Fiber Optic Sensors

Second Edition

OPTICAL SCIENCE AND ENGINEERING

Founding Editor
Brian J. Thompson
University of Rochester
Rochester, New York

1. Electron and Ion Microscopy and Microanalysis: Principles and Applications, *Lawrence E. Murr*
2. Acousto-Optic Signal Processing: Theory and Implementation, *edited by Norman J. Berg and John N. Lee*
3. Electro-Optic and Acousto-Optic Scanning and Deflection, *Milton Gottlieb, Clive L. M. Ireland, and John Martin Ley*
4. Single-Mode Fiber Optics: Principles and Applications, *Luc B. Jeunhomme*
5. Pulse Code Formats for Fiber Optical Data Communication: Basic Principles and Applications, *David J. Morris*
6. Optical Materials: An Introduction to Selection and Application, *Solomon Musikant*
7. Infrared Methods for Gaseous Measurements: Theory and Practice, *edited by Joda Wormhoudt*
8. Laser Beam Scanning: Opto-Mechanical Devices, Systems, and Data Storage Optics, *edited by Gerald F. Marshall*
9. Opto-Mechanical Systems Design, *Paul R. Yoder, Jr.*
10. Optical Fiber Splices and Connectors: Theory and Methods, *Calvin M. Miller with Stephen C. Mettler and Ian A. White*
11. Laser Spectroscopy and Its Applications, *edited by Leon J. Radziemski, Richard W. Solarz, and Jeffrey A. Paisner*
12. Infrared Optoelectronics: Devices and Applications, *William Nunley and J. Scott Bechtel*
13. Integrated Optical Circuits and Components: Design and Applications, *edited by Lynn D. Hutcheson*
14. Handbook of Molecular Lasers, *edited by Peter K. Cheo*
15. Handbook of Optical Fibers and Cables, *Hiroshi Murata*
16. Acousto-Optics, *Adrian Korpel*
17. Procedures in Applied Optics, *John Strong*
18. Handbook of Solid-State Lasers, *edited by Peter K. Cheo*
19. Optical Computing: Digital and Symbolic, *edited by Raymond Arrathoon*
20. Laser Applications in Physical Chemistry, *edited by D. K. Evans*
21. Laser-Induced Plasmas and Applications, *edited by Leon J. Radziemski and David A. Cremers*

22. Infrared Technology Fundamentals, *Irving J. Spiro and Monroe Schlessinger*
23. Single-Mode Fiber Optics: Principles and Applications, Second Edition, Revised and Expanded, *Luc B. Jeunhomme*
24. Image Analysis Applications, *edited by Rangachar Kasturi and Mohan M. Trivedi*
25. Photoconductivity: Art, Science, and Technology, *N. V. Joshi*
26. Principles of Optical Circuit Engineering, *Mark A. Mentzer*
27. Lens Design, *Milton Laikin*
28. Optical Components, Systems, and Measurement Techniques, *Rajpal S. Sirohi and M. P. Kothiyal*
29. Electron and Ion Microscopy and Microanalysis: Principles and Applications, Second Edition, Revised and Expanded, *Lawrence E. Murr*
30. Handbook of Infrared Optical Materials, *edited by Paul Klocek*
31. Optical Scanning, *edited by Gerald F. Marshall*
32. Polymers for Lightwave and Integrated Optics: Technology and Applications, *edited by Lawrence A. Hornak*
33. Electro-Optical Displays, *edited by Mohammad A. Karim*
34. Mathematical Morphology in Image Processing, *edited by Edward R. Dougherty*
35. Opto-Mechanical Systems Design: Second Edition, Revised and Expanded, *Paul R. Yoder, Jr.*
36. Polarized Light: Fundamentals and Applications, *Edward Collett*
37. Rare Earth Doped Fiber Lasers and Amplifiers, *edited by Michel J. F. Digonnet*
38. Speckle Metrology, *edited by Rajpal S. Sirohi*
39. Organic Photoreceptors for Imaging Systems, *Paul M. Borsenberger and David S. Weiss*
40. Photonic Switching and Interconnects, *edited by Abdellatif Marrakchi*
41. Design and Fabrication of Acousto-Optic Devices, *edited by Akis P. Goutzoulis and Dennis R. Pape*
42. Digital Image Processing Methods, *edited by Edward R. Dougherty*
43. Visual Science and Engineering: Models and Applications, *edited by D. H. Kelly*
44. Handbook of Lens Design, *Daniel Malacara and Zacarias Malacara*
45. Photonic Devices and Systems, *edited by Robert G. Hunsberger*
46. Infrared Technology Fundamentals: Second Edition, Revised and Expanded, *edited by Monroe Schlessinger*
47. Spatial Light Modulator Technology: Materials, Devices, and Applications, *edited by Uzi Efron*
48. Lens Design: Second Edition, Revised and Expanded, *Milton Laikin*
49. Thin Films for Optical Systems, *edited by Francoise R. Flory*
50. Tunable Laser Applications, *edited by F. J. Duarte*
51. Acousto-Optic Signal Processing: Theory and Implementation, Second Edition, *edited by Norman J. Berg and John M. Pellegrino*

52. Handbook of Nonlinear Optics, *Richard L. Sutherland*
53. Handbook of Optical Fibers and Cables: Second Edition, *Hiroshi Murata*
54. Optical Storage and Retrieval: Memory, Neural Networks, and Fractals, *edited by Francis T. S. Yu and Suganda Jutamulia*
55. Devices for Optoelectronics, *Wallace B. Leigh*
56. Practical Design and Production of Optical Thin Films, *Ronald R. Willey*
57. Acousto-Optics: Second Edition, *Adrian Korpel*
58. Diffraction Gratings and Applications, *Erwin G. Loewen and Evgeny Popov*
59. Organic Photoreceptors for Xerography, *Paul M. Borsenberger and David S. Weiss*
60. Characterization Techniques and Tabulations for Organic Nonlinear Optical Materials, *edited by Mark G. Kuzyk and Carl W. Dirk*
61. Interferogram Analysis for Optical Testing, *Daniel Malacara, Manuel Servin, and Zacarias Malacara*
62. Computational Modeling of Vision: The Role of Combination, *William R. Uttal, Ramakrishna Kakarala, Spiram Dayanand, Thomas Shepherd, Jagadeesh Kalki, Charles F. Lunskis, Jr., and Ning Liu*
63. Microoptics Technology: Fabrication and Applications of Lens Arrays and Devices, *Nicholas Borrelli*
64. Visual Information Representation, Communication, and Image Processing, *edited by Chang Wen Chen and Ya-Qin Zhang*
65. Optical Methods of Measurement, *Rajpal S. Sirohi and F. S. Chau*
66. Integrated Optical Circuits and Components: Design and Applications, *edited by Edmond J. Murphy*
67. Adaptive Optics Engineering Handbook, *edited by Robert K. Tyson*
68. Entropy and Information Optics, *Francis T. S. Yu*
69. Computational Methods for Electromagnetic and Optical Systems, *John M. Jarem and Partha P. Banerjee*
70. Laser Beam Shaping, *Fred M. Dickey and Scott C. Holswade*
71. Rare-Earth-Doped Fiber Lasers and Amplifiers: Second Edition, Revised and Expanded, *edited by Michel J. F. Digonnet*
72. Lens Design: Third Edition, Revised and Expanded, *Milton Laikin*
73. Handbook of Optical Engineering, *edited by Daniel Malacara and Brian J. Thompson*
74. Handbook of Imaging Materials: Second Edition, Revised and Expanded, *edited by Arthur S. Diamond and David S. Weiss*
75. Handbook of Image Quality: Characterization and Prediction, *Brian W. Keelan*
76. Fiber Optic Sensors, *edited by Francis T. S. Yu and Shizhuo Yin*
77. Optical Switching/Networking and Computing for Multimedia Systems, *edited by Mohsen Guizani and Abdella Battou*
78. Image Recognition and Classification: Algorithms, Systems, and Applications, *edited by Bahram Javidi*

79. Practical Design and Production of Optical Thin Films: Second Edition, Revised and Expanded, *Ronald R. Willey*
80. Ultrafast Lasers: Technology and Applications, *edited by Martin E. Fermann, Almantas Galvanauskas, and Gregg Sucha*
81. Light Propagation in Periodic Media: Differential Theory and Design, *Michel Nevière and Evgeny Popov*
82. Handbook of Nonlinear Optics, Second Edition, Revised and Expanded, *Richard L. Sutherland*
83. Polarized Light: Second Edition, Revised and Expanded, *Dennis Goldstein*
84. Optical Remote Sensing: Science and Technology, *Walter Egan*
85. Handbook of Optical Design: Second Edition, *Daniel Malacara and Zacarias Malacara*
86. Nonlinear Optics: Theory, Numerical Modeling, and Applications, *Partha P. Banerjee*
87. Semiconductor and Metal Nanocrystals: Synthesis and Electronic and Optical Properties, edited by *Victor I. Klimov*
88. High-Performance Backbone Network Technology, *edited by Naoaki Yamanaka*
89. Semiconductor Laser Fundamentals, *Toshiaki Suhara*
90. Handbook of Optical and Laser Scanning, *edited by Gerald F. Marshall*
91. Organic Light-Emitting Diodes: Principles, Characteristics, and Processes, *Jan Kalinowski*
92. Micro-Optomechatronics, *Hiroshi Hosaka, Yoshitada Katagiri, Terunao Hirota, and Kiyoshi Itao*
93. Microoptics Technology: Second Edition, *Nicholas F. Borrelli*
94. Organic Electroluminescence, *edited by Zakya Kafafi*
95. Engineering Thin Films and Nanostructures with Ion Beams, *Emile Knystautas*
96. Interferogram Analysis for Optical Testing, Second Edition, *Daniel Malacara, Manuel Sercin, and Zacarias Malacara*
97. Laser Remote Sensing, *edited by Takashi Fujii and Tetsuo Fukuchi*
98. Passive Micro-Optical Alignment Methods, *edited by Robert A. Boudreau and Sharon M. Boudreau*
99. Organic Photovoltaics: Mechanism, Materials, and Devices, *edited by Sam-Shajing Sun and Niyazi Serdar Saracftci*
100. Handbook of Optical Interconnects, *edited by Shigeru Kawai*
101. GMPLS Technologies: Broadband Backbone Networks and Systems, *Naoaki Yamanaka, Kohei Shiomoto, and Eiji Oki*
102. Laser Beam Shaping Applications, *edited by Fred M. Dickey, Scott C. Holswade and David L. Shealy*
103. Electromagnetic Theory and Applications for Photonic Crystals, *Kiyotoshi Yasumoto*
104. Physics of Optoelectronics, *Michael A. Parker*
105. Opto-Mechanical Systems Design: Third Edition, *Paul R. Yoder, Jr.*
106. Color Desktop Printer Technology, *edited by Mitchell Rosen and Noboru Ohta*
107. Laser Safety Management, *Ken Barat*

108. Optics in Magnetic Multilayers and Nanostructures, *Štefan Višňovský*
109. Optical Inspection of Microsystems, *edited by Wolfgang Osten*
110. Applied Microphotonics, *edited by Wes R. Jamroz, Roman Kruzelecky, and Emile I. Haddad*
111. Organic Light-Emitting Materials and Devices, *edited by Zhigang Li and Hong Meng*
112. Silicon Nanoelectronics, *edited by Shunri Oda and David Ferry*
113. Image Sensors and Signal Processor for Digital Still Cameras, *Junichi Nakamura*
114. Encyclopedic Handbook of Integrated Circuits, *edited by Kenichi Iga and Yasuo Kokubun*
115. Quantum Communications and Cryptography, *edited by Alexander V. Sergienko*
116. Optical Code Division Multiple Access: Fundamentals and Applications, *edited by Paul R. Prucnal*
117. Polymer Fiber Optics: Materials, Physics, and Applications, *Mark G. Kuzyk*
118. Smart Biosensor Technology, *edited by George K. Knopf and Amarjeet S. Bassi*
119. Solid-State Lasers and Applications, *edited by Alphan Sennaroglu*
120. Optical Waveguides: From Theory to Applied Technologies, *edited by Maria L. Calvo and Vasudevan Lakshiminarayanan*
121. Gas Lasers, *edited by Masamori Endo and Robert F. Walker*
122. Lens Design, Fourth Edition, *Milton Laikin*
123. Photonics: Principles and Practices, *Abdul Al-Azzawi*
124. Microwave Photonics, *edited by Chi H. Lee*
125. Physical Properties and Data of Optical Materials, *Moriaki Wakaki, Keiei Kudo, and Takehisa Shibuya*
126. Microlithography: Science and Technology, Second Edition, *edited by Kazuaki Suzuki and Bruce W. Smith*
127. Coarse Wavelength Division Multiplexing: Technologies and Applications, *edited by Hans Joerg Thiele and Marcus Nebeling*
128. Organic Field-Effect Transistors, *Zhenan Bao and Jason Locklin*
129. Smart CMOS Image Sensors and Applications, *Jun Ohta*
130. Photonic Signal Processing: Techniques and Applications, *Le Nguyen Binh*
131. Terahertz Spectroscopy: Principles and Applications, *edited by Susan L. Dexheimer*
132. Fiber Optic Sensors, Second Edition, *edited by Shizhuo Yin, Paul B. Ruffin, and Francis T. S. Yu*

Fiber Optic Sensors

Second Edition

Edited by
Shizhuo Yin
Paul B. Ruffin
Francis T. S. Yu

CRC Press
Taylor & Francis Group
Boca Raton London New York

CRC Press is an imprint of the
Taylor & Francis Group, an **informa** business

CRC Press
Taylor & Francis Group
6000 Broken Sound Parkway NW, Suite 300
Boca Raton, FL 33487-2742

First issued in paperback 2019

ISBN-13: 978-1-4200-5365-4 (hbk)
ISBN-13: 978-0-367-38756-3 (pbk)

Library of Congress Cataloging-in-Publication Data

Fiber optic sensors / Shizhuo Yin, Paul B. Ruffin, Francis T.S. Yu, eds. -- 2nd ed.
p. cm. -- (Optical science and engineering)
Includes bibliographical references and index.
ISBN 978-1-4200-5365-4 (hardback : alk. paper) 1. Optical fiber detectors. I. Yin, Shizhuo, 1963- II. Ruffin, Paul B. III. Yu, Francis T. S., 1932-

TA1815.F527 2008
681'.25--dc22 2007049892

Visit the Taylor & Francis Web site at
http://www.taylorandfrancis.com

and the CRC Press Web site at
http://www.crcpress.com

Contents

Preface

In recent years, fiber optic sensors have developed from the laboratory research and development stage to practical applications. The market for fiber optic sensor technology may be divided into two broad categories of sensors: intrinsic and extrinsic. Intrinsic sensors are used in medicine, defense, and aerospace applications, and they can be used to measure temperature, pressure, humidity, acceleration, and strain. Extrinsic sensors are used in telecommunications to monitor the status and performance of the optical fibers within a network.

The purpose of this updated book is to provide a tutorial overview on fiber optic sensor principles and applications. In particular, the updated and new chapters reflect both the recent advances in fiber optic sensor technology itself (such as the application of photonic crystal fibers to fiber optic gyroscopes and fiber optic grating inscription by femtosecond laser illumination) and new application opportunities that have great potential (e.g., fiber optic sensors provide for medical treatment that is minimally invasive).

This text covers a wide range of topics in fiber optic sensors, although it is by no means complete. All chapters are written by experts in the field. Nine chapters were included in the previous version of the book, but have been updated. Chapter 5 and Chapter 11 are newly added chapters. Chapter 5 (harsh environment fiber optic grating sensors inscribed by femtosecond laser illumination) introduces state-of-the-art fiber optic grating sensor technology and Chapter 11 (fiber optic chemical/biological sensors) reviews the recent advances in this fast growing application sector.

Chapter 1 gives an overview of fiber optic sensors that includes the basic concepts, historical development, and some of the classic applications. This overview provides the essential background material needed to facilitate the objectives of later chapters.

Chapter 2 deals with fiber optic sensors based on Fabry–Perot interferometers. The major merits of this type of sensor include high sensitivity, compact size, and no need for fiber couplers. The high sensitivity and multiplexing capabilities of this type of fiber optic sensor make it particularly well suited for smart structure monitoring applications.

Chapter 3 introduces a polarimetric fiber optic sensor. The polarization state of light that propagates in an optical fiber can be changed through external perturbation. By employing polarization-maintaining fiber, the effect of polarization changes induced by external perturbation can be exploited for sensing applications. One of the major features of this type of sensor is that it offers an excellent trade-off between sensitivity and robustness.

Chapter 4 reviews fiber-grating-based fiber optic sensors. Fiber grating technology (Bragg and long-period gratings) is a very powerful tool for high-sensitivity, quasi-distributed sensing.

Chapter 5 is a newly added chapter (replacing the original Chapter 5 on distributed fiber optic sensors) that introduces a new type of fiber grating inscribed by femtosecond laser irradiation. This type of fiber grating sensor offers the advantage of harsh environment sensing because the gratings are not erased at high temperatures. Additionally, the fibers do not need to be doped with Ge as they are when a grating is written using UV. As a result, these new gratings can be produced as almost any type of fiber (such as photonic crystal fibers and sapphire fibers), which greatly increases the number of applications to which they can be applied.

Chapter 6 discusses fiber optic specklegram sensors. A fiber specklegram is formed by the interference between different modes that propagate in multimode optical fibers. Since the specklegram is formed by common-mode interference, it can have a very high sensitivity to some environmental factors (such as bending) and less sensitivity to others (such as temperature fluctuations). Thus, it is a very unique type of fiber optic sensor.

Chapter 7 introduces interrogation techniques for fiber optic sensors. This chapter emphasizes the physical effects in optic fibers when a fiber is subjected to external perturbations.

Chapter 8 focuses on fiber gyroscope sensors. First, the basic concepts are introduced. Fiber gyroscope sensors are based on the interference between two light beams that propagate in opposite directions in a fiber loop. Since a large number of turns are used, a very high sensitivity can be realized. Second, practical issues related to fiber optic gyroscopes, such as modulation and winding techniques, are reviewed. The content of this chapter has been substantially updated in this new version to include (1) polarization analysis of a fiber optic gyroscope (FOG) sensor coil and (2) recent advances in winding technology.

Chapter 9 introduces a fiber optic hydrophone system. This chapter deals with several key issues, such as interferometer configuration, interrogation/demodulation schemes, multiplexing architecture, polarization fading mitigation, and system integration. It also includes discussions on related technologies, such as fiber optic amplifiers, wavelength division multiplexing components, optical isolators, and circulators.

Chapter 10 discusses the applications of fiber optic sensor technology to structural health monitoring, including bridges, dams, the electric power industry, etc.

Chapter 11 is a newly added chapter that provides a review on fiber optic chemical and biomedical sensors, which represent a fast growing market for fiber optic sensing technology.

This text will be a useful reference for researchers and technical staffs engaged in the field of fiber optic sensors. The book can also serve as a viable text or reference book for engineering students and professors who are interested in fiber optic sensors.

Stuart (Shizhuo) Yin
Paul Ruffin
Francis T. S. Yu

Contributors

I. Bennion Aston University, Birmingham, England

P. L. Chu City University of Hong Kong, Kowloon, Hong Kong

Shanglian Huang Chongqing University, Chongqing, China

Yoonchan Jeong Seoul National University, Seoul, Korea

Byoungho Lee Seoul National University, Seoul, Korea

Craig Michie University of Strathclyde, Glasgow, Scotland

G. D. Peng The University of New South Wales, Sydney, Australia

Y. J. Rao Chongqing University, Chongqing, China

Paul B. Ruffin U.S. Army Research, Development, and Engineering Command, Redstone Arsenal, Alabama

Henry F. Taylor Texas A&M University, College Station, Texas

Eric Udd Blue Road Research, Fairview, Oregon

Shizhuo Yin The Pennsylvania State University, University Park, Pennsylvania

Francis T. S. Yu The Pennsylvania State University, University Park, Pennsylvania

Chun Zhan The Pennsylvania State University, University Park, Pennsylvania

Lin Zhang Aston University, Birmingham, England

W. Zhang Aston University, Birmingham, England

1

Overview of Fiber Optic Sensors

Eric Udd

CONTENTS

1.1 Introduction

Over the past 20 years two major product revolutions have taken place due to the growth of the optoelectronics and fiber optic communications industries. The optoelectronics industry has brought about such products as compact disc players, laser printers, bar code scanners, and laser pointers. The fiber optic communications industry has revolutionized the telecommunications industry by providing higher performance, more reliable telecommunication links with ever decreasing bandwidth cost. This revolution is bringing about the benefits of high-volume production to component users and a true information superhighway built of glass.

In parallel with these developments, fiber optic sensor technology [1–6] has been a major user of technology associated with the optoelectronic and fiber optic communications industries. Many of the components associated with these industries were often developed for fiber optic sensor applications. Fiber optic sensor technology, in turn, has often been driven by the development and subsequent mass production of components to support these industries. As component prices have fallen and quality improvements have been made, the ability of fiber optic sensors to displace traditional sensors for rotation, acceleration, electric and magnetic field measurement, temperature, pressure, acoustics, vibration, linear and angular position, strain, humidity, viscosity, chemical measurements, and a host of other sensor applications has been enhanced. In the early days of fiber optic sensor technology, most commercially successful fiber optic sensors were squarely targeted at markets where existing sensor technology was marginal or, in many cases, nonexistent. The inherent advantages of fiber optic sensors, which include their (1) ability to be lightweight, of very small size, passive, low power, and resistant to electromagnetic interference; (2) high sensitivity; (3) bandwidth; and (4) environmental ruggedness, were heavily used to offset their major disadvantages of high cost and end-user unfamiliarity.

The situation is changing. Laser diodes that cost $3000 in 1979 with lifetimes measured in hours now sell for a few dollars in small quantities, have reliability of tens of thousands of hours, and are widely used in compact disc players, laser printers, laser pointers, and bar code readers. Single-mode optical fiber that cost $20/meter in 1979 now costs less than $0.10/meter, with vastly improved optical and mechanical properties. Integrated optical devices that were not available in usable form at that time are now commonly used to support production models of fiber optic gyros. Also, they could drop in price dramatically in the future while offering ever more sophisticated optical circuits. As these trends continue, the opportunities for fiber optic sensor designers to produce competitive products will increase and the technology can be expected to assume an ever more prominent position in the sensor marketplace. In the following sections the basic types of fiber optic sensors being developed are briefly reviewed, followed by a discussion of how these sensors are and will be applied.

1.2 Basic Concepts and Intensity-Based Fiber Optic Sensors

Fiber optic sensors are often loosely grouped into two basic classes referred to as *extrinsic*, or hybrid, fiber optic sensors and *intrinsic*, or all-fiber, sensors. Figure 1.1 illustrates the case of an extrinsic fiber optic sensor.

In this case an optical fiber leads up to a "black box" that impresses information onto the light beam in response to an environmental effect. The information could be impressed in terms of intensity, phase, frequency, polarization, spectral content, or other methods. An optical fiber then carries the light with the environmentally impressed information back to an optical and/or electronic processor. In some cases the input optical fiber also acts as the output fiber. The intrinsic or all-fiber sensor shown in Figure 1.2 uses an optical fiber to carry the light beam, and the environmental effect impresses information onto the light beam while it is in the fiber. Each of these classes

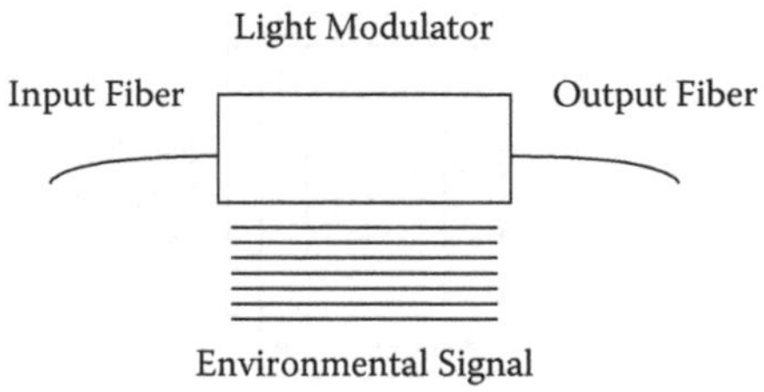

FIGURE 1.1
Extrinsic fiber optic sensors consist of optical fibers that lead up to and out of a "black box" that modulates the light beam passing through it in response to an environmental effect.

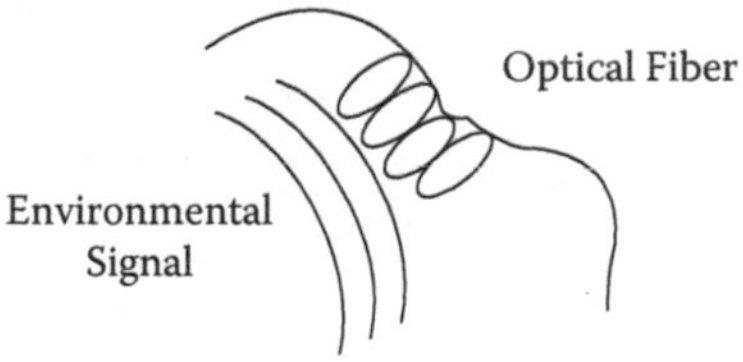

FIGURE 1.2
Intrinsic fiber optic sensors rely on the light beam propagating through the optical fiber being modulated by the environmental effect either directly or through environmentally induced optical path length changes in the fiber itself.

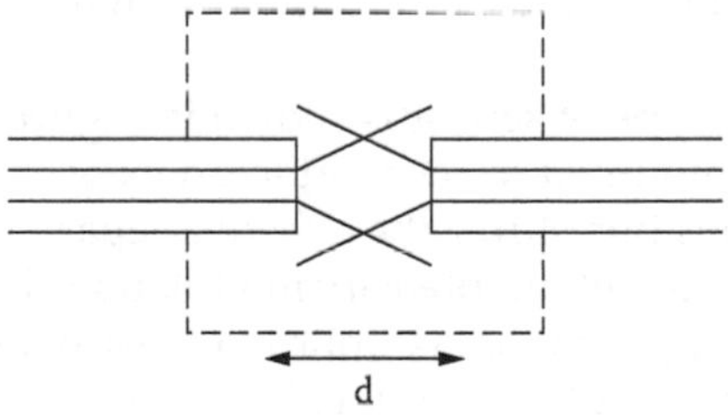

FIGURE 1.3
Closure and vibration fiber optic sensors based on numerical aperture can be used to support door closure indicators and measure levels of vibration in machinery.

of fibers in turn has many subclasses with, in some cases, sub-subclasses [1] that consist of large numbers of fiber sensors.

In some respects the simplest type of fiber optic sensor is the hybrid type that is based on intensity modulation [7,8]. Figure 1.3 shows a simple closure or vibration sensor that consists of two optical fibers held in close proximity to each other. Light is injected into one of the optical fibers; when it exits, the light expands into a cone of light whose angle depends on the difference between the index of refraction of the core and cladding of the optical fiber. The amount of light captured by the second optical fiber depends on its acceptance angle and the distance d between the optical fibers. When the

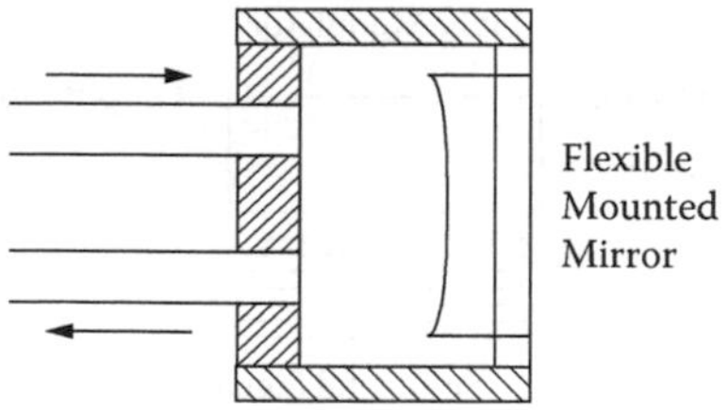

FIGURE 1.4
A numerical aperture fiber sensor based on a flexible mirror can be used to measure small vibrations and displacements.

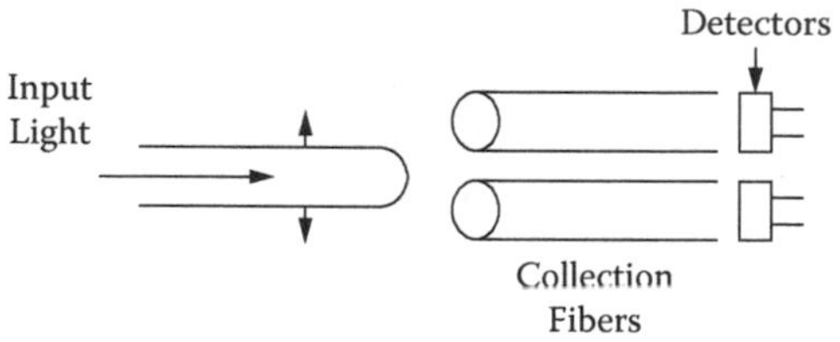

FIGURE 1.5
A fiber optic translation sensor based on numerical aperture uses the ratio of the output on the detectors to determine the position of the input fiber.

distance *d* is modulated, it in turn results in an intensity modulation of the light captured.

A variation on this type of sensor is shown in Figure 1.4. Here a mirror is used that is flexibly mounted to respond to an external effect such as pressure. As the mirror position shifts, the effective separation between the optical fibers shifts with a resultant intensity modulation. These types of sensors are useful for such applications as door closures where a reflective strip, in combination with an optical fiber acting to input and catch the output reflected light, can be used.

With two optical fibers arranged in a line, a simple translation sensor can be configured as in Figure 1.5. The output from the two detectors can be proportioned to determine the translational position of the input fiber.

Several companies have developed rotary and linear fiber optic position sensors to support applications such as fly-by-light [9]. These sensors attempt to (1) eliminate electromagnetic interference susceptibility to improve safety, and (2) lower shielding needs to reduce weight. Figure 1.6 shows a rotary position sensor [10] that consists of a code plate with variable reflectance patches placed so that each position has a unique code. A series of optical fibers is used to determine the presence or absence of a patch.

An example of a linear position sensor using wavelength division multiplexing (WDM) [11] is illustrated by Figure 1.7. Here a broadband light source, which might be a light-emitting diode, is used to couple light into the system. A single optical fiber is used to carry the light beam up to a WDM

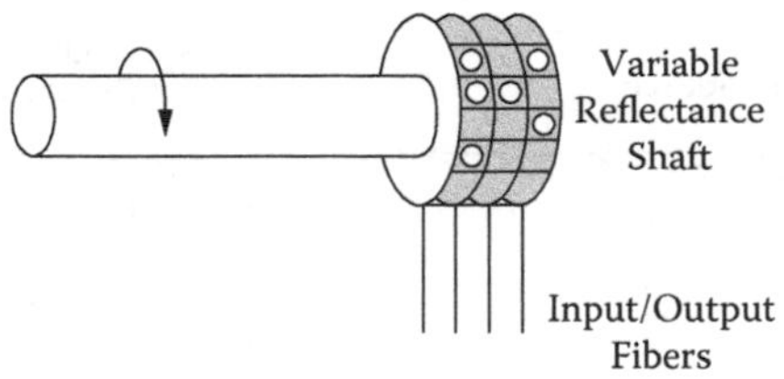

FIGURE 1.6
Fiber optic rotary position sensor based on reflectance used to measure the rotational position of the shaft via the amount of light reflected from dark and light patches.

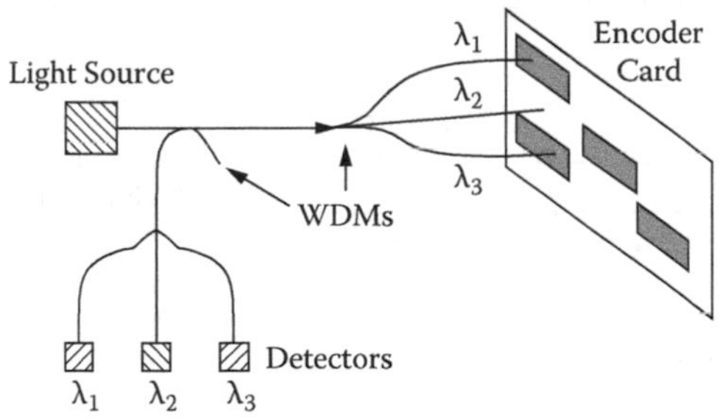

FIGURE 1.7
A linear position sensor using wavelength division multiplexing decodes position by measuring the presence or absence of a reflective patch at each fiber position as the card slides by via independent wavelength separated detectors.

element that splits the light into separate fibers that are used to interrogate the encoder card and determine linear position. The boxes on the card of Figure 1.7 represent highly reflective patches, while the rest of the card has low reflectance. The reflected signals are then recombined and separated by a second wavelength division multiplexing element so that each interrogating fiber signal is read out by a separate detector.

A second common method of interrogating a position sensor using a single optical fiber is to use time division multiplexing methods [12]. In Figure 1.8 a light source is pulsed. The light pulse then propagates down the optical fiber and is split into multiple interrogating fibers. Each of these fibers is arranged so that the fibers have delay lines that separate the return signal from the encoder plate by a time that is longer than the pulse duration. When the returned signals are recombined onto the detector, the net result is an encoded signal burst corresponding to the position of the encoded card.

These sensors have been used to support tests on military and commercial aircraft that have demonstrated performance comparable to conventional electrical position sensors used for rudder, flap, and throttle positions [9]. The principal advantages of the fiber position sensors are immunity to electromagnetic interference and overall weight savings.

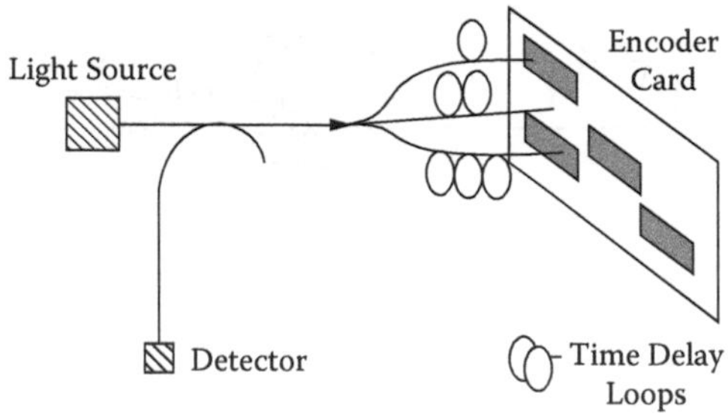

FIGURE 1.8
A linear position sensor using time division multiplexing measure decodes card position via a digital stream of ons and offs dictated by the presence or absence of a reflective patch.

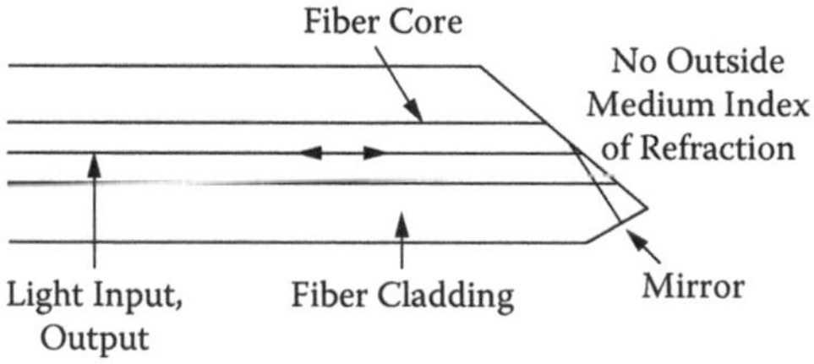

FIGURE 1.9
Fiber sensor using critical angle properties of a fiber for pressure/index of refraction measurement via measurements of the light reflected back into the fiber.

Another class of intensity-based fiber optic sensors is based on the principle of total internal reflection. In the case of the sensor in Figure 1.9, light propagates down the fiber core and hits the angled end of the fiber. If the medium into which the angled end of the fiber is placed has a low enough index of refraction, then virtually all the light is reflected when it hits the mirrored surface and returns via the fiber. If, however, the medium's index of refraction starts to approach that of the glass, some of the light propagates out of the optical fiber and is lost, resulting in an intensity modulation.

This type of sensor can be used for low-resolution measurement of pressure or index of refraction changes in a liquid or gel with 1 to 10% accuracy. Variations on this method have also been used to measure liquid level [13], as shown by the probe configuration of Figure 1.10. When the liquid level hits the reflecting prism, the light leaks into the liquid, greatly attenuating the signal.

Confinement of a propagating light beam to the region of the fiber cores and power transfer from two closely placed fiber cores can be used to produce a series of fiber sensors based on evanescence [14–16]. Figure 1.11 illustrates two fiber cores that have been placed in close proximity to one another. For single-mode optical fiber [17], this distance is on the order of 10 to 20 microns.

When single-mode fiber is used, there is considerable leakage of the propagating light beam mode beyond the core region into the cladding or medium

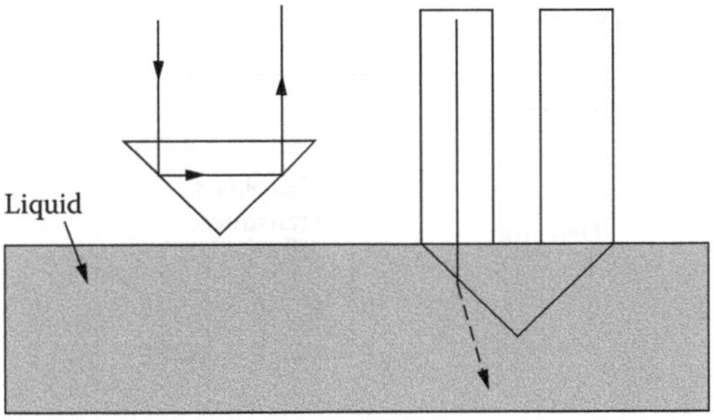

FIGURE 1.10
A liquid-level sensor based on the total internal reflection detects the presence or absence of liquid by the presence or absence of a return light signal.

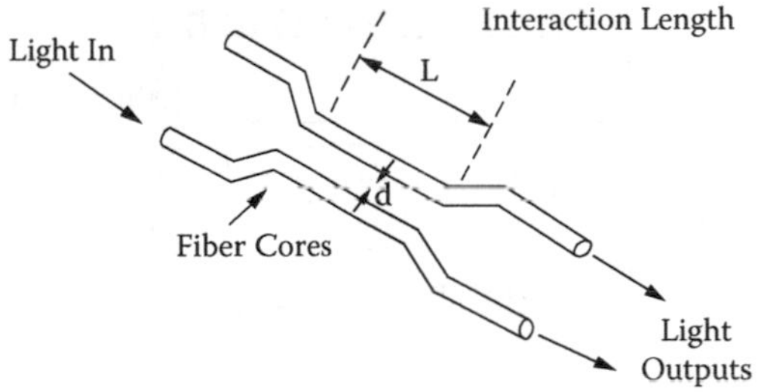

FIGURE 1.11
Evanescence-based fiber optic sensors rely on the cross-coupling of light between two closely spaced fiber optic cores. Variations in this distance due to temperature, pressure, or strain offer environmental sensing capabilities.

around it. If a second fiber core is placed nearby, this evanescent tail will tend to cross-couple to the adjacent fiber core. The amount of cross-coupling depends on a number of parameters, including the wavelength of light, the relative index of refraction of the medium in which the fiber cores are placed, the distance between the cores, and the interaction length. This type of fiber sensor can be used for the measurement of wavelength, spectral filtering, index of refraction, and environmental effects acting on the medium surrounding the cores (temperature, pressure, and strain). The difficulty with this sensor, which is common to many fiber sensors, is optimizing the design so that only the desired parameters are sensed.

Another way that light may be lost from an optical fiber is when the bend radius of the fiber exceeds the critical angle necessary to confine the light to the core area and there is leakage into the cladding. Local microbending of the fiber can cause this to occur, with resultant intensity modulation of light propagating through an optical fiber. A series of microbend-based fiber sensors has been built to sense vibration, pressure, and other environmental effects [18–20]. Figure 1.12 shows a typical layout of this type of device consisting of a light source, a section of optical fiber positioned in a microbend

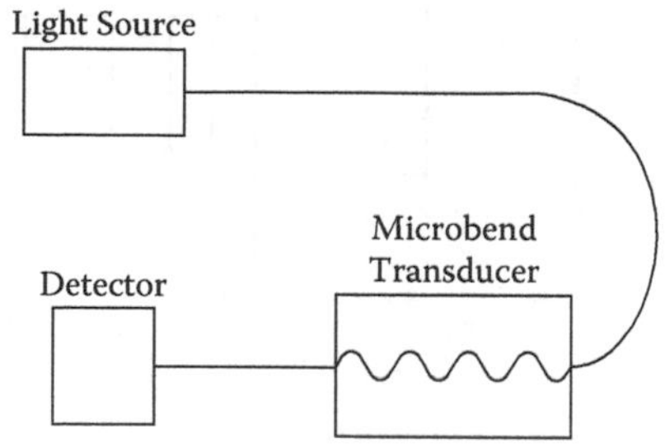

FIGURE 1.12
Microbend fiber sensors are configured so that an environmental effect results in an increase or decrease in loss through the transducer due to light loss resulting from small bends in the fiber.

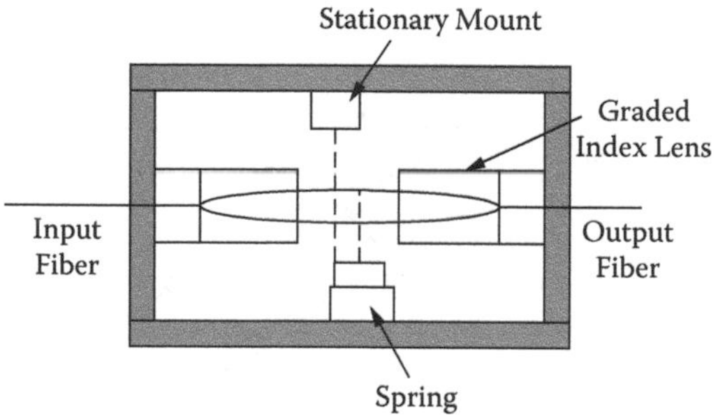

FIGURE 1.13
Grating-based fiber intensity sensors measure vibration or acceleration via a highly sensitive shutter effect.

transducer designed to intensity modulate light in response to an environmental effect, and a detector. In some cases the microbend transducer can be implemented by using special fiber cabling or optical fiber that is simply optimized to be sensitive to microbending loss.

One last example of an intensity-based sensor is the grating-based device [21] shown in Figure 1.13. Here an input optical light beam is collimated by a lens and passes through a dual grating system. One of the gratings is fixed while the other moves. With acceleration the relative position of the gratings changes, resulting in an intensity-modulated signal on the output optical fiber.

One of the limitations of this type of device is that, as the gratings move from a totally transparent to a totally opaque position, the relative sensitivity of the sensor changes, as Figure 1.14 shows. For optimum sensitivity the gratings should be in the half-open/half-closed position. Increasing sensitivity means finer and finer grating spacings, which in turn limit dynamic range.

To increase sensitivity without limiting dynamic range, multiple-part gratings that are offset by 90° should be used, as shown in Figure 1.15. If two outputs are spaced in this manner, the resulting outputs are in quadrature, as shown in Figure 1.16.

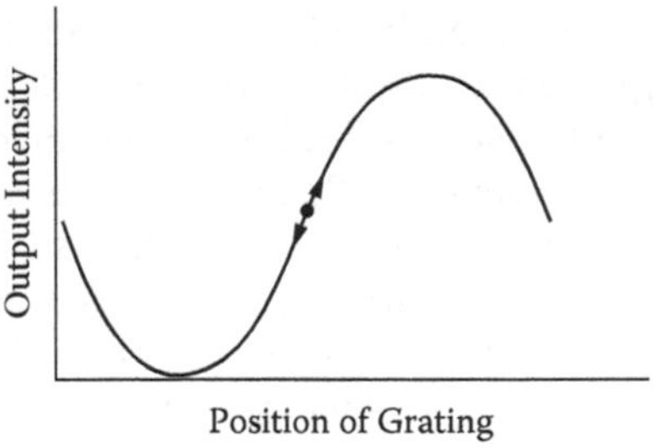

FIGURE 1.14
Dynamic range limitations of the grating-based sensor of Figure 1.13 are due to smaller grating spacing increasing sensitivity at the expense of range.

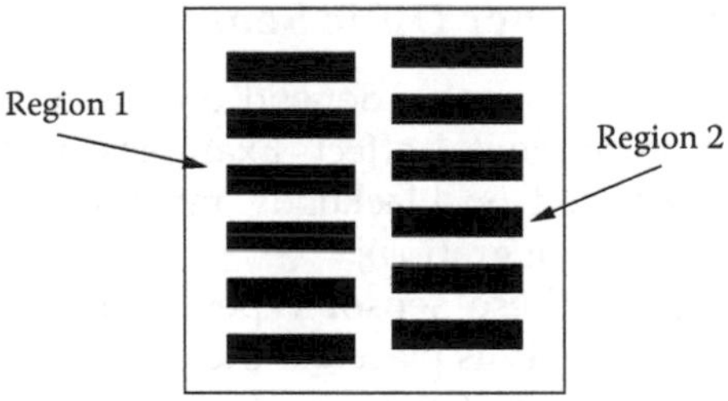

FIGURE 1.15
Dual grating mask with regions 90° out of phase to support quadrature detection, which allows grating-based sensors to track through multiple lines.

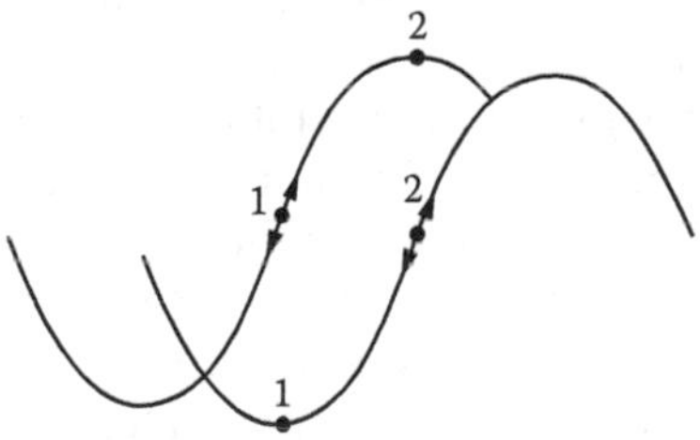

FIGURE 1.16
Diagram of a quadrature detection method that allows one area of maximum sensitivity while the other reaches a minimum, and vice versa, allowing uniform sensitivity over a wide dynamic range.

When one output is at optimal sensitivity, the other is at its lowest sensitivity, and vice versa. By using both outputs for tracking, one can scan through multiple grating lines, enhancing dynamic range and avoiding the signal fadeout associated with positions of minimal sensitivity.

Intensity-based fiber optic sensors have a series of limitations imposed by variable losses in the system that are not related to the environmental effect to be measured. Potential error sources include variable losses due to connectors and splices, microbending loss, macrobending loss, and mechanical creep and misalignment of light sources and detectors. To circumvent these problems, many of the successful higher performance, intensity-based fiber sensors employ dual wavelengths. One of the wavelengths is used to calibrate out all of the errors due to undesired intensity variations by bypassing the sensing region. An alternative approach is to use fiber optic sensors that are inherently resistant to errors induced by intensity variations. The next section discusses a series of spectrally based fiber optic sensors that have this characteristic.

1.3 Spectrally Based Fiber Optic Sensors

Spectrally based fiber optic sensors depend on a light beam modulated in wavelength by an environmental effect. Examples of these types of fiber sensors include those based on blackbody radiation, absorption, fluorescence, etalons, and dispersive gratings.

One of the simplest of these sensor types is the blackbody sensor of Figure 1.17. A blackbody cavity is placed at the end of an optical fiber. When the cavity rises in temperature, it starts to glow and act as a light source.

Detectors in combination with narrow-band filters are then used to determine the profile of the blackbody curve and, in turn, the temperature, as in Figure 1.18. This type of sensor has been successfully commercialized and used to measure temperature to within a few degrees Celsius under intense radio frequency (RF) fields. The performance and accuracy of this sensor are better at higher temperatures and fall off at temperatures on the order of 200°C because of low signal-to-noise ratios. Care must be taken to ensure that the hottest spot is the blackbody cavity and not on the optical fiber lead itself, as this can corrupt the integrity of the signal.

Another type of spectrally based temperature sensor, shown in Figure 1.19, is based on absorption [22]. In this case a gallium arsenide (GaAs) sensor probe is used in combination with a broadband light source and input/output optical fibers. The absorption profile of the probe is temperature dependent and may be used to determine temperature.

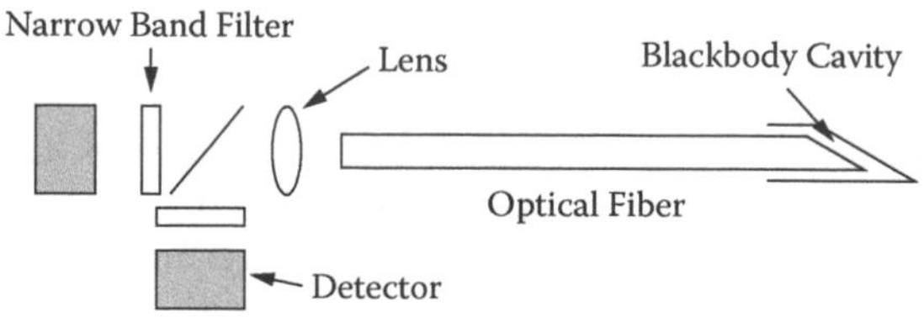

FIGURE 1.17
Blackbody fiber optic sensors allow the measurement of temperature at a hot spot and are most effective at temperatures of higher than 300°C.

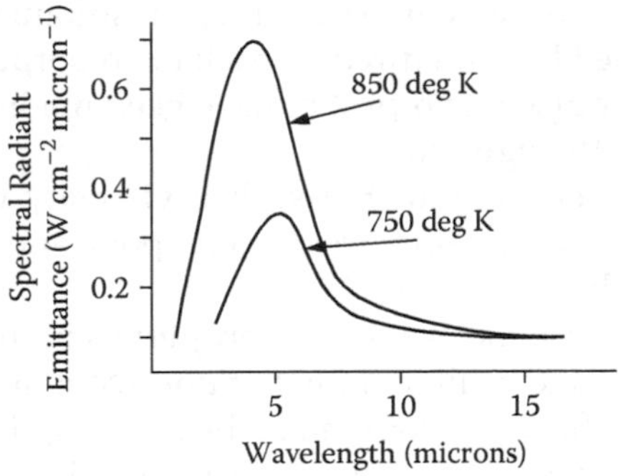

FIGURE 1.18
Blackbody radiation curves provide unique signatures for each temperature.

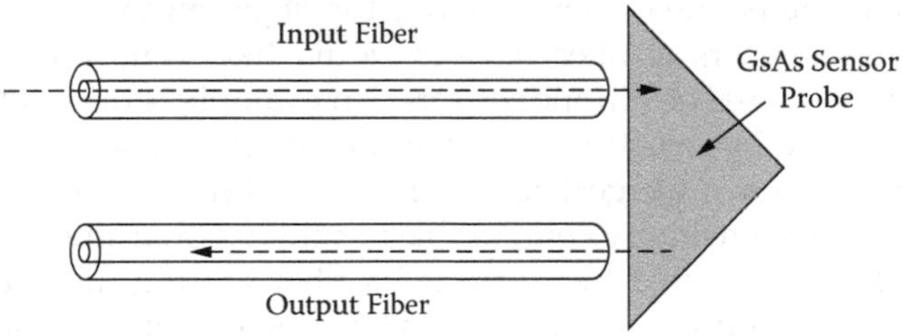

FIGURE 1.19
Fiber optic sensor based on variable absorption of materials such as GaAs allows the measurement of temperature and pressure.

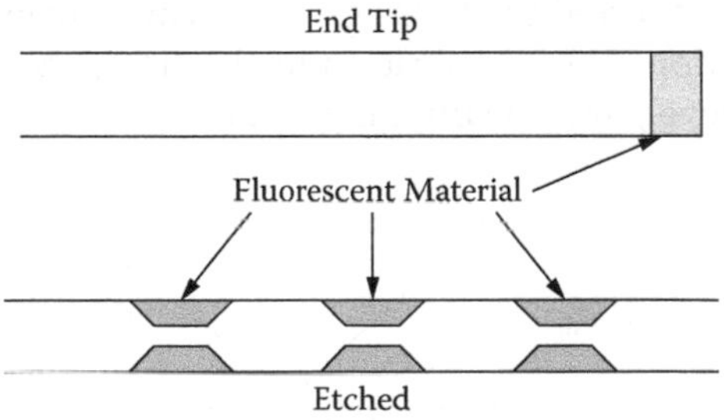

FIGURE 1.20
Fluorescent fiber optic sensor probe configurations can be used to support the measurement of physical parameters as well as the presence or absence of chemical species. These probes may be configured to be single ended or multipoint by using side etch techniques and attaching the fluorescent material to the fiber.

Fluorescent-based fiber sensors [23,24] are widely used for medical applications and chemical sensing and can also be used for physical parameter measurements such as temperature, viscosity, and humidity. There are a number of configurations for these sensors; Figure 1.20 illustrates two of the most common. In the case of the end-tip sensor, light propagates down the fiber to a probe of fluorescent material. The resultant fluorescent signal is captured by the same fiber and directed back to an output demodulator. The light sources can be pulsed, and probes have been made that depend on the time rate of decay of the light pulse.

In the continuous mode, parameters such as viscosity, water vapor content, and degree of cure in carbon fiber reinforced epoxy and thermoplastic composite materials can be monitored.

An alternative is to use the evanescent properties of the fiber, etch regions of the cladding away, and refill them with fluorescent material. By sending a light pulse down the fiber and looking at the resulting fluorescence, a series of sensing regions may be time division multiplexed.

It is also possible to introduce fluorescent dopants into the optical fiber itself. This approach causes the entire optically activated fiber to fluoresce. By using time division multiplexing, various regions of the fiber can be used to make a distributed measurement along the fiber length.

In many cases, users of fiber sensors would like to have the fiber optic analog of conventional electronic sensors. An example is the electrical strain gauge widely used by structural engineers. Fiber grating sensors [25–28] can be configured to have gauge lengths from 1 millimeter to approximately 1 centimeter, with sensitivity comparable to conventional strain gauges.

This sensor is fabricated by "writing" a fiber grating into the core of a germanium-doped optical fiber. This can be done in a number of ways. One method, illustrated by Figure 1.21, uses two short-wavelength laser beams that are angled to form an interference pattern through the side of the optical fiber. The interference pattern consists of bright and dark bands that represent local changes in the index of refraction in the core region of the fiber. Exposure time for making these gratings varies from minutes to hours, depending on the dopant concentration in the fiber, the wavelengths used, the optical power level, and the imaging optics.

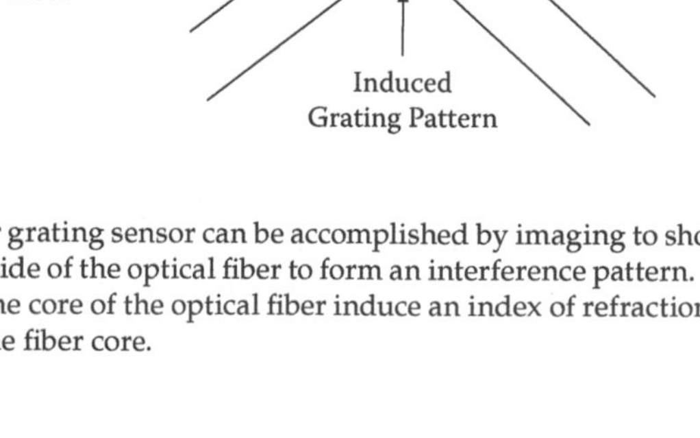

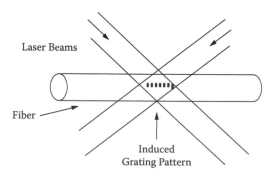

FIGURE 1.21
Fabrication of a fiber grating sensor can be accomplished by imaging to short-wavelength laser beams through the side of the optical fiber to form an interference pattern. The bright and dark fringes imaged on the core of the optical fiber induce an index of refraction variation resulting in a grating along the fiber core.

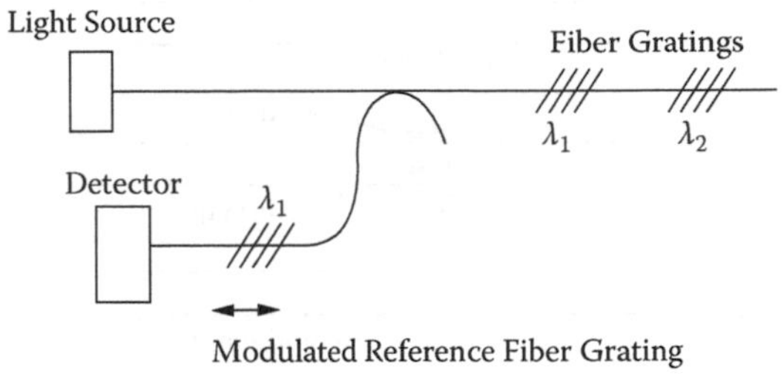

FIGURE 1.22
Fiber grating demodulation systems require very high-resolution spectral measurements. One way to accomplish this is to beat the spectrum of light reflected by the fiber grating against the light transmission characteristics of a reference grating.

Other methods that have been used include the use of phase masks as well as interference patterns induced by short, high-energy laser pulses. The short duration pulses have the potential to be used to write fiber gratings into the fiber as it is being drawn.

Substantial efforts are being made by laboratories around the world to improve the manufacturability of fiber gratings because they have the potential to be used to support optical communication as well as sensing technology.

Once the fiber grating has been fabricated, the next major issue is how to extract information. When used as a strain sensor, the fiber grating is typically attached to, or embedded in, a structure. As the fiber grating is expanded or compressed, the grating period expands or contracts, changing the grating's spectral response.

For a grating operating at 1300 nanometers, the change in wavelength is about 10^{-3} nanometers per microstrain. This type of resolution requires the use of spectral demodulation techniques that are much better than those associated with conventional spectrometers. Several demodulation methods have been suggested using fiber gratings, etalons, and interferometers [29,30]. Figure 1.22 illustrates a system that uses a reference fiber grating. The reference fiber grating acts as a modulator filter. By using similar gratings for the reference and signal gratings and adjusting the reference grating to line up with the active grating, one may implement an accurate closed-loop demodulation system.

An alternative demodulation system would use fiber etalons such as those shown in Figure 1.23. One fiber can be mounted on a piezoelectric transducer and the other moved relative to a second fiber end. The spacing of the fiber ends as well as their reflectivity in turn determines the spectral filtering action of the fiber etalon, illustrated by Figure 1.24.

The fiber etalons in Figure 1.23 can also be used as sensors [31–33] for measuring strain, as the distance between mirrors in the fiber determines their transmission characteristics. The mirrors can be fabricated directly into the fiber by cleaving the fiber, coating the end with titanium dioxide, and then resplicing. An alternative approach is to cleave the fiber ends and insert them

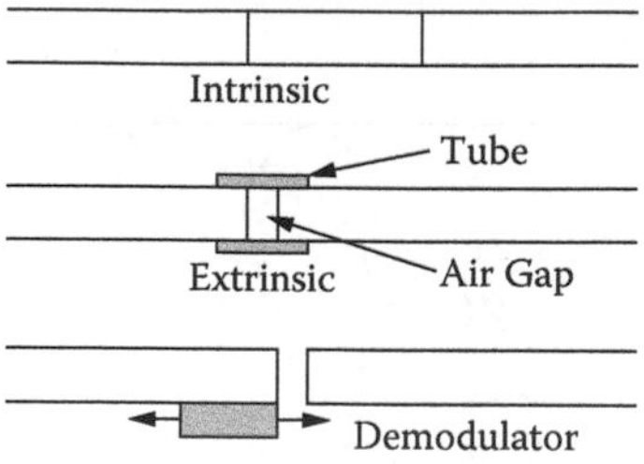

FIGURE 1.23
Intrinsic fiber etalons are formed by in-line reflective mirrors that can be embedded into the optical fiber. Extrinsic fiber etalons are formed by two mirrored fiber ends in a capillary tube. A fiber etalon-based spectral filter or demodulator is formed by two reflective fiber ends that have a variable spacing.

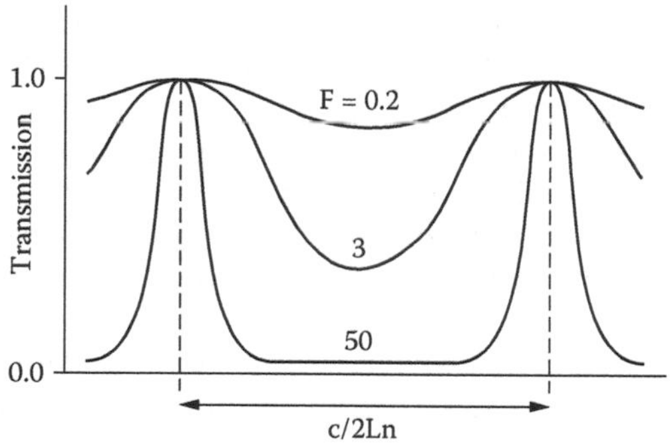

FIGURE 1.24
The transmission characteristics of a fiber etalon as a function of finesse, which increases with mirror reflectivity.

into a capillary tube with an air gap. Both of these approaches are being investigated for applications where multiple in-line fiber sensors are required.

For many applications a single point sensor is adequate. In these situations an etalon can be fabricated independently and attached to the end of the fiber. Figure 1.25 shows a series of etalons that have been configured to measure pressure, temperature, and refractive index, respectively.

In the case of pressure, the diaphragm has been designed to deflect. Pressure ranges of 15 to 2000 pounds per square inch can be accommodated by changing the diaphragm thickness with an accuracy of about 0.1% full scale [34]. For temperature the etalon has been formed by silicon–silicon dioxide interfaces. Temperature ranges of 70 to 500 kelvins can be selected, and for a range of about 100 kelvins a resolution of about 0.1 kelvin is achievable [34]. For refractive index of liquids, a hole has been formed to allow the flow of the liquid to be measured without the diaphragm deflecting. These devices have been commercialized and are sold with instrument packages [34].

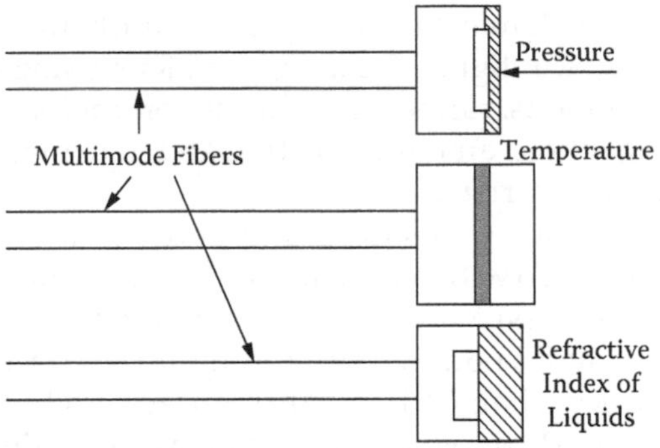

FIGURE 1.25
Hybrid etalon-based fiber optic sensors often consist of micromachined cavities that are placed on the end of optical fibers and can be configured so that sensitivity to one environmental effect is optimized.

1.4 Interferometric Fiber Optic Sensors

One of the areas of greatest interest has been in the development of high-performance interferometric fiber optic sensors. Substantial efforts have been undertaken on Sagnac interferometers, ring resonators, and Mach–Zehnder and Michelson interferometers, as well as dual-mode, polarimetric, grating, and etalon-based interferometers. This section briefly reviews the Sagnac, Mach–Zehnder, and Michelson interferometers.

1.4.1 Sagnac Interferometer

The Sagnac interferometer has been principally used to measure rotation [35–38] and is a replacement for ring laser gyros and mechanical gyros. It may also be employed to measure time-varying effects such as acoustics, vibration, and slowly varying phenomena such as strain. By using multiple interferometer configurations, it is possible to employ the Sagnac interferometer as a distributed sensor capable of measuring the amplitude and location of a disturbance.

The single most important application of fiber optic sensors in terms of commercial value is the fiber optic gyro. It was recognized very early that the fiber optic gyro offered the prospect of an all-solid-state inertial sensor with no moving parts, unprecedented reliability, and a potential of very low cost.

The potential of the fiber optic gyro is being realized as several manufacturers worldwide are producing them in large quantities to support automobile navigation systems, pointing and tracking of satellite antennas, inertial measurement systems for commuter aircraft and missiles, and as the backup guidance system for the Boeing 777. They are also being baselined for such future programs as the Comanche helicopter and are being developed to support long-duration space flights. Other applications using fiber optic gyros include mining operations, tunneling, attitude control for a radio-controlled helicopter, cleaning robots, antenna pointing and tracking, and guidance for unmanned trucks and carriers.

Two types of fiber optic gyros are being developed. The first type is an open-loop fiber optic gyro with a dynamic range on the order of 1000 to 5000 (dynamic range is unitless), with a scale factor accuracy of about 0.5% (this accuracy number includes nonlinearity and hysteresis effects) and sensitivities that vary from less than 0.01 to 100°/hour and higher [38]. These fiber gyros are generally used for low-cost applications where dynamic range and linearity are not the crucial issues. The second type is the closed-loop fiber optic gyro that may have a dynamic range of 10^6 and scale factor linearity of 10 parts per million or better [38]. These types of fiber optic gyros are primarily targeted at medium- to high-accuracy navigation applications that have high turning rates and require high linearity and large dynamic ranges.

The basic open-loop fiber optic gyro is illustrated by Figure 1.26. A broadband light source such as a light-emitting diode is used to couple light into an input/output fiber coupler. The input light beam passes through a polarizer that is used to ensure the reciprocity of the counterpropagating light beams through the fiber coil. The second central coupler splits the two light beams into the fiber optic coil, where they pass through a modulator used to generate a time-varying output signal indicative of rotation. The modulator is offset from the center of the coil to impress a relative phase difference between the counterpropagating light beams. After passing through the fiber coil, the two light beams recombine the pass back though the polarizer and are directed onto the output detector.

When the fiber gyro is rotated clockwise, the entire coil is displaced, slightly increasing the time it takes light to traverse the fiber optic coil. (Remember that the speed of light is invariant with respect to the frame of reference; thus, coil rotation increases path length when viewed from outside the fiber.) Thus, the clockwise propagating light beam has to go through a slightly longer optical path length than the counterclockwise beam, which is moving in

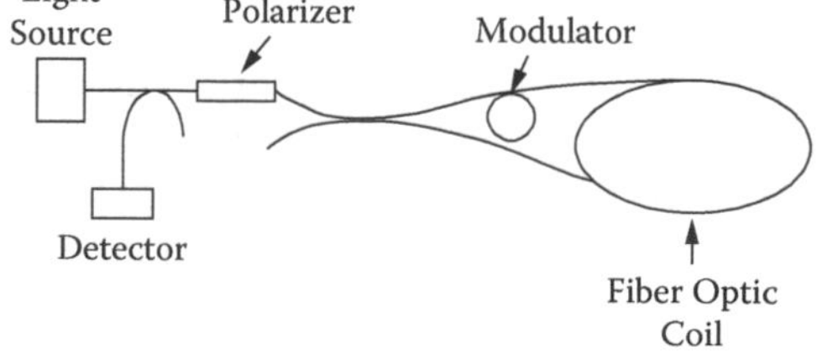

FIGURE 1.26
Open-loop fiber optic gyros are the simplest and lowest cost rotation sensors. They are widely used in commercial applications where their dynamic range and linearity limitations are not constraining.

a direction opposite to the motion of the fiber coil. The net phase difference between the two beams is proportional to the rotation rate.

By including a phase modulator loop offset from the fiber coil, a time difference in the arrival of the two light beams is introduced, and an optimized demodulation signal can be realized. The right side of Figure 1.27 shows this. In the absence of the loops the two light beams traverse the same optical path and are in phase with each other, shown on the left-hand curve of Figure 1.27. The result is that the first or a higher order odd harmonic can be used as a rotation rate output, resulting in improved dynamic range and linearity, shown in Figure 1.28.

Further improvements in dynamic range and linearity can be realized by using a "closed-loop" configuration where the phase shift induced by rotation is compensated by an equal and opposite artificially imposed phase shift. One way to accomplish this is to introduce a frequency shifter into the loop, shown in Figure 1.29.

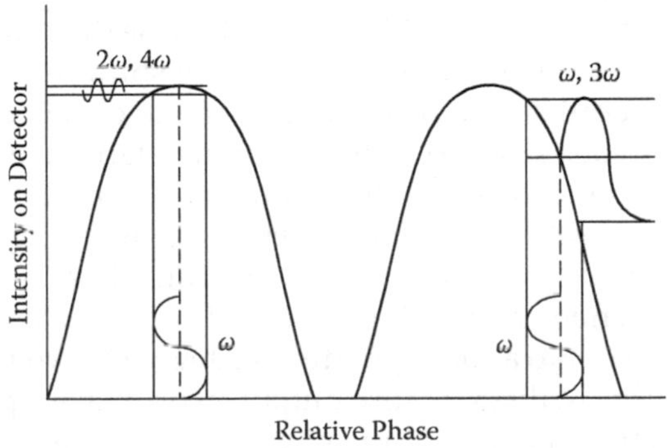

FIGURE 1.27
An open-loop fiber optic gyro has predominantly even-order harmonics in the absence of rotation. Upon rotation, the open-loop fiber optic gyro has an odd harmonic output whose amplitude indicates the magnitude of the rotation rate and whose phase indicates direction.

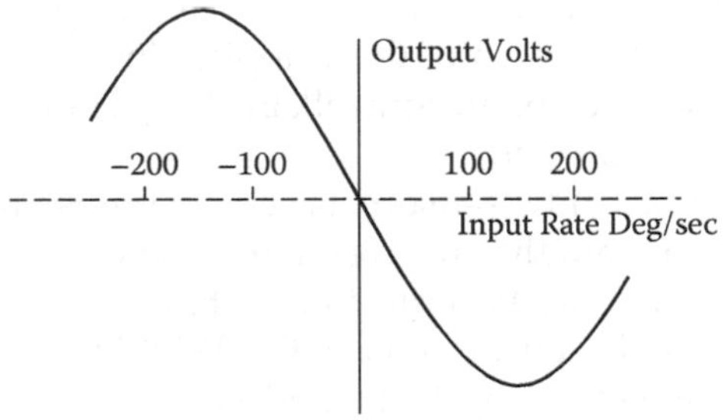

FIGURE 1.28
A typical open-loop fiber optic gyro output, obtained by measuring one of the odd harmonic output components' amplitude and phase, results in a sinusoidal output that has a region of good linearity centered about the zero rotation point.

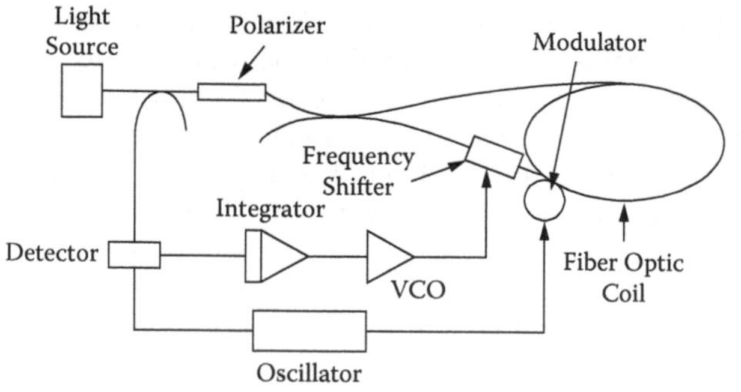

FIGURE 1.29
Closed-loop fiber optic gyros use an artificially induced nonreciprocal phase between counterpropagating light beams to counterbalance rotationally induced phase shifts. These fiber gyros have the wide dynamic range and high linearity needed to support stringent navigation requirements.

The relative frequency difference of the light beams propagating in the fiber loop can be controlled, resulting in a net phase difference proportional to the length of the fiber coil and the frequency shift. In Figure 1.29, this is done by using a modulator in the fiber optic coil to generate a phase shift at a rate ω. When the coil is rotated, a first harmonic signal at *w* is induced with phase that depends on rotation rate in a manner similar to that described previously with respect to open-loop fiber gyros. By using the rotationally induced first harmonic as an error signal, one can adjust the frequency shift by using a synchronous demodulator behind the detector to integrate the first harmonic signal into a corresponding voltage. This voltage is applied to a voltage-controlled oscillator whose output frequency is applied to the frequency shifter in the loop so that the phase relationship between the counterpropagating light beams is locked to a single value.

It is possible to use the Sagnac interferometer for other sensing and measurement tasks. Examples include slowly varying measurements of strain with 100-micron resolution over distances of about 1 kilometer [39], spectroscopic measurements of wavelength of about 2 nanometers [40], and optical fiber characterization such as thermal expansion to accuracies of about 10 parts per million [40]. In each of these applications frequency shifters are used in the Sagnac loop to obtain controllable frequency offsets between the counterpropagating light beams.

Another class of fiber optic sensors, based on the Sagnac interferometer, can be used to measure rapidly varying environmental signals such as sound [41,42]. Figure 1.30 illustrates two interconnected Sagnac loops [42] that can be used as a distributed acoustic sensor. The WDM in the figure is a device that either couples two wavelengths (λ_1 and λ_2 in this case) together or separates them.

The sensitivity of this Sagnac acoustic sensor depends on the signal's location. If the signal is in the center of the loop, the amplification is zero because both counterpropagating light beams arrive at the center of the loop at the

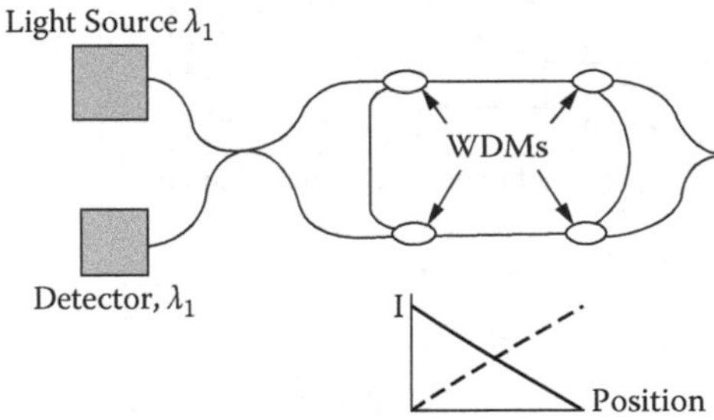

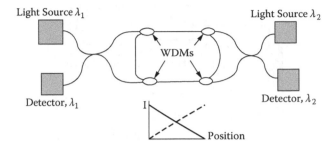

FIGURE 1.30
A distributed fiber optic acoustic sensor based on interlaced Sagnac loops allows the detection of the location and the measurement of the amplitude along a length of optical fiber that may be many kilometers long.

same time. As the signal moves away from the center, the output increases. When two Sagnac loops are superposed, as in Figure 1.30, the two outputs may be summed to give an indication of the amplitude of the signal and ratioed to determine position.

Several other combinations of interferometers have been tried for position and amplitude determinations, and the first reported success consisted of a combination of the Mach–Zehnder and Sagnac interferometers [41].

1.4.2 Mach–Zehnder and Michelson Interferometers

One of the great advantages of all-fiber interferometers, such as Mach–Zehnder and Michelson interferometers [43] in particular, is that they have extremely flexible geometries and a high sensitivity that allow the possibility of a wide variety of high-performance elements and arrays, as shown in Figure 1.31.

Figure 1.32 shows the basic elements of a Mach–Zehnder interferometer: a light source/coupler module, a transducer, and a homodyne demodulator. The light source module usually consists of a long coherence length isolated

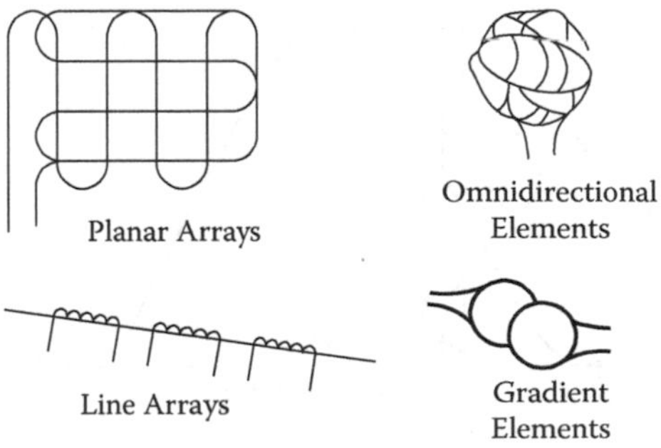

FIGURE 1.31
Flexible geometries of interferometric fiber optic sensors' transducers are one of the features of fiber sensors attractive to designers configuring special-purpose sensors.

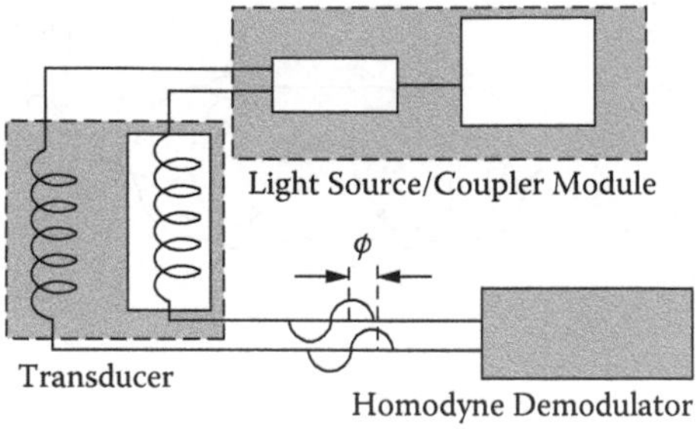

FIGURE 1.32
The basic elements of the fiber optic Mach–Zehnder interferometer are a light source module to split a light beam into two paths, a transducer used to cause an environmentally dependent differential optical path length between the two light beams, and a demodulator that measures the resulting path length difference between the two light beams.

laser diode, a beamsplitter to produce two light beams, and a means of coupling the beams to the two legs of the transducer. The transducer is configured to sense an environmental effect by isolating one light beam from the environmental effect; using the action of the environmental effect on the transducer induces an optical path length difference between the two light beams. Typically, a homodyne demodulator is used to detect the difference in optical path length (various heterodyne schemes have also been used) [43].

One of the basic issues with the Mach–Zehnder interferometer is that the sensitivity varies as a function of the relative phase of the light beams in the two legs of the interferometer, as shown in Figure 1.33. One way to solve the signal fading problem is to introduce a piezoelectric fiber stretcher into one of the legs and adjust the relative path length of the two legs for optimum sensitivity. Another approach has the same quadrature solution as the grating-based fiber sensors discussed earlier.

Figure 1.34 illustrates a homodyne demodulator. The demodulator consists of two parallel optical fibers that feed the light beams from the transducer into a graded index (GRIN) lens. The output from the GRIN lens is

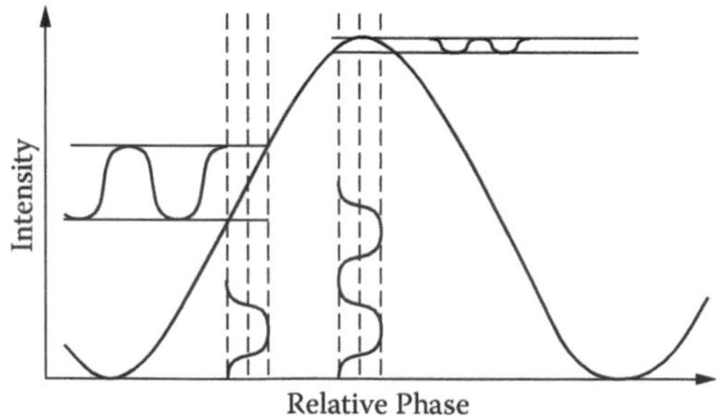

FIGURE 1.33
In the absence of compensating demodulation methods, the sensitivity of the Mach–Zehnder varies with the relative phase between the two light beams. It falls to low levels when the light beams are completely in or out of phase.

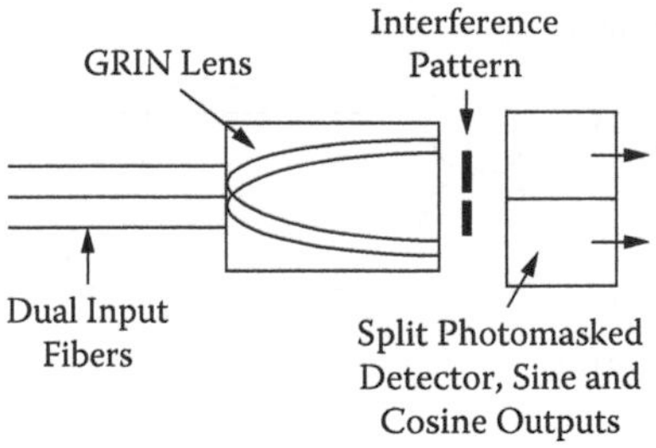

FIGURE 1.34
Quadrature demodulation avoids signal fading problems. The method shown here expands the two beams into an interference pattern that is imaged onto a split detector.

an interference pattern that "rolls" with the relative phase of the two input light beams. If a split detector is used with a photomask arranged so that the opaque and transparent line pairs on the mask in front of the split detector match the interference pattern periodicity and are 90° out of phase on the detector faces, sine and cosine outputs result. These outputs may be processed using quadrature demodulation electronics, as shown in Figure 1.35. The result is a direct measure of the phase difference.

Further improvements on these techniques have been made: notably, the phase-generated carrier approach shown in Figure 1.36. A laser diode is current modulated, resulting in the output frequency of the laser diode being frequency modulated as well. If a Mach–Zehnder interferometer is arranged so that its reference and signal leg differ in length by an amount $(L_1 - L_2)$, then the net phase difference between the two light beams is $2\pi F(L_1 - L_2)n/c$, where n is an index of refraction of the optical fiber and c is the speed of light in vacuum. If the current modulation is at a rate ω, then relative phase differences are modulated at this rate and the output on the detector will be odd and even harmonics of it. The signals riding on the carrier harmonics of ω and 2ω are in quadrature with respect to each other and can be processed using electronics similar to those of Figure 1.35.

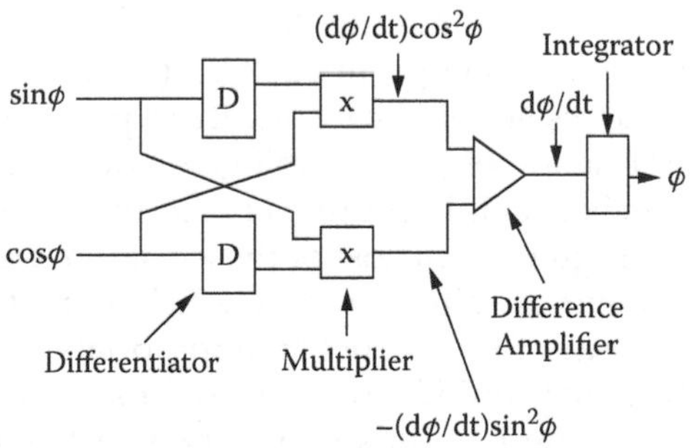

FIGURE 1.35
Quadrature demodulation electronics take the sinusoidal outputs from the split detector and convert them via cross-multiplication and differentiation into an output that can be integrated to form the direct phase difference.

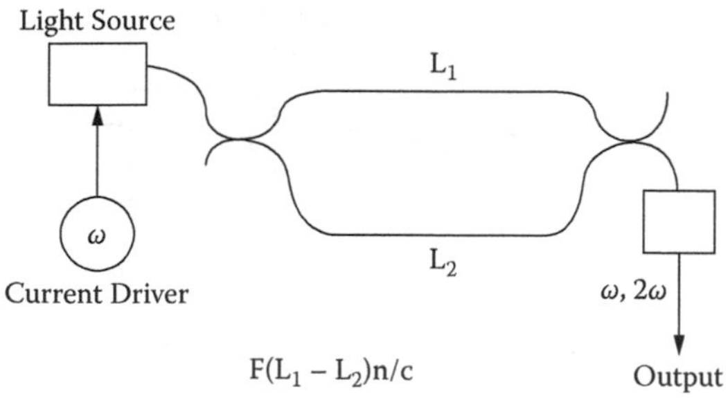

FIGURE 1.36
The phase-generated carrier technique allows quadrature detection via monitoring even and odd harmonics induced by a sinusoidally frequency-modulated light source used in combination with a length offset Mach–Zehnder interferometer to generate a modulated phase output whose first and second harmonics correspond to sine and cosine outputs.

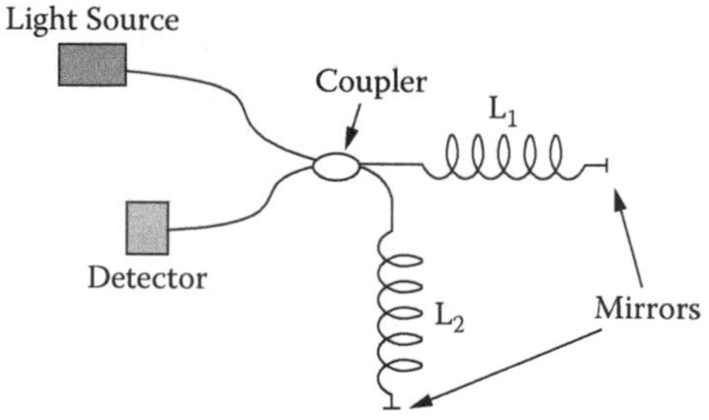

FIGURE 1.37
The fiber optic Michelson interferometer consists of two mirrored fiber ends and can utilize many of the demodulation methods and techniques associated with the Mach–Zehnder.

The Michelson interferometer in Figure 1.37 is in many respects similar to the Mach–Zehnder. The major difference is that mirrors have been put on the ends of the interferometer legs. This results in very high levels of back reflection into the light source, greatly degrading the performance of early systems. Using improved diode pumped YAG (yttrium aluminum garnet) ring lasers as light sources largely overcame these problems. In combination with the recent introduction of phase conjugate mirrors to eliminate polarization fading, the Michelson is becoming an alternative for systems that can tolerate the relatively high present cost of these components.

In order to implement an effective Mach–Zehnder or Michelson-based fiber sensor, it is necessary to construct an appropriate transducer. This can involve a fiber coating that could be optimized for acoustic, electric, or magnetic field response. Figure 1.38 illustrates a two-part coating that consists of a primary and secondary layer. These layers are designed for optimal response to pressure waves and for minimal acoustic mismatches between the medium in which the pressure waves propagate and the optical fiber. These coated fibers are often used in combination with compliant mandrills

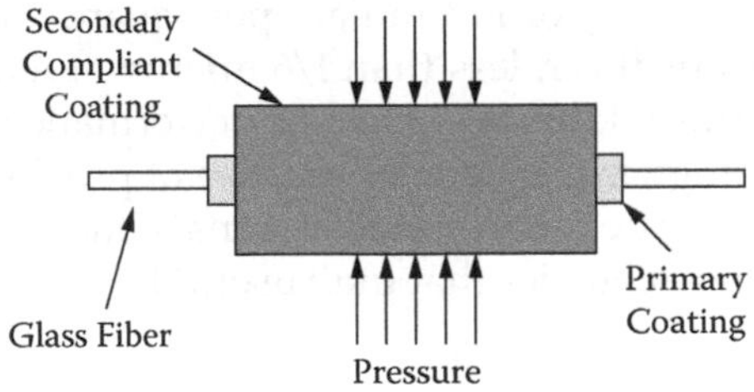

FIGURE 1.38
Coatings can be used to optimize the sensitivity of fiber sensors. An example would be to use soft and hard coatings over an optical fiber to minimize the acoustic mismatch between acoustic pressure waves in water and the glass optical fiber.

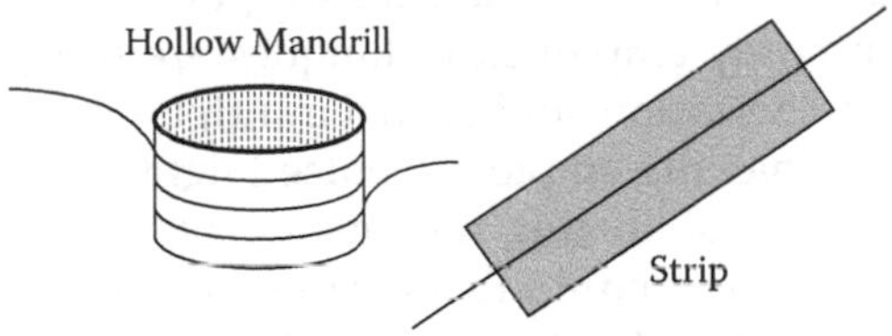

FIGURE 1.39
Optical fiber bonded to hollow mandrills and strips of environmentally sensitive material are common methods used to mechanically amplify environmental signals for detection by fiber sensors.

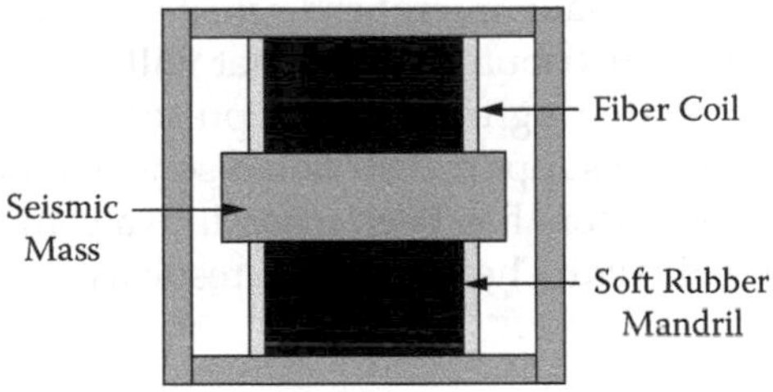

FIGURE 1.40
Differential methods are used to amplify environmental signals. In this case a seismic/vibration sensor consists of a mass placed between two fiber coils and encased in a fixed housing.

or strips of material as in Figure 1.39 that act to amplify the environmentally induced optical path length difference.

In many cases the mechanical details of the transducer design are critical to good performance such as the seismic/vibration sensor of Figure 1.40. Generally, the Mach–Zehnder and Michelson interferometers can be configured with sensitivities that are better than 10^{-6} radians per square root hertz. For optical receivers, the noise level decreases as a function of frequency. This phenomenon results in specifications in radians per square root hertz. As an example, a sensitivity of 10^{-6} radians per square root hertz at 1 hertz means a sensitivity of 10^{-6} radians, while at 100 hertz the sensitivity is 10^{-7} radians.

As an example, a sensitivity of 10^{-6} radians per square root hertz means that for a 1-meter-long transducer, less than 1/6 micron of length change can be resolved at 1-hertz bandwidths [44]. The best performance for these sensors is usually achieved at higher frequencies because of problems associated with the sensors also picking up environmental signals due to temperature fluctuations, vibrations, and acoustics that limit useful low-frequency sensitivity.

1.5 Multiplexing and Distributed Sensing

Many of the intrinsic and extrinsic sensors may be multiplexed [45], offering the possibility of large numbers of sensors supported by a single fiber optic line. The most commonly employed techniques are time, frequency, wavelength, coherence, polarization, and spatial multiplexing.

Time division multiplexing employs a pulsed light source, launching light into an optical fiber and analyzing the time delay to discriminate between sensors. This technique is commonly employed to support distributed sensors where measurements of strain, temperature, or other parameters are collected. Figure 1.41 illustrates a time division multiplexed system that uses microbend-sensitive areas on pipe joints.

As the pipe joints are stressed, microbending loss increases and the time delay associated with these losses allows the location of faulty joints. The entire length of the fiber can be made microbend sensitive and Rayleigh scattering loss is used to support a distributed sensor that will predominantly measure strain. Other types of scattering from optical pulses propagating down optical fiber have been used to support distributed sensing; notably, Raman scattering for temperature sensors has been made into a commercial product by York Technology and Hitachi. These units can resolve temperature changes of

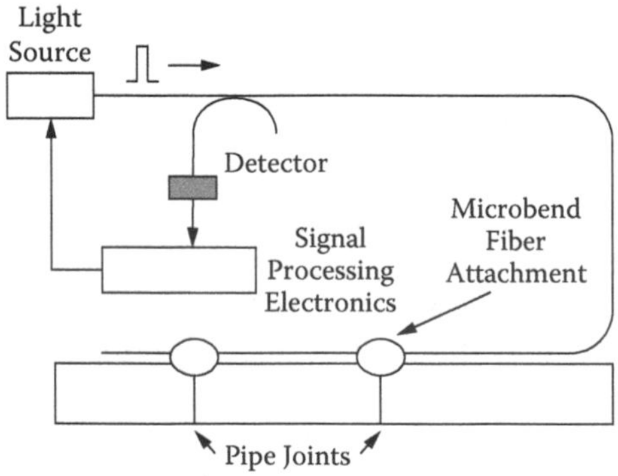

FIGURE 1.41
Time division multiplexing methods can be used in combination with microbend-sensitive optical fiber to locate the position of stress along a pipeline.

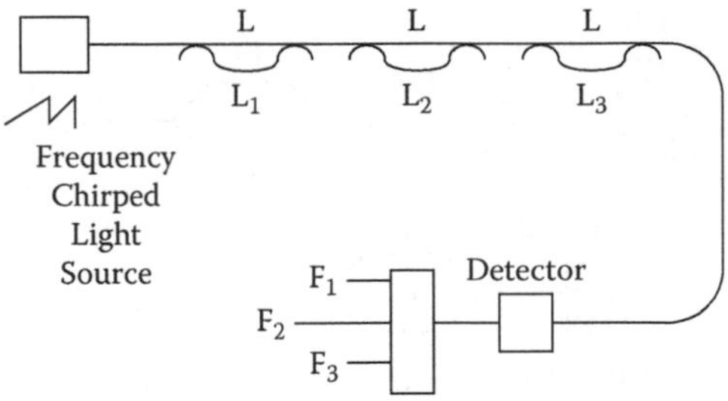

FIGURE 1.42
Frequency division multiplexing can be used to tag a series of fiber sensors. In this case the Mach–Zehnder interferometers are shown with a carrier frequency on which the output signal rides.

about 1°C with spatial resolution of 1 meter for a 1-kilometer sensor using an integration time of about 5 minutes. Brillouin scattering has been used in laboratory experiments to support both strain and temperature measurements.

A frequency division multiplexed system is shown in Figure 1.42. In this example a laser diode is frequency chirped by driving it with a sawtooth current drive. Successive Mach–Zehnder interferometers are offset with incremental lengths $(L - L_1)$, $(L - L_2)$, and $(L - L_3)$, which differ sufficiently so that the resultant carrier frequency of each sensor $(dF/dt)(L - L_n)$ is easily separable from the other sensors via electronic filtering of the output of the detector.

Wavelength division multiplexing is one of the best methods of multiplexing as it uses optical power very efficiently. It also has the advantage of being easily integrated into other multiplexing systems, allowing the possibility of large numbers of sensors supported in a single fiber line. Figure 1.43 illustrates a system where a broadband light source, such as a light-emitting diode, is coupled into a series of fiber sensors that reflect signals over wavelength bands that are subsets of the light source spectrum. A dispersive element, such as grating or prism, is used to separate the signals from the sensors onto separate detectors.

Light sources can have widely varying coherence lengths depending on their spectrum. By using light sources that have coherence lengths that are short compared to offsets between the reference and signal legs in

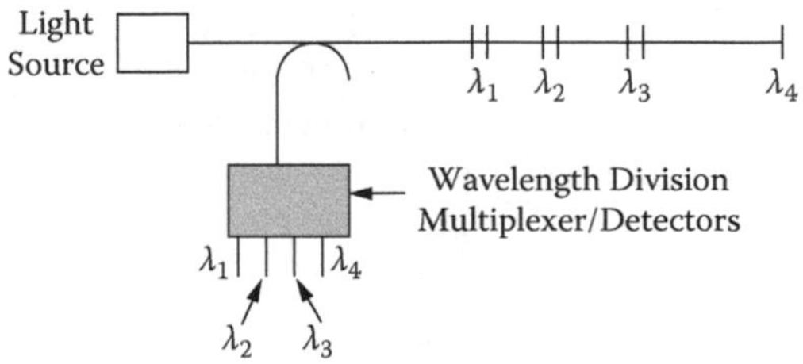

FIGURE 1.43
Wavelength division multiplexing is often very energy efficient. A series of fiber sensors is multiplexed by being arranged to reflect in a particular spectral band that is split via a dispersive element onto separate detectors.

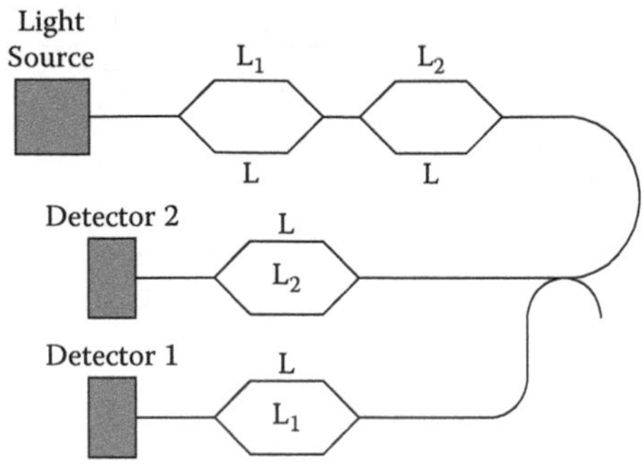

FIGURE 1.44
A low-coherence light source is used to multiplex two Mach–Zehnder interferometers by using offset lengths and counterbalancing interferometers.

Mach–Zehnder interferometers and between successive sensors, a coherence multiplexed system similar to Figure 1.44 may be set up. The signal is extracted by putting a rebalancing interferometer in front of each detector so that the sensor signals may be processed. Coherence multiplexing is not used as commonly as time, frequency, and wavelength division multiplexing because of optical power budgets and the additional complexities in setting up the optics properly. It is still a potentially powerful technique and may become more widely used as optical component performance and availability continue to improve, especially in the area of integrated optic chips, where control of optical path length differences is relatively straightforward.

One of the least commonly used techniques is polarization multiplexing. In this case the idea is to launch light with particular polarization states and extract each state. A possible application is shown in Figure 1.45, where light is launched with two orthogonal polarization modes; preserving fiber and

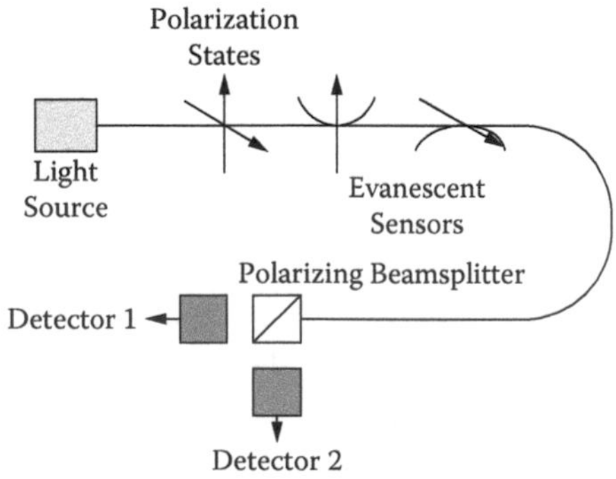

FIGURE 1.45
Polarization multiplexing is used to support two fiber sensors that access the cross-polarization states of polarization-preserving optical fiber.

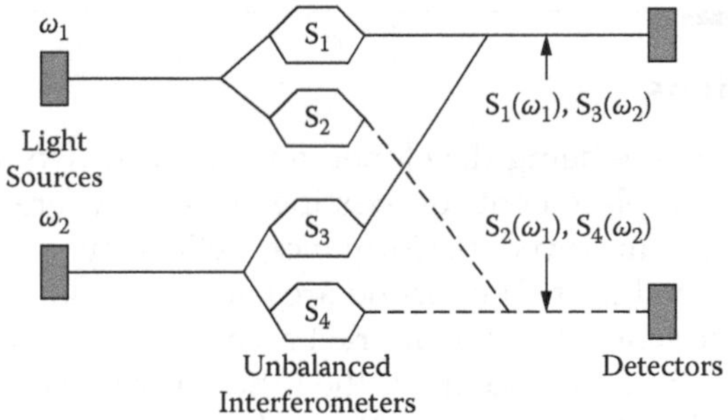

FIGURE 1.46
Spatial multiplexing of four fiber optic sensors may be accomplished by operating two light sources with different carrier frequencies and cross-coupling the sensor outputs onto two output fibers.

evanescent sensors have been set up along each of the axes. A polarizing beamsplitter is used to separate the two signals. There is recent interest in using polarization-preserving fiber in combination with time domain techniques to form polarization-based distributed fiber sensors. This has the potential to offer multiple sensing parameters along a single fiber line.

Finally, it is possible to use spatial techniques to generate large sensor arrays using relatively few input and output optical fibers. Figure 1.46 shows a 2 by 2 array of sensors where two light sources are amplitude modulated at different frequencies. Two sensors are driven at one frequency and two more at the second. The signals from the sensors are put onto two output fibers, each carrying a sensor signal from two sensors at different frequencies. This sort of multiplexing is easily extended to *m* input fibers and *n* output fibers to form *m* by *n* arrays of sensors, as in Figure 1.47.

All of these multiplexing techniques can be used in combination with one another to form extremely large arrays.

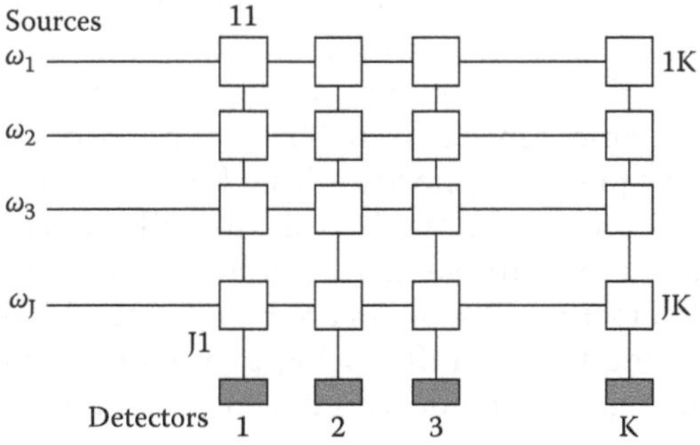

FIGURE 1.47
Extensions of spatial multiplexing the *JK* sensors can be accomplished by operating *J* light sources at *J* different frequencies and cross-coupling to *K* output fibers.

1.6 Applications

Fiber optic sensors are being developed and used in two major ways. The first is as a direct replacement for existing sensors where the fiber sensor offers significantly improved performance, reliability, safety, and/or cost advantages to the end user. The second area is the development and deployment of fiber optic sensors in new market areas.

For the case of direct replacement, the inherent value of the fiber sensor, to the customer, has to be sufficiently high to displace older technology. Because this often involves replacing technology the customer is familiar with, the improvements must be substantial.

The most obvious example of a fiber optic sensor succeeding in this arena is the fiber optic gyro, which is displacing both mechanical and ring laser gyros for medium-accuracy devices. As this technology matures, it can be expected that the fiber gyro will dominate large segments of this market.

Significant development efforts are underway in the United States in the area of fly-by-light [9], where conventional electronic sensor technologies are targeted to be replaced by equivalent fiber optic sensor technology that offers sensors with relative immunity to electromagnetic interference, significant weight savings, and safety improvements.

In manufacturing, fiber sensors are being developed to support process control. Often the selling points for these sensors are improvements in environmental ruggedness and safety, especially in areas where electrical discharges could be hazardous.

One other area where fiber optic sensors are being mass produced is the field of medicine [46–49], where they are being used to measure blood-gas parameters and dosage levels. Because these sensors are completely passive, they pose no electrical-shock threat to the patient and their inherent safety has led to a relatively rapid introduction.

The automotive industry, construction industry, and other traditional sensor users remain relatively untouched by fiber sensors, mainly because of cost considerations. This can be expected to change as the improvements in optoelectronics and fiber optic communications continue to expand along with the continuing emergence of new fiber optic sensors.

New market areas present opportunities where equivalent sensors do not exist. New sensors, once developed, will most likely have a large impact in these areas. A prime example of this is in the area of fiber optic smart structures [50–53]. Fiber optic sensors are being embedded into or attached to materials (1) during the manufacturing process to enhance process control systems, (2) to augment nondestructive evaluation once parts have been made, (3) to form health and damage assessment systems once parts have been assembled into structures, and (4) to enhance control systems. A basic fiber optic smart structure system is shown in Figure 1.48.

Fiber optic sensors can be embedded in a panel and multiplexed to minimize the number of leads. The signals from the panel are fed back to an

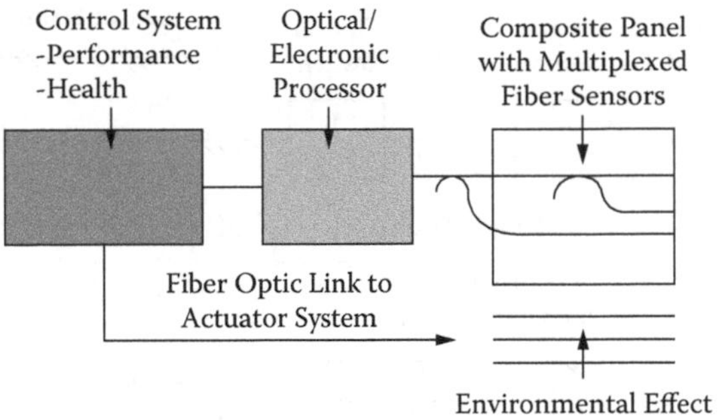

FIGURE 1.48
Fiber optic smart structure systems consist of optical fiber sensors embedded or attached to parts sensing environmental effects that are multiplexed and directed down. The effects are then sent through an optical/electronic signal processor that in turn feeds the information to a control system that may or may not act on the information via a fiber link to an actuator.

optical/electronic processor for decoding. The information is formatted and transmitted to a control system that could be augmenting performance or assessing health. The control system would then act, via a fiber optic link, to modify the structure in response to the environmental effect.

Figure 1.49 shows how the system might be used in manufacturing. Here fiber sensors are attached to a part to be processed in an autoclave. Sensors could be used to monitor internal temperature, strain, and degree of cure. These measurements could be used to control the autoclaving process, improving the yield and quality of the parts.

Interesting areas for health and damage assessment systems are on large structures such as buildings, bridges, dams, aircraft, and spacecraft. In order to support these types of structures, it will be necessary to have very large numbers of sensors that are rapidly reconfigurable and redundant. It will also be absolutely necessary to demonstrate the value and cost effectiveness of these systems to the end users.

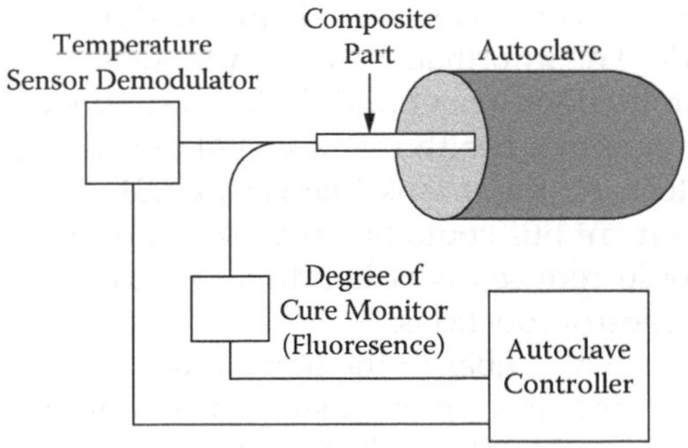

FIGURE 1.49
Smart manufacturing systems offer the prospect of monitoring key parameters of parts as they are being made, which increases yield and lowers overall costs.

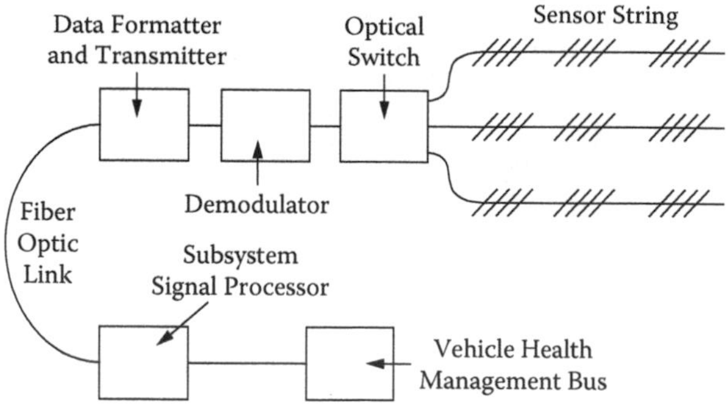

FIGURE 1.50
A modular architecture for a large smart structure system would consist of strings of fiber sensors accessible via an optical switch and demodulator system that could select key sensors in each string. The information would then be formatted and transmitted after conditioning to a vehicle health management bus.

One approach to this problem is to use fiber sensors that have the potential to be manufactured cheaply in very large quantities while offering superior performance characteristics. Two candidates under investigation are the fiber gratings and etalons described earlier. Both offer the advantages of spectrally based sensors and have the prospect of rapid in-line manufacture. In the case of the fiber grating, the early demonstration of fiber being written into it as it is being pulled has been especially impressive. These fiber sensors could be folded into the wavelength and time division multiplexed modular architecture shown in Figure 1.50. Here sensors are multiplexed along fiber strings and an optical switch is used to support the many strings. Potentially, the fiber strings could have tens or hundreds of sensors, and the optical switches could support a like number of strings. To avoid overloading the system, the output from the sensors could be slowly scanned to determine status in a continuously updated manner.

When an event occurred that required a more detailed assessment, the appropriate strings and the sensors in them could be monitored in a high-performance mode. The information from these sensors would then be formatted and transmitted via a fiber optic link to a subsystem signal processor before introduction onto a health management bus. In the case of avionics, the system architecture might look like Figure 1.51. The information from the health management bus could be processed and distributed to the pilot or, more likely, could reduce his or her direct workload, leaving more time for the necessary control functions.

As fiber to the curb and fiber to the home move closer to reality, there is the prospect of merging fiber optic sensor and communication systems into very large systems capable of monitoring the status of buildings, bridges, highways, and factories over widely dispersed areas. Functions such as fire, police, maintenance scheduling, and emergency response to earthquakes,

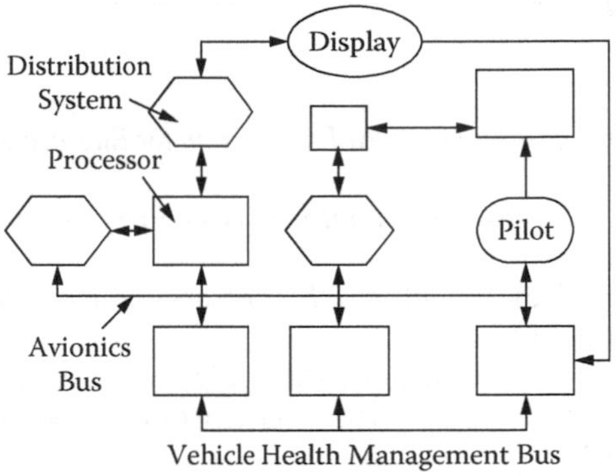

FIGURE 1.51
A typical vehicle health management bus for an avionics system would be the interface point for the fiber optic smart structure modules of Figure 1.50.

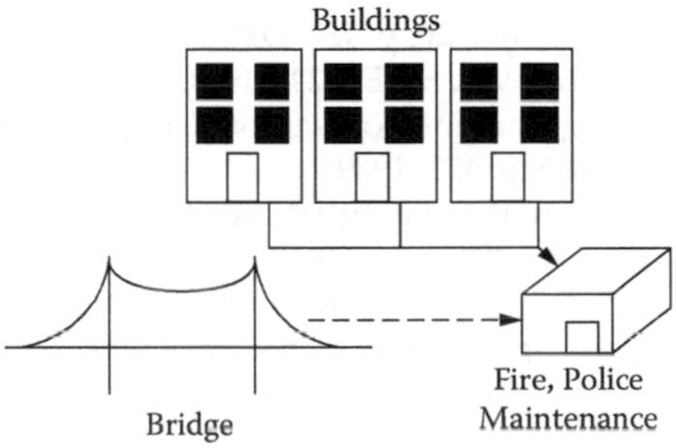

FIGURE 1.52
Fiber optic sensor networks to monitor the status of widely dispersed assets as buildings, bridges, and dams could be used to augment fire, police, and maintenance services.

hurricanes, and tornadoes could be readily integrated into very wide area networks of sensors, as in Figure 1.52.

It is also possible to use fiber optic sensors in combination with fiber optic communication links to monitor stress buildup in critical fault locations and dome buildup of volcanoes. These widely dispersed fiber networks may offer the first real means of gathering information necessary to form prediction models for these natural hazards.

Acknowledgment

Figure 1.1 through Figure 1.52 are drawn from the *Fiber Optic Sensor Workbook*, copyright Eric Udd/Blue Road Research, and are used with permission.

References

1. E. Udd, ed., *Fiber Optic Sensors: An Introduction for Engineers and Scientists*, Wiley, New York, 1991.
2. J. Dakin and B. Culshaw, *Optical Fiber Sensors: Principles and Components*, Vol. 1, Artech, Boston, 1988.
3. B. Culshaw and J. Dakin, *Optical Fiber Sensors: Systems and Applications*, Vol. 2, Artech, Norwood, MA, 1989.
4. T. G. Giallorenzi, J. A. Bucaro, A. Dandridge, G. H. Sigel, Jr., J. H. Cole, S. C. Rashleigh, and R. G. Priest, Optical fiber sensor technology, *IEEE J. Quant. Elec.*, **QE-18**, p. 626, 1982.
5. D. A. Krohn, *Fiber Optic Sensors: Fundamentals and Applications*, Instrument Society of America, Research Triangle Park, NC, 1988.
6. E. Udd, ed., Fiber optic sensors, *Proc. SPIE*, **CR-44**, 1992.
7. S. K. Yao and C. K. Asawa, Fiber optical intensity sensors, *IEEE J. Sel. Areas Commun.*, **SAC-1**, 3, 1983.
8. N. Lagokos, L. Litovitz, P. Macedo, and R. Mohr, Multimode optical fiber displacement sensor, *Appl. Opt.*, **20**, p. 167, 1981.
9. E. Udd, ed., Fly-by-light, *Proc. SPIE*, **2295**, 1994.
10. K. Fritsch, Digital angular position sensor using wavelength division multiplexing, *Proc. SPIE*, **1169**, p. 453, 1989.
11. K. Fritsch and G. Beheim, Wavelength division multiplexed digital optical position transducer, *Opt. Lett.*, **11**, p. 1, 1986.
12. D. Varshneya and W. L. Glomb, Applications of time and wavelength division multiplexing to digital optical code plates, *Proc. SPIE*, **838**, p. 210, 1987.
13. J. W. Snow, A fiber optic fluid level sensor: Practical considerations, *Proc. SPIE*, **954**, p. 88, 1983.
14. T. E. Clark and M. W. Burrell, Thermally switched coupler, *Proc. SPIE*, **986**, p. 164, 1988.
15. Y. F. Li and J. W. Lit, Temperature effects of a multimode biconical fiber coupler, *Appl. Opt.*, **25**, p. 1765, 1986.
16. Y. Murakami and S. Sudo, Coupling characteristics measurements between curved waveguides using a two core fiber coupler, *Appl. Opt.*, **20**, p. 417, 1981.
17. D. A. Nolan, P. E. Blaszyk, and E. Udd, Optical fibers, in *Fiber Optic Sensors: An Introduction for Engineers and Scientists*, E. Udd, ed., Wiley, New York, 1991.
18. J. W. Berthold, W. L. Ghering, and D. Varshneya, Design and characterization of a high temperature, fiber optic pressure transducer, *IEEE J. Lightwave Tech.*, **LT-5**, p. 1, 1987.
19. D. R. Miers, D. Raj, and J. W. Berthold, Design and characterization of fiberoptic accelerometers, *Proc. SPIE*, **838**, p. 314, 1987.
20. W. B. Spillman and R. L. Gravel, Moving fiber optic hydrophone, *Opt. Lett.*, **5**, p. 30, 1980.
21. E. Udd and P. M. Turek, Single mode fiber optic vibration sensor, *Proc. SPIE*, **566**, p. 135, 1985.
22. D. A. Christensen and J. T. Ives, Fiberoptic temperature probe using a semiconductor sensor, *Proc. NATO Advanced Studies Institute*, Dordrecht, The Netherlands, p. 361, 1987.

23. S. D. Schwab and R. L. Levy, In-service characterization of composite matrices with an embedded fluorescence optrode sensor, *Proc. SPIE,* **1170**, p. 230, 1989.
24. K. T. V. Gratten, R. K. Selli, and A. W. Palmer, A miniature fluorescence referenced glass absorption thermometer, *Proc. 4th Int. Conf. Opt. Fiber Sensors,* Tokyo, p. 315, 1986.
25. W. W. Morey, G. Meltz, and W. H. Glenn, Bragg-grating temperature and strain sensors, *Proc. Opt. Fiber Sensors,* '89, Springer–Verlag, Berlin, p. 526, 1989.
26. G. A. Ball, G. Meltz, and W. W. Morey, Polarimetric heterodyning Bragg-grating fiber laser, *Opt. Lett.,* **18**, p. 1976, 1993.
27. J. R. Dunphy, G. Meltz, F. P. Lamm, and W. W. Morey, Multi-function, distributed optical fiber sensor for composite cure and response monitoring, *Proc. SPIE,* **1370**, p. 116, 1990.
28. W. W. Morey, Distributed fiber grating sensors, *Proc. 7th Opt. Fiber Sensor Conf., IREE Australia, Sydney,* p. 285, 1990.
29. A. D. Kersey, T. A. Berkoff, and W. W. Morey, Fiber-grating based strain sensor with phase sensitive detection, *Proc. SPIE,* **1777**, p. 61, 1992.
30. D. A. Jackson, A. B. Lobo Ribeiro, L. Reekie, and J. L. Archambault, Simple multiplexing scheme for a fiber optic grating sensor network, *Opt. Lett.,* **18**, p. 1192, 1993.
31. E. W. Saaski, J. C. Hartl, G. L. Mitchell, R. A. Wolthuis, and M. A. Afromowitz, A family of fiber optic sensor using cavity resonator microshifts, *Proc. 4th Int. Conf. Opt. Fiber Sensors,* Tokyo, 1986.
32. C. E. Lee and H. F. Taylor, Interferometeric optical fiber sensors using internal mirrors, *Electron. Lett.,* **24**, p. 193, 1988.
33. C. E. Lee and H. F. Taylor, Interferometeric fiber optic temperature sensor using a low coherence light source, *Proc. SPIE,* **1370**, p. 356, 1990.
34. Private communication, Elric Saaski, Research International, Woodinville, WA.
35. H. Lefevre, *The Fiber Optic Gyroscope,* Artech, Norwood, MA, 1993.
36. W. K. Burns, ed., *Optical Fiber Rotation Sensing,* Academic Press, San Diego, 1994.
37. R. B. Smith, ed., *Selected Papers on Fiber Optic Gyroscopes,* SPIE Milestone Series, **MS 8**, 1989.
38. S. Ezekial and E. Udd, ed., *Fiber Optic Gyros: 15th Anniversary Conf., Proc. SPIE,* **1585**, 1991.
39. R. J. Michal, E. Udd, and J. P. Theriault, Derivative fiber-optic sensors based on the phase nulling optical gyro, *Proc. SPIE,* **719**, 1986.
40. E. Udd, R. J. Michal, J. P. Theriault, R. F. Cahill, High accuracy light source wavelength and optical fiber dispersion measurements using the Sagnac interferometer, *Proc. 7th Opt. Fiber Sensors Conf.,* IREE Australia, Sydney, p. 329, 1990.
41. J. P. Dakin, D. A. J. Pearce, A. P. Strong, and C. A. Wade, A novel distributed optical fiber sensing system enabling the location of disturbances in a Sagnac loop interferometer, *Proc. SPIE,* **838**, p. 325, 1987.
42. E. Udd, Sagnac distributed sensor concepts, *Proc. SPIE,* **1586**, p. 46, 1991.
43. A. Dandridge, Fiber optic sensors based on the Mach–Zehnder and Michelson interferometers, in *Fiber Optic Sensors: An Introduction for Engineers and Scientists,* E. Udd, ed., Wiley, New York, 1991.
44. F. Bucholtz, D. M. Dagenais, and K. P. Koo, High frequency fiber-optic magnetometer with 70 ft per square root hertz resolution, *Electron. Lett.,* **25**, p. 1719, 1989.
45. A. D. Kersey, Distributed and multiplexed fiber optic sensors, in *Fiber Optic Sensors: An Introduction for Engineers and Scientists,* E. Udd, ed., Wiley, New York, 1991.

46. O. S. Wolfbeis and P. Greguss, eds., Biochemical and medical sensors, *Proc. SPIE*, **2085**, 1993.
47. A. Katzir, ed., Optical fibers in medicine VIII, *Proc. SPIE*, **1893**, 1993.
48. F. P. Milanovich, ed., Fiber optic sensors in medical diagnostics, *Proc. SPIE*, **1886**, 1993.
49. R. A. Lieberman, ed., Chemical, biochemical, and environmental fiber sensors V, *Proc. SPIE*, **1993**.
50. E. Udd, Fiber optic smart structures, in *Fiber Optic Sensors: An Introduction for Engineers and Scientists*, Wiley, New York, 1991.
51. R. Clauss and E. Udd, eds., Fiber optic smart structures and skins IV, *Proc. SPIE*, **1588**, 1991.
52. J. S. Sirkis, ed., Smart sensing, processing and instrumentation, *Proc. SPIE*, **2191**, 1994.
53. E. Udd, ed., *Fiber Optic Smart Structures*, Wiley, New York, 1995.

2

Fiber Optic Sensors Based upon the Fabry–Perot Interferometer

Henry F. Taylor

CONTENTS

2.1 Introduction

The Fabry–Perot interferometer (FPI), sometimes called the Fabry–Perot etalon, consists of two mirrors of reflectance R_1 and R_2 separated by a cavity

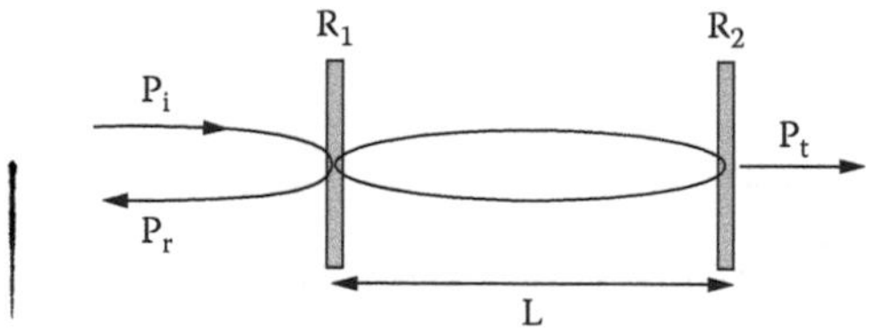

FIGURE 2.1
Fabry–Perot interferometer, with P_i, P_r, and P_t the incident, reflected, and transmitted optical power, respectively.

of length L, as in Figure 2.1. Since its invention in the late 19th century [1], the bulk-optics version of the FPI has been widely used for high-resolution spectroscopy. In the early 1980s, the first results on fiber optic versions of the FPI were reported. In the late 1980s, fiber Fabry–Perot interferometers began to be applied to the sensing of temperature, strain, and ultrasonic pressure in composite materials. This early work laid the foundation for extensive research and development as well as commercialization, which followed during the 1990s.

Fiber Fabry–Perot interferometers are extremely sensitive to perturbations that affect the optical path length between the two mirrors. The sensing region can be very compact—equivalent to a "point" transducer in some applications. Unlike other fiber interferometers (Mach–Zehnder, Michelson, Sagnac) used for sensing, the Fabry–Perot contains no fiber couplers—components that can complicate the sensor's deployment and the interpretation of data. The fiber FPI would appear to be an ideal transducer for many smart structure sensing applications, including those in which the sensor must be embedded in a composite or metal. Finally, these versatile measurement devices are amenable to the application of space division, time division, frequency division, and coherence multiplexing techniques for reducing the cost of multipoint monitoring.

Later sections in this chapter review the theory of the FPI; describe a number of configurations for fiber FPIs; review optical monitoring and multiplexing methods; discuss the embedding of the sensors in materials of technological interest; summarize results achieved in measuring temperature, strain, pressure, and several other measurands; and speculate briefly on future directions for research and development.

2.2 Theory of the Fabry–Perot Interferometer

Mathematical analyses developed decades ago for the bulk FPI also apply to the fiber optic interferometers of interest here. In this section, general expressions for the transmittance and reflectance of the FPI that are applicable in characterizing the performance of the fiber optic sensors are introduced.

The individual mirrors in the FPI can be characterized by transmittances T_i and reflectances R_i, $i = 1, 2$, such that $R_i + T_i = 1$. The excess loss, which corresponds to the portion of the incident power absorbed or scattered out of the beam by the mirror, is neglected in this analysis. The Fabry–Perot reflectance R_{FP} and transmittance T_{FP} are found to be [2]

$$R_{FP} = \frac{R_1 + R_2 + 2\sqrt{R_1 R_2}\cos\phi}{1 + R_1 R_2 + 2\sqrt{R_1 R_2}\cos\phi} \tag{2.1}$$

$$T_{FP} = \frac{T_1 T_2}{1 + R_1 R_2 + 2\sqrt{R_1 R_2}\cos\phi} \tag{2.2}$$

where R_{FP} represents the ratio of the power reflected by the FPI P_r to the incident power P_i, T_{FP} is the ratio of the transmitted power P_t to the incident power, and ϕ, the round-trip propagation phase shift in the interferometer, is given by

$$\phi = \frac{4\pi n L}{\lambda} \tag{2.3}$$

with n the refractive index of the region between the mirrors and λ the freespace optical wavelength. It has been assumed that the light experiences a $\pi/2$ phase shift at each reflection, as appropriate for dielectric mirrors, which is added to the propagation phase shift of Eq. (2.3).

It is evident from Eq. (2.2) that T_{FP} is a maximum for $\cos\phi = -1$ or $\phi = (2m + 1)\pi$, with m an integer. If we define $\Delta = \phi - (2m + 1)\pi$, then near a maximum in T_{FP}, $\cos\phi \approx -(1 - \Delta^2/2)$, with $\Delta \ll 1$. In the case that the mirror reflectances are equal and approach unity, then Eq. (2.2) simplifies to

$$T_{FP} = \frac{T^2}{(1-R)^2 + R\Delta^2} \tag{2.4}$$

where $R = R_1 = R_2$ and $T = 1 - R$. The maximum transmittance occurs when $\Delta = 0$. The finesse F, a frequently used figure of merit for the FPI, is defined as the ratio of the phase change between adjacent transmittance peaks to the phase change between half-maximum points on either side of a peak. From Eq. (2.2) it follows that T_{FP} is a periodic function of ϕ with period 2π, so that a phase change of 2π radians is required to tune from one peak to the next. But it follows from Eq. (2.4) that, for high-reflectance mirrors, T_{FP} is half its maximum value for $\Delta = \pm(1-R)/\sqrt{R}$. This implies that the finesse can be written as

$$F = \pi\sqrt{R}/(1-R) \tag{2.5}$$

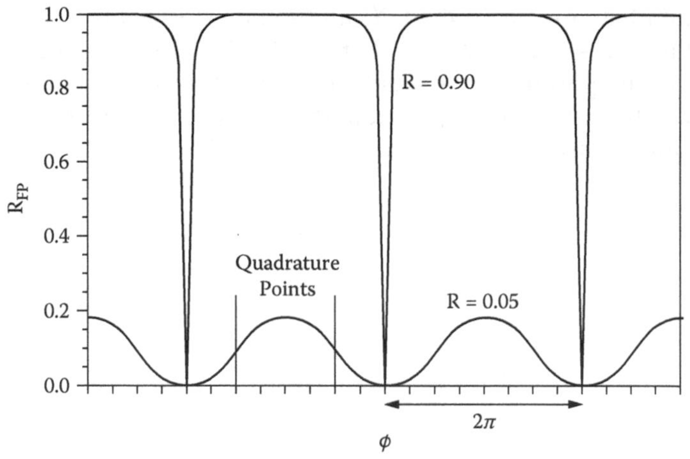

FIGURE 2.2
Dependence of reflectance R_{FP} of a lossless FPI on round-trip phase shift for two values of the individual mirror reflectance, $R = 0.9$ and 0.05. For $R = 0.05$, two quadrature points corresponding to maximum sensitivity of R_{FP} to ϕ are indicated.

Thus, in an interferometer with lossless mirrors, $F = 29.8$ for $R = 0.9$ and $F = 312.6$ for $R = 0.99$.

Another limiting case where the mirror reflectances are low is of particular interest in the case of fiber Fabry–Perot sensors. Assuming once again that the mirrors have equal reflectances, with $R = R_1 = R_2$, it follows from Eqs. (2.1) and (2.2) that if $R \ll 1$, then

$$R_{FP} \cong 2R(1 + \cos\phi) \tag{2.6}$$

$$T_{FP} \cong 1 - 2R(1 + \cos\phi) \tag{2.7}$$

It should be noted that the concept of finesses is not intended to apply to FPIs with $R \ll 1$. In fact, it follows from our definition of finesse and Eq. (2.2) that $F = 1$ for $R = 0.172$ and F is undefined for $R < 0.172$.

The reflectance for a lossless FPI given by Eq. (2.1) is plotted as a function of round-trip phase shift in Figure 2.2 for $R = 0.9$ and $R = 0.05$. The approximate expression from Eq. (2.6) closely follows the exact curve in the latter case.

2.3 Fiber Fabry–Perot Sensor Configurations

The first reports on fiber Fabry–Perot interferometers began to appear in the early 1980s. In these early experiments the interferometer cavity was a single-mode fiber with dielectric mirrors [3,4] or cleaved fiber ends [5] to serve as mirrors. Reflectance or transmittance was monitored as light from a HeNe laser was focused into the fiber interferometer. The expected high sensitivity of these interferometers to strain and temperature was confirmed.

The fiber Fabry–Perot sensors that have evolved from this earlier work are generally classified as intrinsic or extrinsic. In both intrinsic and extrinsic sensors, a fiber (in most cases, single mode) transports light from an emitter to the interferometer and from the interferometer to a photodetector. In an intrinsic fiber Fabry–Perot interferometric sensor, generally termed an "FFPI" sensor, the two mirrors are separated by a length of single-mode fiber and the measurand affects the optical path length of the light propagating in the fiber itself. In an extrinsic fiber-based Fabry–Perot sensor, generally referred to as an "EFPI" sensor, the two mirrors are separated by an air gap or by some solid material other than the fiber. Thus, in the EFPI sensor the measurand affects the optical path length in a medium other than the fiber that transports the monitoring light to and from the interferometer. Both FFPIs and EFPIs are designed such that a measurand affects the optical length of the cavity, and light reflected or transmitted by the interferometer is converted by a photodetector to an electrical signal that is processed electronically to evaluate the measurand.

2.3.1 Intrinsic Fiber Optic Fabry–Perot Interferometer (FFPI) Sensors

Three configurations for FFPI sensors are shown in Figure 2.3. In the simplest case (Figure 2.3(a)) a cleaved or polished end of the fiber forms one mirror and the second mirror is internal to the fiber. The intrinsic version most widely studied and used (Figure 2.3(b)) has two internal mirrors followed by a "nonreflecting" fiber end. A broken or cut fiber end generally shows very low reflectance; alternatively, the end can be polished at an angle. Finally, the interferometer cavity can be formed by fiber Bragg reflectors (Figure 2.3(c)).

The use of internal fiber mirrors in FFPIs was first reported in 1987. Internal mirrors are reflectors formed as an integral part of a continuous length of fiber. They have been produced by "bad" fusion splices between uncoated fibers [6,7] and by fusion splicing of an uncoated fiber to a fiber with a thin dielectric [8,9] or metallic [10] coating on the end.

Internal mirrors formed from dielectric coatings have shown the best mechanical properties, lowest excess optical loss, and widest range of reflectance values. The most commonly used mirror material is TiO_2, which has a refractive index of 2.4 (vs. 1.46 for fused silica). The reflection results from the refractive index discontinuities at the two film–fiber interfaces. The TiO_2 films have been produced by sputtering in an rf planar magnetron system or by electron beam evaporation. Typical film thicknesses are in the neighborhood of 100 nm. The fusion splicer is operated at a lower arc current and duration than for a normal splice, and several splicing pulses are used to form a mirror [8]. The mirror reflectance generally decreases monotonically as a function of the number of splicing pulses, making it possible to select a desired reflectance over the range from much less than 1% to about 10%

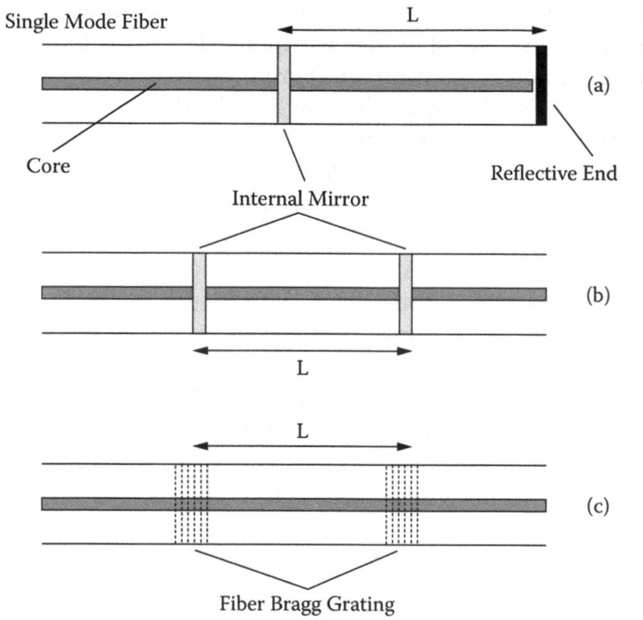

FIGURE 2.3
Intrinsic FFPT sensor configurations: (a) cavity formed by an internal mirror and fiber end; (b) cavity formed by two internal mirrors; (c) cavity formed by two fiber Bragg gratings. In each case *L* is the physical length of the optical cavity.

during the splicing process. With careful fabrication, excess losses in these mirrors can be less than 1% (0.05 dB). To achieve internal mirror reflectances greater than 10%, multilayer TiO_2/SiO_2 films produced by magnetron sputtering have been used. A reflectance of 86% was achieved using multilayer mirrors in an FFPI with a finesse of 21 at a wavelength of 1.3 μm [11].

The cavity for an FFPI sensor can also be formed from fiber Bragg grating (FBG) mirrors [12,13] (Figure 2.3(c)). These mirrors, which consist of alternating low- and high-refractive index regions with spatial period Λ, are written by interfering two ultraviolet laser beams in a single-mode fiber that may be doped with hydrogen to enhance writing sensitivity. The FBG mirrors are spatially distributed over lengths of the order of millimeters, several orders of magnitude greater than the extent of the internal dielectric mirrors (Figure 2.3(b)). FBG mirrors have high reflectance over a narrow spectral region (spectral width of the order of nanometers) centered at a wavelength $2n\Lambda$, with n the fiber refractive index. They also generally have very low excess loss ($\ll$0.1 dB).

Since the light in the sensing region is confined by the fiber core, diffraction does not limit the cavity length for an FFPI sensor. Cavity lengths from 100 μm to 1 m have been demonstrated, with lengths in the vicinity of 1 cm commonly used.

2.3.2 Extrinsic Fiber Fabry–Perot Interferometer (EFPI) Sensors

One of earliest—and still one of the most useful—EFPI configurations (Figure 2.4(a)) makes use of a diaphragm positioned near the cleaved or polished end of a fiber [14,15]. The air-gap cavity is bounded by the reflecting surfaces of the diaphragm and the end of the fiber. After the fiber is positioned to achieve the desired cavity length (typically, microns), it is permanently bonded to the supporting structure. Such short cavity lengths make it possible to operate these sensors with multimode fiber and low-coherence LED light sources. Another configuration (Figure 2.4(b)) makes use of a film of a transparent solid material on the end of the fiber, such that the cavity is in the film, bounded by the fiber–film and film–air interfaces [16,17]. Another widely used EFPI configuration (Figure 2.4(c)) makes use of an air-gap cavity formed between two cleaved or polished fiber surfaces, where the fibers are aligned end to end in a hollow tube [18]. Finally, an EFPI termed the "in-line fiber etalon" (ILFE) (Figure 2.4(d)) makes use of an air-gap cavity in a section of hollow-core fiber spliced between two single-mode fibers [19].

Since the light in the EFPI cavity is not confined, optical loss in the interferometer due to diffraction effectively limits the practical length of the optical cavity to a few hundred microns for most applications.

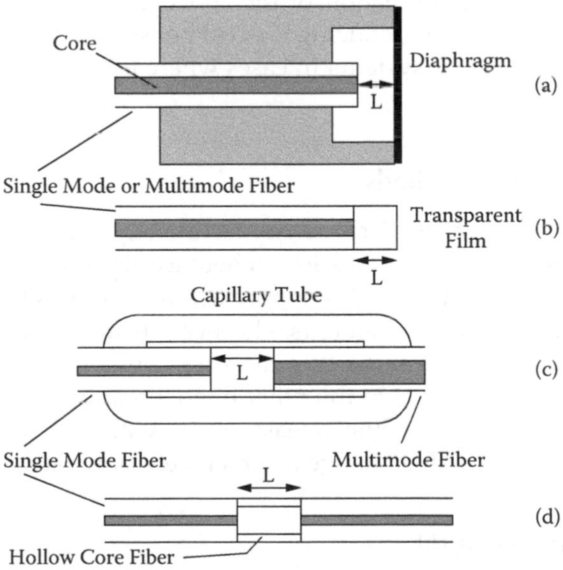

FIGURE 2.4
Extrinsic EFPI sensor configurations: (a) cavity formed by fiber end and diaphragm; (b) cavity formed by surfaces of a film on the end of a fiber; (c) cavity formed by end of a single-mode fiber and a multimode fiber aligned in a capillary tube; (d) cavity formed by single-mode fiber ends spliced to a hollow-core fiber. In each case *L* is the physical length of the optical cavity, which is an air gap in configurations (a), (c), and (d).

2.4 Optical Interrogation Methods and Multiplexing Techniques

In most electrical sensors (e.g., thermocouples for temperature measurement, piezoelectric pressure sensors) the raw signal is a monotonic and fairly linear function of the measurand. The situation with interferometric optical sensors is not so straightforward. It is evident from Figure 2.2 that the reflected (or transmitted) optical signal is a periodic function of optical phase shift in the cavity. Since the phase shift is close to a linear function of the measurand in almost all interferometric sensors, the unprocessed optical signal is a highly nonlinear function of the measurand value. Determining the measurand-induced phase shift from the optical signal is a major challenge—in some cases, the major challenge—in the engineering of an interferometric measurement system.

Multiplexing is another critical issue in fiber sensor design. Here, multiplexing is broadly defined as the use of one optical source to apply light to multiple sensors, the use of one fiber to access multiple sensors, the use of one photodetector to convert the optical signal from multiple sensors, the use of one electronic signal processor to compute measurand values for multiple sensors, or any combination of the above. Multiplexing is important because it opens the way to reducing the cost per sensor, thus improving the cost effectiveness of fiber systems in cases where more than one point is to be monitored.

2.4.1 Interrogation Methods

In today's fiber optic Fabry–Perot sensor technology, interrogation methods using high-coherence (laser) and low-coherence light sources both play a prominent role. As a rule, the laser provides higher sensitivity and faster response for dynamic measurements. However, using a single laser it is not possible to measure phase shifts that can vary over a range greater than 2π radians from a "cold start." On the other hand, broadband sources are well adapted to cases where "absolute" measurement with high dynamic range is needed for measurands that change at a relatively slow rate.

Laser (Single Wavelength)

A typical experimental arrangement for monitoring the reflectance of an FFPI or EFPI sensor with a single laser is shown in Figure 2.5. Light from a laser diode passes through a fiber coupler and is reflected by the fiber interferometer. After passing through the coupler again, the reflected light is converted by a photodiode to an electrical signal, which is processed electronically and/or displayed. Some typical waveforms obtained using a pulsed 1.3-μm semiconductor laser with an FFPI sensor are shown in Figure 2.6 [8]. A Faraday

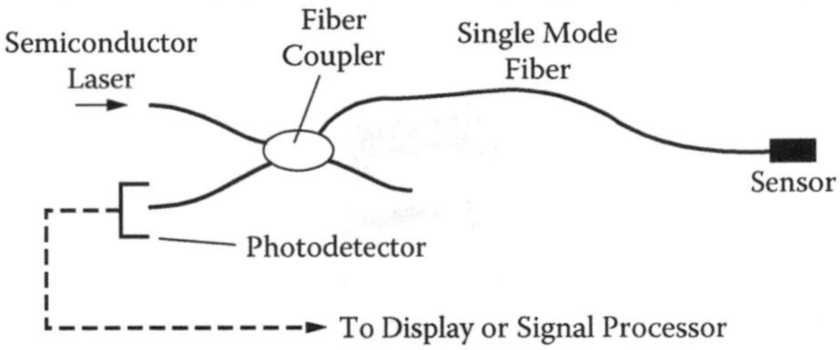

FIGURE 2.5
Arrangement for monitoring the reflected power from a Fabry–Perot fiber sensor.

isolator was placed in series with the laser to suppress destabilizing optical feedback. If the laser pulse is short, the driving current amplitude small, and the interferometer cavity short, the reflected amplitude is almost constant for the duration of the pulse, as in Figure 2.6(a). Increasing the pulse width, driving current amplitude, and interferometer length cause the reflected signal from the FFPI to vary with time due to chirping of the laser. As the laser heats during the pulse, its frequency—and consequently, the reflected power from the interferometer—changes with time, as in Figure 2.6(b). Thus, for the duration of the pulse the reflected optical signal sweeps through fringes of the interferometer.

A laser operated at constant bias current (cw condition) can be used for monitoring an interferometric fiber sensor. In this case, it is desirable to operate the interferometer at the quadrature point to get a quasi-linear dependence of reflected optical power on phase shift. Such an arrangement was used in the measurement of dynamic in-cylinder pressure variations in a diesel engine using an FFPI sensor [20]. A feedback arrangement for locking onto the quadrature condition by adjusting the laser frequency to maintain a constant average value for the photodetector current was implemented in a system for monitoring ultrasonic pressure [21].

In the single-laser schemes [20,21] the linear dynamic range region is limited to $\ll\pi$ rad. A passive scheme that makes it possible to overcome this limitation has been implemented using two simultaneously monitored EFPI sensors [18]. The interferometers are exposed to the same measurand and use the same light source, but they are fabricated so that their round-trip optical path lengths differ by an odd integral multiple of a quarter wavelength. By suitable processing of the two "quadrature-shifted" optical signals, it is possible to extract information on the measurand-induced phase shift while avoiding sensitivity nulls and ambiguity in direction of phase change, which result if a single cw laser and a single sensing interferometer are used. Referring to Figure 2.2, these sensitivity null/ambiguity points occur at maxima and at minima in the reflected (or transmitted) optical power. Thus, when one interferometer is at a "zero-sensitivity" point, the other is at quadrature, where sensitivity is greatest and where direction of phase change can be unambiguously determined.

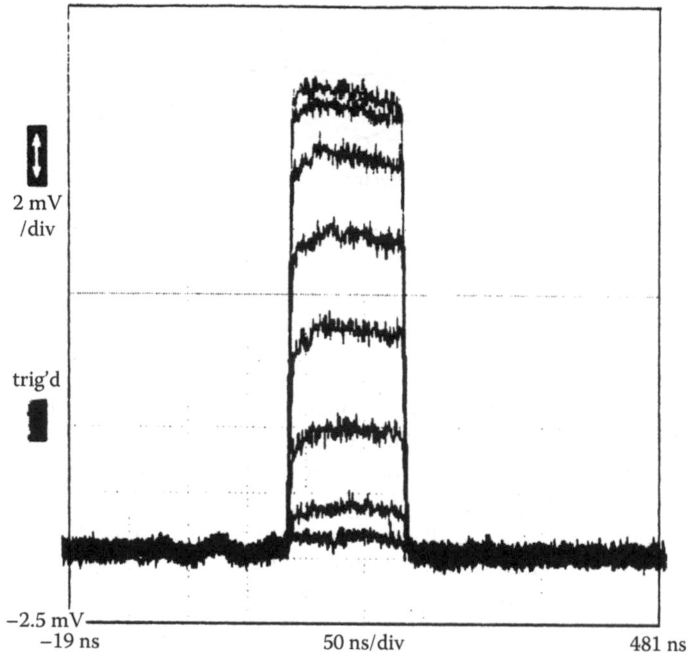

(a)

FIGURE 2.6
Oscilloscope tracings showing reflected power from FFPI at different temperatures for a pulsed laser input. In (a), the temperature was changed from 77 to 56°C in 3°C increments. The interferometer was short (1.5 mm), the pulse width was short (100 nsec), and the modulating pulse amplitude was small so that the effect of laser chirping was small. In (b), the temperature was changed from 29 to 23°C in 1.5°C increments. A longer (1 cm) interferometer, longer pulse duration (800 ns), and larger modulating pulse amplitude cause the interferometer output to sweep through fringes in response to chirping of the laser.

Another way to overcome the inherent problems of sensitivity nulls and direction-of-change ambiguities in interferometric sensors is to modulate the frequency of the light source. In one case, the bias current to a 1.3-μm distributed feedback laser was repetitively ramped to give a linear dependence of optical frequency on time (linear "chirp") during each cycle. Thus, the reflected optical signal from an FFPI sensor is a temporal fringe pattern. The interferometer phase shift was determined digitally by "counting" the time from the initiation of a ramp until the sensor signal crosses a "threshold" proportional to the laser output power [22]. With this system and a microcontroller-based digital processor, it is possible to accurately measure a phase shift while following a fringe pattern through many π radians.

Multiple Wavelengths

Multiple wavelengths from one or more cw light sources can be employed as a way of overcoming the sensitivity nulls and direction of change ambiguities

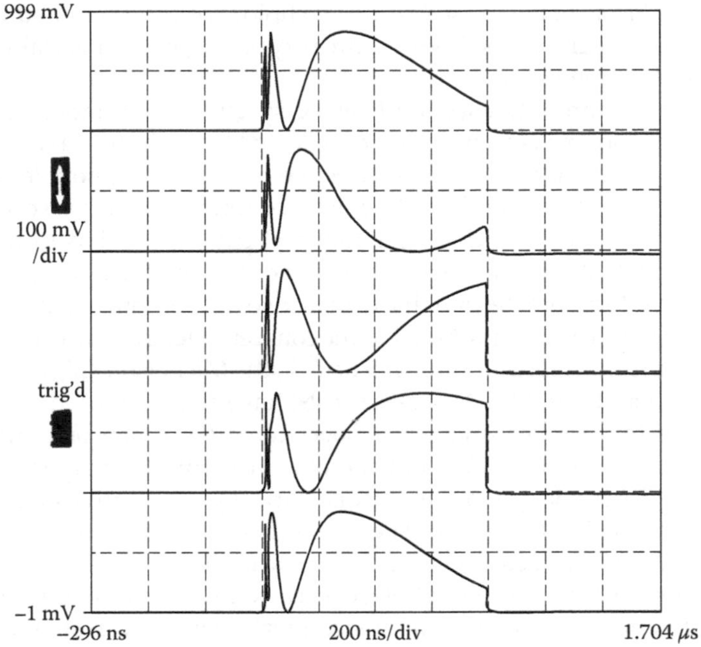

(b)

FIGURE 2.6 (continued).

that occur when interferometers are interrogated using a single coherent source. Using a 3-nm wavelength separation between two modes of an AlGaInP laser, Potter et al. [23] achieved quadrature optical outputs from a 20-μm-long EFPI cavity. A similar approach using spectrally filtered optical signals from EFPI sensors monitored with a low-coherence superluminescent diode (SLD) light source has been reported in which two wavelengths [24] and three wavelengths [25] are simultaneously monitored. In another experiment, two wavelengths from an Er:fiber broadband source were selected by fiber Bragg grating after reflection from an EFPI sensor to provide quadrature optical signals [26].

Broadband Light Source

"White light interferometry" (WLI) refers to interferometric measurement using a broadband light source. In the present context "white light" does not imply that the source emits in the visible region of the spectrum—indeed, almost all WLI systems for monitoring fiber optic sensors use infrared light. As applied to Fabry–Perot sensors, the WLI system is designed to determine the optical path length of the region between the mirrors. The implication of the term "white light" is that the spectrum is sufficiently broad that the coherence length of the light source is much less than the round-trip optical path length of the FPI. Suitable light sources for WLI include semiconductor

SLDs, LEDs, laser diodes biased near threshold, optically pumped Er:doped fibers, and tungsten lamps. These sources typically have spectral widths of a few tens of nanometers.

In the most commonly used WLI sensor system configuration, light from the broadband source is transmitted or reflected by both the sensing interferometer and a reference interferometer before reaching the photodetector [27]. When the optical path length of the reference interferometer is scanned, the photodetector output is a fringe (interference) pattern that has its maximum amplitude (peak of central fringe) where the optical path length difference of the reference interferometer and that of the sensor interferometer are equal. The width of the pattern is proportional to the coherence length of the light source. The most common reference interferometer is the Michelson [27], although the Mach–Zehnder [28] and the Fabry–Perot [29] have also been used.

A WLI system for reading out a fiber Fabry–Perot sensor is illustrated schematically in Figure 2.7(a), and the optical monitoring signal is shown in Figure 2.7(b). A change in the sensor's optical path length causes a lateral translation of the fringe pattern. The output reading is obtained by applying a suitable calibration factor relating fringe location to the measurand value.

The fringe data are processed electronically to determine the exact path length difference of the scanned reference interferometer corresponding to the central fringe peak. Identification of the central fringe is a key issue, because even in the absence of noise, adjacent fringes have amplitudes almost as high. A mistaken central fringe identification leads to an error in round-trip optical path length of at least one wavelength. An approach that greatly reduces the chance of central fringe error by improving the fringe amplitude contrast uses two broadband sources with a wide wavelength separation [30,31].

Another "white light" interrogation scheme does not use a scanned reference interferometer but instead measures the spectrum of the light transmitted (or reflected) by the fiber interferometer [32]. The length L of an EFPI is given by

$$L = \frac{\lambda_1 \lambda_2}{2(\lambda_2 - \lambda_1)} \tag{2.8}$$

where λ_1 and λ_2 represent adjacent peaks in the transmittance (or reflectance) spectrum.

2.4.2 Multiplexing Methods

Fiber Fabry–Perot sensors are amenable to several multiplexing approaches as a means of decreasing the number of expensive components and hence reducing the overall system cost.

Space Division Multiplexing

A multiplexing scheme in which two broadband light sources operating at different wavelengths and one electronic signal processor can monitor up to

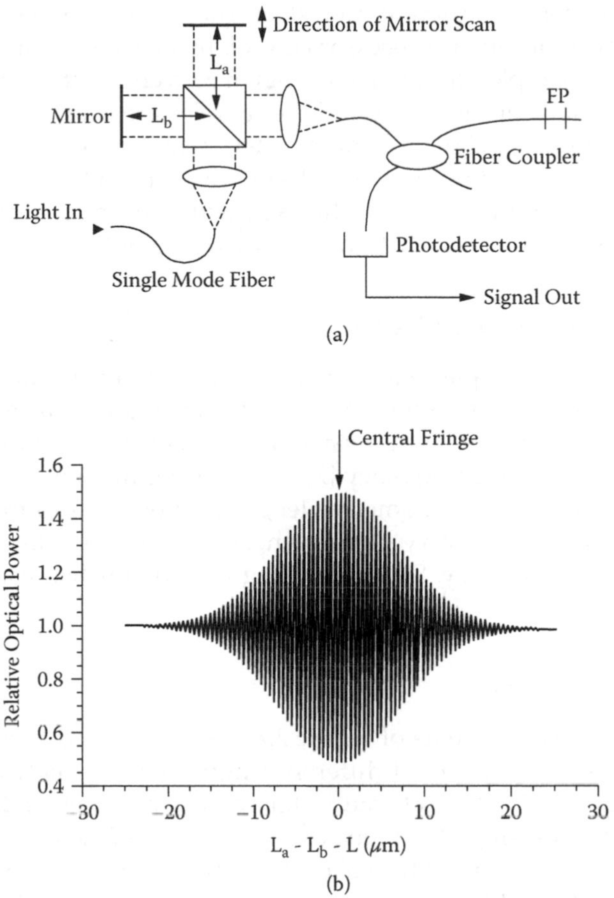

FIGURE 2.7
Measurement of interferometer phase shift using low-coherence (white light) interferometry: (a) optical configuration; (b) fringe pattern. Here *L* represents the optical length of the FP.

32 EFPI sensors accessed optically via a fiber optic distribution network has been described [33]. Light from each sensor is directed by a four-port directional coupler to a separate photodetector. The photodetector signals are processed digitally to determine measurand values for each sensor. Another scheme with a similar physical configuration in which a single distributed feedback laser and one microcontroller-based processor can monitor up to 24 FFPI sensors has also been reported [22].

Time Division Multiplexing

The reflectively monitored Fabry–Perot sensor is well suited for time division multiplexing using a pulsed light source [8]. In such a system, fiber delay lines of different lengths are provided between transmitter and receiver such that, for each laser pulse, the photodetector receives reflected pulses from each of the sensors in different time slots. The signals can be processed by digital means under microprocessor control. In such a system, the reflected waveforms are sampled for analog-to-digital conversion at fixed time delays relative to the start of the pulse, and the samples are averaged digitally. In-phase and quadrature signals from the interferometers needed for high sensitivity can be obtained by adjusting the dc bias current to the laser between pulses. A reference FFPI time-multiplexed with the sensors can be used to correct for fluctuations and drift in the laser wavelength.

Frequency Division Multiplexing

A multiplexing technique that allows for several FFPI sensors of different lengths to be located close together on a linear bus uses a semiconductor laser light source driven with a sawtooth current waveform to produce a linear chirp in the laser frequency [7]. By choosing the sensor lengths to be integral multiples of a fundamental length proportional to the chirp rate, the FFPI sensor signals vary at frequencies that are also linear multiples of the sawtooth frequency. Demultiplexing is accomplished using electrical bandpass filters.

Coherence Multiplexing

The WLI monitoring scheme of Figure 2.7 can be extended to the interrogation of Fabry–Perot sensors of different length in a fiber network, as illustrated in Figure 2.8. This scheme is known as *coherence multiplexing* [34]. Coherence multiplexing also requires the use of a reference interferometer matched in length to within the coherence length of the light source (≅10 μm for a typical LED or SLD) to the sensor FPI being interrogated. Light from the broadband source must be transmitted or reflected by both the sensor and reference interferometer before reaching the photodetector. If the reference interferometer is scanned, one fringe peak is observed for each sensing interferometer. In one experiment using a quartz halogen lamp as the light source and a scanned Michelson reference interferometer, six EFPI strain sensors of different optical cavity length were multiplexed in series along the length of a single-mode fiber [35]. In another case, two FFPI strain sensors were multiplexed in serial and parallel arrangements using a multimode laser diode as the light source [36].

Another coherence multiplexing demonstration utilized two EFPI sensors with different cavity lengths in a serial arrangement on a single-mode fiber, with spectral analysis using Eq. (2.8) to determine the interferometer lengths. The coherence length of the light source is less than the length of either sensor cavity. The demultiplexing scheme utilized a scannable EFPI reference cavity so that the interference pattern from one of the sensor EFPIs is seen only when its length is closely matched to that of the reference EFPI [37].

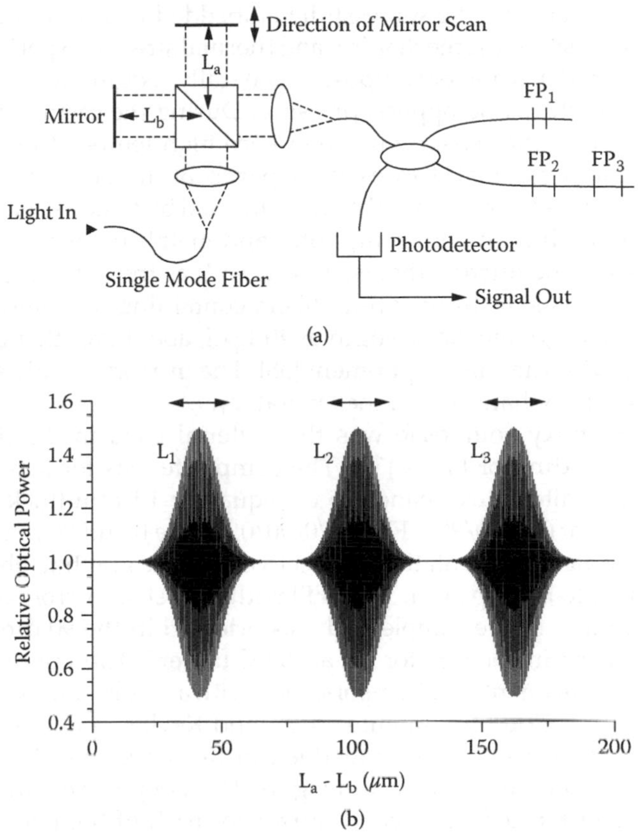

FIGURE 2.8
Arrangement for coherence multiplexing: (a) optical system for sensor interrogation; (b) calculated fringe patterns from three Fabry–Perot interferometers of lengths L_1, L_2, and L_3.

2.5 Embedded Sensors

One of the most desirable attributes of a sensor technology for smart structure applications is the ability to embed the sensor within the structural material. Embedding makes it possible to measure parameters at locations not accessible to ordinary sensors, which must be attached to the surface. In some cases the embedded sensor can continue to function properly under conditions (e.g., elevated temperatures) that epoxies for bonding a sensor to the structural material will not survive. The embedded sensor is protected from damage and isolated from extraneous environmental effects by the structure itself.

In order for a sensor to be successfully embedded in a composite or metal part, it must withstand the mechanical and thermal stresses experienced when the part is formed. Curing of composites generally requires a combination of elevated temperature and applied pressure. During the casting of metals of structural interest, the sensor will experience high temperatures combined with severe compressional stresses as the part cools to room temperature.

The FFPI with dielectric internal mirrors formed by fusion splicing is a candidate for embedding in both composites and metals because the mechanical properties of the mirrors themselves, as well as those of the silica fibers, are excellent. Tensile tests on several fibers containing the internal mirrors indicated an average tensile strength at 40 kpsi, about half that of ordinary splices made with the same equipment [38]. The mirrors readily survive the stresses of ordinary handling in the laboratory.

A graphite–epoxy composite was the material used in the first experiments on embedding of FFPIs [39]. The composite was formed from eight coupons of graphite–epoxy panel, 15 cm square × 1.1 mm thick. The panel had a sequence of 0/90/0/90/FFPI/90/0/90/0, where 0 and 90 indicate parallel and perpendicular orientation, respectively, of the graphite fibers in each coupon relative to the top layer. The FFPI with dielectric mirrors was embedded in the middle of the sample and was oriented in the 90 direction. The sample was cured in vacuum for 2 h at 180°C under 5.3 atm pressure.

In another experiment, FFPI sensors, each with a single aluminum internal mirror, were embedded in graphite–PEEK and Kevlar–epoxy coupons, and a three-axis strain rosette was embedded in Kevlar–epoxy [40]. The strain rosette configuration is illustrated in Figure 2.9. A deposited aluminum film on the fiber end formed the second mirror for each of the interferometers. The sensors were tested by monitoring the reflectance as a flexural force was applied to the coupon. The phase change in a number of sensors tested was found to be a linear function of strain over the range 0–500 με, and no significant hysteresis was observed.

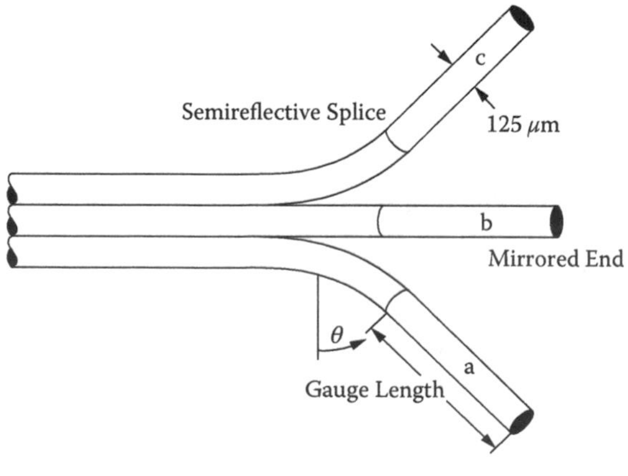

FIGURE 2.9
Three-axis strain rosette with FFPI sensors.

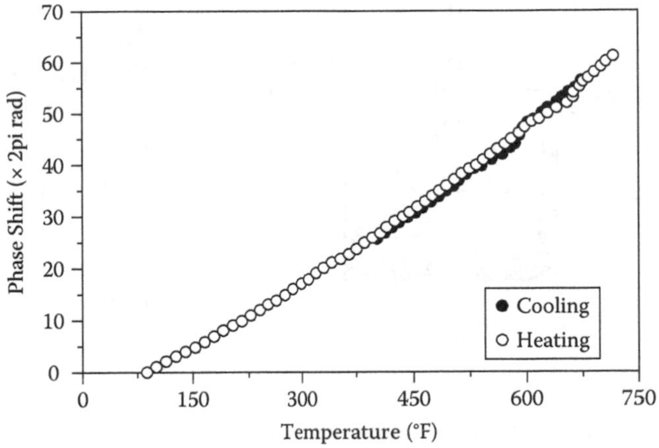

FIGURE 2.10
Dependence of interferometer phase shift ϕ on temperature during the process of curing a graphite–PEEK composite.

The embedding of EFPI three-axis strain rosettes in neat resin and graphite–epoxy composites has also been reported [41]. Strains measured in the graphite–epoxy sample were within 5% of those seen with surface-mounted electrical strain gages for orientations parallel and perpendicular to the graphite fibers, but some anomalies were seen at the 45° orientation. Embedding of an EFPI sensor with the ILFE configuration in a graphite–epoxy composite plate has also been reported [42]. The ILFE is much smaller than other EFPI sensors and thus represents a relatively minor perturbation in the mechanical structure of the composite matrix. Furthermore, it is less sensitive to lateral strains than are FFPI sensors.

In another experiment, an FFPI with dielectric mirrors and a thermocouple were embedded between the second and third layers of a 16-layer graphite–PEEK panel [43]. The panel was then cured in a hot press using standard temperature and pressure profiles. Figure 2.10 shows the dependence of round-trip optical phase shift in the interferometer on temperature (determined using the embedded thermocouple) during the curing process under a constant pressure of 16.9 atm. An abrupt change in the slope of the curve is observed around 700°F for heating and 600°F for cooling. This phenomenon is thought to be associated with the strain that results from decrystallization (for heating) or crystallization (for cooling) of the polymer in the coupon. This kind of information is important in optimizing the curing process, since the crystallization and decrystallization temperatures can be adjusted by changing the temperature and pressure profiles.

FFPI sensors have also been embedded in aluminum [44]. The parts were cast in graphite molds in air, as illustrated in Figure 2.11. Breakage of the fibers at the air–metal interface during the casting process was avoided by passing the fiber through stainless-steel stress-relief tubes, which extended a short distance into the finished part. Thermal expansion of the aluminum

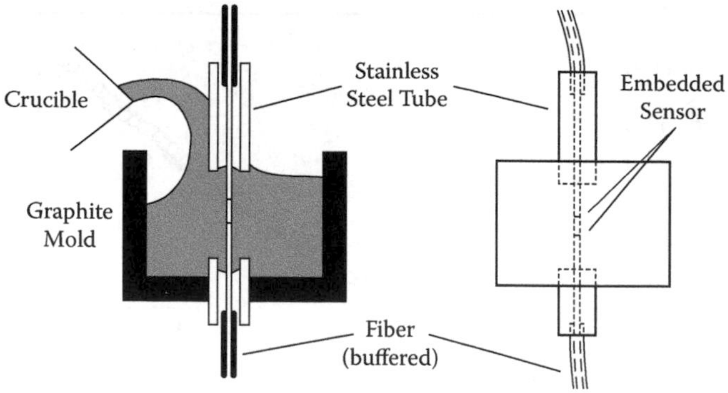

FIGURE 2.11
Process for embedding an FFPI in cast aluminum part.

caused the optical phase in the embedded FFPI to be 2.9 times more sensitive to temperature changes than for the same interferometer in air.

The embedding of EFPIs in reinforced concrete has also been reported [45]. Strain measurements using the fiber optics sensors agreed to within about 5% of data obtained with foil (electrical) gages.

2.6 Applications

Most of the effort to date in fiber optic Fabry–Perot sensing has been devoted to the measurement of temperature, strain, and pressure. Other demonstrated measurands include displacement, humidity, magnetic field, and liquid flow rate.

2.6.1 Temperature Measurement

The sensitivity of an interferometric fiber optic sensor to temperature is determined by the rate of change of optical path length nL with temperature. For an intrinsic sensor in a fused silica fiber, the optical path length change is dominated by the temperature coefficient of refractive index, which is over an order of magnitude greater than the thermal expansion coefficient. In fact, the thermal expansion coefficient in fused silica at room temperatures is lower than that of almost all materials of technological interest and becomes negative at cryogenic temperatures.

An FFPI with internal mirror reflectances of about 2% was used to sense temperature over the range –200 to 1050°C [38]. No hysteresis or change in the mirror reflectances was observed over the temperature range of the experiment, although it was noted that the fiber became brittle at the higher temperatures. As illustrated by the data of Figure 2.12, the sensitivity is low-

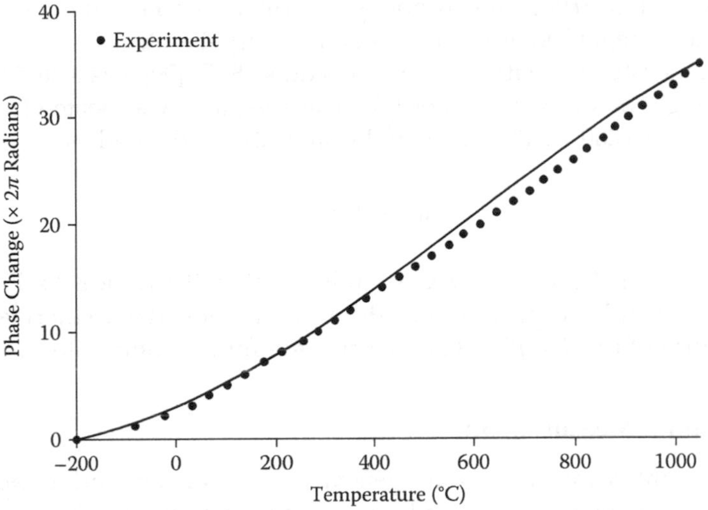

FIGURE 2.12
Phase shift in FFPI determined by fringe counting as a function of the temperature as measured with a thermocouple. Each dot represents a 2π-radian phase shift. The solid line is a theoretical plot determined from data on the temperature dependence of refractive index and thermal expansion coefficient for fused silica.

est at cryogenic temperatures and the response is fairly linear at room temperature and above.

An FFPI interrogated with WLI was used to measure temperature over the range from 20 to 800°C with a resolution of 0.025°C [28]. A piezoelectrically scanned fiber Mach–Zehnder served as the reference interferometer. The FFPI used in the experiment had been annealed to remove a slight birefringence that was observed to adversely affect the sensor's performance.

One of the earliest EFPI temperature sensor designs utilizes a thin (≅8 μm) layer of silicon bonded on one side to the end of a multimode fiber and on the other to a glass plate [14]. The cavity is formed by the two surfaces of the silicon layer, and the temperature sensitivity results primarily from the silicon's large temperature coefficient of refractive index. The sensor is monitored with an LED light source by measuring the ratio of optical power reflected at two different wavelengths within the LED spectral range. Linear response with a resolution of 0.02°C (with sample averaging) was observed over the range from 10 to 60°C.

Another EFPI temperature sensor used a cavity formed by the cleaved ends of two fibers bonded end to end with an air gap in a stainless-steel tube. The sensor was monitored using WLI, and a temperature resolution of 0.006°C over the range 27.3 to 62.5°C was reported [46]. The temperature sensitivity in this case is due to the thermal expansion of the stainless steel in the region between the fiber bonds. The cavity was so long (≅500 μm) that, due to diffraction, only a small part of the light leaving the input fiber was captured by the core of that fiber after reflection and thus contributed to the interference.

Technically, this structure is properly described as a Fizeau interferometer, which here is regarded as a subclass of the Fabry–Perot.

The temperature sensitivity of an embedded FFPI [39] was tested by monitoring fringes at a wavelength of 1.3 μm as the panel was heated from room temperature to 200°C. The value of the quantity Φ_T, defined as

$$\Phi_T = d\phi/\phi dT, \tag{2.9}$$

was measured to be 8.0×10^{-6}/°C, slightly less than the value 8.3×10^{-6}/°C for the same sensor in air prior to embedding. From these data, a thermal expansion coefficient of 2.1×10^{-7}/°C was estimated for the composite.

2.6.2 Strain Measurement

Surface-mounted transducers are generally used for monitoring parameters such as strain, temperature, and acoustic pressure in structural materials. Although embedding is the ultimate goal in some smart materials development efforts, externally mounted fiber sensors can also be very useful in this field. The surface-mounted sensor can be readily repositioned or replaced. Furthermore, surface-mounted sensors can be used with materials that must be processed at such high temperatures that embedding may be difficult or impossible.

The conventional way to monitor strain in a structural part is to bond electrical strain gages to its surface. The high sensitivity of the fiber interferometer to longitudinal strain can be utilized in a similar manner. In the first experiment of this type, an FFPI with a cavity formed with a silver internal mirror and a silvered fiber end was bonded to the surface of a cantilevered aluminum beam [10]. A conventional resistive foil strain gage was also bonded to the beam as a reference. Strain was introduced by loading the beam in flexure. The optical phase change in the FFPI, determined by monitoring the reflected power and counting fringes, was found to be a linear function of the strain reading from the resistive device over the range 0 to 1000 με.

For strain measurements using FFPI sensors embedded in a graphite–PEEK panel, two electric strain gages (ESGs) were bonded to opposite sides of the completed coupon, above and below the embedded FFPI [43]. A strain was induced by applying load to the center of the coupon, which was supported at opposite ends. The strain sensitivity of the FFPI, $\Delta\phi/\Delta L$, was determined to be 9.1×10^6 rad/m by comparing the optical phase shift with the ESG readings. This is about 18% less than the value of $\Delta\phi/\Delta L$ measured for a similar fiber in air. Strain measurements were also made at elevated temperature. Both the fiber optic and ESG sensors showed good linearity at 200°F. However, the ESG response was unstable at 300°F, while the same linear load profiles were observed at 200 and 300°F for the FFPI sensor.

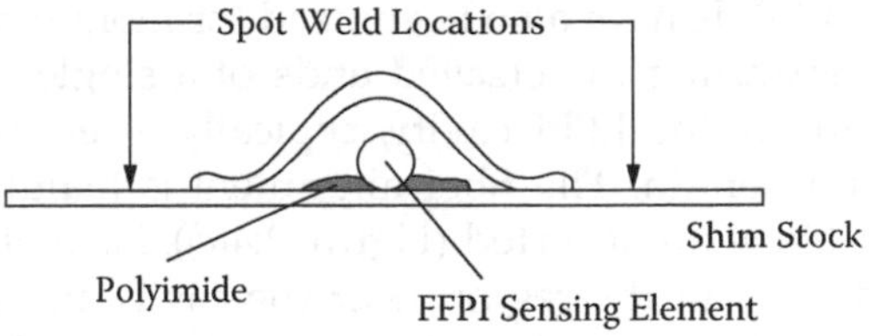

FIGURE 2.13
Strain sensor design for mounting to a metal surface by spot welding.

The packaging method of Figure 2.13 in which the fiber interferometer is bonded to a thin, flat strip of stainless-steel shim stock was developed for mounting the FFPI onto a metal structure by spot welding. Unlike epoxy bonding, this technique can be applied to a wet surface in rainy weather. Furthermore, calibration of the shim-stock-mounted sensor is more reliable than if the gage is directly epoxied to the structure to be monitored. Twelve of these sensors were mounted on load-bearing elements in a railroad bridge and on a rail, where they collected data on strains induced by passing trains for over a year [47]. Excellent agreement with data from co-located electrical strain gages was observed.

An experiment designed to achieve very high strain sensitivity with temperature compensation [48] utilized two long-cavity (≅1 m), co-located FFPIs interrogated using a frequency-scanned, 1.3-μm Nd:YAG laser. The cavity for one of the FFPIs was a fused silica fiber, while the cavity for the second interferometer was a fluoride glass fiber. The temperature coefficients of phase shift are very different in these fiber materials, so that the data from the two FFPIs can be processed to achieve a high degree of temperature compensation. A strain sensitivity of 1.5 nε was achieved with this technique.

An FFPI strain sensor using fiber Bragg grating (FBG) mirrors has been reported, in which two chirped gratings were co-located in the single-mode fiber [12], The spatial frequency of one grating increased with increasing displacement along the fiber axis and decreased for the other. At any given optical frequency, a Fabry–Perot interferometer was formed between the sections of the two gratings that exhibited reflectance peaks at that frequency. Because the gratings were chirped, the Fabry–Perot free spectral range was a function of the optical frequency of the tunable laser used for interrogating the sensor within a 30-nm spectral range. Strain was unambiguously measured over the range from 3 to 1300 με.

Another FFPI with FBG mirrors was interrogated using WLI with an Er:fiber broadband source and a scanned fiber Mach–Zehnder reference interferometer [13]. Polarization maintaining single-mode fiber was used for the sensor. Coarse strain information was obtained by measuring the wavelength of maximum reflectance as determined from the FBG, while fine resolution was obtained from the phase shift measured for the FFPI. A strain sensitivity of 33 nε with an unambiguous measurement range of 800 με was reported.

Surface-mounted FFPIs have also been used for monitoring strain. In this case, an air gap separating the cleaved ends of a single-mode fiber and a multimode fiber forms the EFPI cavity, typically a few microns to a few hundred microns in length. The fiber alignment is maintained by a silica tube in which the fibers are inserted (Figure 2.4(c)). Each of the fibers is free to move longitudinally in the region near the cavity but is constrained at some point along its length by bonding either to the silica tube or to the sample being monitored. The distance between bonding points is the length of the region over which a perturbation affects the sensor output, known as the gage length. The EFPI is usually monitored in reflection, with the input and reflected light carried by the single-mode fiber.

In one experiment, two of the EFPIs were bonded on the surface of a ceramic material in close proximity to one another [49]. The cavity lengths in the two sensors were slightly different, so that the round-trip phase shifts were different by about $\pi/2$ radians. By simultaneously monitoring the two sensor outputs, it is possible to determine the direction of change of the phase shift and avoid sensitivity nulls. When the ceramic was temperature cycled from 25 to 600°C, the sensor data provided information on the expansion of a crack in the material.

EFPIs with the ILFE configuration have also been used for strain measurement. In one experiment, an EFPI embedded in a graphite–epoxy plate measured dynamic strain subsequent to the impact of a weight dropped on the sample [42]. In another experiment, an EFPI sensor bonded to a metal bar measured the longitudinal strain response to a projectile impact [26]. The signal processing scheme was designed to respond to strain rates as high as 10 ε/s. Displacement measurements carried out to determine the noise floor of the dual-wavelength readout system yielded an estimated $0.8\ \mathrm{n}\varepsilon/\sqrt{\mathrm{Hz}}$ for the strain measurement sensitivity.

2.6.3 Pressure Measurement

The high strain sensitivity of FFPI sensors has been utilized in the measurement of gas pressure in internal combustion engines [20]. Many engines are constructed so that an element such as a fuel injector valve, which is exposed to the combustion chamber pressure, is bolted to the cylinder head. The variation in longitudinal strain on the bolts during an engine cycle is approximately proportional to the combustion chamber pressure. An FFPI epoxied into a hole drilled in one of these bolts can thus be used for engine pressure measurements. More recently an in-cylinder sensor with the configuration of Figure 2.14 has been developed in which the FFPI sensing element is embedded along the axis of a metal rod, which is then inserted into a metal housing with a 0.5-mm-thick lower wall (diaphragm) [22,50]. A nut at the top of the housing is torqued to produce a slight compression of the metal rod. The sensor is mounted in a threaded port in the cylinder head. An adapter attenuates the effect of rapid temperature changes that occur during a combustion cycle by allowing the gas to cool before it reaches the

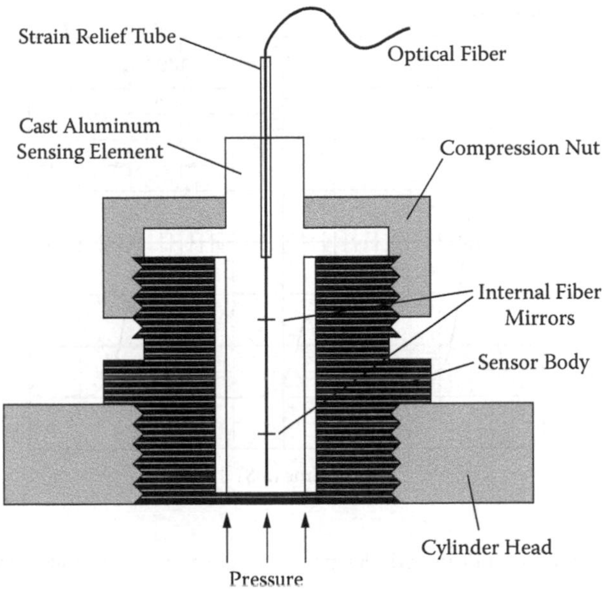

FIGURE 2.14
Sensor for measuring in-cylinder pressure in an internal combustion engine.

diaphragm, so that thermal compensation is not needed. None of the FFPI combustion sensors is provided with any form of external cooling. These sensors have operated for several hundred thousand sensor-hours in large, natural gas-fueled engines, where they are exposed to temperatures in the 200–300°C range. Agreement with water-cooled piezoelectric reference sensors is consistently better than 1%, as in the data of Figure 2.15. Up to 24 of the sensors can be monitored with a single signal conditioning unit containing a distributed feedback laser as the light source.

The first reported pressure measurement with a fiber optic Fabry–Perot interferometer utilized the EFPI–diaphragm configuration of Figure 2.4(a) [14]. The diaphragm was a thin (≅4 μm) silicon membrane bonded to a glass substrate in which a multimode fiber was mounted. The cavity length was in the 1.4–1.7 μm range. The resolution and accuracy in this sensor were reported to be 1 mm Hg over the pressure range from 750–1050 mm Hg. This sensor was developed for biomedical applications, such as intracranial pressure monitoring in patients with brain trauma.

Another sensor developed for biomedical purposes, a long-cavity (500 μm) diaphragm-based EFPI interrogated using low-coherence interferometry, achieved a mean pressure resolution of 0.06 mm Hg over the range 0–360 mm Hg [51]. In this sensor, the 5-μm-thick stainless-steel diaphragm was bonded to a Pyrex tube. Temperature compensation was provided by a co-located EFPI temperature sensor isolated from the pressure environment.

Another application of diaphragm-based EFPI sensors is the optical multichannel transducer array designed to measure pressure profiles with high

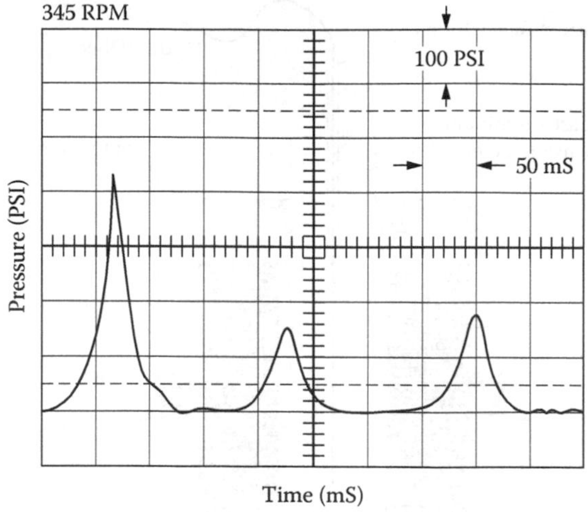

FIGURE 2.15
Pressure versus time for FFPI in-cylinder pressure sensor, measured for a misfiring single-cylinder engine.

time and spatial resolution for wind tunnel application [52]. The silicon diaphragms were about 20 μm thick, and the cavity length was 55 μm. A dual-laser interrogation technique provided quadrature signals for each of the 32 elements in the array. The pressure resolution was 0.1 psi, with a frequency response to 50 kHz.

Diaphragm-based EFPIs have also been used as sensitive microphones. In one case, with a thin metallized Mylar membrane as the diaphragm, flat response over the frequency range 20 Hz–3 kHz with >65 dB dynamic range was achieved, using a HeNe laser as the monitoring light source [53]. Another EFPI microphone utilized a 50-μm-thick polyethylene terepthalate (PET) film on the end of a multimode fiber as the pressure-sensitive cavity [17]. A HeNe laser was used to interrogate the sensor, which was biased near quadrature. The acoustic noise floor was 15 kPa over a 25-MHz bandwidth.

A diaphragm-based FFPI pressure sensor of Figure 2.16 in which the end of a fiber containing the FFPI is bonded to a thin (≅50 μm) stainless-steel diaphragm has also been reported [54]. The fiber is bonded under tension so that pressure-induced deflection of the diaphragm acts to reduce the strain on the FFPI. A sensitivity of 0.4 torr over the range 0 to 100 torr was reported.

The ability to embed FFPIs in structural materials suggests application as a pressure transducer in ultrasonic nondestructive testing (NDT). In conventional NDT studies, piezoelectric transducers (PZTs) are positioned on the surface of the sample to launch and detect ultrasonic waves. The ability to locate the receiving transducer deep within the material opens the possibility of obtaining new information on the properties of a bulk sample. For example, it may be possible to better detect and locate delamination sites in composites or cracks in metals. Experiments with a PZT transducer

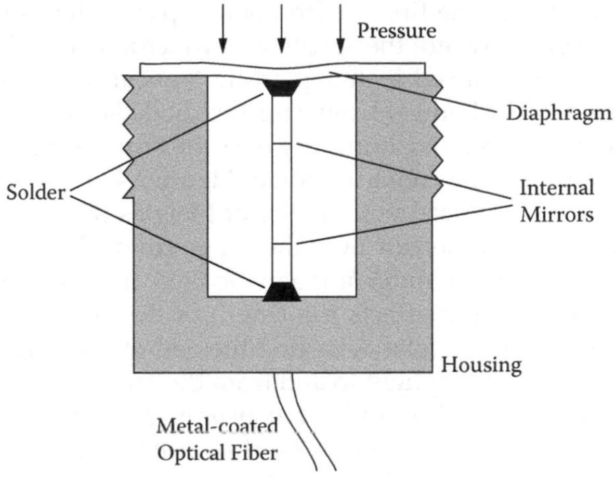

FIGURE 2.16
Diaphragm-mounted FFPI pressure sensor.

positioned on the surface of a graphite–epoxy composite sample yielded an acoustic response from an embedded FFPI over the frequency range 100 kHz to 5 MHz [55]. In another experiment acoustic waves at a frequency of 5 MHz were monitored by an FFPI sensor embedded in an epoxy plate [21]. The compressional acoustic wave generated by the transducer interacts with light in the fiber to produce a phase shift via the strain-optic effect. Only the acoustic wave in the region between the mirrors contributes to the sensor output.

An aluminum-embedded FFPI has also been used to detect ultrasonic waves launched by a surface-mounted PZT transducer over the frequency range from 0.1 to 8 MHz [44]. In another experiment, an EFPI bonded to the surface of an aluminum block was used to detect surface acoustic waves launched by a piezoelectric transducer at a frequency of 1 MHz [56]. A dual-wavelength scheme with lasers at 0.78 and 0.83 μm was used to obtain in-phase and quadrature signals from the sensor. The EFPI also detected the acoustic noise burst produced by breaking a pencil lead on the surface of the aluminum sample.

2.6.4 Other Applications

A displacement sensor utilized a 0.83-nm superluminescent diode with WLI in measuring the length of a cavity formed by the end of a single-mode fiber and a movable mirror [57]. The static sensitivity was estimated to be 6.1 nm $Hz^{-1/2}$. In another displacement measurement experiment using quadrature-shifted EFPIs with WLI interrogation, a repeatability of 0.5 μm and an accuracy of 1.5 μm over a 150-μm range of motion of the target [58] were obtained. In another case a displacement sensitivity of 30 nm for measuring the physiological motion of a ciliary bundle in the inner ear of an anesthetized turtle was achieved [59].

Humidity was one of the first environmental parameters sensed with an EFPI in an experiment where the sorption of water in a TiO_2 film on the tip of a fused silica fiber caused a change in the optical thickness of the film [16]. The reflected optical power from a 0.8-μm diode laser was found to be a monotonic function of relative humidity over the humidity range 0–80%.

An EFPI sensor with the configuration of Figure 2.4(c) in which the multimode fiber has been replaced with a wire of Metglas ($Fe_{77.5}B_{15}Si_{7.5}$) has been used as a magnetic field sensor over the range from 100 to 35,000 nT [60]. In response to a change in ambient magnetic field, the Metglas experiences a dimensional change that affects the length of the air-gap cavity formed between the end of the Metglas wire and the end of the single-mode fiber, which transmits monitoring light to and from the sensing head.

An FFPI sensor mounted in a pipe has been used to measure the rate of liquid flow [61]. A flexible polymer tube that protects the fiber serves as a nucleation center for the formation of vortices, a process that causes transient strains in the fiber. Experimentally, the frequency of the signal measured by this "vortex-shedding" fiber optic flow meter was found to be proportional to liquid velocity over the range of 0.14 to 3.0 m/s.

2.7 Conclusions

A number of configurations for intrinsic (FFPI) and extrinsic (EFPI) fiber optic Fabry–Perot sensors have been developed for a variety of applications. Both FFPI and EFPI sensors can be optically interrogated using either coherent (laser) or low-coherence light sources, and both are amenable to space division, time division, frequency division, and coherence multiplexing techniques. They can be embedded in materials of technological interest, including composites and metals. They can provide high sensitivity, dynamic range, and response speed for measurement of temperature, strain, pressure, displacement, magnetic field, and flow rate.

Fiber sensor applications have not developed as quickly as some had expected a decade ago. Nevertheless, the availability of commercial Fabry–Perot sensor products is expanding and a strong technological foundation for the future has been laid. Biomedical applications, smart military and commercial structures, industrial equipment monitoring, and "downhole" applications in the oil and gas industry are four areas where rapid growth can be anticipated during the next decade. Advances in technology, increased volume production, and the widespread use of multiplexing should make fiber Fabry–Perot sensors cost competitive in an increasing number of markets. Optical networks in which sensors for different measurands are interconnected on a common single-mode fiber bus are on the horizon. As in the past, fiber sensor technology will continue to benefit from rapid advances in componentry spurred by the explosive growth in fiber telecommunications.

References

1. C. Fabry and A. Perot, Théorie et applications d'une nouvelle méthode de Spectroscopie Interférentielle, *Ann. Chem. Phys.* **16**, p. 115, 1899.
2. C. E. Lee and H. F. Taylor, Sensors for smart structures based upon the Fabry–Perot interferometer, in *Fiber Optic Smart Structures*, E. Udd, ed., Wiley, New York, pp. 249–269, 1995.
3. S. J. Petuchowski, T. G. Giallorenzi, and S. K. Sheem, A sensitive fiber-optic Fabry–Perot interferometer, *IEEE J. Quantum Electron*, **17**, p. 2168, 1981.
4. T. Yoshino, K. Kurosawa, and T. Ose, Fiber–optic Fabry–Perot interferometer and its sensor applications, *IEEE J. Quantum Electron*, **18**, p. 1624, 1982.
5. A. D. Kersey, D. A. Jackson, and M. Corke, A simple fibre Fabry–Perot sensor, *Opt. Commun.*, **45**, p. 71, 1983.
6. P. A. Leilabady and M. Corke, All-fiber-optic remote sensing of temperature employing interferometric techniques, *Opt. Lett.*, **12**, p. 773, 1987.
7. F. Farahi, T. P. Newson, P. A. Leilabady, J. D. C. Jones, and D. A. Jackson, A multiplexed remote fiber optic Fabry–Perot sensing system, *Int. J. Optoelectronics*, **3**, p. 79, 1988.
8. C. E. Lee and H. F. Taylor, Interferometric optical fibre sensors using internal mirrors, *Electron. Lett.*, **24**, p. 193, 1988.
9. M. N. Inci, S. R. Kidd, J. S. Barton, and J. D. C. Jones, Fabrication of single-mode fibre optic Fabry–Perot interferometers using fusion spliced titanium dioxide optical coatings, *Meas. Sci. Tech.*, **3**, p. 678, 1992.
10. T. Valis, D. Hogg, and R. M. Measures, Fiber optic Fabry–Perot strain gauge, *IEEE Photonics Tech. Lett.*, **2**, p. 227, 1990.
11. C. E. Lee, W. N. Gibler, R. A. Atkins, and H. F. Taylor, In-line fiber Fabry–Perot interferometer with high-reflectance internal mirrors, *J. Lightwave Tech.*, **10**, p. 1376, 1992.
12. K. P. Koo, M. LeBlanc, T. E. Tsai, and S. T. Vohra, Fiber-chirped grating Fabry–Perot sensor with multiple-wavelength-addressable free-spectral ranges, *IEEE Photon, Tech. Lett.*, **10**, p. 1006, 1998.
13. Y. J. Rao, M. R. Cooper, D. A. Jackson, C. N. Pannell, and L. Reekie, Absolute strain measurement using an in-fibre-Bragg-grating-based Fabry–Perot sensor, *Electron. Lett.*, **36**, p. 708, 2000.
14. R. A. Wolthuis, G. L. Mitchell, E. Saaski, J. C. Hartl, and M. A. Afromowitz, Development of medical pressure and temperature sensors employing optical spectrum modulation, *IEEE Trans. Biomed, Engineering*, **38**, p. 974, 1991.
15. G. L. Mitchell, Intensity-based and Fabry–Perot interferometer sensors, in *Fiber Optic Sensors: An Introduction for Engineers and Scientists*, E. Udd, ed., Wiley, New York, p. 139, 1991.
16. F. Mitschke, Fiber optic sensor for humidity, *Opt. Lett.*, **14**, p. 967, 1989.
17. P. C. Beard and T. N. Mills, Miniature optical fibre ultrasonic hydrophone using a Fabry–Perot polymer film interferometer, *Electron. Lett.*, **33**, p. 801, 1997. Also P. C. Beard and T. C. Mills, Extrinsic optical-fiber ultrasound sensor using a thin polymer film as low-finesse Fabry–Perot interferometer, *Appl. Opt.*, **35**, p. 663, 1996.
18. K. A. Murphy, M. F. Gunther, A. M. Vengsarkar, and R. O. Claus, Quadrature phase-shifted, extrinsic Fabry–Perot optical fiber sensors, *Opt. Lett.*, **16**, p. 273, 1991.

19. J. Sirkis, T. A. Berkoff, R. T. Jones, H. Singh, A. D. Kersey, E. J. Friebele, and M. A. Putnam, In-line fiber etalon (ILFE) fiber optic strain sensors, *J. Lightwave Tech.*, **13**, p. 1256, 1995.
20. R. A. Atkins, J. H. Gardner, W. H. Gibler, C. E. Lee, M. D. Oakland, M. O. Spears, V. P. Swenson, H. F. Taylor, J. J. McCoy, and G. Beshouri, Fiber optic pressure sensors for internal combustion engines, *Appl. Opt.*, **33**, p. 1315, 1994.
21. J. F. Dorighi, S. Krishnaswamy, and J. D. Achenbach, Stabilization of an embedded fiber optic Fabry–Perot sensor for ultrasound detection, *IEEE Trans. Ultrason., Ferroelectron, and Freq. Contr.*, **42**, p. 820, 1995.
22. R. Sadkowski, C. E. Lee, and H. F. Taylor, Multiplexed fiber-optic sensors with digital signal processing, *Appl. Opt.*, **34**, p. 5861, 1995.
23. J. Potter, A. Ezbiri, and R. P. Tatam, A broad band signal processing technique for miniature low-finesse Fabry–Perot interferometric sensors, *Opt. Commun.*, **140**, p. 11, 1997.
24. N. Furstenau and M. Schmidt, Fiber-optic extrinsic Fabry–Perot interferometer vibration sensor with two-wavelength passive quadrature readout, *IEEE Trans. Instrumentation Measurement*, **47**, p. 143, 1998.
25. M. Schmidt and N. Furstenau, Fiber-optic extrinsic Fabry–Perot interferometer sensors with three-wavelength digital phase demodulation, *Opt. Lett.*, **24**, p. 599, 1999.
26. Y. L. Lo, J. S. Sirkis, and C. C. Wang, Passive signal processing of in-line fiber etalon sensors for high strain-rate loading, *J. Lightwave Tech.*, **15**, p, 1578, 1997.
27. G. Beheim, Remote displacement measurement using a passive interferometer with a fiber optic link, *Appl. Opt.*, **24**, p. 2335, 1985.
28. H. S. Choi, C. E. Lee, and H. F. Taylor, High-performance fiber optic temperature sensor using low-coherence interferometry, *Opt. Lett.*, **22**, p. 1814, 1997.
29. C. E. Lee and H. F. Taylor, Fiber-optic Fabry–Perot temperature sensor using a low-coherence light source, *J. Lightwave Tech.*, **9**, p. 129, 1991.
30. S. Chen, K. T. V. Grattan, B. T. Meggitt, and A. W. Palmer, Instantaneous fringe-order identification using dual broadband sources with widely spaced wavelengths, *Electron. Lett.*, **29**, p. 334, 1993.
31. Y. J. Rao, Y. N. Ning, and D. A. Jackson, Synthesized source for white-light systems, *Opt. Lett.*, **18**, p. 462, 1993.
32. V. Bhatia, M. B. Sen, K. A. Murphy, and R. O. Claus, Wavelength-tracked white light interferometry for highly sensitive strain and temperature measurements, *Electron. Lett.*, **32**, p. 248, 1996.
33. Y. J. Rao and D. A. Jackson, A prototype multiplexing system for use with a large number of fiber-optic-based extrinsic Fabry–Perot sensors exploiting coherence interrogation, *Proc. SPIE*, **2507**, p. 90, 1995.
34. C. M. Davis, C. J. Zarobila, and J. D. Rand, Fiber-optic temperature sensors for microwave environments, *Proc. SPIE*, **904**, p. 114, 1988.
35. M. Singh, C. J. Tuck, and G. F. Fernando, Multiplexed optical fiber Fabry–Perot sensors for strain metrology, *Smart Mater. Struct.*, **8**, p. 549, 1999.
36. S. C. Kaddu, S. F. Collins, and D. J. Booth, Multiplexed intrinsic optical fibre Fabry–Perot temperature and strain sensors addressed using white light interferometry, *Meas. Sci. Tech.*, **10**, p. 416, 1999.
37. V. Bhatia, K. A. Murphy, R. O. Claus, M. E. Jones, J. L. Grace, T. A. Tran, and J. A. Greene, Optical fibre based absolute extrinsic Fabry–Perot interferometric sensing system, *Meas. Sci. Tech.*, **7**, p. 58, 1996.

38. C. E. Lee, R. A. Atkins, and H. F. Taylor, Performance of fiber-optic temperature sensor from –200 to 1050°C, *Opt. Lett.*, **13**, p. 1038, 1988.
39. C. E. Lee, H. F. Taylor, A. M. Markus, and E. Udd, Optical-fiber Fabry–Perot embedded sensor, *Opt. Lett.*, **14**, p. 1225, 1989.
40. T. Valis, D. Hogg, and R. M. Measures, Composite material embedded fiber-optic Fabry–Perot strain gauge, *Proc. SPIE*, **1370**, p, 154, 1990.
41. S. W. Case, J. J. Lesko, B. R. Fogg, and G. P. Carman, Embedded extrinsic Fabry–Perot fiber optic strain rosette sensors, *J. Intelligent Mat. Sys. Struct.*, **5**, p. 412, 1994.
42. C. C. Chang and J. L. Sirkis, Design of fiber optic sensor systems for low velocity impact detection, *Smart Mater. Struct.*, **7**, p. 166, 1997.
43. C. E. Lee, W. N. Gibler, R. A. Atkins, J. J. Alcoz, H. F. Taylor, and K. S. Kim, Fiber optic Fabry–Perot sensors embedded in metal and in a composite, *8th Optical Fiber Sensors Conf.*, Monterey, CA, Jan. 1992.
44. C. E. Lee, W. N. Gibler, R. A. Atkins, J. J. Alcoz, and H. F. Taylor, Metal-embedded fiber optic Fabry–Perot sensors, *Opt. Lett.*, **16**, p. 1990, 1991.
45. S. F. Masri, M. S. Agbabian, A. M. Abdel-Ghaffar, M. Higazy, R. O. Claus, and M. J. deVries, Experimental study of embedded fiber-optic strain gauges in concrete structures, *J. Eng. Mech.*, **120**, p. 1696, 1994.
46. Y. J. Rao and D. A. Jackson, Recent progress in fibre optic low-coherence interferometry, *Meas. Sci. Tech.*, **7**, p. 981, 1996.
47. W. Lee, J. Lee, C. Henderson, H. F. Taylor, R. James, C. E. Lee, V. Swenson, R. A. Atkins, and W. G. Gemeiner, Railroad bridge instrumentation with fiber optic sensors, *Appl. Opt.*, **38**, p. 1110, 1999.
48. E. J. Friebele, M. A. Putnam, H. J. Patrick, A. D. Kersey, A. S. Greenblatt, G. P. Ruthven, M. H. Krim, and K. S. Gottschalck, Ultra-high-sensitivity fiber-optic strain and temperature sensor, *Opt. Lett.*, **23**, p. 222, 1998.
49. K. A. Murphy, C. E. Kobb, A. J. Plante, S. Desu, and R. O. Claus, High temperature sensing applications of silica and sapphire optical fibers, *Proc. SPIE*, **1370**, p. 169, 1990.
50. C. E. Lee and H. F. Taylor, A fiber optic pressure sensor for internal combustion engines, *Sensors*, **15**, 3, p. 20, Mar. 1998.
51. Y. J. Rao and D. A. Jackson, A prototype fibre-optic-based Fizeau medical pressure and temperature sensor system using coherence reading, *Meas. Sci. Tech.*, **5**, p. 741, 1994.
52. J. Castracane, L. P. Clow, and G. Seidler, Optical multichannel transducer array for wind tunnel applications, *Opt. Eng.*, **35**, p. 2627, 1996.
53. C. Zhou and S. V. Letcher, Fiber optic microphone based on a combination of Fabry–Perot interferometry and intensity modulation, *J. Acoust. Soc. Amer.*, **98**, p. 1042, 1995.
54. T. W. Kao and H. F. Taylor, High-sensitivity intrinsic fiber-optic Fabry–Perot pressure sensor, *Opt. Lett.*, **21**, p. 615, 1996.
55. J. J. Alcoz, C. E. Lee, and H. F. Taylor, Embedded fiber-optic Fabry–Perot ultrasound sensor, *IEEE Trans. Ultrasonics, Ferroelectrics, Freq. Control*, **37**, p. 302, 1990.
56. T. A. Tran, W. V. Miller III, K. A. Murphy, A. M. Vengsarker, and R. O. Claus, Stabilized extrinsic fiber optic Fabry-Perot sensor for surface acoustic wave detection, *Proc. SPIE*, **1584**, p. 178, 1991.
57. L. A. Ferreira, A. B. L. Ribeiro, J. L. Santos, and F. Farahi, Simultaneous displacement and temperature sensing using a white light interrogated low finesse cavity in line with a fiber Bragg grating, *Smart Mater. Struct.*, **7**, p. 189, 1998.

58. T. Li, A. Wang, K. Murphy, and R. Claus, White-light scanning fiber Michelson interferometer for absolute position-distance measurement, *Opt. Lett.*, **20**, p. 785, 1995.
59. M. D. Barrett, E. H. Peterson, and J. W. Grant, Extrinsic Fabry–Perot interferometer for measuring the stiffness of ciliary bundles on hair cells, *IEEE Trans. Biomed. Eng*, **46**, p. 331, 1999.
60. K. D. Oh, J. Ranade, V. Arya, A. Wang, and R. O. Claus, Optical fiber Fabry–Perot interferometric sensor for magnetic field measurement, *IEEE Photon. Tech. Lett.*, **9**, p. 797, 1997.
61. J. X. Fang and H. F. Taylor, Fiber optic Fabry–Perot flow sensor, *Microwave Opt. Tech. Lett.*, **18**, p. 209, 1998.

3

Polarimetric Optical Fiber Sensors

Craig Michie

CONTENTS

3.1 Introduction to Polarimetric Optical Fiber Sensors

The past 20 years have witnessed an intense international research effort devoted to the use of optical fiber-based sensors designed to measure a wide range of physical and chemical parameters. More recently, advances in component technologies and the growth of cost-effective processing power have enabled many of these sensing methods to mature from laboratory systems into commercial products. The operational details of many of these sensors can be found elsewhere. This chapter focuses principally on polarimetric measurement techniques, which relate changes induced in the polarization state of light traveling within an optical fiber to measurands of interest.

The chapter describes the basic physical properties and techniques exploited to produce polarimetric sensors and the limitations of these sensors within practical environments. During the course of the chapter the reader is introduced to a number of measurement schemes that have been investigated across a range of industrial applications, from civil engineering to the electricity supply industry. Through the course of the chapter the reader should become familiar with a range of polarimetric measurement methods and gain an appreciation of the practical limitations of some of the measurement schemes.

3.2 Propagation of Light Waves

Before proceeding with the detailed discussion of the fiber polarimeter, we provide a brief review of some of the fundamentals of wave propagation within optical fibers. Fuller details of this material can be found elsewhere [1–3] but nonetheless it is useful to review some essential background to aid in understanding the sensor operation.

Light is a transverse electromagnetic wave. The electric (E) and magnetic (H) fields are coupled and time varying—the E field generates the H field (and vice versa). Both fields are orthogonal to each other and their propagation in a single direction (the z-direction) can be described by Maxwell's equations (in two dimensions) as

$$\frac{\partial^2 E}{\partial z^2} = \varepsilon_0 \mu_0 \frac{\partial^2 E}{\partial t^2} \tag{3.1}$$

$$\frac{\partial^2 H}{\partial z^2} = \varepsilon_0 \mu_0 \frac{\partial^2 H}{\partial t^2} \tag{3.2}$$

where z represents the coordinate, along the direction of propagation, ε_0 the electric permittivity, μ_0 the magnetic permeability, and t time. The general form for a wave in three dimensions is written as

$$\nabla^2\phi = \frac{1}{\upsilon^2} = \frac{\partial^2\phi}{\partial t^2} \tag{3.3}$$

where υ, the velocity, is equal to

$$\varepsilon_0\mu_0 = \frac{1}{\upsilon^2} \quad \text{or} \quad \upsilon = \frac{1}{\sqrt{\varepsilon_0\mu_0}} \tag{3.4}$$

which evalutes to $3 \times 10^{-8}\ \mathrm{ms^{-1}}$ in a vacuum. In the case of the electric field this expression is written as

$$\nabla^2 \mathrm{E} = \varepsilon_0\mu_0 \frac{\partial^2 \mathrm{E}}{\partial t^2} \tag{3.5}$$

A similar equation applies for the magnetic field vector **H**.

The general solution to Eq. (3.5) describes a sinusoid of angular frequency ω:

$$E = E_0 e^{i(k\cdot r - \omega t)} \tag{3.6}$$

Here E and E_0 are, in general, complex vectors and $k \cdot r = xk_x + yk_y + zk_z$, where (k_x, k_y, k_z) are the components of the propagation direction and (x, y, z) are the components of the point in the space where the field E is evaluated.

Considering, for simplicity, only plane monochromatic (single-frequency) waves propagating in free space in the direction Oz, the general solution to the wave equation for the electric field can be expressed in the form

$$\mathrm{E}_x = e_x \cos(\omega \mathrm{t} - \mathrm{kz} + \delta_x) \tag{3.7}$$

$$\mathrm{E}_y = \mathrm{e}_y \cos(\omega \mathrm{t} - \mathrm{kz} + \delta_y) \tag{3.8}$$

where δ_x and δ_y are arbitrary phase angles. Thus, it is possible to describe this solution completely by means of two waves: one in which the electric field lies entirely in the xz-plane, and the other in which it lies entirely in the yz-plane. If these waves are observed at a particular value of z, say z_0, they take the oscillatory form:

$$\mathrm{E}_x = \mathrm{e}_x \cos(\omega \mathrm{t} + \delta'_x), \qquad \delta'_x = \delta_x - \mathrm{kz}_0 \tag{3.9}$$

$$\mathrm{E}_y = \mathrm{e}_y \cos(\omega \mathrm{t} + \delta'_y), \qquad \delta'_y = \delta_y - \mathrm{kz}_0 \tag{3.10}$$

and the tip of each vector appears to oscillate sinusoidally with time along a line. E_x is said to be linearly polarized in the direction Ox, and E_y is said to be linearly polarized in the direction Oy.

The tip of the vector, which is the sum of E_x and E_y, will in general describe an ellipse whose Cartesian equation in the xy-plane at the chosen z_0 will be given by

$$\frac{E_x^2}{e_x^2}+\frac{E_y^2}{e_y^2}+2\frac{E_x E_y}{e_x e_y}\cos\delta=\sin^2\delta \tag{3.11}$$

where $\delta=\delta_y'-\delta_x'$.

Now, the state of polarization (SOP) of an electromagnetic wave propagating along the z-axis may be ascertained from the path traced by the tip of the electric field vector **E** in the xy-plane. Three states of polarization can be defined: linear, circular, and elliptical.

3.2.1 Polarization

Linear Polarization

In a linearly polarized wave, traveling in the z-direction, the electric field vector oscillates along a straight line in xy-plane. The wave can be resolved into two orthogonal components along the x- and y-axes (E_x and E_y), with a path difference of $\delta = m\pi$, where m is an integer. If m is 0 or an even integer, then the two components are in phase. If m is odd, then the wave is still linearly polarized, but it is oriented orthogonally (see Figure 3.1).

Circular Polarization

Circular polarization arises from the superposition along the x- and y-axes of two linearly polarized waves of equal amplitude but differing in phase by $\delta = \pm\pi/2$. In this case the tip of the electric field vector describes a circle that rotates clockwise (or counterclockwise) according to the phase relationship. A phase difference of $\delta = -\pi/2 + 2m\pi$, where $m = 0, \pm1, \pm2$ …, gives rise to right circular polarized light (i.e., the electric field vector rotates clockwise

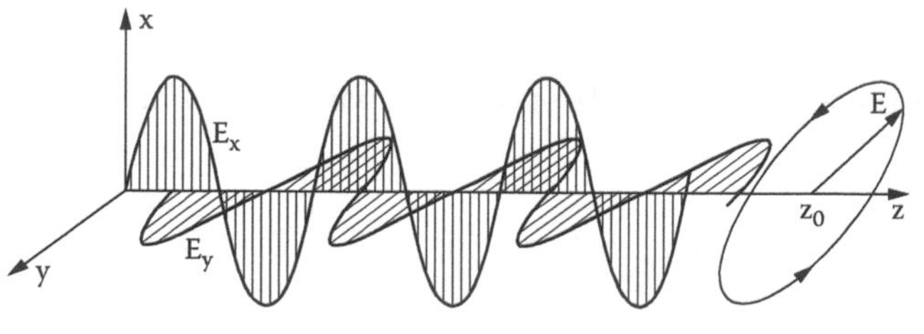

FIGURE 3.1
Polarization components of *E* wave.

when veiwed looking toward the source), and a phase difference of $\delta = \pi/2 + 2m\pi$ gives rise to left circular polarized light. It is useful to note that linearly polarized waves can be synthesized from two opposite polarized circular waves of equal amplitudes. The orientation of the x- and y-axes is immaterial, and circular polarized light can be specified by its amplitude and whether it is left or right polarized.

Elliptical Polarization

In all other circumstances the tip of the electric field vector rotates, tracing an ellipse in the xy-plane, and the electric vector also changes in magnitude. Elliptical polarized light results from the superposition along the x- and y-axes of two linearly polarized waves of arbitrary (and different) amplitudes and phase difference. Elliptical polarization is fully specified by its amplitude, ellipticity, and the orientation of the major axes of the ellipse with respect to some convenient axes. Linear and circular polarization can be considered as special cases of elliptically polarized light.

There are several formal ways of representing polarization and modeling the transmission of polarized light through polarizing media such as birefringent materials. Among these, Jones vectors and matrices [1] and the Poincaré sphere construction [4] are commonly used. The Jones vectors provide a mathematical treatment of polarized light, while the Poincaré sphere creates a geometrical representation of polarization. More details on the Jones vector matrix manipulation follow.

Many optical sensors, including the majority of optical voltage and current sensors, measure changes in polarization state and relate these to the parameter under investigation. It is important, therefore, that the evolution of the polarization state under the application of these fields is well understood.

3.2.2 Jones Matrix Algebra

The Jones matrix formulism enables the polarization state of light traveling through a complex transmission path to be evaluated by using matrix algebra [1]. The polarization state of a wave is represented by a two-component complex number of the form

$$a = \begin{bmatrix} E_x e^{j\partial_x} \\ E_y e^{j\partial_y} \end{bmatrix} \tag{3.12}$$

where E_i describes the amplitude of the electric field vector and δ_i its phase. The preceding expression describes the general case for an elliptically polarized wave. In the case of a linearly polarized light wave with the direction of the electric field at an angle θ to the x-axis, the Jones matrix can be simplified to

$$a = \begin{bmatrix} E_x \cos\theta \\ E_x \sin\theta \end{bmatrix} \tag{3.13}$$

The circular polarized wave can be described as

$$a = \begin{bmatrix} E\cos(\omega t) + i\sin(\omega t) \\ -E\sin(\omega t) + i\cos(\omega t) \end{bmatrix} \tag{3.14}$$

Taking the real components of each wave reveals that $E_x = E\cos(\omega t)$ and $E_y = -E\sin(\omega t)$. At $t = 0$, the x-component is given by E and the y-component is zero. As time progresses, the component of the electric field vector in the x-direction diminishes as the y-component grows. Thus, as stated earlier, the field of the electric vector appears to rotate clockwise by an observer looking into the forward-traveling wave.

A simple example of a Jones matrix is that of an absorbing medium that attenuates the light transmission but does not alter the polarization orientation. In this case the medium can be described as

$$A = \begin{bmatrix} \alpha & 0 \\ 0 & \alpha \end{bmatrix} \tag{3.15}$$

where α represents the attenuation of the light traveling through that medium. An ideal polarizer will block light where the electric field is oriented on one direction and will transmit all of the light in the orthogonal polarization state. The transfer function for such a polarizer can be written as

$$A = \begin{bmatrix} 1 & 0 \\ 0 & 0 \end{bmatrix} \tag{3.16}$$

and light traveling through this polarizer will be modified; thus, y:

$$\begin{bmatrix} 1 & 0 \\ 0 & 0 \end{bmatrix} \begin{bmatrix} E_x \\ E_y \end{bmatrix} = \begin{bmatrix} E_x \\ 0 \end{bmatrix} \tag{3.17}$$

The matrix for a linear polarizer with its transmission axis at an angle θ to the x-axis in the xy-plane is written as [1]

$$P = \begin{bmatrix} \cos^2\theta & \cos\theta\sin\theta \\ \cos\theta\sin\theta & \sin^2\theta \end{bmatrix} \tag{3.18}$$

The output polarization state is obtained by multiplying the input polarization state by a series of matrix elements that are used to describe each of the individual elements within the optical path.

3.2.3 Optical Retarders

Optical retarders are components made of birefringent materials used to change the polarization state of an incident wave. Light traveling through one of the principal axes of the retarder is delayed with respect to the light traveling in the orthogonal axes. On emerging from the retarder, the relative phase of the two components is different from the initial phase and so, therefore, is the state of polarization.

The relative phase difference or retardance $\Delta\Phi$ between the two constituent components' ordinary and extraordinary waves (or components along the fast and slow axes of birefringent material) is given by

$$\Delta\phi = \frac{2\pi}{\lambda} d\left(\left|n_o - n_e\right|\right) \tag{3.19}$$

where d is the thickness of the material, λ is the wavelength of the propagating light, and n_o and n_e are the refractive indices of the ordinary and extraordinary wave components, respectively.

The thickness of the birefringent material d is chosen so as to introduce the required relative phase difference between the two components. Some common retarders are *half-wave plate* and *quarter-wave plate.* The half-wave plate rotates the plane of linearly polarized light by 90°. A phase difference of $\pi/2$ is introduced between the light components propagating along fast and slow axes.

Quarter-Wave Plate

The quarter-wave plate changes linearly polarized light incident at an angle 45° to the optic axis into circularly polarized light. In this case the birefringent material introduces a phase difference of $\pi/4$ between the two component waves, which combine to give circularly polarized light.

3.2.4 Birefringent Optical Fiber

A range of optical fibers has been devised for measurement applications where it is essential to ensure that the polarization state of the propagating wave is preserved. Generally, these are anisotropic fibers that display two distinct refractive indices depending on the electric field orientation of the light entering into them. The two orthogonal principal axes of these birefringent fibers are described as the fast and slow axes, referring to the phase velocity of the light traveling within them. A beam guided in the axis with the higher index (slow axis) will have a lower velocity than a beam at the orthogonal axis (the fast axis). In the special case of light launched with its electric field aligned with the principal axis of the material, the light is guided with its polarization state undisturbed.

The birefringence of the fiber is given by the difference in the refractive index of the two optic axes as

$$B = n_s - n_f \tag{3.20}$$

where n_s and n_f are the indices of refraction of the slow axis and the fast axis, respectively. Birefringence is usually defined in terms of the fiber beat length, L_B, the length of fiber over which the difference in phase shift between the orthogonal polarizations amounts to 2π:

$$L_B = \frac{\lambda}{B} \tag{3.21}$$

If the fiber is subjected to a mechanical perturbation with a period comparable to L_B, strong power coupling between the two orthogonal polarizations will occur. Therefore, the value of L_B should be smaller than the perturbation periods introduced in the drawing process as well as the physical bends and the twists the fiber suffers. Consequently, polarization preserving will be achieved in fibers with short beat lengths, much shorter than the 10-cm value of conventional circular fibers.

The birefringence of a fiber can be determined experimentally through measurement of the beat length [5]. The simplest such method is visual observation of the pattern of the light emitted from the fiber due to Rayleigh scattering. Each scattering particle forms a radiating dipole excited by the incident light. Radiation is maximum orthogonal to, and zero in line with, the direction of the electrical field.

For visible light propagating in the fundamental mode with both orthogonal polarizations equally excited, the intensity of scattered light (observed at an angle of 90° to the direction the polarization launched in the fiber) can be seen to change periodically. Radiation is maximum when the two orthogonal modes are in phase and is zero when the modes are out of phase by π. Consequently, the beat length equals the distance between two points of maximum or minimum intensity, as illustrated in Figure 3.2.

High-Birefringence Fibers and Their Applications

There are two principal methods for introducing birefringence into optical fibers. The first of these relies on modifying the waveguiding characteristics of the core region by altering its geometry such that the circular symmetry is lost and two axes with different refractive indices are produced. These so-called geometrical birefringent fibers, commonly known as *elliptical core fibers*, have beat lengths in the order of a few millimeters [6]. A cross-section of an elliptical core fiber is shown in Figure 3.3(a).

In an alternative design, linear birefringence can be introduced into the fiber by applying asymmetric stress to the core, which modifies the core refractive index profile. This is achieved either by making the area surrounding the

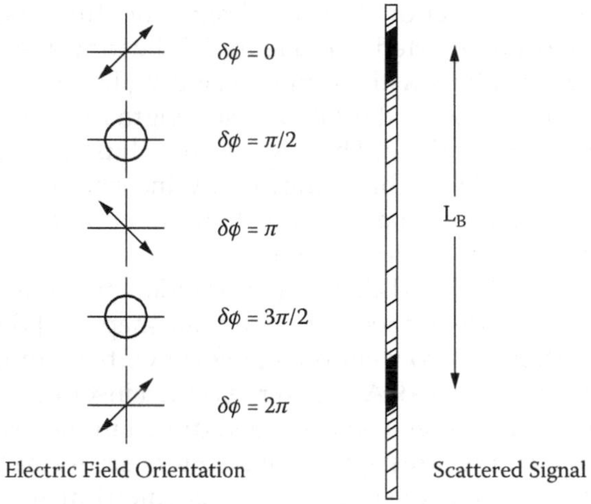

FIGURE 3.2
Visual determination of the beat length, L_B, with $\delta\phi$ the optical phase difference between the two orthogonal polarization states.

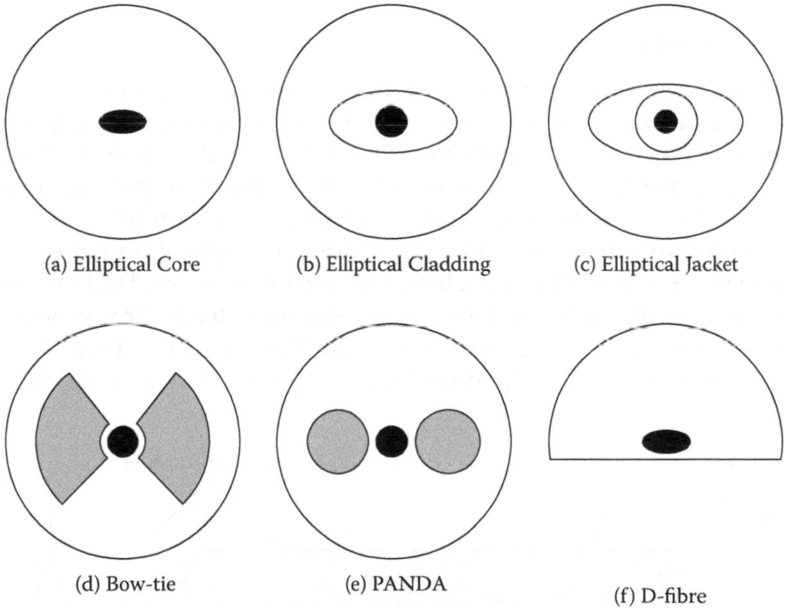

FIGURE 3.3
High-birefringence fiber designs.

core noncircular or by introducing highly doped regions around the core. Birefringence is induced elasto-optically due to the different thermal expansion coefficients of the material surrounding the core combined with the asymmetric fiber cross-section. Various designs of stress-induced birefringence fibers have been reported, and Figure 3.3 illustrates some of them.

High-birefringence fibers with elliptical cladding [7] or an elliptical jacket [8] have been reported; with the latter, heat lengths in the order of a few millimeters have been obtained. The bow-tie fiber, illustrated in Figure 3.3(d), exhibits the highest value of birefringence by incorporating highly doped regions in the area surrounding the core. With this approach, beat lengths as low as 0.5 mm have been reported [9,10].

In a similar design, the PANDA fiber (*p*olarization maintaining *and a*bsorption reducing) has two doped regions of circular shape [10] (Figure 3.3(e)). To fabricate a PANDA fiber a conventional preform with a pair of holes drilled on each side of the core is used. A doped rod is then inserted in each hole, and the composite preform is then collapsed and drawn in the usual way to produce a fiber in which the stress-producing sectors are formed by the doped rods. Designs that combine the two effects of geometrical and stress birefringence have also been reported (e.g., bow-tie fiber with an elliptical core) [11].

3.3 Polarimetric Sensors

In a typical fiber polarimetric sensor, linearly polarized light is launched at 45° to the principal axes of a birefringent fiber such that both eigenmodes are equally excited. The polarization state at the output is converted to intensity using a polarizer analyzer oriented at 90° to the input polarization state (see Figure 3.4). Since the determination of the polarization rotation is essentially reduced to a measurement of intensity of the optical transmission, the arrangement thus described is vulnerable to errors associated with variations in the source output or within the fiber sensor input leads. This problem can be overcome by using a Wollaston prism and two detectors. Using the Wollaston prism, the signals at the detectors are equivalent to those produced by

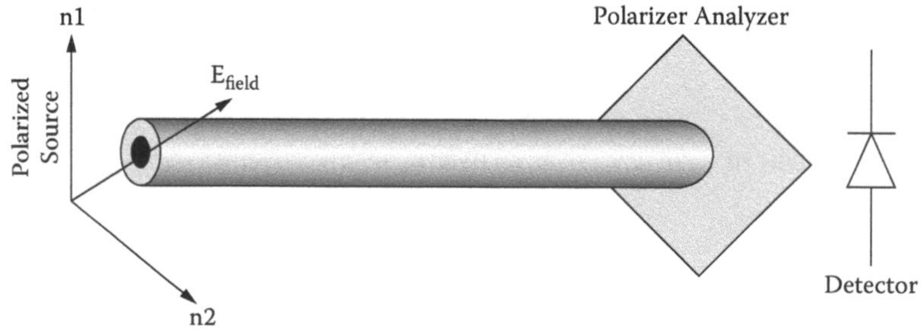

FIGURE 3.4
Schematic representation of a fiber polarimeter.

polarizing analyzers oriented at +45° and –45° to the fiber's birefringent axes. Processing results in a signal independent of the optical power. This advantage comes, however, with the additional difficulty of correctly aligning the Wollaston prism and the two detectors with respect to the fiber birefringent axes.

The polarimeter can be readily analyzed using the Jones matrix algebra as follows. Assuming that the input beam is linearly polarized, both orthogonal polarization modes of the high-birefringence fiber can be equally excited if a half-wave plate is used to align the azimuth of the polarization at 45° to the axes of the fiber. In that case, using the Jones matrix notation, one can express the electric field of the light as

$$E_0 = \frac{1}{\sqrt{2}}\begin{bmatrix}1\\1\end{bmatrix} \tag{3.22}$$

The electric field of the light after traveling through the birefringent section of fiber and the polarizing analyzer at the fiber output can be calculated as

$$E = ABE_0 \tag{3.23}$$

where A and B are the Jones matrices of the polarization analyzer and the high-birefringence fiber, respectively. If we regard the fiber as a phase plate, B can be written as

$$B = \begin{bmatrix} e^{i(\phi_1+\phi_2/2)} & 0 \\ 0 & e^{j(\phi_1-\phi_2/2)} \end{bmatrix} \tag{3.24}$$

where Φ_1 is the mean phase retardance and Φ_2 is the induced phase retardance between the orthogonal polarizations caused by propagation through the fiber.

For the special case of a polarization analyzer positioned with its axis at an angle of 45° with the axes of the fiber, A is given by

$$A = \frac{1}{2}\begin{bmatrix}1 & 1\\1 & 1\end{bmatrix} \tag{3.25}$$

The intensity detected after the polarization analyzer will be

$$I = E^2 = \frac{I_o}{2}[1+\cos\phi_2] \tag{3.26}$$

where I_o is the total output power. Therefore, the change in polarization will be observed as a change of intensity after the analyzer, producing a sinusoidal signal.

The sensitivity of a sensor is then dependent on the degree of polarization rotation introduced by a particular measurand and the minimum resolvable rotation that can be detected. The influence of the measurand on the emergent polarization state can be calculated as follows.

3.3.1 The Optical Phase Change Mechanism

The phase of a light wave guided by a fiber of length L is given by [12]

$$\phi = \beta L = n_{\text{eff}} k_0 L \tag{3.27}$$

where β is the propagation constant of the mode, n_{eff} is its effective index, and k_0 is the free-space wave number of the light equal to $2\pi/\lambda_0$, with λ_0 the wavelength in vacuum. For any two modes guided by the fiber, the difference in their optical phases after length L can be expressed as

$$\Delta\phi = \Delta\beta L = \Delta n_{\text{eff}} k_0 L \tag{3.28}$$

where Δn_{eff} is the difference in effective indices between the two modes.

If the fiber is subjected to an external strain, ε, the differential phase will be affected according to Eq. (3.25). The overall phase response is proportional to the length of the fiber exposed to the external field. Therefore, it is customary to consider the response of the sensor in relation to a unit sensor length. Hence,

$$\frac{1}{L}\frac{\partial(\Delta\phi)}{\partial\varepsilon} = k\frac{\partial(\Delta n_{\text{eff}})}{\partial n}\frac{\partial n}{\partial\varepsilon} + k\frac{\partial(\Delta n_{\text{eff}})}{\partial D}\frac{\partial D}{\partial\varepsilon} + k\Delta n_{\text{eff}}\frac{1}{L}\frac{\partial L}{\partial\varepsilon} \tag{3.29}$$

where D is the fiber core diameter and n the index of the core, or the cladding.

The first term in Eq. (3.25) describes the photoelastic effect that relates mechanical strain to the optical index of refraction of the strained material. For the case of normal circular core fiber, the resulting change of the refractive index of the fiber is calculated according to Eq. (3.26) (assuming that the core and cladding indices are almost equal):

$$\frac{\partial n}{\partial\varepsilon} = -\frac{n^3}{2}[p_{12} - v(p_{11} + p_{12})] \tag{3.30}$$

where v is Poisson's ratio taken to be the same in both core and cladding, and p_{11}, p_{12} are the strain optic coefficients [24].

The second term in Eq. (3.30) reflects the change in the differential effective index due to the longitudinal strain on the diameter of the fiber with $\partial D/\partial\varepsilon$ equal to $-vD$. This term has been shown to have a very small contribution and can generally be neglected. Finally, the last term of Eq. (3.30) is the result of the physical change in length due to strain.

3.3.2 Temperature Sensing

A similar expression can be derived to describe the case when a fiber experiences a change in temperature (*T*):

$$\frac{1}{L}\frac{\partial(\Delta\phi)}{\partial T} = k\frac{\partial(\Delta n_{\text{eff}})}{\partial n}\frac{\partial n}{\partial T} + k\frac{\partial(\Delta n_{\text{eff}})}{\partial L}\frac{\partial L}{\partial T} + k\Delta n_{\text{eff}}\frac{1}{L}\frac{\partial L}{\partial T} \tag{3.31}$$

The first term reflects the change in the refractive index of the fiber due to the pure temperature effect, while the second term reflects the changes in the refractive index due to the changes in the fiber dimensions via the photoelastic effect. The change in fiber length due to thermal expansion or contraction is expressed by the third term with $\partial L/\partial T$ equal to αL, where α is the thermal coefficient of linear expansion. It has been shown [4,6] that the change in phase with temperature is dominated by the change in index rather than by the change in length.

Equations (3.25) and (3.27) have been used to calculate the strain [12] and temperature [13] sensitivity of circular core fibers with relatively good accuracy. In the case of high-birefringence fibers, the calculation is more complicated due to the complexity of the shape and the material consistency of the fiber; the strain and temperature sensitivities of high-birefringence fibers are usually determined experimentally with calibration experiments.

3.3.3 Overmoded Sensors

Dual-mode, or overmoded, sensors operate using a similar principle to polarimeters except that the propagation paths through the fiber are separated spatially. In a slab waveguide, the solutions to Maxwell's equations that describe the propagation of the optical wave fall into two classes and are represented as TE and TM, depending on the field that is oriented perpendicular to the direction of propagation. In a cylindrical waveguide such as an optical fiber, two classes of solutions also exist, but these are not true transverse waves but hybrids and are designated EH or HE—the fundamental mode being the HE11 mode. Higher order modes are generally combinations of other modes and are consequently more difficult to analyze. The linearly polarized (LP) mode approximation [5] has been developed to simplify the analysis of mode propagation in fibers and is widely used. In this approximation the fundamental mode HE11 is described as the LP01 mode. The mode adjacent to it, known as the LP11 mode, is a combination of the HE01 and the EH01, both of which have even and odd solutions. The LP notation significantly eases the complexity of analyzing (and describing) the operation of modal interference sensors and is used here.

Dual-mode sensors [11,14–17], or twin-moded sensors, take advantage of the fact that the LP_{01} and LP_{11} modes travel along the fiber with different velocities. Furthermore, the electric field distribution across the fiber core differs for each mode. The electric field of the fundamental mode (the LP_{01}

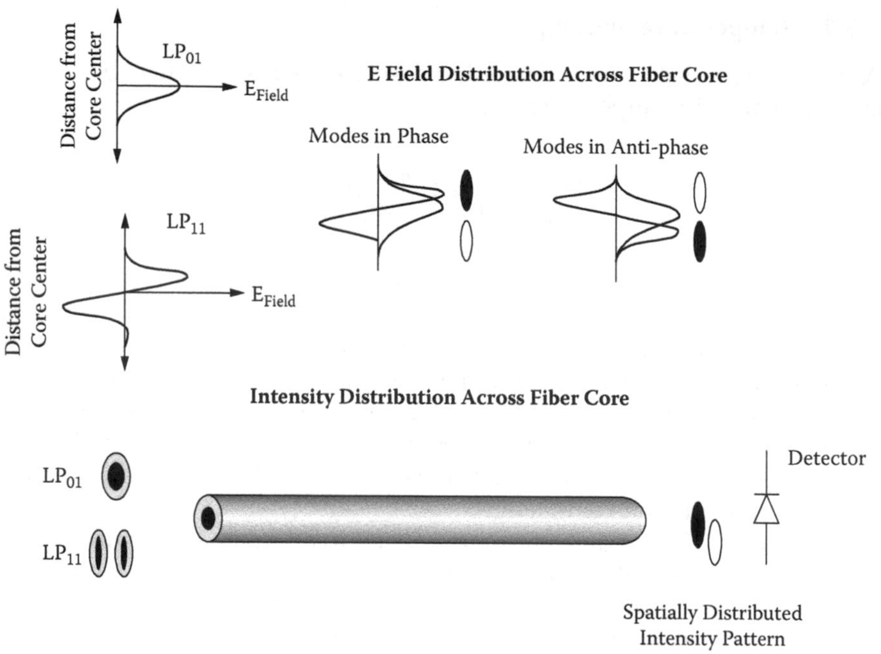

FIGURE 3.5
Schematic representation of dual-mode sensor.

mode) describes a Gaussian profile, whereas the field shape of the LP_{11} mode has two maxima with opposing signs (see Figure 3.5). The light intensity observed across the core diameter is described by the superposition of these two fields. Consequently, the near-field pattern can be observed as a pair of oscillating lobes as the phase relationship between the modes changes. Temperature or strain fields applied to the fiber directly modulate the phase relationship and consequently change the output pattern. This can be measured and directly related to the parameter causing the effect. The difference in effective mode index is much higher for spatial modes than it is for polarization modes; therefore, these sensors are generally more sensitive by one to two orders of magnitude.

These sensors have been researched widely as strain sensors and have been combined with polarimeters to produce sensors that can simultaneously measure two or more parameters such as temperature and strain.

Combined Strain and Temperature Measurement

In cases where the target measurement parameter is dynamic, such as a vibration-induced strain change, it is relatively easy to decouple the influence of simultaneously occurring fields such as temperature and strain. However, when the temporal characteristics of the measurands are similar, for example, quasi-static strain measurements, then some means of recovering the strain and temperature information from a single-sensing fiber is required. Generally, the approach that is adopted is to interrogate the sensor with two different measurement approaches at the same time. This allows the temperature and strain to be recovered by solving a set of simultaneous equations [17,18,20–22]. The measurement parameters Φ_1 and Φ_2 of the sensing schemes used are both functions of strain changes (ε), and temperature change (T) is related using matrix formulation as follows:

$$\begin{bmatrix} \phi_1 \\ \phi_2 \end{bmatrix} = \begin{bmatrix} K_{1T} & K_{1\varepsilon} \\ K_{2T} & K_{2\varepsilon} \end{bmatrix} \begin{bmatrix} T \\ \varepsilon \end{bmatrix} \tag{3.32}$$

or

$$\Phi = K\Omega \tag{3.33}$$

K, the transfer matrix from Ω to Φ, is the characteristic matrix of the sensor under investigation. Provided that the determinant $\Delta = K_{1t}K_2\varepsilon - K_{2t}K_1\varepsilon$ is not equal to zero, this equation may be inverted and T and ε can be calculated as

$$\Omega = K^{-1}\Phi \tag{3.34}$$

where K^{-1} represents the inverse of matrix K.

It has been shown that combinations of polarimetry and dual-mode interferometry [17], or polarimetric measurements on the LP_{01} and LP_{11} modes, are well suited to this type of measurement, enabling temperature and strain to be readily recovered since the relative sensitivities of the two measurement methods can be radically different. However, such sensors are difficult to implement since the launch conditions for the dual-mode signal are different from that of the polarimeter. Consequently, only laboratory demonstrations of such sensors have been reported.

An alternative sensor can be constructed by implementing two polarimeters [18,19] on the same fiber, one on the LP_{01} mode and another on the LP_{11}. This produces the following reasonably well-conditioned matrix:

$$\begin{bmatrix} \delta\Delta\phi_{LP_{01}} \\ \delta\Delta\phi_{LP_{11}} \end{bmatrix} = \begin{bmatrix} -4.87\times10^{-3} & -0.882 \\ 7.2\times10^{-3} & -0.846 \end{bmatrix} \begin{bmatrix} \delta\varepsilon\,(\text{rad}/\mu\varepsilon\,\text{m}) \\ \delta T\,(\text{rad}/^\circ\text{C m}) \end{bmatrix} \tag{3.35}$$

The advantage of a sensor of this type is that it is compatible with coherence multiplexing methods that allow quasi-distributed measurements to be implemented. This is illustrated in the following examples.

3.3.4 Coherence

The coherence length, or coherence times, of an optical source describes the extent to which it can be represented by a pure sine wave. In sensor applications this can have important implications. Highly coherent sources can often cause problems from cross-talk arising from interference from parts of the sensor network that are displaced from the sensor but within the coherence length of the source used to address the system. Short coherence-length sources are often deliberately used in order to restrict the field of view to one single sensor at a time, thus enabling many sensors to be addressed at using a single source and receiver (more details are provided later). The importance of the source coherence length on the performance of an optical sensor network warrants that some time be spent on the basic physical principles.

A beam of light can be considered as comprised of a discrete collection of photons, each with its own characteristic lifetime. The photon is essentially a wave train where the electric field exists for a short period of time, or propagation distance (see Figure 3.6).

The optical wave from a conventional light source is a collection of randomly generated photons with arbitrary phase relationships. The maximum distance over which the phase of a light source can be predicted (the source coherence length) is directly related to the lifetime of the photon and can be written as

$$L = c\partial T \tag{3.36}$$

where c is the speed of light and δT *is* the photon lifetime. The wave train associated with a single photon is not a pure sinusoid but can be considered as a series of sine waves of different frequencies and amplitudes. The distribution of these frequencies is generally characterized by a near-Gaussian or Lorentzian distribution, as shown in Figure 3.7, where the full-width, half-maximum (FWHM) line width of the distribution is given by $\delta v = 1/\delta T$.

The coherence length of a source, the length over which the phase of the light can be related to the spectral spread of the source (its line width) by differentiating the expression $c = v\lambda$ gives

$$\delta \mathrm{v} = \frac{c}{\lambda^2}\delta\lambda. \tag{3.37}$$

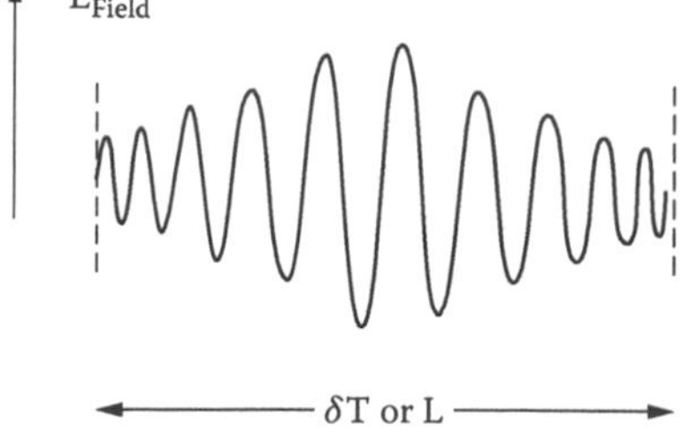

FIGURE 3.6
Wave train representation of a photon.

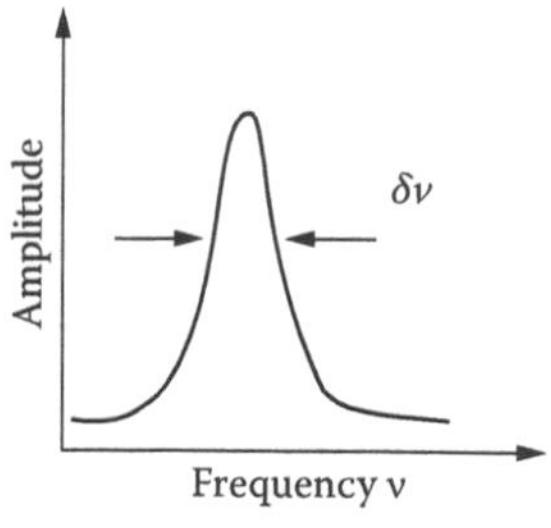

FIGURE 3.7
Spectral spread of laser emission.

Since we know that

$$L = c\delta T$$
$$L = \frac{c}{\delta v} \tag{3.38}$$

then

$$L = \frac{\lambda^2}{\partial \lambda} \tag{3.39}$$

Hence, for a light-emitting diode with a source wavelength of 850 nm and a line width of 50 nm, the coherence length would be approximately 15 μm. This extremely short distance over which the phase can be predicted or measured can be used effectively to multiplex several interferometers.

3.3.5 Coherence Multiplexed Sensors

Time division multiplexing (TDM) a network of optical sensors is an obvious means of sharing the resource of transmitter and detector among a large number of sensor elements. There are obvious advantages to this approach since it harnesses the power and flexibility of silicon-based processing techniques and it is robust. However, it is intrinsically limited by the processing speed of the associated electronics and for high-bandwidth measurements this usually implies a significant cost penalty. Much of this expense can be avoided if the signals can be preprocessed optically such that the only conversion to an electrical signal takes place at the measurand processing stage. Coherence-based multiplexing is one possible route to achieve this goal [19,23,24].

Phase-based sensors and polarimeters measure the relative phase delay introduced between light traveling in two arms of an interferometer. This information is recovered by processing the interference signal derived from beating the output of both arms either in the plane of a polarizer or at the surface of a detector. Consequently, this information can be recovered only if the two signals are coherent—that the relative time delay between the transmission paths does not exceed the coherence time of each individual source. Coherence multiplexed systems exploit this constraint to give each sensor a uniquely identifiable signature.

An array of interferometers can be constructed. Providing that the relative path imbalance (between the sensing and reference arms) differs between each interferometer by more than the coherence length of the interrogating source (typically 20 μm for a light-emitting diode), then each interferometer can be uniquely identified and its signal recovered using a compensating interferometer at the receiver. This concept is illustrated in Figure 3.8.

The primary disadvantage with constructing such arrays is that a great deal of the energy is lost. Furthermore, a series of unbalanced interferometers and matched receiving interferometers must be constructed in order

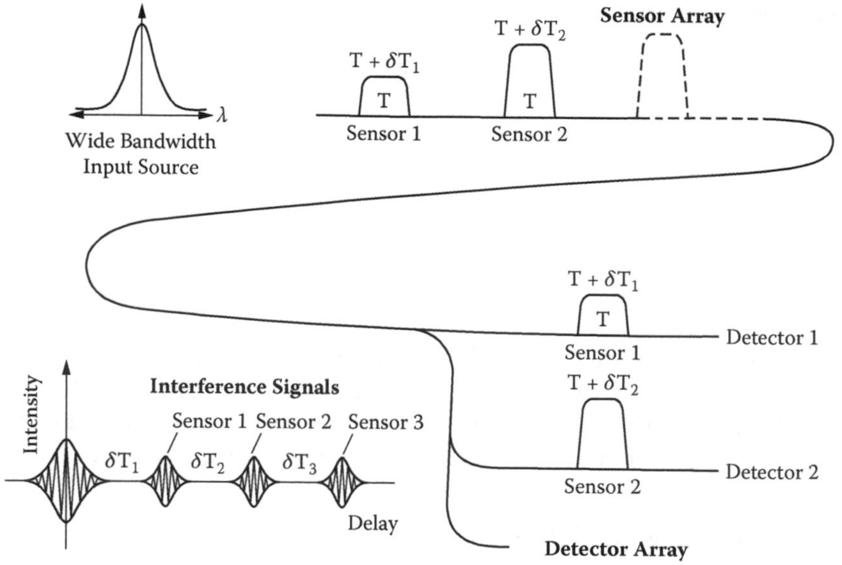

FIGURE 3.8
Coherence multiplexing of interferometric sensors.

to identify each sensor uniquely. A consequence of all this is that the optical fiber network's inherent simplicity is lost. The measurement and compensating arrays are as complicated as, if not more complicated than, a comparable electrical wiring circuit. However, variations on the preceding approach, which use the differential delay experienced by light traveling through the orthogonal axes of polarization-maintaining fiber, recover many of the advantages associated with the optical measurement schemes.

Figure 3.9 illustrates how a single fiber length can be delineated into several measurement sections and the measurement information relating to each section recovered at the receiver. The measurement principle is known as *quasi-distributed polarimetry.* Polarized light from a low-coherence source is launched into one of the principal axes of a polarization-maintaining fiber. At preselected locations along the fiber a small proportion of this light is coupled across to the orthogonal axes. This forms a sensor section, and thermal or mechanical effects in this vicinity will modulate the differential delay, which will be observed as a polarization rotation. As the light travels down the fiber, the light in the slow axis will be delayed with respect to that in the fast axes. At some point this delay will exceed the coherence length of the source and the relative phases between the two axes and the polarization information will no longer be recoverable. At this point an additional cross-coupling point can be introduced and another sensor location identified. The process can be repeated until the end of the fiber length. At the fiber exit a scanning interferometer is used to compensate for the delays introduced by the fiber and hence recover the interference signals. In this way each

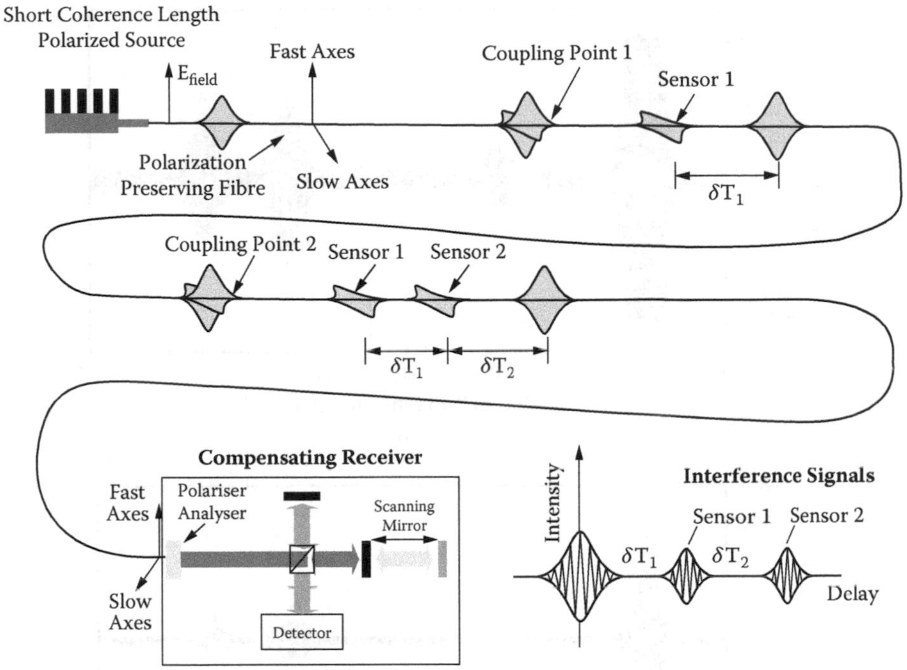

FIGURE 3.9
Quasi-distributed polarimetric sensor.

measurement section can be sequentially addressed and the target measurement parameter recovered as a function of position along the fiber length.

Measurement systems of this type and derivatives of them have been pioneered by Bertin Co. (France) among others and applied to a range of applications including monitoring stress in composite materials and monitoring the temperature of the stators in a nuclear plant. In the latter example the measurement instrument addressed up to 128 channels, and although the channels were separated spatially, the measurement principles are essentially the same [25,26]. Each probe had a temperature accuracy of better than 0.5°C and could operate up to a temperature of 200°C.

Quasi-Distributed Polarimetric Strain and Temperature Sensors

An extension of the principle just illustrated has been shown [19] to permit the simultaneous measurement of temperature and strain on a quasi-distributed basis by exploiting differences in sensitivity of the LP_{10} and the LP_{11} to temperature and strain fields. Implementation of this system required the development of a novel mode-splitting device that physically separated each mode into two distinct fibers that could be directed to a pair of Michelson interferometers to compensate for accumulated path imbalance along the sensing fiber length. Figure 3.10 shows an implementation with two coupling points

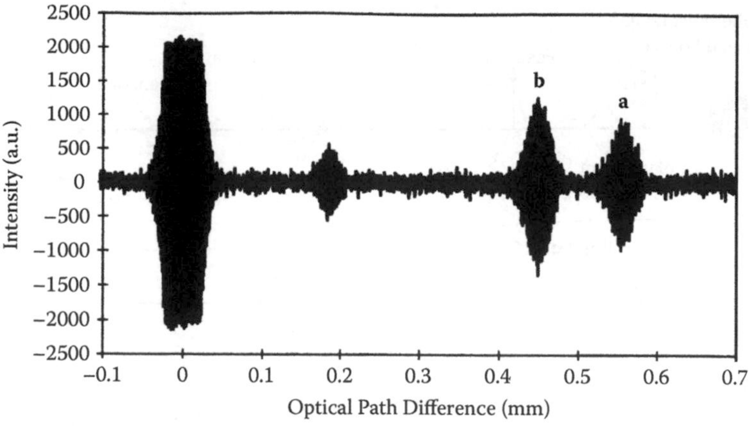

(a)

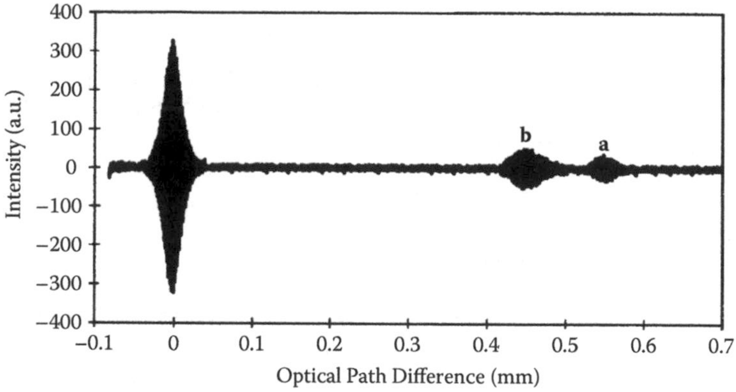

(b)

FIGURE 3.10
Signal of the Michelson interferometers for (a) the LP_{01} mode and (b) the LP_{11} mode. (a) and (b): Interferograms due to the first and second coupling points, respectively.

induced: The first is 50 cm from the fiber input and the second is 120 cm, forming two sensing lengths (point a to the end of the fiber, and point b to the end of the fiber, respectively). The fringes from each mode correspond to the position of the coupling points.

For the first part of the strain and temperature cycle, both the strain and the temperature reach high values. The strain error increases slowly, reaching the highest value of 200 με after 50 sec, when both strain and temperature have a high value. After this point the strain starts to decrease and the error assumes an almost constant value. For the temperature data the error increases rapidly after 50 sec and continues to increase for the whole cycle, reaching the highest value of –6°C when the temperature of the fiber is 70°C. In the second part of Figure 3.11, where the same strain cycle is applied but the temperature in both sensing regions is kept low (approximately 38°C), the errors in both strain and temperature are low (2°C and 50 με).

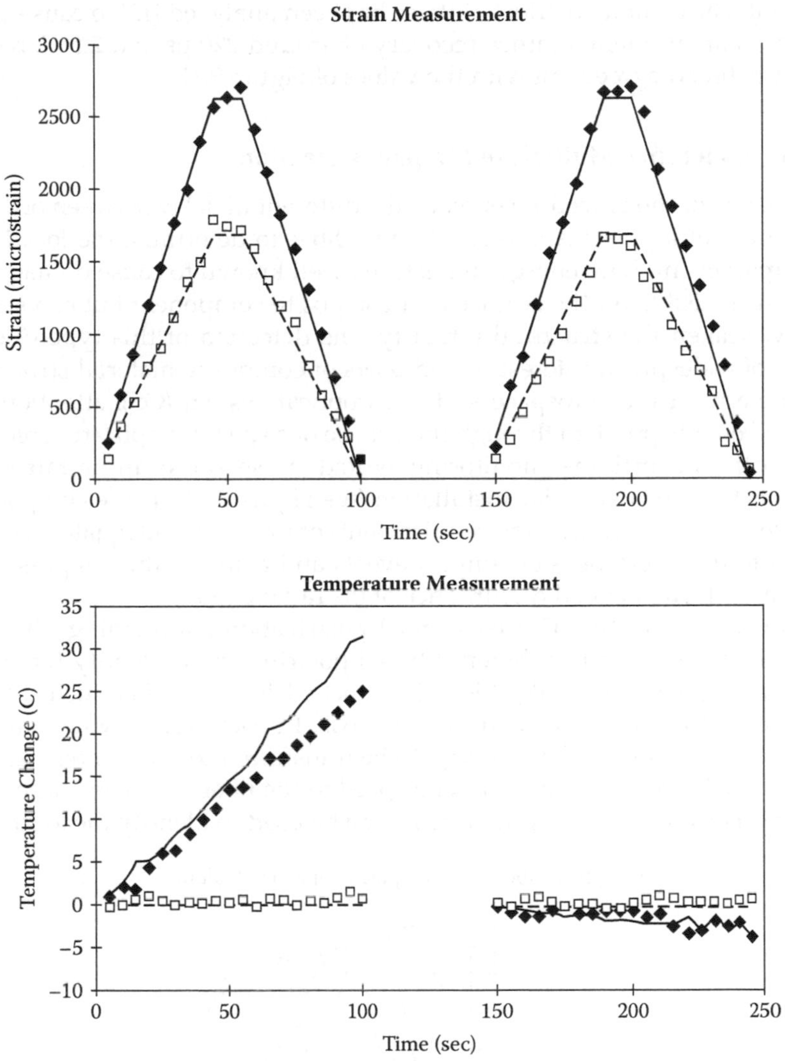

FIGURE 3.11
Simultaneous strain and temperature measurement over two sections of a single fiber.

The experimental performance of the preceding measurement is not particularly impressive, but the principle is demonstrated. First, errors in the measurement process arise from systematic experimental error (the reference thermocouple only samples the temperature at one point, whereas the optical fiber integrates the temperature measurement over its entire length). Second, the phase recovery process on the polarization rotation measurement is accurate to $\pi/10$. This influences not only the measurement but also the matrix calibration process and produces errors of up to 2% in each matrix element. The combined effect of these has been analyzed [13] to cause errors in the strain and temperature recovery of around 220 με and 2.5°C, respectively, in broad agreement with the values of Figure 3.11.

3.3.6 Coherence Multiplexed Impact Detection

A variation of the preceding exploits the differential delay between orthogonal modes of a polarization-maintaining fiber to determine the location of an impact event. Low-energy impacts are well known to cause damage that remains invisible on the surface of a composite component but can significantly weaken the structural integrity. The detection of this type of damage is of widespread interest to end users of composite material structures, particularly in the aerospace sector. In optical system (OS), attention was focused on this problem through the vehicle of a specific application requirement of the consortium—monitoring the Radome section of an aircraft, which protects the sensitive radar installation (see Figure 3.12). These components are pressed by the competing requirements of providing adequate mechanical protection from possible impact events and being as thin as possible to minimize the associated disturbance of the radar signal.

A pressure-sensitive [25] (side hole) polarization-maintaining fiber was used. This was used in coherence-based polarimetry to identify the occurrence of impact events [18]. Light is launched into the fiber such that the electric field is coincident with the fast axis. Impact events, which damage the fiber, cause some of the energy to be transferred to the orthogonal axis, where it will propagate at a reduced speed to the fiber end. The phase delay induced between the two propagating can be used to identify the location of

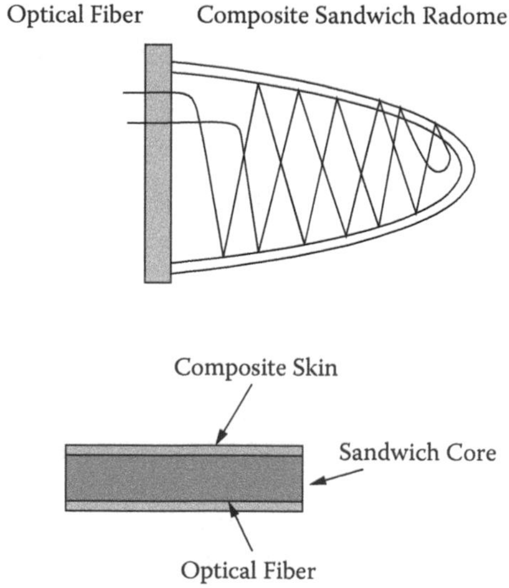

FIGURE 3.12
Optical fiber sensor integrated in Radome structure.

the impact. The delay induced between the impact and the distal end of the fiber is

$$\Delta\Phi = (2\pi/\lambda)\Delta n L \tag{3.40}$$

where Δn is the difference in refractive index between the two axes, L is the distance between the cross-coupling event and the distal end of the fiber, and λ is the operational wavelength. The magnitude of the impact event can be deduced from the amount of cross-coupling that has taken place, and the location of the impact is determined from measurements of the phase delay using a Michelson interferometer.

The Radome was constructed from a sandwich of 1-mm-thick D glass epoxy skins bonded to a 10-mm-thick, high-density foam material, as shown in Figure 3.12. To accommodate the sensor, grooves were machined in the foam. Impact events on the surface are transferred to the fiber and are thus detected. Although the operation of the sensor depends on the orientation of the fiber with respect to the impact event, the fiber has shown to be sensitive to impacts of around 5 J. The spatial location of an impact can be identified to within 1 cm without difficulty.

Similar investigations have been reported for monitoring impact damage more generally within composite materials using other fiber types. Figure 3.13 shows the signal generated from a 50-J impact onto a composite laminate structure [23,24]. The sensitivity of the measurement depends on the interface where the fiber is positioned. Placing the sensor between two 0° plies produces the minimum effect on structural integrity but limits the

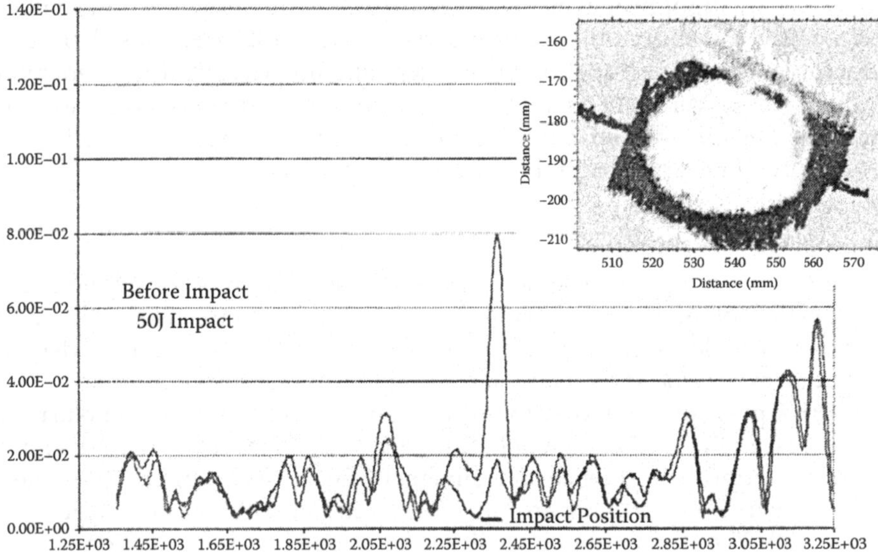

FIGURE 3.13
Impact detection using bow-tie fiber.

sensitivity because the interface effectively disappears after the composite structure is cured.

3.4 Optical Current Measurement

3.4.1 Principle of Operations

So far our treatment of polarimetric sensors has focused entirely on sensors for physical parameters that would commonly be used within structural monitoring applications (for example, temperature and strain measurements). In addition to these an important use of polarimetry has been made within the electricity supply industry to monitor voltage and current variations within the transmission and distribution networks for both metering and protection purposes. This is arguably the most technically developed field of fiber-based polarimetric sensors: Many industrial research laboratories have produced prototype systems that are now moving toward commercialization.

The optical current transducer (OCT) determines the current flow in an electrical conductor by measuring the magnetic flux density within the vicinity of the conductor. If the optical sensor completely encloses the conductor, then a true reading of the current can be obtained; otherwise the reading reflects the magnetic flux density at the measurement point and has to be scaled accordingly.

The magnetic flux density is determined by the polarization rotation induced as the light propagates through a guiding medium (the optical fiber or a sensitive crystal) in the vicinity of the conductor [27–29]. This change in polarization state is a function of the magnetic flux density, the interaction length, and the Verdet constant (V, rads T^{-1} m^{-1}) of the material used to construct the device. The angular rotation θ (measured in degrees) experienced by the light passing through the sensor is described as

$$\theta = VBl \tag{3.41}$$

where B is the magnetic field strength (tesla), a function of the applied current and the conductor geometry, and l is the length of the sensor exposed to the magnetic field. A great many methods can be used to measure the degree of rotation experienced by the light traveling through the magnetic field. The basic principle is illustrated in, for example, in Ning et al. [27,28]. Unpolarized light is first polarized using a polarizing film or crystal. The polarized light is rotated under the influence of the magnetic field; this rotation is converted to an intensity change using an analyzing polarizer at the sensor output (see Figure 3.14). The transmitted intensity is described by the expression

$$I = I_0\left(1 + \sin(2\theta)\right)/2 \tag{3.42}$$

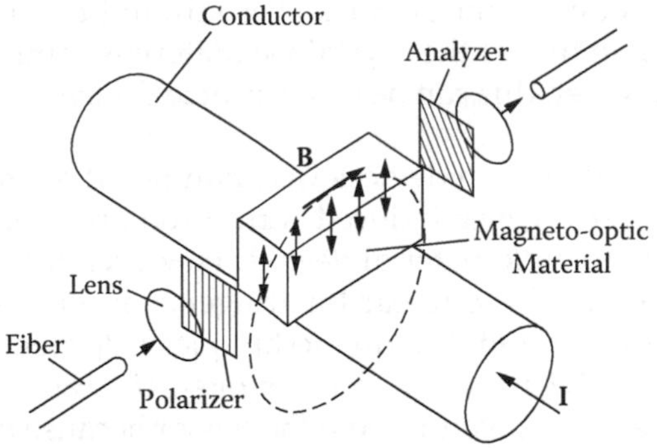

FIGURE 3.14
Schematic of point current sensor.

where I_0 is the input intensity of the light delivered to the sensor. The preceding measurement is sensitive to the absolute value of the steady-state light transmission (I_0), but this sensitivity can be eliminated by normalizing the modulated signal to the dc component.

Although the underlying concept is straightforward, the practical implementation of these devices is far from trivial; many different approaches have been investigated to produce an effective current sensor. In the following section we present a summary of some of the key methods in order to illustrate the diversity of solutions that have been devised.

Ideally, the OCT should use the optical fiber itself as the medium for detecting the magnetic field associated with the current. If this can be performed without the need for excessively complex processing, then it provides the simplest and ultimately most cost-effective solution. Furthermore, this intrinsic sensor design, shown in Figure 3.15, totally encloses the magnetic path, hence providing isolation from other potentially corrupting magnetic

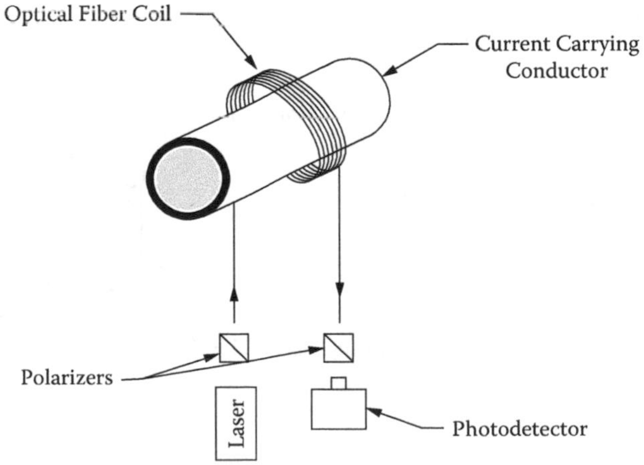

FIGURE 3.15
Wound-fiber type OCT.

sources. The Verdet constant of silica fiber (around 8 × 10^{-6} rad/A) is significantly lower than that of the crystalline materials commonly used in the point sensors. However, this can be compensated for by using multiple turns [27–29].

Early efforts to realize such devices were hampered by instability derived from the linear birefringence induced within the fiber coils, and the focus of effort moved on to bulk material devices. However, more recent research has virtually eliminated the linear birefringence in optical fibers through novel fiber composition and drawing techniques [32]. Fiber with very small linear birefringence (<1 m^{-1}) was demonstrated in 1986. The birefringence increases once the fiber is wound onto the sensor because of the anisotropic stress profile produced across the fiber core, but high-temperature annealing has proven a very effective means of reducing this influence. The annealing process destroys the protective fiber coating; however, a new coating can be applied following the annealing process provided that one takes care to ensure that residual linear birefringence is not reintroduced at this stage. Several major manufacturers, including Toshiba [30,31] and ABB [33], have produced and tested wound-fiber OCT devices in field environments.

Reflective arrangements of this type of sensor (see Figure 3.16) enable the fiber coil to be wrapped around the conductor, obviating the need to disassemble the conductor for fitting [30,31]. In addition, the nonreciprocal nature of the Faraday effect serves to double the sensing length of fiber and can be used to suppress the influence of vibration-induced birefringence since the influence on the clockwise-propagation beam will be in anti-phase to the effect on the counterclockwise-propagation beam.

Toshiba developed a silica fiber OCT based on this design that has been demonstrated to meet the JEC 1201 IPS metering specification. Although the authors do not give details of any specific features implemented to account for vibration-induced error, the influence of mechanical shock from a circuit breaker vibration of 8 G was recorded as less than the rms value of the electronic noise (3.8 A).

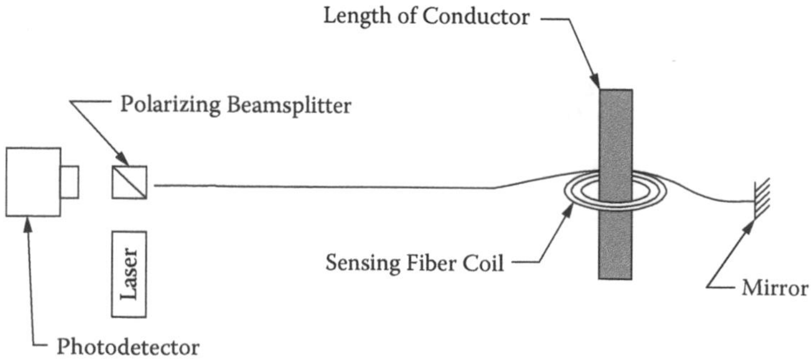

FIGURE 3.16
Reflective-type wound-fiber OCT.

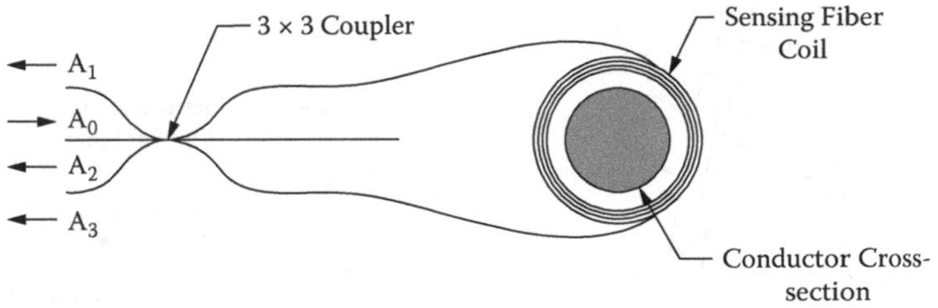

FIGURE 3.17
Sagnac interferometer-based OCT.

Capitalizing on the research that has gone into optical fiber gyroscopes, current measurement devices based on a Sagnac interferometer have been developed [33–36]. The sensor measures nonreciprocal phase changes that arise between light propagating clockwise and counterclockwise through a measurement coil. In the gyroscope, the phase shift gives a measure of the Earth's rotation. In a current meter the phase shift is produced by the influence of the magnetic field.

The Sagnac-based device shown in Figure 3.17 employs a 3 × 3 fiber optic coupler that, from a single input light source, gives three output signal phases separated by 120°. This arrangement enables the interferometer to be optimally biased at all times and allows extremely sensitive measurements to be made over a wide dynamic range. A Sagnac interferometer CT built using York low-birefringence fiber LB 800 and a commercially available gyroscope was evaluated in a 1994 field trial by ABB. These trials show the maximum error in sensor output (compared to a conventional metering class CT) to be less than 0.2%.

As an alternative to the all-fiber current sensor, many people turned their attention to bulk devices built from doped flint glass material [37–39]. Several pieces were assembled to form a simple glass ring structure analogous to a conventional CT iron core (see Figure 3.18).

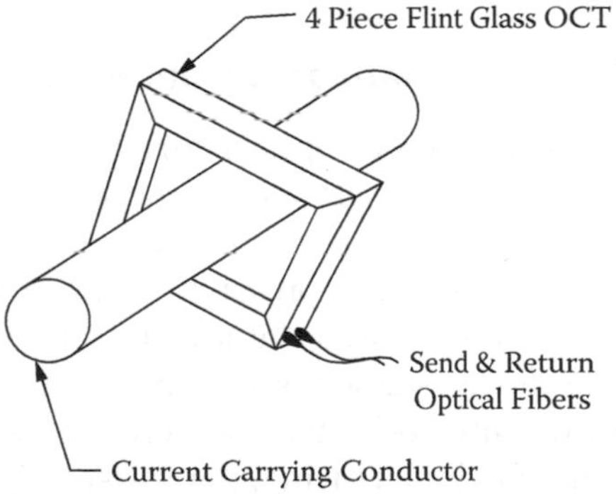

FIGURE 3.18
Simple bulk-optic OCT.

The mode of operation of the bulk-optic OCT is similar to that of a wound-fiber OCT. The light travels around the ring by continual reflection at the "corners" of the structure. The manufacture of these devices is relatively straightforward, but at each reflection the light experiences a slight polarization shift, which can corrupt the measurement process. Efforts to minimize this influence have produced structures that cancel these effects or eliminate them altogether by

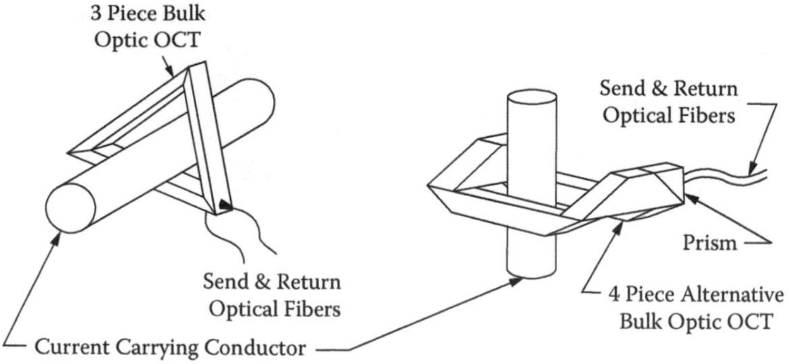

FIGURE 3.19
Revised bulk-optic structures.

forcing all reflections to occur at angles that do not introduce a linear phase shift (see Figure 3.19).

In contrast to wound fiber, the bulk-optic sensors are relatively heavy and require robust housing to support, protect, and mount the assembly on a conductor. However, despite the intrinsic weight of these devices, which can be up to 100 kg, they are considerably lighter than their electrical counterparts, which can weigh up to 7000 kg. Several companies are now marketing metering-class optical current transformers that are produced in this way, mostly for high-voltage applications (>100 kV) where they are competitive in cost. Accurate information on the market size for these instruments is not readily available, but a 1996 report indicated that 50 to 100 sensors of this type are sold annually worldwide [29].

3.4.2 Crystal-Based Optical Current Transducers

A major class of optical current measurement devices is based on the use of magneto-optic crystals as the sensing medium. The first of this class is really an amalgam of conventional and optical technologies known as a *hybrid* or *flux concentrator* consisting of an iron, or ferrite, core with an air gap into which a Faraday rotator cell is placed [29,30]. In this way the magnetic field the device senses is increased, introducing a marked degree of cross-talk immunity and effectively creating a current-measuring, rather than a magnetic field-measuring, device (see Figure 3.20).

If the Verdet coefficient of the material used to construct the OCT is sufficiently high, then the iron core can be eliminated. This arrangement has also received some attention since it allows very compact sensors, which can be readily retrofitted, to be produced. These designs are called "unlinked" because the sensing medium does not enclose the conductor; as such they provide a measure of the magnetic field and not a direct measure of current.

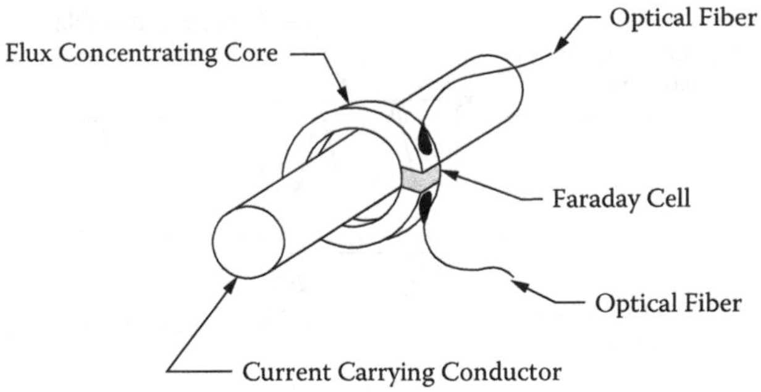

FIGURE 3.20
Hybrid or flux concentrator OCT.

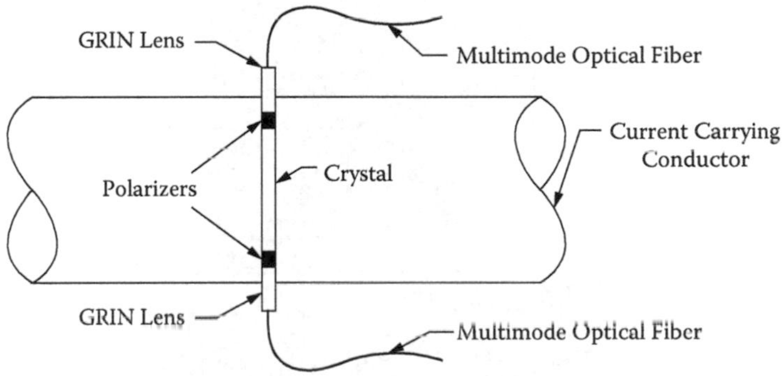

FIGURE 3.21
Single-pass, crystal-based OCT.

Single-pass devices are conceptually the simplest of these (see Figure 3.21), but the sensitivity may be significantly increased [41–46], without substantially complicating the design, using arrangements that force the light to make multiple crossings over the measurement path (see Figure 3.22).

All these devices suffer from the relatively strong dependency of Faraday materials, particularly garnets, to temperature. However, several methods have been reported whereby this influence can be negated or integrated into the postprocessing electronics [44,45].

Temperature and Vibration Compensation Schemes for Crystal-Based OCTs

The sensitivity of OCTs to temperature variations and mechanical disturbances has forced researchers in these fields to find methods to minimize these influences or to compensate for their effect. In the case of mechanically induced transmission changes, the processing required is relatively

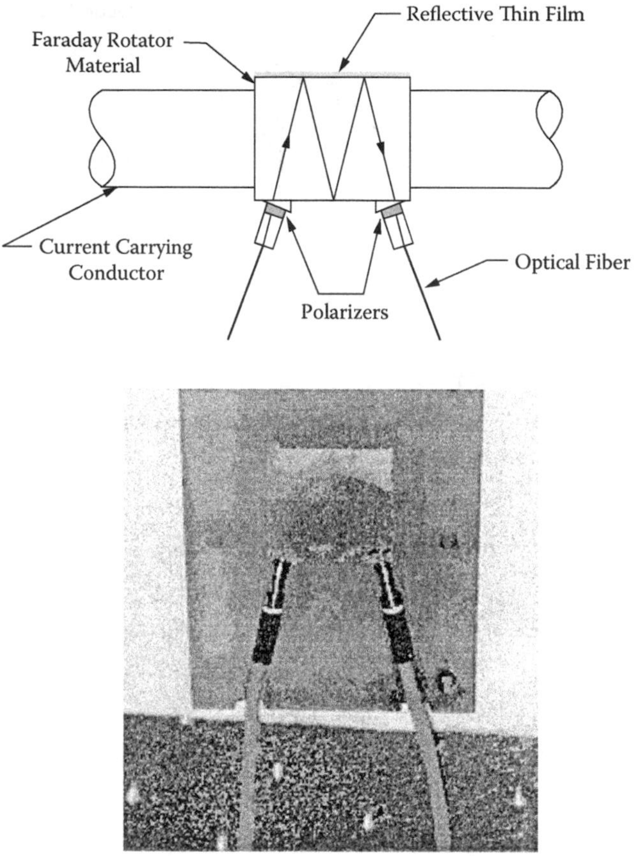

FIGURE 3.22
Relative-type, crystal-based OCT.

straightforward since the nonreciprocal nature of the current measurement process can be employed to great effect.

Two independent light sources that propagate through the optical circuit in opposing directions (see Figure 3.23) are used to interrogate the OCT. The rotation in the optical polarization state depends on the direction of light propagation through the OCT, whereas the intensity changes do not. The two counterpropagating signals can therefore be processed to negate the influence of vibration. Both ac and dc measurements can be corrected in this way, and the effectiveness of the process is clearly demonstrated in Figure 3.24, which shows compensated and uncompensated signals. It should be noted that these signals were recorded as part of an experiment designed to induce artificially high levels of distortion into the measurement signal to illustrate the efficiency of the compensation scheme. They are not representative of normal vibration levels.

Figure 3.25 displays the errors associated with a compensated and uncompensated current during the preceding measurement. Despite extreme

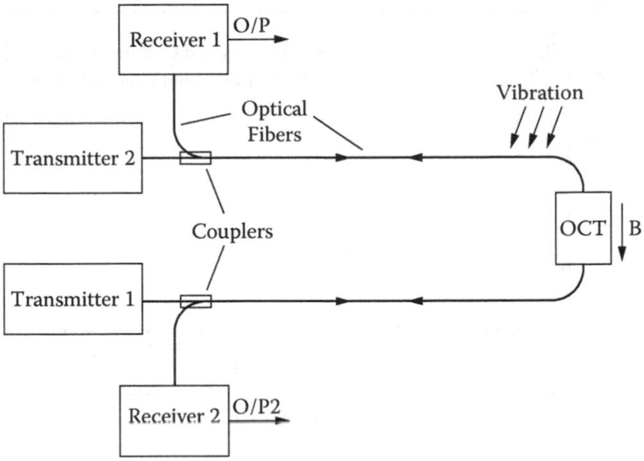

FIGURE 3.23
Configuration of the bidirectional light propagation in the optical system.

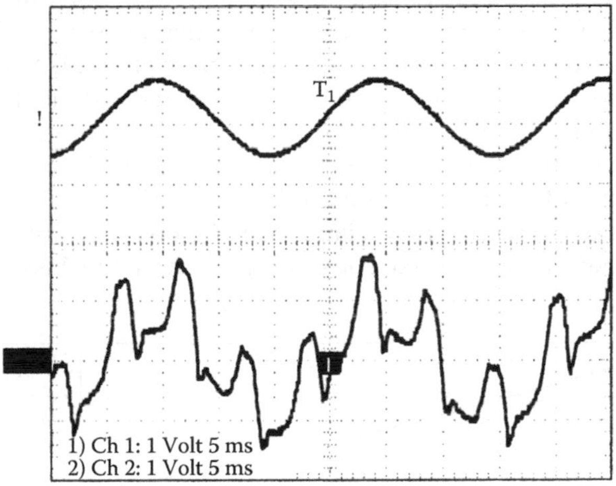

FIGURE 3.24
Oscilloscope traces of the two OCT outputs during the vibration test (vibration: 180 Hz, 8 g, 4 amps test current, connectors vibrated).

mechanical disturbance, the errors of the compensated channel remain similar to those measured in a steady-state condition, while the errors of the uncompensated channel reach very high levels.

Temperature-induced errors arise from the fact that the Verdet constant of many materials (particular garnets) is strongly sensitive to operating temperature. For paramagnetic materials, the variation of the Verdet constant with temperature can be approximated as [44,47,48]

$$V = \alpha / T + \beta \tag{3.43}$$

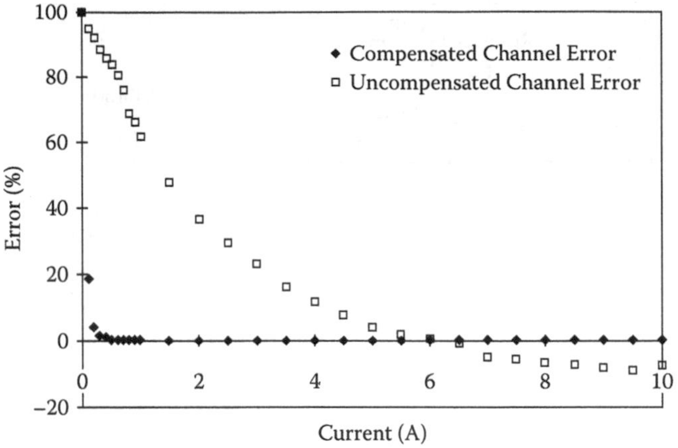

FIGURE 3.25
OCT output errors (vibration: 180 Hz, 8 g, connectors vibrated).

The coefficients α and β are specific to each material and the wavelength used to interrogate the sensor. The sensitivity of two materials, terbium gallium garnet (TGG) and Faraday rotator-5 (FR5) glass, are shown for illustrative purposes in Figure 3.26 because both have relatively high Verdet constants and are readily available commercially. The Verdet constants were estimated from data supplied by the manufacturer and verified by experiment and are displayed in Figure 3.26 at two optical wavelengths (633 and 850 nm) over a typical operational temperature range (250 to 350 K).

The impact of temperature variation can be obtained by evaluating the difference between the Verdet constant at *T*1 and the Verdet constant at

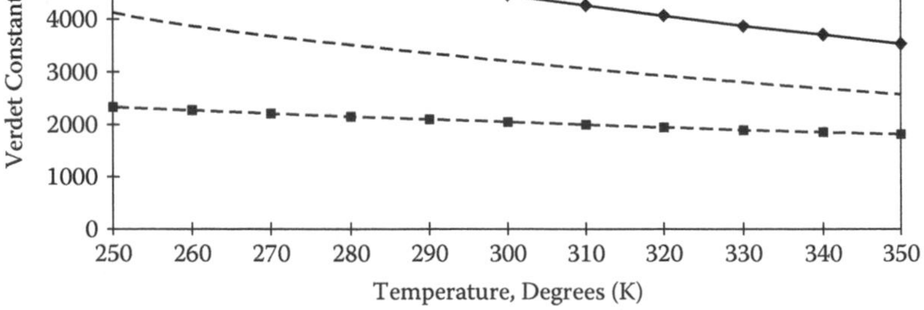

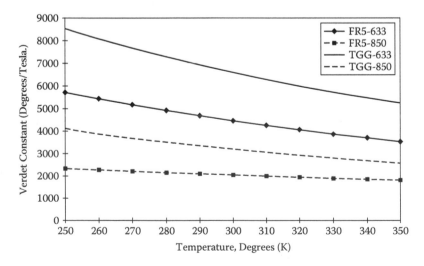

FIGURE 3.26
Variation of Verdet constant with temperature and wavelength.

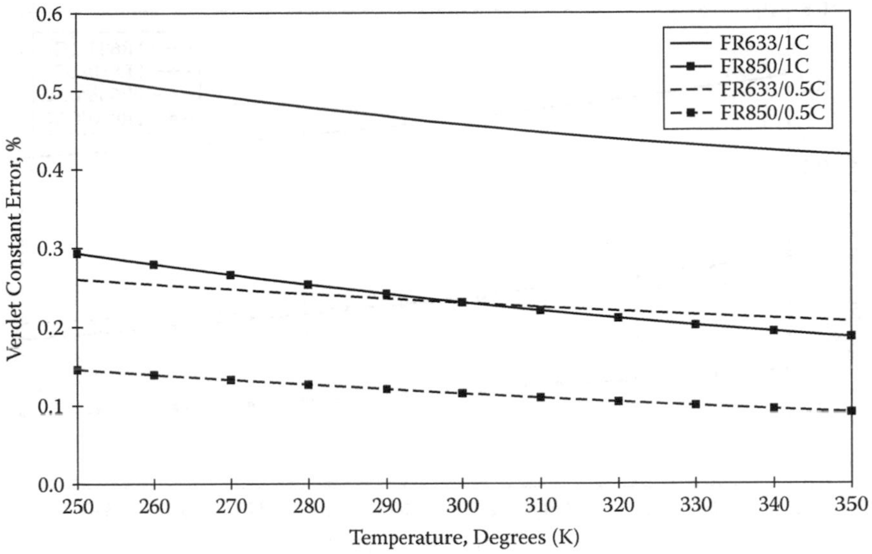

(a)

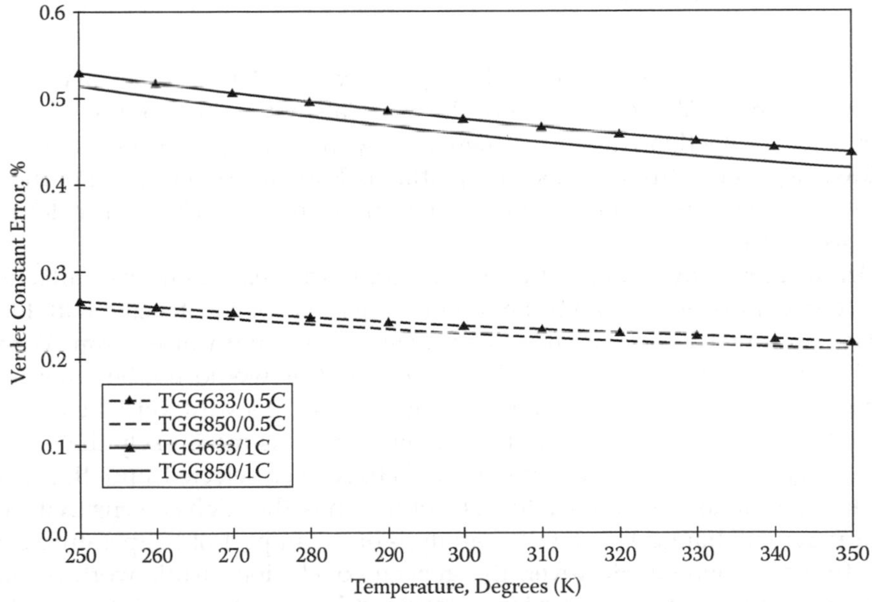

(b)

FIGURE 3.27
Error in Verdet constant (TGG).

$T_{1+\delta T}$ (where δT represents a small perturbation in the temperature measurement) for any given material and wavelength combination. This is shown in Figure 3.27 and Figure 3.28, where, for illustration purposes, δT was set

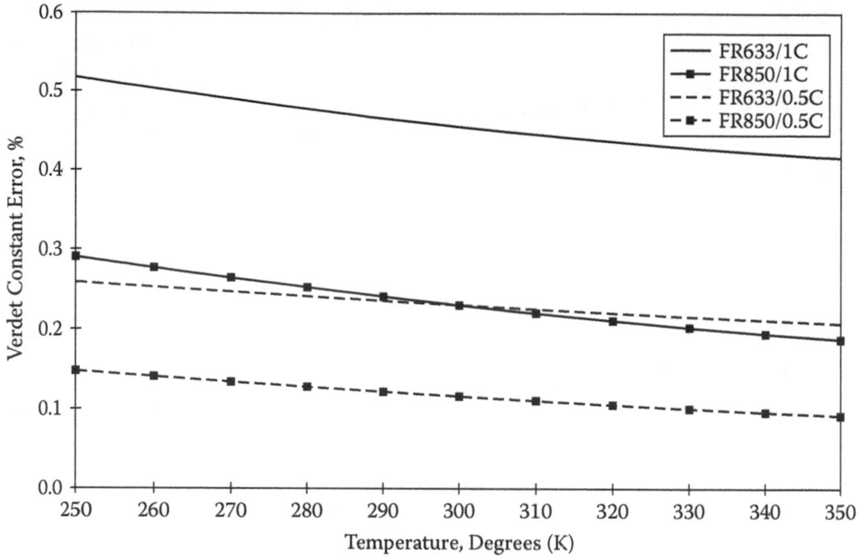

FIGURE 3.28
Error in Verdet constant (FR5).

to 0.5 and 1°C. Knowing the local temperature with 1°C accuracy allows us to ascertain the Verdet constant of the materials with sufficient precision to construct protection-class instruments using any of the preceding materials. However, even if the sensor's temperature is known to 0.5°C, none of these material/wavelength combinations can be used to obtain Class 0.1 metering performance.

In the preceding example, the best performance (i.e., the one least affected by temperature errors) can be obtained using a sensor constructed with FR5 glass and interrogated with an operational wavelength of 850 nm. Without considering the influence of temperature, this would not be a natural choice of sensor parameters since FR5 glass has the lowest Verdet constant of all of the material/wavelength combinations. It is more likely that TGG, interrogated with a 633-nm source, would have been selected since this provides the highest sensitivity to current (but has the highest sensitivity to temperature changes). A detailed analysis of the impact of temperature can be found elsewhere. However, the general conclusion of this work is that to effectively compensate for temperature-induced measurement error, the operating temperature of the measurement crystal must be known to a high degree of accuracy: on the particular cases investigated, 0.1°C for metering-class operation.

Many researchers have attempted to resolve the issue of temperature compensation by using indirect temperature measurement methods where the temperature of the device is derived from current measurements made at two different wavelengths or other similar techniques. However, detailed

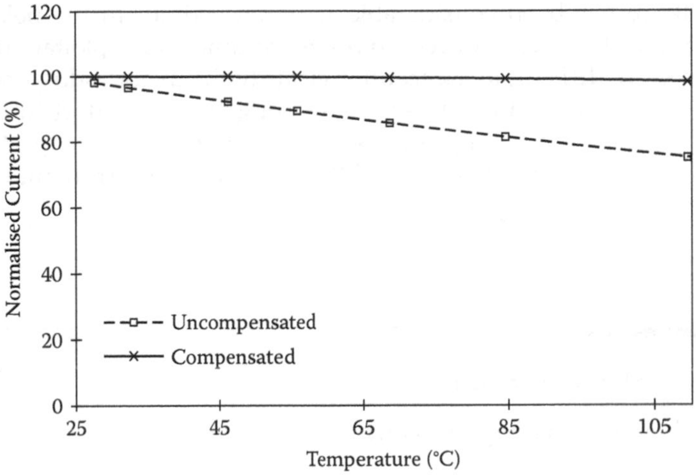

FIGURE 3.29
Temperature-compensated current measurement.

analysis of this and similar dual-parameter measurements has shown that, in general, the temperature recovery process is not robust over the entire measurement range. The only truly effective means to obtain the temperature information is by direct measurement using procedures that are not influenced by current, A recent investigation [44] using a fiber Bragg grating as a temperature probe demonstrated a thermally compensated measurement over the temperature range of 20 to 120°C (see Figure 3.29).

Commercial OCT Devices

Examples of field trials of voltage and current transformers can be found for every major industrial player in this area. Interest in optical technology seems particularly strong in Japan, where companies such as Toshiba, Panasonic, and the Tokyo Electric Company have published widely on the subject. Late 1990s estimates of the market size for optical current transducers suggested that several thousand devices were sold annually within Japan [29,40]. Panasonic alone reported total shipments of several thousands of optical sensor heads for fault detection and metering purposes. Its garnet crystal-based OCT can be obtained with a permalloy core that acts as a flux concentrator. These devices have a 25-mm air gap to house the OCT and are specified as having a noise equivalent current of around 2 mA over a 500-Hz bandwidth. The same device without the flux concentrator is capable of detecting currents of approximately 10 A. Toshiba has concentrated its efforts toward an all-fiber device and reported field trials showing operation within the specified metering class JEC1201. The measurement sensitivity of the wound-fiber current transducer is around 3 A, and the sensor has a long-term stability of better than 0.15% in situ in GIS equipment.

While there has been considerable research activity in the field of current sensing in Europe, information on the commercial exploitation of these devices is more difficult to ascertain. Certainly the major industrial companies—Siemens, ABB, and GEC—have a strong research effort in this field, and this work is producing devices that meet the required performance specification. It is likely that sales of these devices will grow into a steady market over the coming years.

3.5 Optical Voltage Sensor

3.5.1 Polarimeric Voltage Sensors

Polarimetric measurement schemes, directly equivalent to those employed within the OCT, can be used to produce optical voltage transducers. The sensor measures the polarization rotation produced in an optically active crystal $Bi_{12}SiO_2$ (BSO) or something similar (e.g., BGO and BTO) in the presence of an electric field. A voltage measurement can be inferred from this if the system has previously been calibrated. The overall system and a breakdown of the components in the sensor head are shown in Figure 3.30.

The physical mechanism used to detect the voltage change is known as the Pockels effect and exploits the sensitivity of the material refractive index to changes in electric field intensity. Two classes of devices are commonly constructed depending on the type of material used [46]. Class 43 m crystals (e.g., $Bi_4Si_3O_{12}$) have no intrinsic circular birefringence, and crystals of type 23 (e.g., $Bi_{12}Si_3O_{20}$) are intrinsically circularly birefringent. Polarimetric

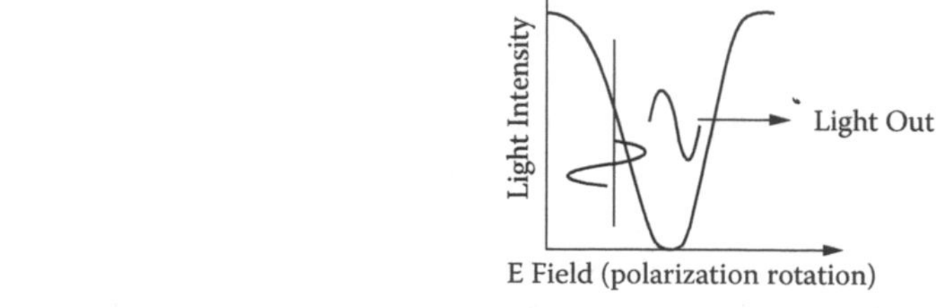

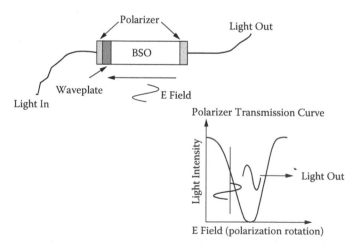

FIGURE 3.30
Schematic of optical voltage sensor.

sensors can be constructed with either class, but the circularly birefringent crystal offers, in principle, the potential to compensate for the sensor's residual temperature response.

Alternative forms of voltage transducers (VT) that have been studied include integrated optic devices. These devices are usually constructed in $LiNbO_3$ using photolithographic processing methods [49]. Integrated optic fabrication methods enable complex optical structures to be integrated into a single substrate forming a compact and potentially rugged device. However, as with all integrated optic component technology, the technology is expensive and the coupling process from the fiber to the device and back into the fiber is not trivial. Several researchers have demonstrated laboratory versions of integrated optic VTs, but to date there has been limited interest within the electricity supply industry.

The optical voltage sensor is in effect an electric field sensor. Ideally, the entire electric field should be dropped across the crystal, but clearly this is not practical. Care must be taken, therefore, to ensure that the electric field developed across the crystal remains constant in relation to the line potential; this is usually achieved with a simple capacitive divider, as shown in Figure 3.31. Generally, this is not possible in the case of electrical VTs because the VT capacitance influences the stability of the divider network. However, an optical VT is equivalent to a VT with an infinite impedance, and usual loading effects can therefore be disregarded.

The implementation of the capacitive divider arrangement is shown in Figure 3.32, and its physical realization is shown in Figure 3.33.

The corona rings represent the capacitor C2 as described in reference 50. C1 is formed by using an additional metal plate (at line potential) and the top corona ring. The copper plate used to form the top plate of C1 is more readily observed in Figure 3.34, which shows a close-up view of the installation.

An OCT is visible on top of the conductor, while the optical voltage transducer (OVT) lies under the conductor oriented radially along the electric field lines.

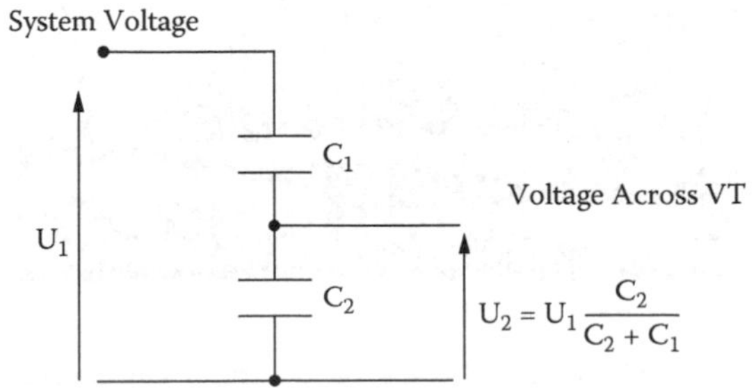

FIGURE 3.31
Capacitive voltage divider.

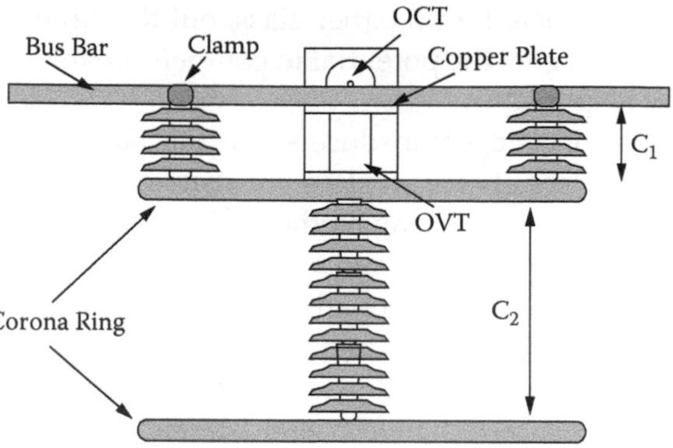

FIGURE 3.32
Implementation of the capacitive voltage divider.

FIGURE 3.33
Field installation of three-phase optical voltage and current sensors showing the capacitive divider implementation.

FIGURE 3.34
Close-up of the OCT/OVT installation.

3.5.2 Optical Network Instability Diagnosis

Two key benefits that optical fiber sensors offer are their immunity to electromagnetic interference and, equally importantly, the fact that they do not influence the electromagnetic circuit. As the use of local/embedded generation systems increases, this feature will become more important. In a 1999 network instability investigation, the task was significantly simplified with the use of optical voltage measurement equipment [50].

During commissioning tests for a 49-MW combined cycle gas turbine (CCGT) system, the generator transformer was back-energized from the 132-kV supply (132/11.5 kV). Each time connection was made to the grid, the steam turbine neutral voltage displacement (NVD) protection relay operated. Optical voltage transducers were used to measure the line voltage and identified that neutral displacement voltage of per-unit level (see Figure 3.35) was present whenever an electrical measurement apparatus was connected

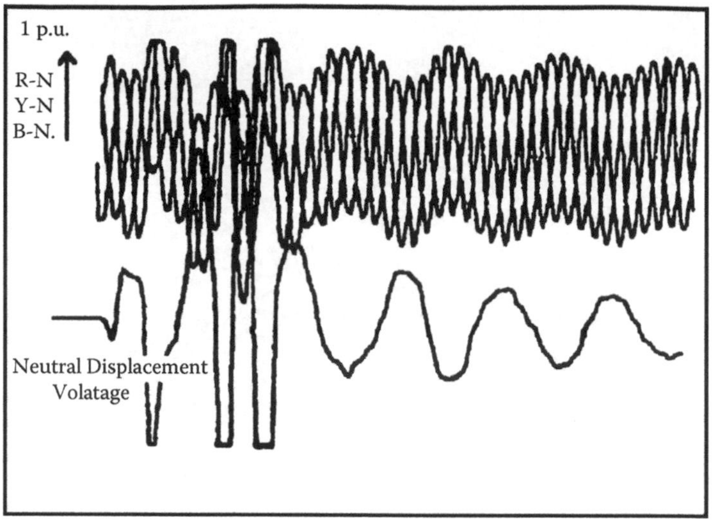

FIGURE 3.35
Voltage traces from steam turbine VT (star winding) RYB voltages and steam turbine open delta voltage, gas turbine neutral displacement voltage.

to the line. The optical transducers did not have this influence since they do not interfere with the electromagnetic circuit. As a result of this intervention, a successful solution to the network instabilities was devised.

3.6 Conclusions

The field of optical fiber sensors has advanced substantially in the past 10 years following the developments made within the supporting component technology driven by the telecommunications industry. Reflecting the relative maturity of the technology, research activity within universities and industrial centers has shifted from demonstrating physical concepts to producing test instruments that can then be engineered into preproduction prototypes. In making this significant shift in emphasis, optical fiber researchers have had to more closely address issues associated with the application of optical fibers in harsh, practical environments.

Fiber optic sensors will undoubtedly have a role to play within a structural monitoring context where much of the research and development effort has been focused. However, there is clearly a significant scope for these sensors elsewhere—for example, within the electricity supply industry, where optical measurement techniques offer sensitivity and performance improvements while reducing costs by minimizing the requirements for insulation. This chapter has focused on polarimetric sensors, which have a major role to play in this context as an enabling technology for voltage and current measurement.

References

1. E. Hecht and A. Zajac, *Optics*, Addison–Wesley, Reading, MA, 1974.
2. Yariv, *Quantum Electronics*, 2nd ed., Wiley, New York, 1975.
3. M. Borne and E. Wolfe, *Principles of Optics*, 6th ed., Pergamon, Oxford, 1986.
4. R. M. A. Azzam and N. M. Bashara, *Ellipsometry and Polarised Light*, North–Holland Elsevier, Amsterdam, 1977.
5. R. B. Dyott, *Elliptical Fiber Waveguides*, Artech House, Norwood, MA, 1995.
6. R. B. Dyott, J. R. Cozens, and D. G. Morris, Preservation of polarization in optical fiber waveguides with elliptical cores, *Electron. Lett.*, **15**, pp. 380–382, 1979.
7. V. Ramaswamy, R. H. Stolen, M. Divino, and W. Pleibel, Birefringence in elliptically clad borosilicate single mode fibers, *Appl. Opt.*, **18**, pp. 40–84, 1979.
8. S. C. Rashleigh and M. J. Marrone, Polarisation holding in high birefringence fiber, *Electron. Lett.*, **18**, pp. 326–327, 1982.
9. R. D. Birch, M. P. Varnham, D. N. Payne, and E. J. Tarbax, Fabrication of polarization-maintaining fibers using gas phase etching, *Electron. Lett.*, **18**, pp. 1036–1038, 1982.
10. N. Shibata, Y. Sasaki, K. Okamoto, and T. Hosaka, Fabrication of polarization maintaining and absorption reducing fibers, *J. Lightwave Technol.*, **LT-1**, pp. 38–43, 1983.
11. B. K. Kim, S. H. Yun, I. K. Huang, and B. Y. Kim, Nonlinear strain response of the two mode fiber optic interferometer, *Opt. Lett.*, **21**, pp. 934–936, 1996.
12. C. D. Butter and G. B. Hocker, Fiber optics strain gauge, *Appl. Opt.*, **17**, pp. 2867–2869, 1978.
13. G. B. Hocker, Fiber-optic sensing of pressure and temperature, *Appl. Opt.*, **18**, pp. 1445–1447, 1979.
14. W. Eickhoff, Temperature sensing by mode–mode interference in birefringence optical fibers, *Opt. Lett.*, **6**, pp. 204–206, 1981.
15. T. A. Eftimov and W. J. Bock, Sensing with a LP_{01}–LP_{02} intermodal interferometer, *J. Lightwave Technol.*, **LT-11**, pp. 2150–2156, 1993.
16. B. Y. Kim, J. N. Blake, S. Y. Huang, and H. J. Shaw, Use of highly elliptical core fibers for two mode fiber devices, *Opt. Lett.*, **12**, pp. 729–731, 1987.
17. A. M. Vengsarkar, W. C. Michie, L. Jankovic, B. Culshaw, and R. O. Claus, Fiber-optic dual-technique sensor for simultaneous measurement of strain and temperature, *J. Lightwave Technol.*, **LT-12**, pp. 170–177, 1994.
18. G. Thursby, W. C. Michie, D. Walsh, M. Konstantaki, and B. Culshaw, Simultaneous recovery of strain and temperature fields by the use of two-moded polarimetry with an in-line mode splitter/analyzer, *Opt. Lett.*, **20**, pp. 1919–1921, 1995.
19. G. Thursby, D. Walsh, W. C. Michie, and B. Culshaw, An in-line mode splitter applied to a dual polarimeter in elliptical core fiber, *Proc. SPIE*, **2360**, 10th Intl. Conf. Optical Fiber Sensors, B. Culshaw and J. D. C. Jones, eds., pp. 339–342, 1994.
20. W. Craig Michie, B. Culshaw, C. Thursby, W. Jin, and M. Konstantaki, Optical fiber sensors for temperature and strain measurement, *Proc. SPIE Smart Structures*, San Diego, 1996.
21. W. Jin, W. C. Michie, G. Thursby, M. Konstantaki, and B. Culshaw, Geometric representation of simultaneous measurement of temperature, *Optical Eng.*, **36**, 8, pp. 2272–2278, 1997.

22. W. Jin, W. C. Michie, G. Thursby, M. Konstantaki, and B. Culshaw, Simultaneous measurement of temperature and strain: Error analysis, *Optical Eng.*, **36**, 2, pp. 506–609, Feb. 1997.
23. I. P. Giles, M. Mondanos, R. Badcock, and P. A. Lloyd, Distributed optical fiber based damage detection in Composites, *Proc. SPIE Conf. Sensory Phenomena and Measurement Instrumentation for Smart Structures and Materials*, **3670**, pp. 311–321, 1999.
24. M. Mondanos, P. A. Lloyd, I. P. Giles, R. Badcock, and K. Weir, Damage detection in composites using polarmetric low coherence polarimetry, *Proc SPIE Int. Soc. Opt. Eng.*, **4185**, pp. 276–763, 2000.
25. C. Clecot, J. J. Guerin, M. Lequime, and M. Rioual, White light fiber-optic sensor network for the thermal monitoring of the stator in a nuclear power plant alternator, *Proc. OFS93*, Florence, Italy, pp. 271–274, 1993.
26. M. N. Charasse, M. Turpin, and J. P. le Pesant, Dynamic pressure sensing with side hole birefringent optical fiber, *Opt. Lett.*, **16**, pp. 1043–1045, 1991.
27. Y. N. Ning, T. Y. Liu, and D. A. Jackson, Two low-cost electrooptic hybrid current sensors capable of operation at extremely high potential, *Rev. Sci. Instrument.*, **63**, pp. 57–71, 1992.
28. Y. N. Ning, Z. P. Wang, A. W. Palmer, and K. T. V. Grattan, Recent progress in optical current sensing techniques, *Rev. Sci. Instrument.*, **66**, 5, pp. 3097–3111, 1995.
29. G. W. Day, K. B. Rochford, and A. H. Rose, Fundamentals and problems of fiber current sensors, *Optical Fiber Sensors*, **11**, pp. 124–128, 1996.
30. M. Takahashi, H. Noda, K. Terai, S. Ikuta, and Y. Mizutani, Optical current sensor for gas insulated switchgear using silica optical fiber, *IEEE Trans. Power Delivery*, **12**, 4, pp. 1422–1428, Oct. 1997.
31. K. Terai, S. Ikuta, Y. Mizutani, M. Takahashi, and H. Noda, Practical optical current transformer for gas insulated switchgear, *Proc. 12th Intl. Conf. Optical Fiber Sensors*, Williamsburg, VA, 1997.
32. A. H. Rose, Z. B. Ken, and G. Day, Improved annealing techniques for optical fibers, *Proc. OFS-10*, Glasgow, pp. 306–311, 1994.
33. Bohnert, H. Brandle, and G. Frosio, Field test of interferometric optical fiber high voltage and current sensors, *Proc. OFS-10*, 1994.
34. J. Barlowe and D. N. Payne, Production of single mode fibers with negligible intrinsic birefringence, *Electronic Lett.*, **17**, p. 725, 1986.
35. T. Bosselmann, Magneto- and electro-optic transformers meet expectations of the electricity supply industry, *Proc. 12th Intl. Conf. Optical Fiber Sensors*, 1997.
36. T. Bosselmann, Comparison of four different optical fiber coil concepts for high voltage magneto-optic current transducers, *Proc. 9th Intl. Conf. Optical Fiber Sensors*, Florence, Italy, 1993.
37. K. Kurosawa, K. Oshida, and O. Sano, Optical current sensor using flint glass fiber, *Proc. OFS-11*, pp. 147–148, 1996.
38. H. Schwarz and M. Hudash, Optical current transformers—A field trial test in a 380 kV system, *ABB Rev.*, **3**, pp. 12–18, 1998.
39. T. Bosslemann and P. Menke, Intrinsic temperature compensation of magneto-optic ac current transformers with glass ring sensor head, *Proc. 10th Intl. Conf. Optical Fiber Sensors*, pp. 20–24, 1994.
40. N. Itoh, H. Minemoto, D. Ishiko, and S. Ishizuka, Commercial current sensor activity in Japan, *Proc. 12th Intl. Conf. Optical Fiber Sensors*, Japan, 1997.
41. P. Niewczas, W. I. Madden, W. C. Michie, A. Cruden, and J. R. McDonald, Progress towards a protection class optical current sensor, *IEEE Power Engineering Rev.*, **20**, 2, pp. 57–59, Feb. 2000.

42. P. Nieweczas, A. Cruden, W. C. Michie, W. I. Madden, and J. R. McDonald, Error analysis of an optical current transducer operating with a digital signal processing system, IMTC99 special issue of the *IEEE Trans. Instrument Measurement*, **9**, 6, pp. 1254–1259, Dec. 2000.
43. A. Cruden, I. Madden, W. C. Michie, P. Niewczas, J. R. McDonald, and I. Andonovic, Optical current measurement system for high voltage applications, International Measurement Committee (IMEKO), *Measurement*, **24**, 2, pp. 97–102, 1998.
44. W. I. Madden, W. C. Michie, A. Cruden, P. Niewczas, J. R. McDonald, and I. Andonovic, Temperature compensation for optical current sensors, *Optical Eng.*, **38**, 10, pp. 1699–1707, 1999.
45. P. Niewczas, A. Cruden, W. C. Michie, W. I. Madden, J. R. McDonald, and I. Andonovic, A vibration compensation technique for an optical current transducer, *Optical Eng.*, **38**, 10, pp. 1708–1714, 1999.
46. A. Cruden, Z. J. Richardson, J. R. McDonald, and I. Andonovic, Optical devices for current and voltage measurement, *IEEE Trans. Power Delivery*, **10**, 3, pp. 1217–1223, July 1995.
47. N. P. Barnes and L. B. Petway, Variation of the Verdet constant with temperature of terbium aluminium garnet, *J. Optical Soc. Am. B*, **9**, 10, pp. 1912–1916, 1992.
48. Product data sheet for FR-5 Glass, HOYA Corporation.
49. N. Jaegerm and F. Rahmatian, Integrated optics Pockels cell as a high voltage sensor, *Proc. OFS-8*, pp. 153–157, 1991.
50. W. C. Michie, A. Cruden, P. Niewczas, W. I. Madden, J. R. McDonald, and A. Kinson, Transient voltage instability investigation using optical voltage sensor, *PE Rev. Lett. IEEE Power Eng. Rev.*, pp. 55–56, Feb. 1999.

4

In-Fiber Grating Optic Sensors

Lin Zhang, W. Zhang, and I. Bennion

CONTENTS

4.1 Introduction

The past decade has witnessed intensive research and development efforts to engineer a new class of fiber optic component—the UV-inscribed in-fiber gratings [1–4]. While the major driving force has come from the need for new, high-performance fiber grating devices, such as wavelength division multiplexing (WDM) filters, erbium doped fiber amplifier (EDFA) gain equalizers, and dispersion compensators to boost the bandwidth for high-speed telecommunication network systems, the utilization of fiber grating technology in it the same optical sensing has increased pace [5,6]. No doubt, the high demand for fiber grating devices in telecommunications will ensure a continuing reduction in the cost of the basic technology, which in turn will encourage the growth of sensing applications using grating-based sensor devices and systems.

In-fiber grating-based sensors have many advantages over conventional electric and alternative fiber optic sensor configurations. They are relatively straightforward, inexpensive to produce, immune to electromagnetic (EM) interference and interruption, lightweight, small in size, and self-referencing with a linear response. Most significantly, their wavelength-encoding multiplexing capability allows tens of gratings in a single piece of fiber to form an optical data-bus network. The combination of their multiplexing capability and inherent compatibility with fiber-reinforced composite materials permits in-fiber gratings to be embedded in a number of important structural materials for smart structure applications. Indeed, the development of structurally integrated fiber optic sensors, using fiber Bragg gratings (FBGs), represents a major contribution to the evolution of smart structures, leading to improvements in both safety and economics in many engineering fields, including major civil works, road and rail bridges, tunnels, dams, maritime structures, airframe sections, projectile delivery systems, and numerous medical appliances.

A list of the extensive reviews of fiber grating principles and their applications in sensors is provided in references 1–6. In this chapter, we attempt to present a brief review of fiber optic sensors based on fiber gratings of short- (Bragg type) and long-period structures, with emphasis on the most updated advances in sensing devices, interrogation techniques, and applications.

4.2 Fiber Bragg Grating Sensors

4.2.1 FBG Theory and Fabrication Technology

The elementary fiber Bragg grating comprises a short section of single-mode optical fiber in which the core refractive index is modulated periodically. As depicted in Figure 4.1, this structure acts as a highly wavelength-selective reflection filter with the wavelength of the peak reflectivity, λ_B, determined by the phase matching condition

$$\lambda_B = 2n_{\text{eff}}\Lambda \tag{4.1}$$

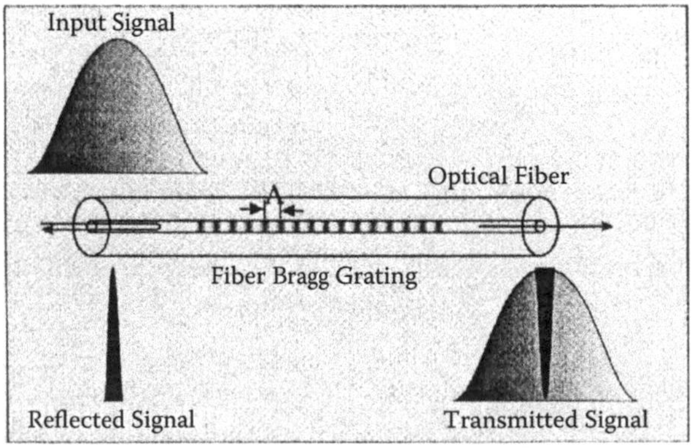

FIGURE 4.1
Schematic diagram of structure and spectral response of fiber Bragg grating.

where n_{eff} is the effective refractive index of the guided mode in the fiber, and Λ is the period of the refractive index modulation with a form of

$$n(z) = n_{co} + \delta n[1 + \cos(2\pi z/\Lambda)] \tag{4.2}$$

where n_{co} is the unexposed core refractive index and δn is the amplitude of the photo-induced index excursion. This periodical index-modulated structure enables the light to be coupled from the forward-propagating core mode into the backward-propagating core mode, generating a reflection response.

The refractive index modulation of an FBG is achieved by exposure of the core to an intense UV interference fringe pattern generated either by the free-space two-beam holographic method or by the diffractive phase mask technique [7,8], as shown in Figure 4.2. The origins of the photosensitivity in optical fibers are now fairly well understood [9], and gratings can be UV written in most fiber types, including standard germanosilicate telecommunications fiber. The achievable UV-induced index change varies from one type of fiber to another, but for inherently less photosensitive fibers, the H_2-loading method is employed to photosensitize the fibers [10]. A typical fiber Bragg grating, used for strain/temperature sensing, has a physical length of a few millimeters and can provide virtually 100% peak reflectivity, with a reflection bandwidth of <0.5 nm. The UV inscription techniques have been refined to the point where FBGs of high-precision profiles and versatile structures are readily produced. Figure 4.3 depicts typical spectral profiles of uniform-period, chirped, resonant, and array gratings that can provide controlling, combining, and routing functions for light signal transmission and processing.

Although the fiber jacket has to be striped to allow UV inscription, with proper handling during the photoinscription phase and by recoating the grating after exposure, the mechanical strength of the fiber is only marginally compromised and maintains its robustness for practical devices [11,12]. A

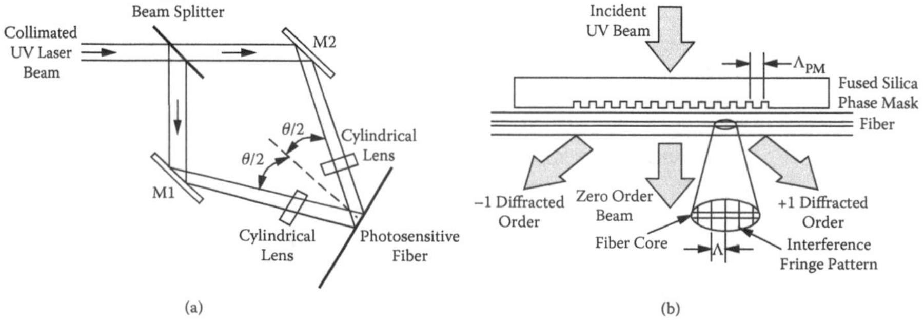

FIGURE 4.2
FBG fabrication techniques: (a) free-space two-beam holographic and (b) diffractive phase mask methods.

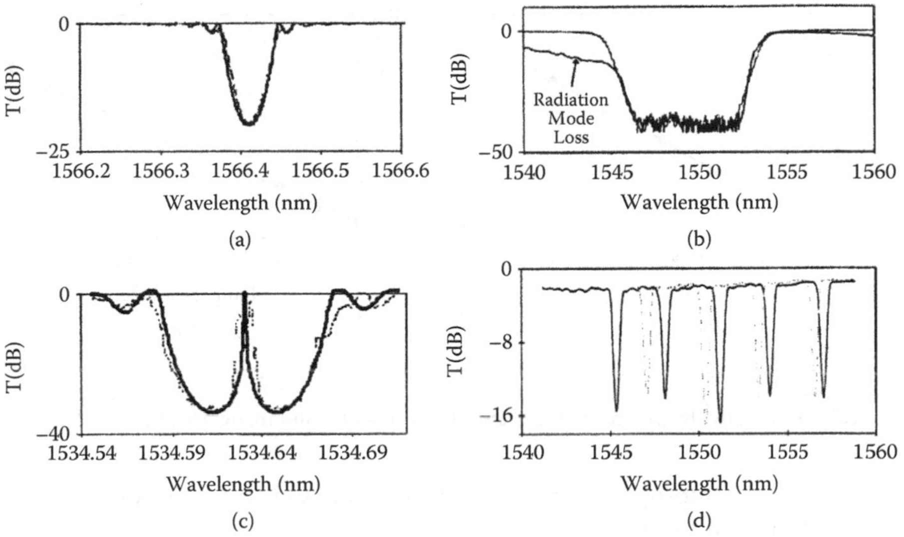

FIGURE 4.3
Experimentally measured and theoretically modeled transmission spectra of FBGs of (a) uniform-period, (b) chirped, (c) phase-shifted, and (d) array structures.

standard annealing process has been identified to treat UV-exposed gratings against subsequent degradation. It has been widely accepted that a lifetime of about 25 years can be expected for commercialized FBG devices; during this period the reflectivity can be expected to decrease by just 1 to 2% [13].

4.2.2 Sensing Principles of FBG

Strain and Temperature Sensing

The sensing function of an FBG derives from the sensitivity of both the refractive index and grating period to externally applied mechanical or thermal perturbations. The strain field affects the response of an FBG directly, through the expansion and compression of grating pitch size and through the strain-optic effect—that is, the strain-induced modification of the refractive index. The temperature sensitivity of an FBG occurs principally through the effect on the induced refractive index change and, to a lesser extent, on the thermal expansion coefficient of the fiber. Thus, the peak reflected wavelength shifts by an amount $\Delta\lambda_B$ in response to strain ε and temperature change ΔT as given by

$$\frac{\Delta\lambda_B}{\lambda_B} = P_e\varepsilon + [P_e(\alpha_s - \alpha_f) + \varsigma]\Delta T \tag{4.3}$$

where P_e is the strain-optic coefficient, α_s and α_f are the thermal expansion coefficients of any fiber bonding material and of the fiber itself, respectively,

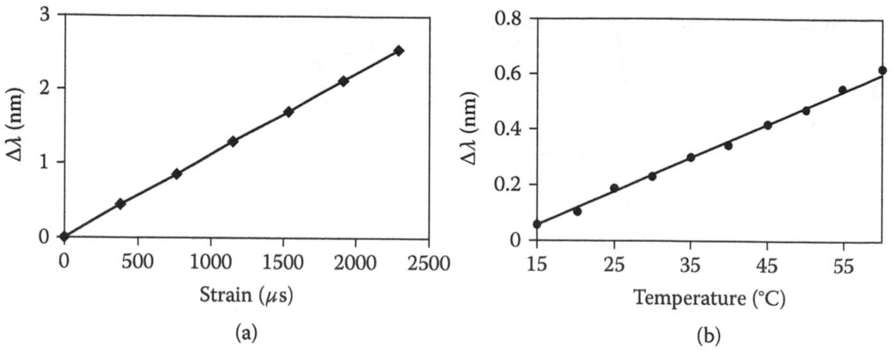

FIGURE 4.4
Typical wavelength-shifting response of an FBG to (a) strain and (b) temperature.

and ς is the thermo-optic coefficient. The normalized strain response at constant temperature is found to be

$$\frac{1}{\lambda_B}\frac{\Delta\lambda_B}{\Delta\varepsilon} = 0.78 \times 10^{-6_{\text{micro-strain}}{}^{-1}}$$

and the normalized thermal responsivity at constant strain is

$$\frac{1}{\lambda_B}\frac{\Delta\lambda_B}{\Delta T} = 6.678 \times 10^{-6}\ {}^{\circ}\text{C}^{-1}$$

For grating produced in silica fiber, representative values of the strain- and temperature-induced wavelength shifts are ~1 pm and ~10 pm/°C at 1300 nm, respectively [5]. Figure 4.4(a) and 4.4(b) show typical Bragg wavelength shift responses to the applied strain and temperature, demonstrating good linear characteristics over practical dynamic ranges.

4.2.3 Interrogation Techniques

FBG optical sensing is based on the principle that the measured information is wavelength-encoded in the Bragg reflection of the grating. The wavelength shift information can be related to the measurand at that sensor position. Therefore, the primary work for the FBG sensor lies in the wavelength interrogation of the Bragg reflection. A simple method to realize this is to use a spectrometer or monochromator, but neither is attractive in practical application due to their bulk-optical nature, size, limited resolution capability, and lack of ruggedness and cost effectiveness.

Precise wavelength measurement using FBG sensors has been a challenging problem since the early stage of FBG sensing work, and many techniques have been proposed for wavelength interrogation. Usually, the wavelength measurement is not very straightforward; thus, the general principle is to convert the wavelength shift to some easily measured parameter, such as amplitude, phase, or frequency.

Wavelength–Amplitude Conversion

Amplitude measurement is the most common and direct way in optical fiber sensors. Converting wavelength shift to amplitude change makes the interrogation simple and cost effective. Several approaches can be related to amplitude measurement.

Edge Filter. An edge filter provides a wavelength-dependent transmittance, offering a linear relationship between the wavelength shift and the output intensity change of the filter. A typical interrogation scheme using an edge filter is shown in Figure 4.5 [14]. The reflection of the FBG sensor is split into two beams of equal intensity. One of the beams is filtered before being detected. The other beam, serving as a reference, is unfiltered and is detected. The two outputs are then amplified and fed to an analog divider. The ratio of the filtered beam to the reference beam provides the wavelength information without suffering any intensity variation from power fluctuation of the source and losses in the optical fiber links.

This approach provides a simple and low-cost solution for the measurement of the wavelength shift caused by FBG sensors. However, the measurement resolution is limited. For the system shown in Figure 4.5, an infrared high-pass filter was used as an edge filter that provides a linear range from 815 to 838 nm. The obtained resolution was approximately 1% over the full-scale measurement range.

As a result of using bulk-optic filters, the alignment is critical and portability is reduced. An all-fiber approach is obviously more attractive. Some fiber devices have been proposed as edge filters for the measurement of wavelength shift. A wavelength demodulation scheme, using a WDM coupler,

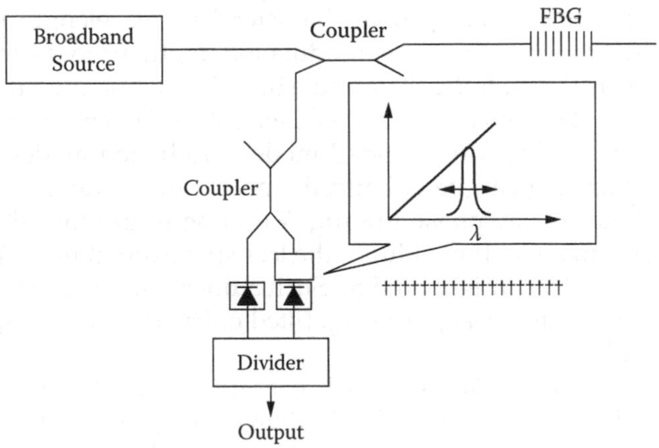

FIGURE 4.5
Schematic of edge filter-based FBG sensor interrogation.

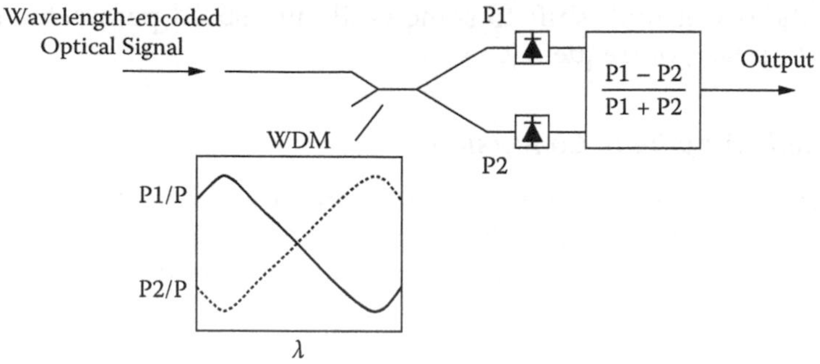

FIGURE 4.6
Wavelength demodulation using a WDM fiber coupler.

was proposed, as shown in Figure 4.6 [15]. The WDM fused-taped coupler shows a monotonic change in the coupling ratio between the two output fiber ports. Its transfer function, as shown in the inset of Figure 4.6, gives a coupling ratio of 0.4 dB/nm over the span of 1520 to 1560 nm. Taking the ratio of the difference and sum of the two outputs of the WDM coupler gives a drift-compensated output for wavelength shift detection. A resolution of ±3 for static strain and ~0.5 for dynamic strain, respectively, has been demonstrated using this approach.

The spectral slope of a WDM coupler is typically less than 10%/nm. The slope steepness determines the sensitivity and minimum detectable wavelength shift. It has been found that by increasing the number of cycles (626 cycles), to make the coupler highly overcoupled, the spectral slope is enhanced up to ~35% at around 1550 nm [16].

A biconical fiber filter has also been proposed as an edge filter [17]. This filter is basically a section of single-mode, depressed-cladding fiber, which consists of a tapered region of decreasing fiber diameter followed by an expanding taper of increasing fiber diameter. The wavelength response of the filter is oscillatory with a large modulation depth, propagating only certain wavelengths through the fiber while heavily attenuating others.

Fiber gratings themselves can act as edge filters. When a fiber grating is written, power coupling from guided mode to radiation modes of the fiber occurs. This coupling can be enhanced, and to some extent controlled, by tilting the fringes of the phase grating. Variation of grating tilt affects the amount of coupling loss, the width of the loss spectrum, thus giving a wavelength-dependent transmission edge. Such a fiber grating edge filter can be realized by using tilted Bragg grating, tilted chirped grating [18], and long-period grating [19].

Matched Fiber Grating Filter. The basic concept of the interrogation scheme using matched fiber gratings [20,21] is shown in Figure 4.7. Light from the broadband source is fed to the sensor grating via the fiber coupler. This propagates back to the receiving grating, which is mounted on a piezoelectric

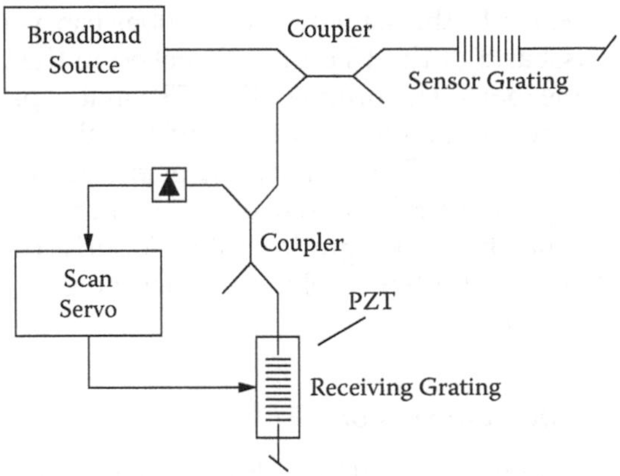

FIGURE 4.7
FBG sensor interrogation employing matched fiber grating filter.

stretcher. The receiving grating is fabricated so that its central reflecting wavelength is identical to the sensor grating. The central reflecting wavelength of the sensor grating will vary in direct proportion to the measurand and generally will not match that of the receiving grating. If the central wavelength of the receiving grating is swept over a defined wavelength range by driving the piezoelectric transducer (PZT), then at one point in the sweep the reflecting wavelengths of both the sensor grating and the receiving grating will exactly match. In this condition a strong signal will be back-reflected from the receiving grating and detected by the photodiode. Providing that the relationship between the driving voltage and the wavelength of the receiving grating is known, the instantaneous wavelength value of the sensor grating can be determined. A closed-loop servo is used to maintain the matched condition and thus track the wavelength shift to the sensor grating. The resolution is dictated by linewidth of the gratings. For gratings with bandwidths of 0.2 and 0.05 nm, a strain resolution of ~4.1 and 1, respectively, has been reported [20]. With open loop this technique can be used to demodulate the multiplexed FBG sensor array.

Another approach has been proposed in which two identical chirped-fiber gratings are employed, but the receiving grating works in transmission mode [22]. If the profiles of these two gratings are identical, then the receiving grating will block the reflection from the sensing grating, and the light received at the photodiode will be minimized. When the sensing grating is stretched or heated, its spectral profile is linearly shifted. This leads to a fraction of the light reflected from the sensing grating falling outside the reflection band of the receiving grating and being transmitted to the photodiode. The quasi-square reflection profiles of the two chirped gratings permit a linear relationship between wavelength change and the light intensity transmitted by the receiving grating. This enables direct measurement of wavelength-encoded reflection, eliminating the need for a piezoelectric tracking system.

Fiber Fabry–Perot Filter. In the scheme using a fiber Fabry–Perot (FFP) filter [23], the narrow passband of the FTP filter is locked to the band of the FBG return signal with piezoelectric tuning of the FTP cavity spacing. In this system, a closed-loop servo is used to track the wavelength shift of a single FBG sensor. With open loop the FFP system can be operated in a scanning mode to track the wavelengths of multiple sensors. The operating principle is the same as that of the matched fiber grating filter. A derivative form of signal detection is used to permit higher resolution measurement of the wavelength shift from the FBG sensor.

Wavelength–Frequency Conversion

An acoustic-optic tunable filter (AOTF) shows wavelength-dependent light transmittance, which is controlled by the radio frequency (RF) applied. By locking the mean optical wavelength of the filter to the instantaneous Bragg wavelength of the FBG sensor, any wavelength shift can be tracked through the deviation of the applied RF frequency [24]. A schematic diagram of this approach is shown in Figure 4.8. If the filter's mean optical wavelength does not match the measured wavelength, the feedback system will pick up an error signal and apply it to a voltage controlled oscillator (VCO) to produce a frequency tuning. This technique can be used to detect the wavelength shift from either the reflection or the transmission of the Bragg grating sensor. In a revised version [25], the wavelength-encoded reflection is split into two parts—one is passed through the AOTF, and the other is fed directly into a second detector. The signals from two detectors are compared to provide an error signal as feedback to the AOTF, so that the AOTF tracks the wavelength. A wavelength resolution of 2.62 pm is reported, corresponding to a strain resolution of 2.24.

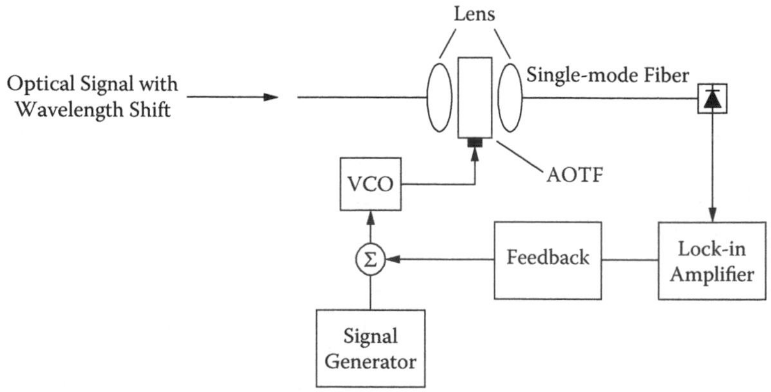

FIGURE 4.8
Interrogation scheme using an acoustic-optic tunable filter.

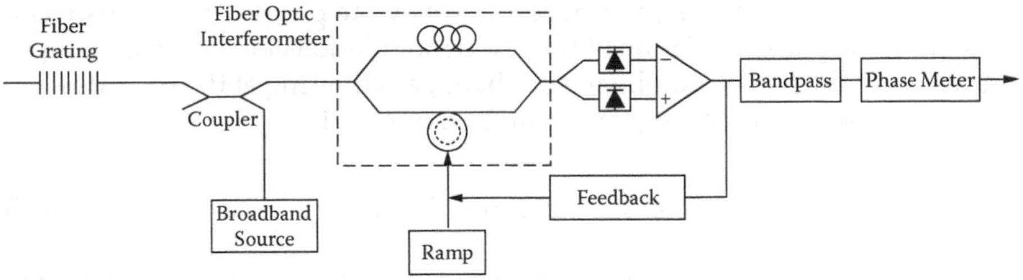

FIGURE 4.9
A typical FBG interrogation scheme employing a fiber optic interferometer.

Wavelength–Phase Conversion

In this approach an unbalanced optical fiber Mach–Zehnder interferometer is used as a wavelength discriminator to detect the wavelength shift of an FBG sensor [26,27]. For a single grating sensor the interrogation system is as shown in Figure 4.9, where a broadband source is used. The wavelength-encoded reflection from the grating is fed to the interferometer through a coupler. Therefore, this Bragg reflection acts as a wavelength-modulated source. The wavelength dependence on the interferometer output can be expressed as

$$I(\lambda) = A[1 + k\cos(\psi(\lambda))] \tag{4.4}$$

where $\psi(\lambda) = 2\pi nd/\lambda$, A is proportional to the input intensity and system loss, d is the length imbalance between two fiber arms, n is the effective index of the fiber core, λ is the wavelength of the return light from the grating sensor, and k is the interference visibility.

For a dynamic wavelength modulation in the Bragg reflection, the change in the phase shift $\Delta\psi(t)$ is

$$\Delta\psi(t) = -\frac{2\pi nd}{\lambda^2}\Delta\lambda \sin \omega t \tag{4.5}$$

Using an interferometric sensor, the high-resolution dynamic phase detection capability of ~10^{-6} rad/$\sqrt{\text{Hz}}$ is achievable [28]. Thus, a length imbalance of 10 mm, an index of 1.46, and a wavelength of 1.55 m would allow a minimum detectable wavelength shift as small as 2.6×10^{-5} pm.

For static or pseudostatic wavelength shifts, a pseudoheterodyne technique is proposed that relies on the generation of an electrical carrier signal at the output by applying a phase ramp to the piezoelectric transducer in one of the fiber arms. Now the change in phase shift can be described as

$$\Delta\psi(t) = -\frac{2\pi nd}{\lambda^2}\Delta\lambda + \frac{2\pi n}{\lambda}d(t) \tag{4.6}$$

where $d(t)$ is a dynamic length imbalance set by the piezoelectric transducer. By setting the phase deviation of the ramp modulation close to 2π, the output signal of the difference amplifier after bandpass filtering at the fundamental of the ramp modulation frequency can be expressed as

$$S(\lambda) = A\cos(\omega t + \Delta\psi(\lambda))] \tag{4.7}$$

A phase meter or a lock-in amplifier usually has a phase resolution of 0.01°, giving rise to a wavelength shift of 4.5×10^{-3} pm for a 10-mm length imbalance.

The demodulation techniques for dynamic and static wavelength shifts are different, although both are realized through fiber optic interferometry. In the measurement of a dynamic wavelength shift, the interferometer converts the wavelength shift into optical phase change; for a static wavelength shift it is the measurement of an electrical phase change. In principle, the latter has more tolerance to polarization fading effects than the former. In general, the fiber optic interferometry offers a much higher wavelength resolution than other interrogation schemes.

Wavelength–Position Conversion

The principle of this approach is the same as that used in the spectrometry. In such a system wavelength interrogation is achieved with a fixed dispersive element (e.g., prism or grating) that distributes different wavelength components at different positions along a line imaged onto an array of detector elements [29–31]. As depicted in Figure 4.10, the reflection from an FBG sensor is launched into a bulk-optical grating. The light, at different wavelengths, will be diffracted in different directions along a straight line. A linear charge coupled device (CCD) camera is used so that light with a different wavelength will be projected to a different position on the CCD. The resolution of the measurement is dependent on the spatial resolution of the bulk grating and the number of the CCD pixels. This technique was first used for monitoring fiber gratings with wavelengths lower than 900 nm, due to the spectral response of the CCD cameras available at that time.

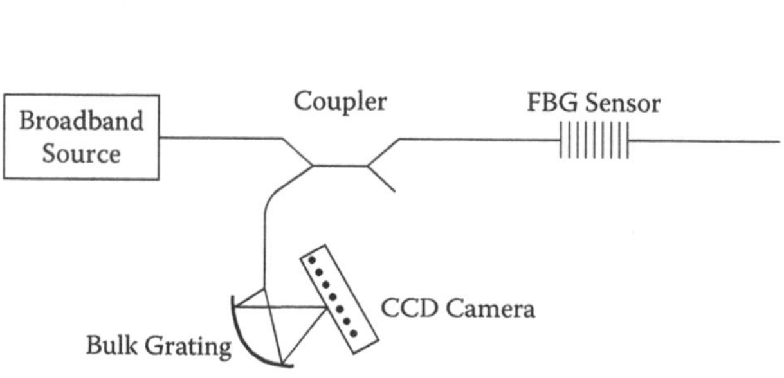

FIGURE 4.10
Schematic of FBG sensor interrogation using dispersive device and CCD camera.

Wavelength–Time Conversion

In this method, wavelength-time conversion is realized by using a dispersive optical delay line [32]. Broadband, ultrafast pulses, generated by a passively mode-locked erbium-fiber laser, are launched into FBG sensors via a dispersive fiber. Reflections from individual gratings propagate back through the dispersive fiber and are monitored by a fast detector and a sampling oscilloscope. The high dispersion of the dispersive fiber converts wavelength shift into a shift in the pulse arrival time at the detector.

Fourier Transform

Several interrogation approaches based on fiber Fourier transform spectrometry have been reported [33,34]. As shown in Figure 4.11, in this approach the reflection from the FBG sensor is coupled into a balanced optical fiber Michelson interferometer. A piezo-driven fiber stretcher is inserted into one of the interferometer arms. A ramp signal is applied to drive the fiber stretcher to provide an optical path difference (OPD) of up to 30 cm, scanning through its zero OPD position. Therefore, an interferogram is generated at the output symmetrically around zero OPD. The width of this interferogram depends on the coherence length of the light input to the interferometer, which is associated with the bandwidth of the Bragg grating. The Fourier transform of the interferogram is realized by feeding the output into an electrical spectrum analyzer, since the frequency at which the fringes appear at the Michelson output is directly proportional to the rate of change of the OPD in the interferometer. By using such a fiber Fourier transform spectrometer, the optical spectrum of the Bragg grating is expressed as an electrical frequency

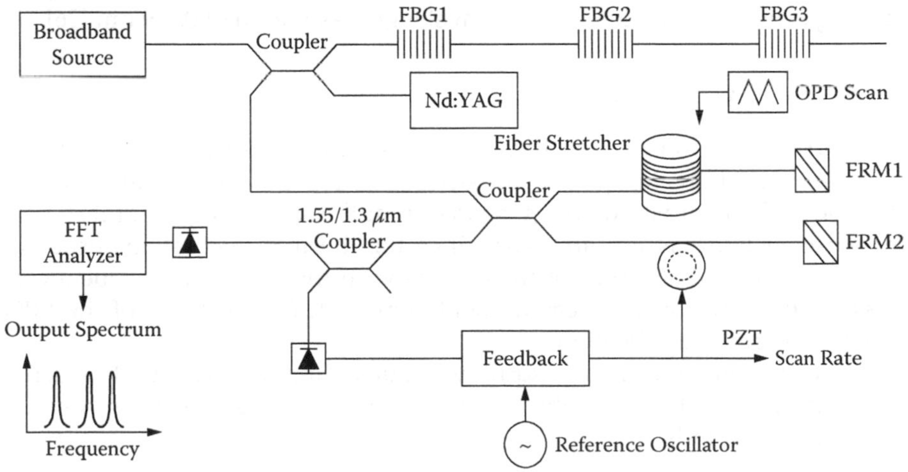

FIGURE 4.11
FBG sensor interrogation based on Fourier transform spectrometer.

spectrum, and thus wavelength shift can be monitored by measuring the corresponding frequency change. An Nd:YAG laser operating at 1.32 mm is used to provide a reference optical signal for a correct scan rate through the feedback system in order to maintain the linearity of the scan. The two Faraday rotator mirrors (FRM) are used to overcome polarization fading effects in the optical fiber interferometer. The resolution of the wavelength measurement is limited by an effective instrumental bandwidth, comparable to that of the FBG. In this case, this value is related to the maximum available scanning OPD from the zero position. The effective coherence length of a typical Bragg grating around 1550 nm is ~0.5 to 1 cm (for a bandwidth of 0.2 to 0.1 nm). Therefore, the OPD must be scanned over a minimum 2-cm range to cover the interferogram. For a system with a maximum OPD scan of 10 cm, a spectral resolution of 0.07 cm^{-1} (i.e., a Bragg wavelength shift of ~0.015 nm in a 1.55-m region) has been reported.

This method allows absolute wavelength measurement. An improved version is reported by Flavin et al. [35] in which a high-coherence source is needed, too, as a reference source. Since the ratio of the optical frequencies of the Bragg reflection and the reference source is equal to the ratio of the two optical phases, the wavelength of the Bragg reflection can be obtained from the ratio of the two optical phases. The phase information is recovered using a technique based on frequency domain processing. By capturing a short interval of each interferogram in a short-scan Michelson and then Fourier transforming it and removing the zero and negative frequencies of the transform, the result is inverse Fourier transformed to yield an analytic signal from which the required phase can be calculated. In this approach only a short optical path scan of 1.2 mm is applied. The measurement is controlled by the wavelength stability of the reference source. A wavelength resolution of 5 pm has been reported. A similar approach using Hilbert transform processing is reported, which gives a wavelength resolution of 0.019 nm [36].

4.2.4 Cross-Sensitivity

The intrinsic response of an FBG to both strain and temperature imposes serious implications for quasi-static strain sensing as any temperature variation along the fiber is indistinguishable from the strain. For dynamic strain measurement this is not an issue, since the thermal fluctuations occur at low frequencies that tend not to coincide with the resonance frequency of interest. For static strain measurement, temperature sensitivity of the FBG can complicate its application.

To recover temperature and strain without ambiguity, a two-grating sensing system is used to describe the relation between the wavelength shifts and the measurands:

$$\begin{pmatrix} \Delta\lambda_1 \\ \Delta\lambda_2 \end{pmatrix} = \begin{pmatrix} K_{1\varepsilon} & K_{1T} \\ K_{2\varepsilon} & K_{2T} \end{pmatrix} \begin{pmatrix} \varepsilon \\ T \end{pmatrix} \tag{4.8}$$

where $\Delta\lambda_1$ and $\Delta\lambda_2$ are the wavelength shifts from two grating sensors, respectively; $K_1\varepsilon$ and K_{1T} represent the response to strain and temperature for grating 1; and $K_2\varepsilon$ and K_{2T} for grating 2. Different cases can be classified as

1. $K_1\varepsilon = K_{2T} = 0$, or $K_2\varepsilon = K_{1T} = 0$, represents an ideal case cross-sensitivity.
2. With one of the elements in the response matrix being zero, one sensor is isolated from either of the perturbations, which can provide a reference for the other sensor.
3. The two sensors have very different responses to strain and temperature so that the determinant of the response matrix is nonzero, and the eigenvalue solution for the matrix equation can therefore be obtained.

A great deal of research effort has been focused on finding possible grating sensor configuration and interrogation techniques that will allow temperature and strain to be measured simultaneously.

Reference Grating

A straightforward solution involves compensating for the temperature/strain influence by using another FBG that is shielded from strain/temperature and measures only one perturbation. Several schemes have been proposed. Among them, the most effective method involves the use of strain-free temperature reference gratings that are co-located with the strain sensors, experiencing the same thermal environment. The compensation is obtained by subtracting the wavelength shift induced by the temperature excursion in the reference grating from the total wavelength shift recorded by the strain sensor [37,38]. Despite its effectiveness, this method is costly as extra gratings and their associated interrogation units are needed in the sensing system.

A passive temperature-compensating package has been proposed to nullify the temperature to wavelength coefficient (i.e., make either K_{1T} or K_{2T} zero). In a typical case [39], the grating is mounted under tension in a package comprising two materials of silica and aluminum, which have different thermal expansion coefficients. As the temperature rises the strain on the grating is released by just the amount necessary to cancel the shift in Bragg wavelength caused by the temperature rise. This grating can be used as a reference grating to the other gratings.

Two FBGs with Very Different Responses to Strain and Temperature

However, a system capable of providing strain and temperature measurements from the same fiber without requiring that a section of the fiber be isolated from temperature or strain is much more desirable. The elimination of cross-sensitivity may be achieved by measurements at two different wavelengths or two different optical modes, for which the strain and temperature responsivity is different.

Due to the wavelength dependence of the photoelastic and thermo-optic coefficients of the fiber glass, there is a small variation in the ratio of responses of FBGs written in different wavelengths. By measuring the responses of two fiber gratings written at 848 and 1298 nm, it has been found that their responses are 6.5% higher for strain and 9.8% lower for temperature at ~850 nm [40]. It gives a nonzero determinant of matrix *K*; thus, changes in both strain and temperature can be determined using the inverse of Eq. (4.8). The two gratings are deliberately superimposed during writing so as to be in contact with the same strain and temperature environment.

Bragg Grating and Fiber Polarization-Rocking Filter

Usually, different fiber grating structures show different responses to temperature and strain. A technique employing two different types of fiber grating has been used to distinguish strain and temperature. Methods employing a uniform fiber grating in combination with different fiber gratings such as fiber polarization-rocking filter [41], sample grating [42], and long-period grating have been proposed [43].

A combination of Bragg grating and polarization-rocking filter is employed to unambiguously determine strain and temperature applied to the fiber (Figure 4.12) [41]. Both gratings are written at the same location in an elliptical-core D-type fiber. Due to the large grating period of the polarization-rocking filter compared to that of the FBG, the resultant bandwidth is so wide (14 nm in reference 40) that it becomes difficult to measure small shifts in the resonant wavelength of this type of grating. A solution of this problem is to write a pair of rocking filters in exactly the same piece of fiber, forming a Mach–Zehnder interferometer. The interferogram obtained from the two rocking filters allows accurate monitoring of the wavelength shifts caused by temperature or strain. By using this configuration, the wavelength shift caused by a temperature change of 150°C is 1.1 nm in the Bragg grating and −30.6 nm in the rocking filter interferogram. Exposed to a strain change of 3000, the wavelength shifts are 2 nm in the Bragg grating and −5.2 nm in the rocking filter interferogram. Compared to the Bragg grating, the rocking filter gives a quite different but much higher response to both strain and temperature. All the elements in the matrix of Eq. (4.8) can be obtained;

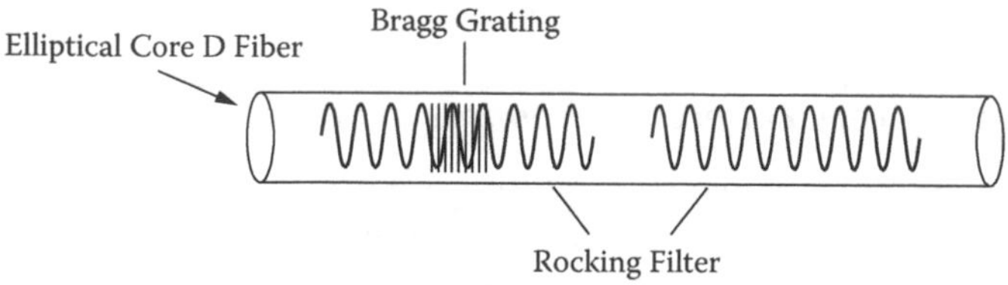

FIGURE 4.12
Bragg grating and polarization-rocking filters for simultaneous measurement of temperature and strain.

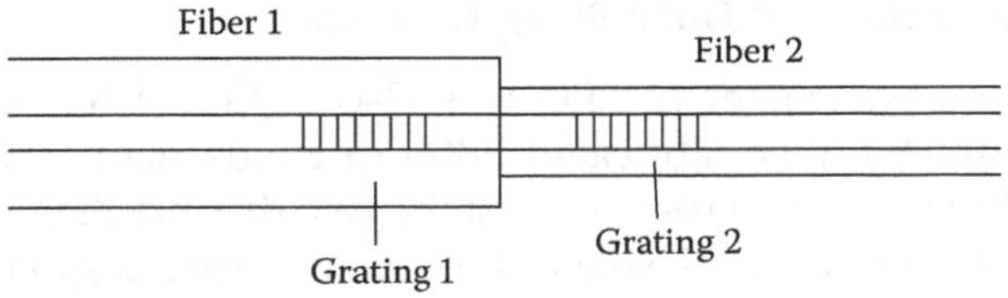

FIGURE 4.13
Two FBGs inscribed on the fibers with different diameters.

thus, an unambiguous measurement for both strain and temperature can be achieved.

The configuration using an FBG in combination with a long-period grating is described in a later section.

FBGs with Different Fiber Diameters

In this approach two fiber Bragg gratings are written on either side of a splice between two fibers with different diameters [44], as shown in Figure 4.13. It is considered that the strain the fiber experiences can be related to the size of its cross-section: The fiber with a larger diameter experiences smaller strain than that with a smaller diameter. Therefore, in such a configuration, a change in temperature will shift the center wavelengths of the two gratings by similar amounts, maintaining their wavelength spacing, while a change in strain results in the center wavelengths' moving by different amounts, changing the wavelength spacing. Eventually, the relative wavelengths provide a measurement of the strain, while the absolute wavelength shifts include information on both the temperature and strain.

Fiber Bragg Grating and Chirped Grating in a Tapered Fiber

In a uniform fiber Bragg grating, both strain and temperature cause a shift in the Bragg wavelength, while effective bandwidth remains almost unaffected. The bandwidth can be made strain dependent if a strain gradient is introduced along the grating length. In the proposed method [45], a chirped grating is written in a tapered fiber. The taper profile can be designed such that the grating becomes linearly chirped when tension is applied, thus creating a strain gradient along the grating. The average strain, which is the integral of local strain divided by the grating length, is measured by the change in the effective bandwidth. Both average strain and temperature still induce a shift in the Bragg wavelength, but the effective bandwidth will only vary with the strain rather than the temperature.

Fundamental and Harmonic Modes of Fiber Bragg Grating

Kalli and coworkers [46] have demonstrated the viability of an approach measuring both fundamental and harmonic modes of an FBG and achieved strain and temperature resolutions of ±17/pm and ±1.7°C/pm.

Polarization-Maintaining Fiber Bragg Gratings

The polarization parameter of FBG was also exploited for –T discrimination. Sudo et al. [47] have fabricated FBGs in Panda fiber and investigated the strain and temperature responses of the two distinct Bragg wavelengths to the fast and slow axes. Wavelength-to-temperature responsivities of 9.5 and 10.1 pm/°C and strain sensitivities of 1.342 and 1.334 pm were measured from the two polarization modes with cross-sensitivity errors of ±2°C and ±20, respectively [47].

4.2.5 Nonstrain/Temperature Sensors

In addition to strain and temperature sensing, fiber Bragg gratings have been used to detect other parameters as well. However, the basic sensing principles for most of those parameters are the same as that for strain and temperature. For instance, FBG-based sensors have been reported to measure parameters such as pressure [48,49], load [50], bending [51], vibration [52], acceleration [53], electrical power [54], electrical current [55], magnetic field [56,57], and hydrogen [58]. Refractive index change in the fiber core causes a Bragg wavelength shift. This has been used to measure high-magnetic fields, where—due to magnetically induced circular birefringence (Faraday effect)—the Bragg resonance is split into two resonances [56]. The surrounding refractive index change of the fiber grating can greatly affect the Bragg reflection. Based on this principle, some chemical sensors have been proposed by writing a Bragg grating on D fiber [59] or side-polished fiber [60]. The surrounding refractive index also plays an important role on the spectral characteristics of long-period fiber grating. This makes it a good candidate for chemical sensing. We will introduce and discuss the detailed information in a later section.

FBGs have been fabricated on multicore fiber for bending sensors [61]. Furthermore, multimode fiber has been considered to make fiber Bragg gratings for sensing purposes [62]. However, due to the complexity of many modes in the reflection of the fiber grating, a stable and accurate measurement is hard to achieve using a multimode fiber grating.

4.3 FBG Sensor Multiplexing Techniques

In many applications a large number of sensors need to be used to achieve a distributed measurement of the parameters. In particular, using sensors in smart structures is of interest where sensor arrays are bonded or embedded into the materials to monitor the health of the structure. FBG sensors have a distinct advantage over other sensors because they are simple, intrinsic sensing elements that can be written into a fiber, and many sensors can be interrogated through a single fiber.

4.3.1 WDM

The most straightforward multiplexing technique for FBG sensors is wavelength division multiplexing (WDM), utilizing the wavelength-encoding feature of an FBG-based sensor. The WDM technique is based on spectral splicing of an available source spectrum. Each FBG sensor can be encoded with a unique wavelength along a single fiber. Since we are operating in the wavelength domain, the physical spacing between FBG sensors can be as short as desired to give accurate distributed information of measurands.

A basic WDM configuration is shown in Figure 4.14. An FBG sensor array is written along a single fiber and illuminated by a broadband source. The optical signals reflected from the FBG sensor array are fed into a wavelength detection scheme. In an early configuration a tunable Fabry–Perot filter was used to distinguish between the wavelength shifts of each FBG sensor [23]. When the tunable filter operates in scanning mode, multiplexed FBG sensors can be interrogated.

A parallel topology is used to allow simultaneous interrogation of all the sensors in WDM, as shown in Figure 4.15(a) [20]. A $1 \times N$ fiber splitter is used to divide the optical reflection into N channels. In each channel a matched fiber grating detects the wavelength shift from a specific FBG sensor (further detail can be seen in the first subsection of Section 4.2.3).

In the parallel scheme each filter receives less than $1/2N$ of optical power as a result of using $1 \times N$ fiber splitter and fiber coupler. More FBG sensors lead to a larger power penalty. An improved scheme using a serial matched FBG array is reported by Brady et al. [63], as shown in Figure 4.15(b). This scheme is claimed to allow the optical power to be used more efficiently than in the parallel topology. As can be seen, however, a large power penalty still exists through the use of the reflection of matched fiber gratings. A revised version of the serial scheme is proposed [64], in which the transmission of the matched FBG is used to monitor the wavelength shift from the corresponding sensing FBG. This reduces the power penalty of 6 dB.

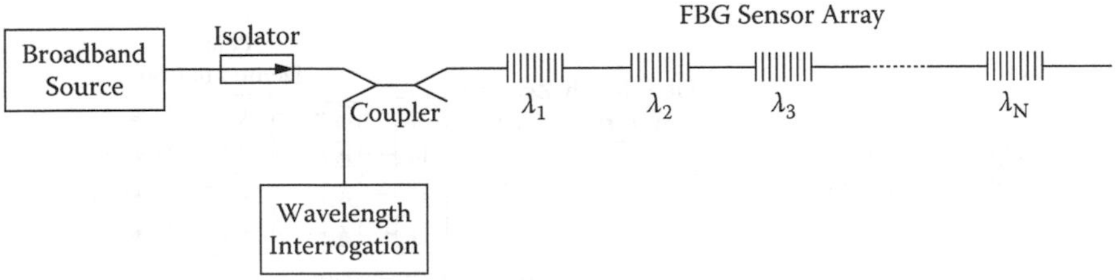

FIGURE 4.14
A sensing system with wavelength division multiplexed FBG sensor array.

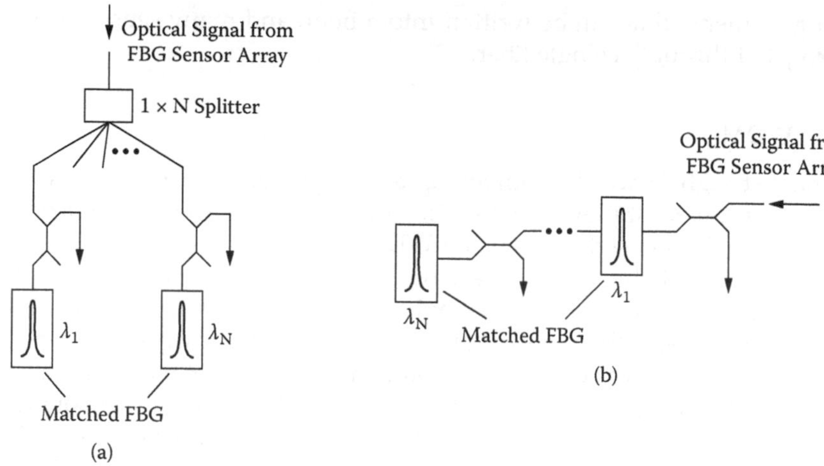

FIGURE 4.15
(a) Parallel topology; (b) serial topology for interrogation of FBG sensors in WDM.

The recent development of WDM devices for the fiber optic communication field allows for the development of new multiplexing schemes. Berkoff and Kersey [65] demonstrated a WDM configuration by using a bandpass wavelength division demultiplexer, as shown in Figure 4.16. A fiber Mach–Zehnder interferometer is used to transform the wavelength-encoded strain information from FBG sensor array signals into a series of interferometric phase-dependent terms. Instead of using a fiber splitter, the output of the interferometer is coupled into a bandpass wavelength division demultiplexer, which is a wavelength splitter and separates the composite interferometer phase signal into discrete channels that correspond to a wavelength window about each of the FBG wavelengths in the array. This technique overcomes the power penalty suffered in the previous parallel and serial schemes. As the dense wavelength division demultiplexer become commercially available, a large number of FBG sensors can be addressed.

There have been many reports on wavelength division multiplexed FBG sensors. The basic concept of these sensor systems is the same as described

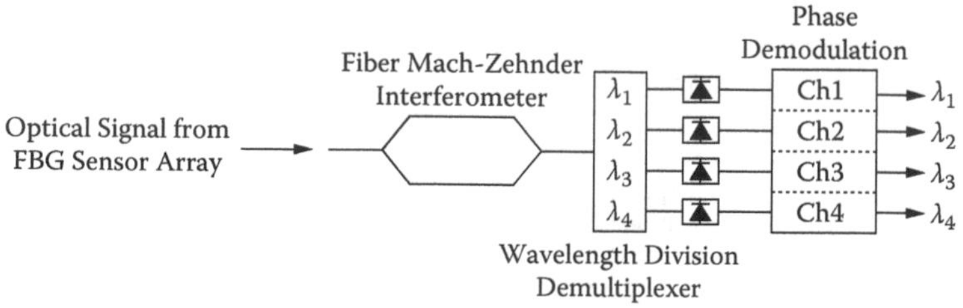

FIGURE 4.16
Interrogation of WDM FBG sensor array using a wavelength division demultiplexer.

previously, except that different interrogation schemes are used. The available number of multiplexed FBG sensors depends on the measurement range of the sensors and the bandwidth of the gratings. Narrowing the grating bandwidth, however, reduces the available optical power from each sensor, degrading the system performance. Fiber Bragg grating laser sensor systems have been proposed to overcome this problem [66,67]. The basic Bragg grating laser sensor employs a Bragg grating as the end reflector of a fiber laser and uses Erdoped fiber as the usual gain medium. Changes in the lasering wavelength occur in response to external perturbations applied on the Bragg reflector. The wavelength addressing capabilities of fiber gratings allows distributed fiber grating sensors to be implemented. However, complicated effects may take place in the output of fiber grating laser sensors, such as mode competition and mode beating [67].

4.3.2 TDM

In time division multiplexed (TDM) FBG sensors, the return pulses between adjacent FBGs are recovered, with the two pulses separated by a distance equal to the round-trip time between gratings. Using the phase-sensitive detection scheme, FBG sensors can be multiplexed using time division addressing, as depicted in Figure 4.17 [68]. Pulsed light from a broadband source is launched into a fiber containing an FBG sensor array, with different peak reflection wavelengths along the fiber length. The input pulse width has to be equal to or less than the round-trip time between two gratings. The return pulse reflections from each grating are then separated in time at the output. The reflected pulses are fed through the Mach–Zehnder interferometer that acts as the wavelength shift detector. Individual sensor pulses can be demultiplexed using an electronic circuit capable of gating out a single pulse within the pulse train signal, allowing phase detection methods to be used to recover sensor information from each sensor channel.

Using the TDM technique described before, an eight-FBG sensor system was demonstrated [69]. The number of multiplexed FBG sensors depends on the pulse width, equal to the separation distance between two adjacent gratings, a figure that could be improved following the recent development

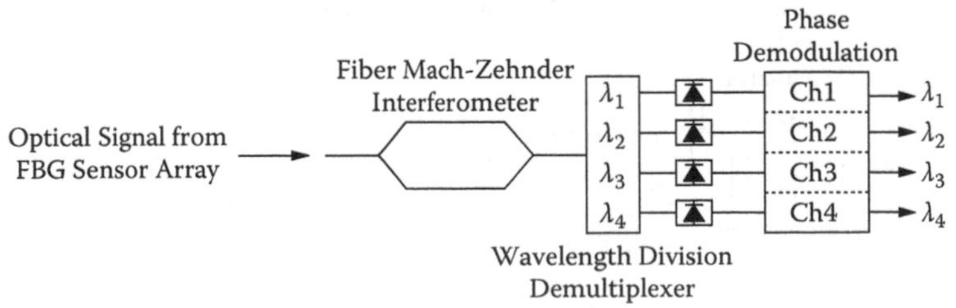

FIGURE 4.16
Interrogation of WDM FBG sensor array using a wavelength division demultiplexer.

of high-speed gating devices. However, as the pulse duty circle increases, the available power for each sensor channel is reduced, causing poor visibility and low signal-to-noise ratio in the detection.

4.3.3 WDM/TDM

WDM provides for tens of FBG sensors per fiber, but TDM can multiply this number several times by reusing the wavelength spectrum of the source.

Mixed WDM/TDM is used in a serial configuration where several WDM grating arrays are concatenated, each at a greater distance along the fiber [69]. The FBGs are fabricated at low reflectivity. By launching a short pulse of light from the broadband source, the reflections will return to the detector at successively later time intervals. The detection electronics are designed to respond to the reflection signals only during a selected time window so that a single WDM set of FBG sensors is selected for detection. However, in this serial configuration cross-talk arises from the effect of multiple reflection between the FBGs, as they possess the same Bragg wavelength.

A parallel configuration, as shown in Figure 4.18, eliminates this effect [70]. The input pulse is split into the FBG sensor arrays. A certain length of fiber delay line separates the time window for each WDM set. But the overall optical efficiency is reduced due to the use of a fiber splitter.

Koo et al. [71] proposed a system that combines code division multiplexing (CDM) with WDM to achieve dense wavelength division multiplexing of FBG sensors [71]. This is an alternative to TDM. It has the advantage of having a larger average sensor output power than the TDM technique. Like the traditional code division multiplexing access technique, this CDM is based on a correlation technique for separating individual sensor outputs from a single multiplexed output of many sensors. In this approach, the optical signal launched into the FBG sensors is modulated with a pseudorandom bit sequence (PRBS). By adjusting the time delay between successive FBG sensors to be equal to one bit length or multiple bit lengths of the PRBS, a given individual sensor can be selected from the entire sensor array response by

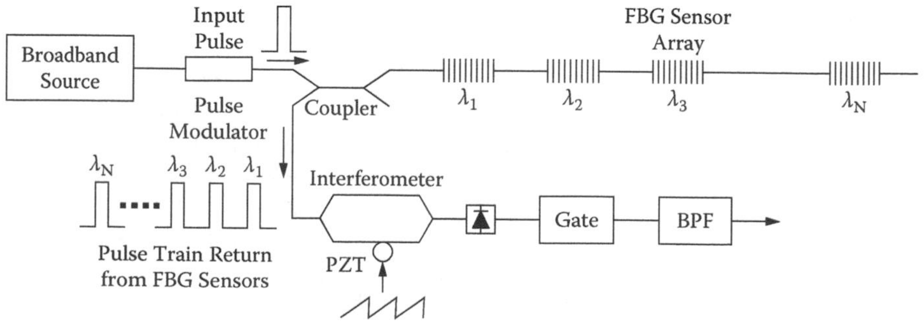

FIGURE 4.17
Schematic of a time division multiplexed FBG sensing system.

correlating its response with a reference PRBS (autocorrelation process). Using this approach, the wavelength separation between sensors can be reduced to the fiber Bragg grating linewidth. The authors claimed that with a reasonable code sequence length of 127, the multiplexing of >100 FBG sensors in a single WDM FBG array is possible, using a source spectral width of 40 nm.

4.3.4 IWDM

In a WDM system, the number of grating sensors to be multiplexed is determined by the required operating wavelength range of each FBG sensor and the total useable bandwidth of the light source. As the amplitude of the reflection is totally ignored, the WDM system does not allow the spectral response of adjacent gratings to cross each other. A technique for addressing an FBG sensor array, called intensity and wavelength dual-coding multiplexing (IWDM), has been proposed [72], as shown in Figure 4.19. In the IWDM technique a set of spectrally distributed FBGs with relatively low reflectivity (low-R) and high reflectivity (high-R) is alternately placed along a fiber. All the low-R gratings have the dual-peak spectral feature, and the bandwidth of each of them is similar to that of the high-R grating. The dual-peaked profile of the lower reflectivity gratings facilitates resolution of the respective strain-encoded wavelength peaks when the higher and lower reflectivity spectra overlap.

Figure 4.20 shows the results of an experiment in which a higher reflectivity grating is kept unstrained while an adjacent lower reflectivity, dual-peaked grating is subjected to increasing strain sufficient for the reflected spectrum of the latter to pass completely through the unstrained grating spectrum. Successive spectral profiles are shown in Figure 4.20(a), and the separate traces in Figure 4.20(b) correspond to wavelength shift measurements on each of the two peaks of the dual-peaked grating. It is clear that continuous strain measurement may be affected. Thus, as illustrated, this concept permits a doubling of the number of strain sensor elements for a given source bandwidth, but it is readily apparent that further intensity levels may be used in the same manner to achieve yet further increases in sensor density. Furthermore, in conjunction with TDM and spatial division multiplexing

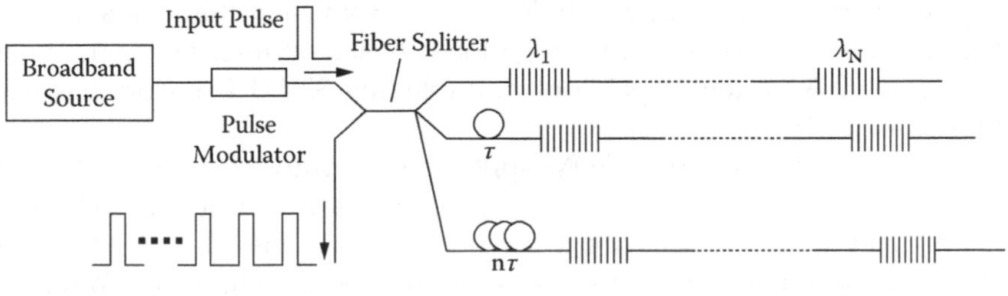

FIGURE 4.18
A parallel configuration of combined WDM/TDM multiplexing topology.

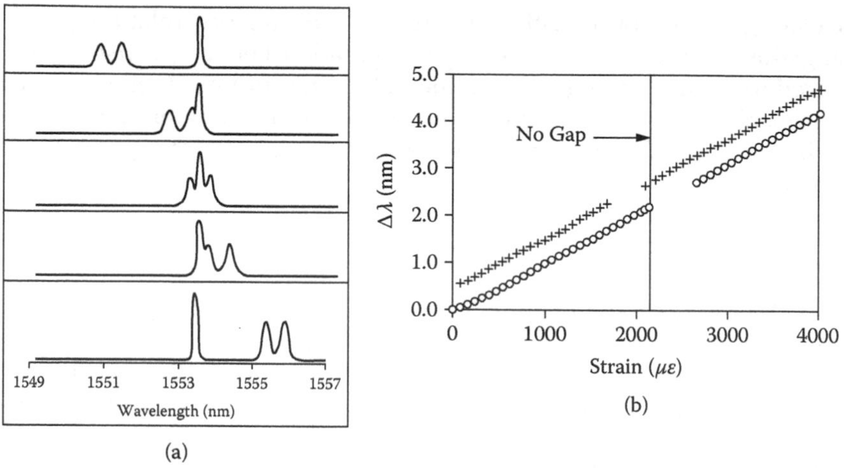

FIGURE 4.20
(a) Successive spectral reflection profiles from a pair of IWDM sensors as the strain on one is increased; (b) wavelength shift $\Delta\lambda$ plotted against applied strain measured for each peak of the dual-peaked grating reflection profile.

(SDM), this technique could lead to interrogation of a substantially larger number of FBG sensors.

4.3.5 SDM

The multiplexing techniques described previously are suitable for applications where distributed sensing is needed. In many situations, however, point measurements are required where FBG sensors in a network must be operated independently and be interchangeable and replaceable without any substantial recalibration. This requires that all FBGs in the network have identical characteristics. Furthermore, it is very desirable to be able to multiplex a large number of identical FBGs, for reasons of economy and multiple-parameter measurements.

A spatial division multiplexing (SDM) configuration is depicted in Figure 4.21. A superluminescent diode is used as a source, a Michelson interferometer as a wavelength scanner. All the FBGs have the same nominal Bragg wavelength and bandwidth. Reflection from each channel is detected separately. Wavelength-shift interrogation is based on interferometric phase measurement. A system capable of multiplexing 32 FBG sensors has been demonstrated [73].

However, the conventional SDM splits sensors into many fiber channels and employs a separate electronic signal processing unit for each channel. This inevitably brings with it the penalties associated with the use of multiple single sensors in terms of cost, robustnesss, weight, etc. To remedy this shortcoming, the most popular format is to combine SDM with other multiplexing techniques [73–75]. Davies et al. [74] employed a combined

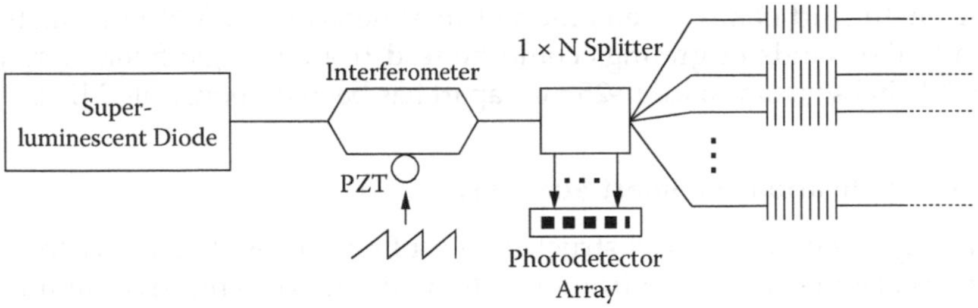

FIGURE 4.21
A spatial division multiplexing configuration using a fiber splitter and photodetector array.

WDM/SDM technique where a multichannel optical fiber switch is used, allowing for separate sensors on multiple fiber channels to share the same processing electronics. In the reported configuration, the system has the capability to monitor 12 FBG sensors along each of five fibers, for a total of 60 sensor elements. Chen et al. have reported an SDM/WDM system based on using a two-dimensional CCD array to generate an $M \times N$ matrix, with spatial positions of the fiber channels encoded along one axis of the CCD, while wavelengths are encoded along the orthogonal axis [30,31].

A novel Bragg grating-based method has been used as an SDM technique, utilizing the fact that the reflective spectrum of a low-reflectivity Bragg grating is the Fourier transform of its spatial structural variation [76]. In an example of this multiplexing technique, as shown in Figure 4.22, Bragg gratings with very low reflectivity and the same Bragg wavelength are inscribed on a single fiber. Optical power from the tunable laser is coupled into the fiber with FBG sensors. Scanning the tunable laser results in a measurement of the reflected power from the sensing fiber, as a function of wavelength. The Fourier transformation of these data gives reflected power as a function of distance from a reference reflector located at the entry point into the sensing fiber. Thus, the location of the Bragg gratings relative to the reference reflector is accurately known. Reflected power as a function of distance information allows for the selection of a particular data set along the length of fiber in the region of a Bragg grating. This data set then is Fourier transformed back into reflected power as a function of wavelength. The peak reflected power is used to determine the wavelength of the grating being interrogated. Together with the initially measured wavelength, the temperature-/strain-induced

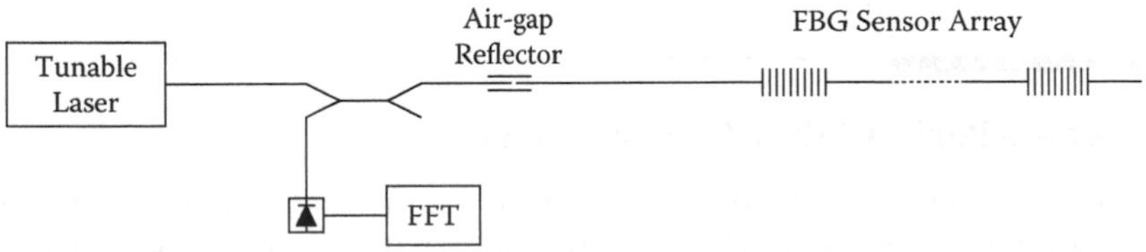

FIGURE 4.22
SDM multiplexing system based on Fourier transform.

wavelength shift of any given grating can be determined. With this method, literally thousands of gratings could be read from a single fiber. A system with 12 FBG sensors spaced ~25 mm apart has been demonstrated [77].

4.3.6 Applications in Smart Structure

Generally speaking, a smart structure must first possess the ability to sense its internal and external environment. It should then be able to communicate the sensory signals via appropriate pathways to one or many signal processing and control modules, where the information is analyzed and appropriate actions decided. If necessary, the decisions must be conveyed to actuators incorporated within the structure, which respond by altering its characteristics such as the shape, size, stiffness, position, or natural frequency [78].

Optical fiber sensors have been recognized as a good candidate for use in smart structures. It is found that an embedded optical fiber with a cladding diameter below a certain value does not degrade the static mechanical properties of the host [79]. A number of benefits in combining optical fibers and composites have been identified [80]. Compared with other sensing technologies, they provide several compelling advantages. One of the most important is their immunity to electromagnetic interference, which eliminates the need for the costly and heavy shielding normally required to minimize electrical noise picked up when using electrically active sensors. They provide a high-bandwidth, high-sensitivity, and high-dynamic-range measurement capability. Additional advantages of this technology include fatigue and corrosion resistance and the inherent strength of glass fibers—features that enable this technology to perform well in the extremely hostile environments of high temperatures, erratic vibrations, and shock loadings. Optical fiber itself can be used as the communication medium. The use of optical fiber technology with smart structures enables both the realization of multiplexed arrays of fiber sensors interconnected by other fibers and the implementation of distributed sensors. The fiber Bragg grating sensor is an excellent example of this. To date, fiber Bragg grating sensors have been demonstrated on civil engineering structures such as dams; bridges [81–84]; oil wells [85]; industry pressure vessels and boilers [86]; composite material structures such as modern yacht masts [87], aircraft [77], spacecraft [75]; and curing and impact detection in composite materials [88–90].

4.4 Long-Period Fiber Grating Sensors

Recently, R&D efforts have been directed to a new class of grating devices—long-period fiber gratings (LPFGs). In contrast to the short-period (~1 m) FBGs, LPFGs have modulated refractive index periods typically in a range of 50–500 m and are capable of coupling light between the forward-propagating core and cladding modes, generating a series of attenuation resonance peaks in transmission. These types of devices, therefore, possess an intrinsically low level of back reflection and are inexpensive to batch-produce. The popularity of LPFGs has increased dramatically over the last few years as notable numbers of LPFG-based devices and systems have been demonstrated for a multitude of applications in optical fiber communications and sensing.

Since LPFGs were first proposed as band rejection filters by Vengsarker et al. in 1996 [91], a variety of applications have been found for them. In optical communication and signal processing, LPFGs have been demonstrated as EDFA gain-flattening filters [92–94], WDM isolation filters [95,96], high extinction ratio polarizers [97,98], and mode converters [99,100]. For optical sensing applications, LPFGs have been implemented as temperature/strain/refractive-index sensors [101–104] as well as sensing demodulators [19,105]. The fact that light coupling involves cladding modes means that the LPFG spectral response is strongly influenced by the optical properties of the cladding and the surrounding medium. This unique feature has yielded several novel sensing concepts and devices. Its sensitivity to the external refractive index has been exploited for chemical sensing applications [106–108]. Other sensing devices with sensitivity advantages include optical load and bend sensors [109–112]. These nonaxial-strain sensors are particularly interesting as the sensing mechanism is based on a perturbation-encoded resonance mode-splitting measurement rather than conventional wavelength-shifting detection. Furthermore, the property of quadratic dispersion of LPFGs [113] has been utilized to implement measurand-induced-intensity and dual-resonance sensors, offering the attractive features of using economical signal demodulaiton schemes and possessing superhigh sensitivities.

It should be pointed out, however, that most of the LPFG-based sensor devices so far are still in their early development stage. Many practical device issues need to be resolved before they can be used in real applications. Problems such as packaging and parameter cross-sensitivity are especially pressing.

4.4.1 LPFG Theory and Fabrication Technology

LPFG Theory

The long periodicity of LPFG results in light coupling from the core mode to the discrete cladding modes. The coupled cladding modes attenuate rapidly as they propagate along the length of the fiber due to the lossy cladding-surrounding interface and bends in the fiber. Consequently, a series of attenuation resonance bands is generated only in transmission. As in the case of the FBG, the resonance of the LPFG, λ_{res}, occurs only at the phase match condition

$$\lambda_{res}\left(n_{co,eff}^{01} - n_{cl,eff}^{m}\right)\Lambda \tag{4.9}$$

where $n_{co,eff}^{01}$ and $n_{cl,eff}^{m}$ are the effective indices of the fundamental core mode and the *m*th cladding mode, respectively, and Λ is the period of grating.

Typical spectral profiles of two LFPGs with a similar period size (~500 m) but different grating lengths (1 and 3 cm) are depicted in Figure 4.23, showing a series of broad attenuation peaks located in the optical fiber communication window ranging from 1.2 to 1.6 m. The contrast in spectral profiles between the two gratings indicates that the bandwidth of an LPFG can be controlled by simply altering the grating length.

The LPFG fabrication technology is now sophisticated enough for a variety of LPFGs of complex structures to be readily produced. Among the demonstrated structures, including phase shift [114,115], cascaded [116–118], and chirped [97,119] LPFGs, the Mach–Zehnder interferometer (MZI) cascade configuration has attracted the most attention [96,114,116–121]. This structure concatenates a pair of identical LPFGs, each of which has a loss band close to 3 dB. As shown in Figure 4.24(a), the first LPFG acts as a beamsplitter coupling 50% light from the core to the cladding, while the second grating combines the lights traveling in the core and the cladding. The intrinsic optical phase difference between the co-propagating core and cladding modes between the two LPFGs results in a set of narrow interference fringes generated inside the original loss band. The free spectral range (FSR)—the spectral spacing between two adjacent resonant peaks—is defined by the following expression:

$$\Delta\lambda \approx \lambda^2 \Big/ \left[\left(n_{\text{co,eff}}^{01} - n_{\text{cl,eff}}^{m}\right)d\right] \tag{4.10}$$

where d is the cavity length, which can be easily tailored to the required specification for FSR. The representative profiles of two cascaded LPFGs with cavity lengths of 50 and 190 mm are shown in Figure 4.24(b). Clearly, the subresonances dramatically enhance the spectral resolution of the device. In addition, the resonant structure also provides an extra parameter FSR-utilizable for multiparameter sensing.

It is interesting to note that a variant configuration of an MZI requiring only one LPFG can be formed and has been demonstrated as a temperature sensor [122]. This configuration works on the principle of the self-interference realized by the LPFG and the fiber-end coating. In comparison with the dual-LPFG interferometer, the single LPFG device offers an added advantage of signal interrogation from the reflection mode.

Fabrication Techniques

The increasing popularity of LPFGs largely stems from the fact that LPFGs are easy and inexpensive to fabricate as compared with FBGs. The large period size allows the UV writing of LPFGs using amplitude masks that are significantly less expensive and less quality critical than the phase masks generally used in fabricating FBGs. Furthermore, the large period size permits the point-by-point direct UV inscription, not just removing the need for amplitude masks but also offering the added advantage of the realization of

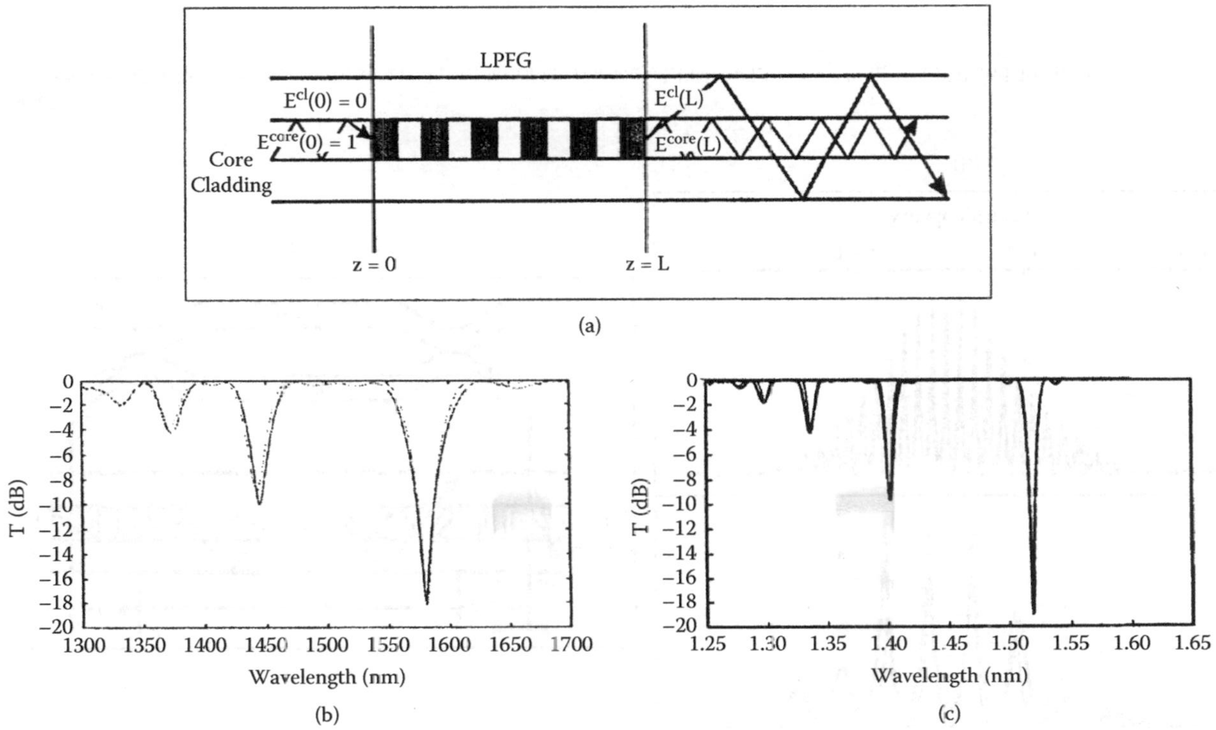

FIGURE 4.23
(a) Mode coupling mechanism of LPFG. Theoretically modeled and experimentally measured transmission spectra of two LPFGs of ~500 m period; (b) ~1 cm length; and (c) ~3 cm length.

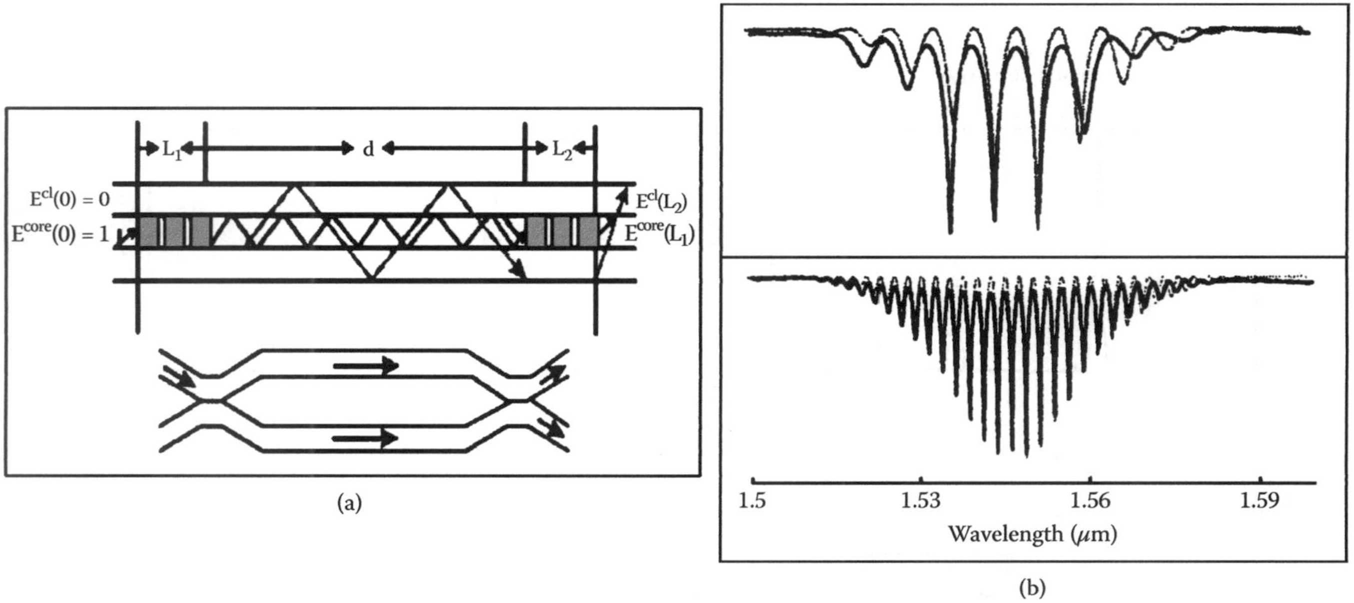

FIGURE 4.24
(a) LPFG Mach–Zehnder interferometer; (b) spectral profiles of two cascade LPFGs with cavity lengths of 50 and 190 mm.

arbitrary periods and profiles [123]. Recently, several non-UV-writing techniques have also been reported, showing that LPFGs can even be produced using a CO_2 laser, a microlens array, and a fusion splicer [124–128].

4.4.2 LPFG Temperature and Strain Sensing

Temperature and Strain Sensing

Any modulation of the core and cladding guiding properties will modify the spectral response of LPFGs, which is essential for sensing purposes. The external perturbations are detected as wavelength-dependent loss modulation of an LPFG. In 1996, Bhatia and Vengsarkar first demonstrated LPFG optical sensors capable of measuring temperature/strain as well as external refractive index [101]. They found that the average temperature sensitivity of LPFGs is much larger than that of FBGs and strongly fiber-type and grating-structure dependent. In their investigation, LPFGs were fabricated in four different types of fiber and their spectral behavior was examined for the cladding modes of different orders. It was noticed that the third-order cladding mode of an LPFG with a 210-m period produced in the 980-nm single-mode standard fiber exhibited the highest temperature sensitivity of 0.154 nm/°C—10 times higher than that of an FBG (~0.014 nm/°C). Figure 4.25(a) presents the comparative results from an LPFG with a 400-m period and a standard FBG, both fabricated in B/Ge co-doped fiber: The amplification of the thermal sensitivity of the LPFG is also approximately 10 times that of the FBG.

The LPFGs investigated by Bhatia and Vengsarkar as strain sensors also exhibited a greater sensitivity than that of FBGs, but the relative difference is not as large as in the temperature case. The axial-strain-encoded wavelength shift of the LPFG produced in the Corning fiber was only twice as much as the value obtained from the FBG produced in the same fiber. However, there has not been a consistent conclusion of the LPFGs' always exhibiting a larger

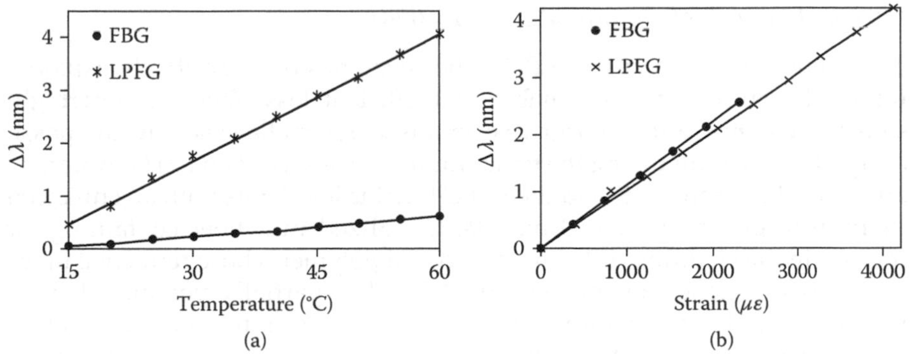

FIGURE 4.25
Comparison of (a) temperature-induced and (b) strain-induced wavelength shifts of LPFG and FBG.

average strain sensitivity than that of FBGs since several subsequent reports claimed that similar or even smaller strain sensitivities were observed for the LPFG structures. Figure 4.25(b) relates to a case where the strain sensitivity of an LPFG is similar to that of an FBG involving the same two gratings as used for thermal response evaluation in Figure 4.25(a).

It should be particularly pointed out that, in LPFG temperature and strain sensing, the wavelength shift of the cladding modes of LPFGs can be toward either short wavelengths (blue shift) or long wavelengths (red shift) [91,129,130], depending on the dispersion property of the waveguide. In general, for sensing in the wavelength region close to the telecommunication window, the blue shift is mostly associated with the use of relative short-period (high-order mode) LPFGs, whereas the red shift takes place when the LPFGs of relative long periods (low-order mode) are used. The wavelength shift direction is also strongly fiber-type dependent.

Temperature-Insensitive LPFG Sensors

The large thermal sensitivity of LPFGs imposes severe temperature cross-sensitivity effects on the nontemperature sensors, affecting the accuracy of the measurement. Issues in relation to thermal instability and temperature cross-sensitivity have to be resolved before practical LPFG devices can be developed for telecommunication and sensing applications. Judkins et al. proposed and demonstrated a temperature-insensitive LPFG fabricated in a specially designed fiber [131]. The principle involves balancing the temperature dependence of the refractive index of the host fiber and that of the grating periodicity. By tailoring the refractive index profile of the host fiber, they successfully reduced the temperature sensitivity of the LPFG to –0.0045 nm/°C, which was even one order of magnitude less than that of FBGs. Strain and refractive index sensors were later produced in this fiber showing no signs of temperature cross-sensitivity [132].

LPFGs of Superhigh-Temperature Sensitivity

The thermal responsivity of LPFGs, however, has been positively explored for implementing widely tunable high-efficient loss filters or superhigh-sensitivity temperature sensors. Several researchers have sought the possibility of further enhancing the temperature sensitivity of an LPFG. Abramov and coworkers demonstrated the LPFG tunable loss filters with an enhanced thermal responsivity of ~80 nm/100°C realized in a fiber containing air regions in the cladding, filled with a special polymer whose refractive index is a strong function of temperature [133]. The partially polymer-cladded LPFG achieved a temperature sensitivity eight times that of the standard LPFG. In addition, the device is insensitive to the external refractive index change because of its buffer effect from the polymer clad. The polymer-cladded LPFGs were then metal coated for electrical tuning and demonstrated themselves as high-efficiency tunable loss filters [134].

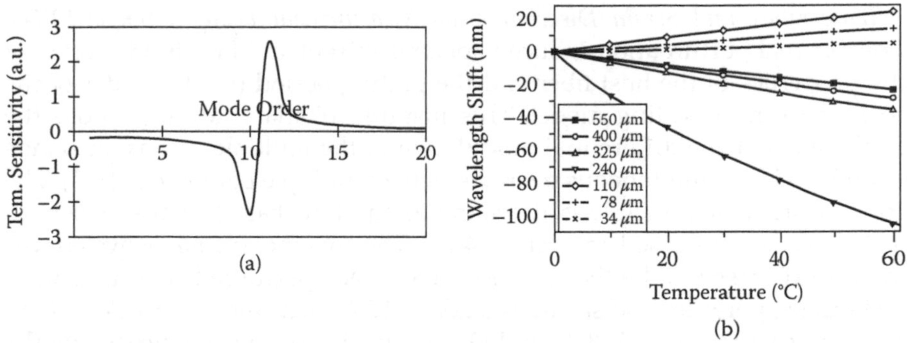

FIGURE 4.26
(a) Theoretically calculated temperature sensitivity as a function of cladding mode order for 1550-nm wavelength region; (b) measured temperature-induced wavelength shifts for seven LPFGs with small, medium, and large periods.

Recently, Shu et al. reported that the temperature sensitivity could be dramatically enhanced by writing the LPFG in B/Ge co-doped fiber and by choosing appropriate cladding modes [130]. The theoretical modeling of their study indicates that the thermal responsivity of LPFGs strongly depends on the order of the cladding mode. The relationship between the sensitivity and mode order, shown in Figure 4.26(a), clearly suggests that there is a special mode-order region. In this region, the temperature-induced wavelength shift changes direction and, more important, the temperature sensitivity of the cladding mode is significantly higher than that of the other modes outside this region. The thermal behavior of seven LPFGs with low, medium, and large periods was evaluated experimentally to test the theoretical findings. Figure 4.26(b) plots the magnitude and direction of the wavelength shifts against the temperature for the seven gratings. The temperature responsivities of the seven gratings clearly follow the theoretical mode-order-dependent trend. The four lower order modes of the gratings with longer periods blue-shift, whereas the three higher order modes of the gratings with shorter periods red-shift. The grating with 240-m period, whose mode order falls in the sensitive mode region, demonstrated a superhigh-temperature sensitivity of 2.75 nm/°C. This value is three times more than that of the partially polymer-cladded LPFGs.

It is interesting to speculate that if such an LPFG is used as a temperature sensor and interrogated by an optical spectrum analyzer with a resolution of 0.1 nm, a temperature change as small as 0.036°C could be detected by this LPFG.

Simultaneous Temperature and Strain Sensing Using LPFG and Hybrid LPFG/FBG

Efforts have also been made to identify techniques capable of co-measuring temperature and other sensing parameters. For this purpose, both the single-structure LPFG and hybrid LPFG/FBG have been investigated.

Temperature and Strain Discrimination Measurement Using a Single LPFG. The external perturbation-induced spectral shift of an LPFG is a function of the properties of the host fiber and the grating period (i.e., the order of the corresponding cladding mode). This inspired Bhatia et al. to propose the exploitation of the differential modulation of the multiple-resonance bands of an LPFG for simultaneous measurement of multiple parameters [135]. The co-measurement principle is the same as the FBG-based dual-wavelength technique, as discussed in Section 4.2.4. The two measurable wavelengths in an LPFG case will be the two loss bands, each perturbed simultaneously by both temperature and strain (i.e., $\Delta\lambda_1 = A\Delta T + B\Delta$, and $\Delta\lambda_2 = C\Delta T + D\Delta$). The four coefficients—*A*, *B*, *C*, and *D*—can be determined by measuring the temperature and strain responses separately for each resonance band. Once they are determined, the standard sensitivity matrix inversion technique can be applied to separate the strain and temperature effects.

However, for large values of ΔT and $\Delta\varepsilon$, the wavelength shifts with strain and temperature may not be linear, making *A* and *C* functions of ΔT and *B* and *D* functions of $\Delta\varepsilon$. Furthermore, the temperature cross-sensitivity could make *B* and *D* functions of ΔT as well, introducing complexity into the equations, imposing difficulty and limitations for real applications. Nevertheless, simultaneous temperature and strain measurement using a single LPFG structure was investigated for small dynamic ranges. The investigative gratings were fabricated in two different types of fiber (Corning SMF-28 and Corning 1060-nm Flexcor) and evaluated for temperature and strain responses independently to determine the coefficients. It was found that the LPFG fabricated in the former fiber showed considerable nonlinear and cross-sensitivity effects, but over the small ranges ($\Delta T < 30$°C and $\Delta\varepsilon < 2000$), the wavelength shifts with the strain and temperature were linear and the cross-sensitivity was negligible. The minimum detectable strain and temperature changes were measured to be 30 and 0.6°C. The gratings fabricated in the second type of fiber, whose intrinsic strain coefficient was very small, demonstrated better applicability, as one of the two resonance bands appeared insensitive to the strain (i.e., $D \approx 0$). Thus, the solution for the sensitivity matrix was significantly simplified. As a result, the temperature dynamic range was notably increased and the maximum errors were only 1°C and 58 over ranges of 125°C and 2100, respectively.

Temperature and Strain Discrimination Measurement Using a Hybrid LPFG/FBG. An alternative method for temperature and strain discrimination measurement employing a hybrid LPFG/FBG was proposed by Patrick et al. [136]. The LPFG and FBGs were fabricated in different types of fiber, ensuring that the LPFG has a much larger thermal coefficient but a smaller strain coefficient than that of the FBG. The FBGs were UV written in Lycom single-mode fiber, exhibiting temperature and strain sensitivities of 0.009 nm/°C and 0.001 nm, respectively. The LPFGs were produced in AT&T 3D 980-nm single-mode fiber with a period of 246 m, yielding a temperature response of 0.06 nm/°C, which is about seven times that of the FBGs, and a strain response of 0.0005 nm, only half the value for the FBG.

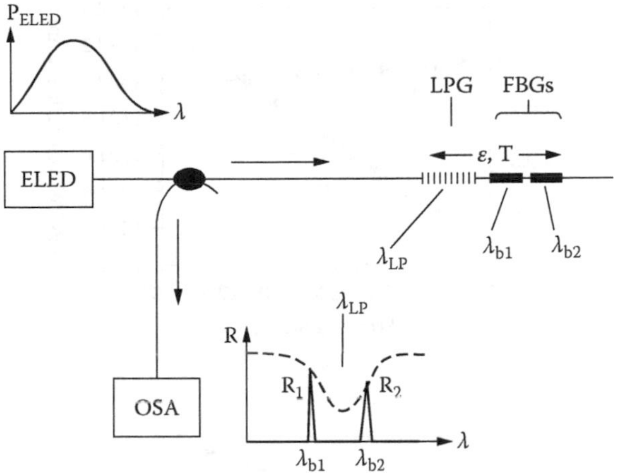

FIGURE 4.27
Schematic of the hybrid FBG/LPFG strain/temperature sensor and detection system. (From Patrick, H. et al., *IEEE Photon. Tech. Lett.*, 8, 9, pp. 1223–1225, 1996.)

The performance of this method was further enhanced by combining two FBGs with one LPFG in order to avoid poor accuracy of wavelength-shift detection from the broad spectral response of the LPFG. The sensor configuration and the interrogation principle are shown in Figure 4.27. The two FBGs were designed to have central Bragg wavelengths spectrally located on both sides of the LPFG transmission loss band. The interrogation is based on the principle that strain-/temperature-induced wavelength shifts of both LPFG and FBGs were measured entirely from the reflection signals of the FBGs. The reflection signal function $F(R_1, R_2)$ is given by

$$F(R_1, R_2) = \frac{\left(\sqrt{R_1} - \sqrt{R_2}\right)}{\left(\sqrt{R_1} + \sqrt{R_2}\right)} \tag{4.11}$$

where R_1 and R_2 are the reflection intensities of the two FBGs. $F(R_1, R_2)$ is the normalized difference in the intensity of the FBG reflections due to the LPFG transmission. The introduction of the $F(R_1, R_2)$ function results in a sensitivity matrix relating the changes in $F(R_1, R_2)$ and in one of the FBG wavelengths to the strain and temperature as

$$\begin{bmatrix} F(R_1, R_2) \\ \Delta\lambda_{b2} \end{bmatrix} = \begin{bmatrix} A & B \\ C & D \end{bmatrix} \begin{bmatrix} \Delta T \\ \Delta\varepsilon \end{bmatrix} \tag{4.12}$$

Similarly, the temperature and strain coefficients of the sensitivity matrix of the hybrid sensor can be determined by measuring the temperature response for one of the FBG shifts and R_1 and R_2 at a constant strain, and vice versa. Figure 4.28 plots the results from the calibration measurement of the

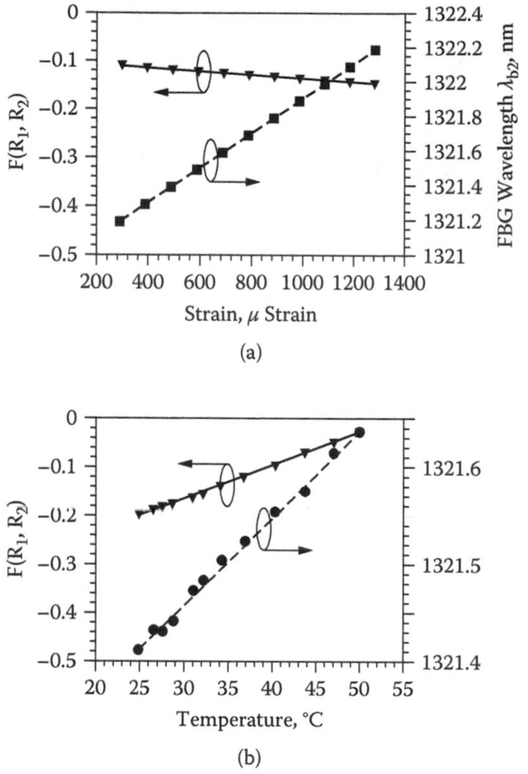

FIGURE 4.28
$F(R_1, R_2)$ and λ_{b2} versus (a) strain when the hybrid sensor was held at a constant temperature of 38°C and (b) temperature while the sensor was held at a constant 59°C. (From Patrick, H. et al., *IEEE Photon. Tech. Lett.*, **8**, 9, pp. 1223–1225, 1996.)

hybrid sensor investigated by Patrick et al., showing good linear responses allowing calculation of the four coefficients from the best-fit technique. Once the sensor has been calibrated, the matrix allows the discrimination between temperature and strain without having to determine λ_{LP} directly.

The LPFG/FBG hybrid sensor was tested under the condition where the strain and temperature changed simultaneously, achieving relatively high dynamic ranges and resolution compared with the single-LPFG sensor. Over the range from 290 to 1270 and 25 to 50°C, the rms deviation was found to be ±9 and ±1.5°C for the applied strain and temperature, respectively.

4.5 Novel Sensing Applications of LPFGS

The unique optical properties of the LPFG underlying the core-cladding, mode-coupling mechanism have given rise to a number of novel sensing concepts and devices. The sensitivity to the external refractive index has not just stimulated intense study of this topic but has also led to optical chemical sensors [107,108]. The birefringence and asymmetry-induced resonance mode-splitting effect has been exploited for transverse strain and structure shape sensing. The study of the waveguide quadratic-dispersion characteristics of LPFG has initiated intensity-measurement-based and dual-resonance sensors offering the attractive advantages of using a low-cost interrogation system [113] and superhigh sensitivity. These new phenomena of LPFG structures are of academic importance but also stimulate interest in the realization of new devices.

4.5.1 Refractive Index Sensing

Patrick and coworkers were among the first to systematically analyze the response of LPFGs to an external refractive index. They delineated the high dependence of LPFG spectral response to the external refractive index on the grating period. The spectral behavior of LPFGs with different periods was examined for an extended external refractive index range from $n = 1$ to 1.7. Figure 4.29 plots the wavelength and strength of a coupled cladding mode of an LPFG with a 275-m period against the applied external refractive index. The centered wavelength exhibited a blue-shift effect when the index changed from 1 to 1.44. When the index was close to 1.45–1.46, the resonance virtually disappeared, as, at this point, the cladding modes were discrete guide modes. With further increases in the index ($n > 1.46$), the LPFG behaves as a "hollow waveguide" and the mode-coupling mechanism switches to a

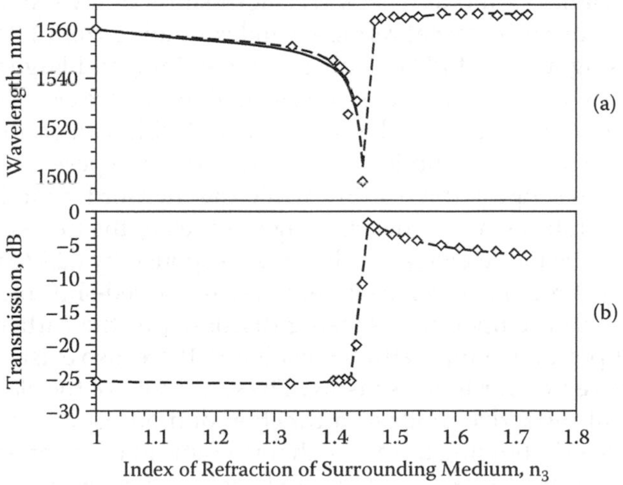

FIGURE 4.29
Experimentally measured (a) center wavelength and (b) transmission of a coupled cladding mode of a 275-mm LPFG against the external refractive index from 1 to 1.7. (From Patrick, H. J. et al., *J. Lightwave Tech.*, **16**, 9, pp. 1606–1611, 1998.)

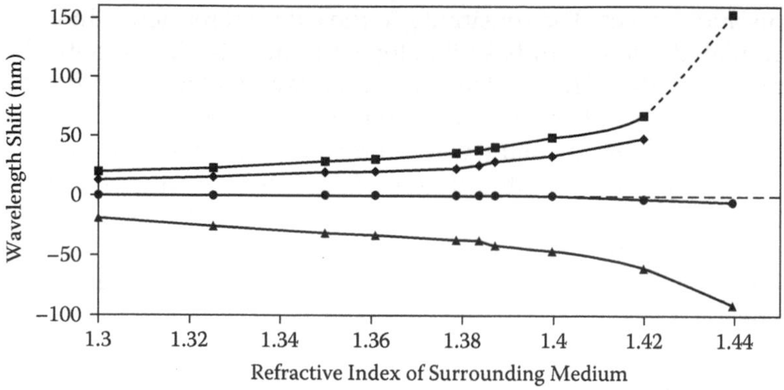

FIGURE 4.30
Experimentally measured external index-induced wavelength shifts of four LPFGs of periods 78, 110, 240, and 550 m.

different regime, where the light couples to the leaky cladding modes [137]. In this regime, the coupled cladding mode reappears with a near-fixed central wavelength, and its strength increases with the index. The spectral response in this regime may be exploited as a refractive index-encoded intensity sensor specifically for applications requiring high index measurement.

It should be pointed out that Patrick's investigation on the spectral response of LPFGs to the external index was mainly in LPFGs with relatively large periods, to the extent that the high-order cladding modes were excluded. Recently, Allsop et al. [108] comprehensively studied the dependence of wavelength shift to the external refractive index on the order of the coupled cladding mode. The wavelength shifts of low- and high-order modes caused by the external refractive index ranging from 1.3 to 1.45 were measured using a set of LPFGs with small, medium, and large periods; the results are shown in Figure 4.30. Two principal trends are clear from this figure. First, the external index-induced wavelength shift can occur in opposite directions: The low-order modes associated with large periods shift toward the shorter wavelengths (blue shift), while the high-order modes from the small-period gratings move toward longer wavelengths (red shift). Second, as the order of the mode increases, the index responsivity ($|\Delta\lambda/\Delta n|$) increases in the blue-shift regime, whereas it decreases in the red-shift regime.

As in the LPFG temperature sensing discussed in the earlier subsection "LPFGs of Superhigh-Temperature Sensitivity," there also exists a super sensitivity mode region for index sensing. The sensitivity of the cladding mode to the external refractive index as a function of mode order has been theoretically modeled, showing a similar characteristic to the temperature sensitivity plot in Figure 4.26(a). The 11th mode of the LPFG with a 78-m period, which falls into the high-sensitivity mode region, has exhibited the highest index sensitivity among the seven investigated gratings. The studies on the dependence of the sensitivity on cladding-mode order offer good guidance for designing appropriate high-performance LPFG sensors.

It is worth noting that altering cladding physical properties can also greatly enhance the index sensitivity. Chiang et al. recently reported the enhancement of the external refractive index sensitivity of an LPFG resulting from a small reduction in the cladding radius via an etching process [138]. They etched a fiber containing an LPFG with a 157-m period using an HF solution. Although the cladding radius was reduced by only ~5% (from 62.5 to 59 m), the index sensitivity of the LPFG increased twofold.

Chemical Sensing. The unique sensitivity to the external refractive index of an LPFG structure has invoked some discussion on using LPFGs for chemical/biochemical sensing. Patrick et al. demonstrated that LPFGs could be used in effectively measuring the concentration of ethylene, a principal compound, in the antifreezer [106]: The 275-m LPFG they had fabricated detected a tiny change (less than 1%) in that concentration. Allsop and coworkers performed the detection of organic aromatic compounds in paraffin by an LPFG sensor with optimized sensitivity. They suggested that an LPFG approach in this application may be more attractive than the conventional high performance liquid chromatography (HPLC) and UV spectroscopy methods, as an LPFG could potentially offer in-situ process control measurements for oil refinery industrial applications [108].

The large inherent temperature and bend cross-sensitivity of an LPFG obviously impose problems for practical chemical sensing applications. Zhang and Sirkis proposed a scheme to compensate the temperature effect on LPFG chemical sensing [107]. In the scheme a configuration concatenating an FBG and an LPFG was employed, enabling elimination of the temperature cross-sensitivity from referencing the temperature using the FBG. Such LPFG/FBG hybrid chemical sensors were used to determine the concentration of sucrose solutions and to distinguish various chemicals and demonstrated to some extent the effectiveness of temperature compensation.

The property of index sensing of LPFG may be exploited for environmental monitoring. So far, few applications have been seen in this area. There is reason to believe that such applications will greatly increase in the future [139].

4.5.2 Optical Load Sensors

It was noted that nonaxial strain could radically change the spectral profile of an LPFG. This property has then been exploited for optical load sensing. The measurement principle is rather different from the conventional wavelength-shifting detection method and is based on measuring the pronounced polarization mode splitting that accompanies the application of the load. With no load applied, the spectral response of an LPFG reveals no polarization dependence due to the degeneracy of orthogonally polarized eigenmodes in the circular symmetric structure. Upon application of the load, however, the degeneracy is broken, leading to the appearance of the two spectrally separated and orthogonally polarized eigenmodes. The spectral separation increases with further loading. Thus, the load can be measured indirectly by measuring the polarization mode splitting.

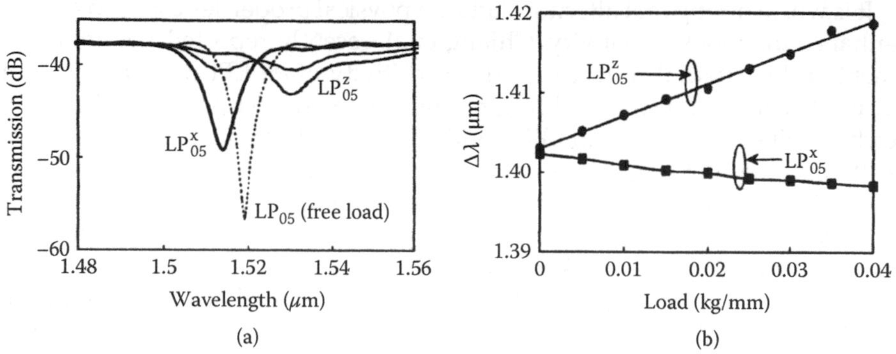

FIGURE 4.31
(a) Transmission spectra of the $LP_{0.5}$ mode with and without loading exited using arbitrary and *x*- and *z*-polarization; (b) wavelength shifts of the two split modes against loading.

The optical load sensor based on measuring the polarization mode splitting of an LPFG was first reported by Liu et al. [109]. A 490-m LPFG fabricated in B/Ge co-doped fiber was found exhibiting resonance mode splitting under a small load, and the splitting was remarkably sensitive to the loading and the polarization state of the probe light. Figure 4.31(a) depicts the transmission spectra of the $LP_{0.5}$ mode of the LPFG when it was free and under loading with the launched light with three different polarization states. The coupled light was found switchable to either polarization mode. A maximum polarization extinction ratio of 11 dB was achieved for *x*-polarization, although only 5 dB was obtained for *z*-polarization. The response of the mode splitting to the load was linear (Figure 4.31(b)) and reached a maximum of ~20 nm under a 4-kg load, giving a transverse strain sensitivity of 250 nm/kg/mm. This value was ~400 times higher than that of the FBG in a similar loading condition [140]. It was proposed that the high transverse strain sensitivity of the LPFG may be combined with its axial strain sensing capability to be developed as a multi-axis sensor.

It is important to point out that under the same loading condition and using the same type of fiber, the transverse strain will induce a similar enhancement to the birefringence to both the FBG and the LPFG fibers. The amplified transverse strain sensitivity by the LPFG originates from the cladding mode-coupling mechanism. The magnitude of the sensitivity amplification is approximately the period size ratio between the LPFG and FBG. This can be clearly seen from the following equation [141]:

$$\frac{\Delta\lambda_{0m}|_{\mathrm{LPG}}}{\Delta\lambda_{B}|_{\mathrm{FBG}}}=\frac{\left(n_{01}^{co,x}-n_{0m}^{cl,x}\right)-\left(n_{01}^{co,y}-n_{0m}^{cl,y}\right)}{2\left(n_{01}^{co,x}-n_{01}^{co,y}\right)}\frac{\Lambda_{\mathrm{LPG}}}{\Lambda_{B}}\sim\frac{1}{2}\frac{\Lambda_{\mathrm{LPG}}}{\Lambda_{B}}\sim 250 \tag{4.13}$$

The value of 500 is a typical example, where $\Lambda_{\mathrm{LPFG}} = 500$ m and $\Lambda_{\mathrm{FBG}} = 1$ m. Clearly, this value is in good agreement with experimental results.

However, it should be stressed that although the overall transverse strain sensitivity of LPFGs is remarkably high, the broad resonance of the LPFGs will limit the load sensing, achieving high resolution. Given a standard LPFG with bandwidth of ~10 nm, a load less than ~0.004 kg may not be detectable using an interrogation system with a resolution of 0.1 nm.

4.5.3 Optic Bend Sensors

LPFGs have also been implemented as optical bend sensors [110,111,142,143]. Two techniques have been demonstrated to measure the bending curvature; one involves detecting the bend-induced cladding-mode resonance wavelength shift [142], and the other is based on measuring the asymmetry-generated cladding-mode resonance splitting [110,111].

Optical Bend Sensing Based on Wavelength Shift Detection

Patrick et al. investigated the possibility of LPFGs being used for structural bend sensing by examining the changes in wavelength and attenuation strength of LPFGs under bending [142]. A correlation between the applied bending curvatures and the resonance wavelength shifts was identified for LPFGs produced in several specific fibers.

Although four different types of fiber were used in their investigation, the reported detail of the results was for an LPFG with a 150-μm period grating written in the photosensitive fiber. Figure 4.32(a) depicts the spectral profiles of this grating when it was free and under bending (the curvature was 2.3/m). Under bending, the resonance peak clearly shifts toward longer wavelength and the attenuation weakens. The measurement also revealed that the bend response was fiber-orientation dependent as the shift and strength of the resonance peak changed differently for different fiber orientations. Figure 4.32(b)

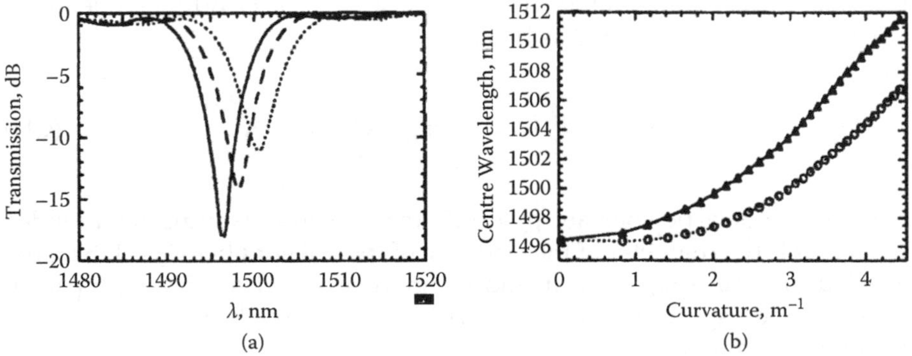

FIGURE 4.32
(a) Spectra of LPFG free and under bending; (b) evolution of LPFG attenuation band center wavelength against curvature for θ = 0° and 180°. (From Patrick, H. J. et al., *Electron. Lett.*, **34**, 18, pp. 1773–1774, 1998.)

shows the overall behavior of this sensor for two orientations ($\theta = 0°$ and $\theta = 180°$) measured over a curvature range from 0 to 4.4/m. Figure 4.32(b) clearly indicates that the wavelength-shift responsivity of the LPFG to the bending is nonlinear: it is less sensitive for small bends, and the sensitivity grows with increasing curvature. The achieved maximum bend sensitivity is ~3.5 nm/m. The authors suggested that the orientation dependence of the bend response could be attributed to the UV exposure-induced concentricity error between the core and cladding modes [144].

The wavelength-shift detection sensing method has obvious limitations associated with nonlinear response, fiber orientation dependence, and decaying of the attenuation band. Given that these limitations are hard to overcome, the implementation of real sensor devices could be highly problematic.

Optical Bending Based on Mode-Splitting Measurement

In comparison, the mode resonance splitting detection technique shows some advantages for bending curvature measurement. Several independent groups of researchers have observed the phenomenon of bend-induced cladding-mode splitting in the LPFGs covering both the normal and special fibers. In the case of Rathje and his colleagues, the LPFGs were produced in the fiber with a large core concentricity error [145]. For Liu [110], Ye [111], and Chen [112] and their coworkers, however, the conventional B/Ge co-doped photosensitive and standard telecom fibers were used.

The mode splitting under bending can be attributed to the symmetry break of the fiber geometry. The imposed curvature in the fiber splits the degeneracy of the cladding mode of the circular fiber, leading to two closely separated cladding modes, one with its field predominantly on the outer side of the cladding relative to the center of the bend, and the other with its field mainly on the inner side of the cladding. These separate cladding modes have correspondingly higher and lower effective index values compared to the value of the straight fiber. The spectral split of the cladding mode to the curvature may be estimated as [146].

$$\Delta\lambda \approx \Delta n_{0m}^{\mathrm{cl}} \cdot \Lambda = \rho_e \cdot n_{0m}^{\mathrm{cl}} \cdot \frac{D_{\mathrm{cl}}}{\sqrt{2}} \cdot K_{\mathrm{bend}} \cdot \Lambda \tag{4.14}$$

For the photoelastic constant $\rho_e \approx 0.22$, the effective cladding mode index $n_{0m}^{\mathrm{cl}} = 1.444$, the diameter of the cladding of the unbent fiber $D_{\mathrm{cl}} = 125$ m, the applied curvature $K_{\mathrm{bend}} = 5/\mathrm{m}$, and the grating period $\Lambda = 500$ m, Eq. (4.14) yields a mode splitting of 70.3 nm, which was in excellent agreement with the experimental results.

The bending experiment conducted by Liu et al. finds that the response of mode splitting to the curvature is near linear over a curvature range from 0 to 5.5/m, but the bend sensitivity is mode-order dependent (Figure 4.33(b)). Figure 4.33(a) shows the spectral profiles of the grating written in the B/Ge

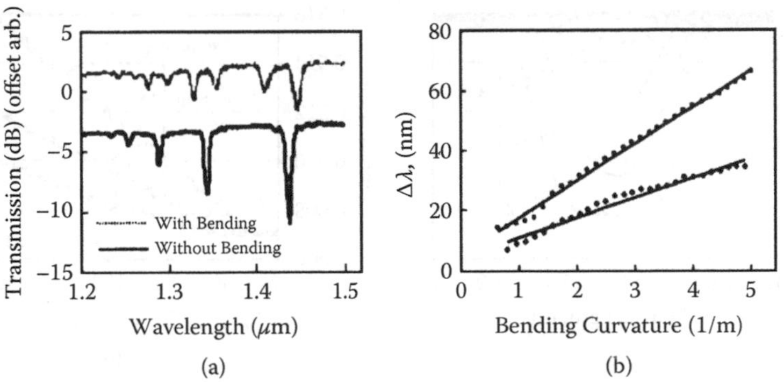

FIGURE 4.33
(a) LPFG spectral profiles with and without bending; (b) resonance mode splitting against applied curvature for LP and LPFG modes.

fiber with and without bending. The splitting of the five coupled cladding modes is very pronounced. Up to 70 nm, splitting was measured for the LP_{05} mode under bend with a curvature of 5/m, giving a bending sensitivity of 14 nm/m^{-1}, which is nearly three times higher than the maximum value demonstrated by the wavelength detection method. The broad spectral resonance of the LPFG also limits high-resolution bend measurement. As a result, a curvature less than 0.4/m may not be detectable.

Ye et al. [111] went further to exploit the possibility of discrimination between temperature and bending curvature by measuring the spectral separation to obtain the curvature information while monitoring the temperature from the central wavelengths of the split peaks. Based on this principle, a bend-temperature sensitivity matrix was generated with the four coefficients determined by performing the bending experiments at several different temperatures. In principle, such a matrix can be used to separate the temperature effect from the bend, providing an independent measurement for each parameter.

4.5.4 Quadratic-Dispersion LPFGS

Intensity-Measurement-Based Sensors

The core-to-cladding mode coupling provides the LPFG structure with versatile waveguide dispersion characteristics. LPFGs can be classified into three dispersion categories: positive linear dispersion ($d\lambda/d\Lambda > 0$); negative linear dispersion ($d\lambda/d\Lambda < 0$); and quadratic dispersion ($d\lambda/d\Lambda = 0$). Most of the demonstrated LPFG-based conventional sensors belong to the first two categories, relying on measuring the external perturbation-induced resonance wavelength shift. In the case of the quadratic-dispersion LPFG, the external perturbation changes only the coupling strength between the core and cladding modes and does not affect the resonant wavelength. In other words, when a quadratic LPFG is subjected to strain or to a change in temperature,

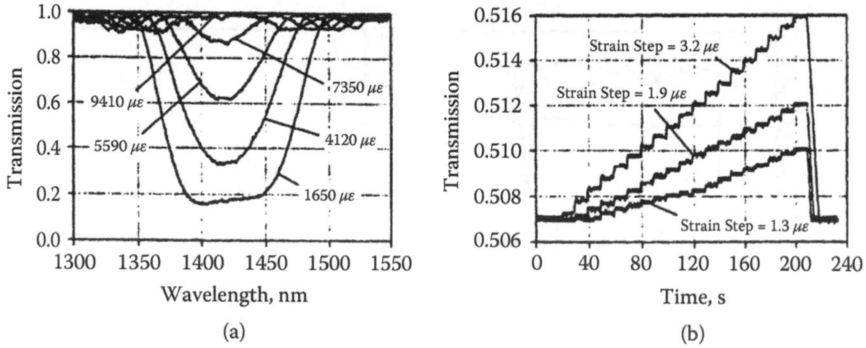

FIGURE 4.34
(a) Transmission spectrum evolution of a quadratic-dispersion LPFG under strain (from 1650 to 9410); (b) change in transmission intensity of a quadratic-dispersion LPFG under increasing strains of three different discrete steps. The strain is increasing linearly with time. (From Grubsky, V. and Feinberg, J., *Opt. Lett.*, **25**, 4, pp. 203–205, 2000.)

its transmission loss changes. This property can be exploited for intensity-measurement strain and temperature sensors. These types of sensors are attractive because there is no need for the costly optical spectrum analyzer or other dispersive devices in their signal interrogation.

The quadratic dispersion LPFGs were first proposed and demonstrated by Grubsky and Feinberg [113]. Two gratings with periods of 50.1 and 69.4 m produced in photosensitive fiber using a 334-nm near-UV laser exhibited a quadratic-dispersion property and were implemented as intensity-measurement sensors. Figure 4.34(a) depicts the transmission profiles for one of these two gratings when it experienced several strains. It is clear from the figure that while the coupling strength falls with each increase in the strain, the central resonant wavelength of the grating barely moves. Figure 4.34(b) shows that the quadratic-dispersion LPFG is capable of detecting the intensity change corresponding to a strain change as small as 1.3. Temperature sensing was also demonstrated using the same grating. According to the authors, a small change of 3% to the transmission was measured, corresponding to a temperature change of 11°C. They speculated on this basis that a 0.04°C temperature change could be detected if a detector of 0.01% resolution was used with this sensor.

Dual-Resonance Sensors

It is also interesting to note that if the phase match condition is slightly off the quadratic-dispersion point, the coupled cladding-mode-resonance splits into two peaks. This phenomenon has been observed and utilized for sensing experiments by Shu and colleagues [147–149]. Since the dual-resonance peaks shift in opposite direction in an external perturbation field, the sensitivity can be significantly improved by measuring the split between the two peaks instead of the shift of one peak. Figure 4.35 presents a picture of

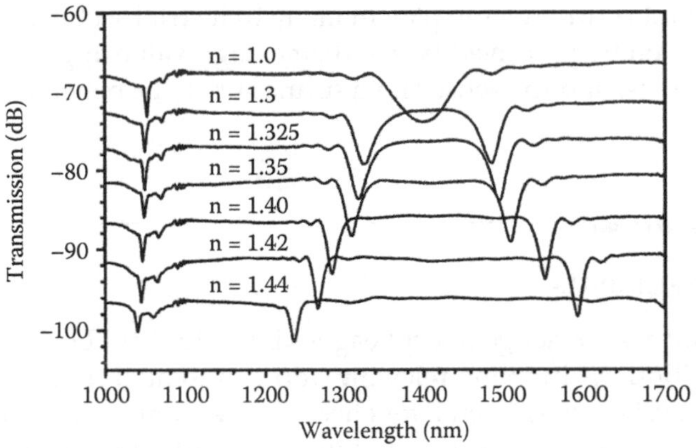

FIGURE 4.35
External refractive index-change-induced spectral evolution of a dual-peak LPFG sensor.

the external refractive index-induced spectral evolution of a dual-resonance LPFG sensor with a relatively short period of 153 m. It can be seen from the figure that if the index sensitivity is just measured from the shift of the blue peak, the value is –33.3 nm/1.42–1.44, whereas if it is measured from the red peak, the value is greater than +107.8 nm/1.42–1.44. (Note: it was not possible to measure the shift to the full range, as it was beyond the light source range.) This is significantly higher than that of the blue peak. However, if the refractive index is measured from the split between the blue and red peaks, the sensitivity will be further dramatically improved.

4.6 Conclusions

In-fiber grating sensing technology has been advanced to the point of commercial reality and offers an attractive approach to quasi-distributed strain measurement and structural health monitoring. Systems employing multiplexed arrays of short, periodic fiber Bragg gratings—surface mounted or embedded in structural materials including carbon fiber composites—have performed successfully in numerous field trials involving a wide variety of structures. But the underlying grating inscription technology continues to advance, and more sophisticated grating structures, offering sensor performance advantages, are now readily producible. It is to be expected that these new grating types will rapidly extend the impressive capabilities already demonstrated. Long-period gratings, rather simpler to produce than Bragg gratings, have distinct sensitivity advantages in some applications, but are, as yet, rather difficult to multiplex. Based on performance, potentially low costs, and architectural flexibility, the increasing emphasis on smart structures suggests that both fiber Bragg gratings and long-period gratings have important future roles to play in many industrial sectors. Grating technology is driven by perceived large-volume, high-value applications in telecommunications, and the sensor field seems sure to be benefit.

Acknowledgments

We gratefully acknowledge our colleagues Dr. Y. Liu and Dr. X. Shu for their direct contributions in many relevant works presented in this chapter. We are also grateful to Dr. L. Everall for checking the manuscript. Our own work on fiber gratings has been supported in major part by the Engineering and Physical Sciences Research Council.

References

1. I. Bennion, J. A. R. Williams, L. Zhang, K. Sugden, and N. J. Doran, UV-written in-fiber Bragg gratings, *Opt. Quantum Electron.*, **28**, pp. 93–135, 1996.
2. R. Kashyap, *Fiber Bragg Gratings*, Academic Press, New York, 1999.
3. A. Othonos and K. Kalli, *Fibre Bragg Gratings: Fundamentals and Applications in Telecommunications and Sensing*, Artech House, London, 1999.
4. K. O. Hill and G. Meltz, Fiber Bragg grating technology fundamentals and overview, *J. Lightwave Tech.*, **15**, pp. 1263–1276, 1997.
5. Y. J. Rao, In-fiber Bragg grating sensors, *Measurement Sci. Tech.*, **8**, pp. 355–375, 1997.
6. A. D. Kerscy, M. A. Davis, H. J. Patrick, M. LeBlanc, K. P. Koo, C. G. Askins, M. A. Putnam, and E. J. Friebeie, Fibre grating sensors, *J. Lightwave Tech.*, **15**, 8, pp. 1442–1463, 1997.
7. G. Meltz, W. W. Morey, and W. H. Glenn, Formation of Bragg gratings in optical fibers by transverse holographic method, *Opt. Lett.*, **14**, pp. 823–825, 1989.
8. K. O. Hill, B. Malo, F. Bilodeau, D. C. Johnson, and J. Albert, Bragg gratings fabricated in monomode photosensitive optical fiber by UV exposure through a phase mask, *Appl. Phys. Lett.*, **62**, pp. 1035–1037, 1993.
9. B. Poumellec, P. Niay, M. Douay, and J. F. Bayon, The UV-induced refractive index grating in Ge:SiO_2 preform: Additional CW experiments and the macroscopic-origin of the change in index, *J. Phys. D. Appl. Phys.*, **29**, pp. 1842–1856, 1996.
10. P. J. Lemaire, R. M. Atkins, V. Mizrahi, and W. A. Reed, High pressure H_2 loading as a technique for achieving ultrahigh UV photosensitivity and thermal sensitivity in GeO_2 doped optical fibers, *Electron. Lett.*, **29**, pp. 1191–1193, 1993.
11. D. Varelas, H. G. Limberger, R. P. Salathe, and C. Kotrotsios, UV-induced mechanical degradation of optical fibers, *Electron. Lett.*, **33**, pp. 804–806, 1997.
12. D. Varelas, H. G. Limberger, and R. P. Salathe, Enhanced mechanical performance of single-mode optical fibers irradiated by a CW UV laser, *Electron. Lett.*, **33**, pp. 704–705, 1997.

13. S. Kannan, J. Z. Y. Guo, and P. J. Lemaire, Thermal stability analysis of UV-induced fiber Bragg gratings, *J. Lightwave Tech.*, **15**, pp. 1478–1483, 1997.
14. S. M. Melle, K. Liu, and M. Measures, A passive wavelength demodulation system for guided-wave Bragg grating sensors, *IEEE Photon. Technol. Lett.*, **4**, 5, pp. 516–518, 1992.
15. M. A. Davis and A. D. Kersey, All-fiber Bragg grating strain-sensor demodulation technique using a wavelength division coupler, *Electron. Lett.*, **30**, pp. 75–77, 1994.
16. Q. Zhang, D. A. Brown, H. Kung, J. E. Townsend, M. Chen, L. J. Reinhart, and T. F. Morse, Use of highly overcoupled couplers to detect shifts in Bragg wavelength, *Electron. Lett.*, **31**, 6, pp. 480–482, 1995.
17. A. B. Lobo Ribeiro, L. A. Ferreira, M. Tsvetkov, and J. L. Santos, All fiber interrogation technique for fiber Bragg sensors using a biconical fiber filter, *Electron. Lett.*, **32**, 4, pp. 382–383, 1996.
18. Y. Liu, L. Zhang, and I. Bennion, Fabricating fiber edge filters with arbitrary spectral response based on tilted chirped grating structures, *Meas. Sci. Tech.*, **10**, pp. L1–L3, 1999.
19. R. W. Falloon, L. Zhang, L. A. Everall, J. A. R. Williams, and I. Bennion, All-fiber optical sensing system: Bragg grating sensor interrogated by a long-period grating, *Meas. Sci. Tech.*, **9**, pp. 1969–1973, 1998.
20. D. A. Jackson, A. B. L. Ribeiro, L. Reekie, and J. L. Archambault, Simple multiplexing scheme for a fiber-optic grating sensor network, *Opt Lett.*, **18**, 14, pp. 1192–1194, 1993.
21. L. A. Ferreira, J. L. Santos, and F. Farahi, Pseudoheterodyne demodulation technique for fiber Bragg grating sensors using two matched gratings, *IEEE Photon. Technol. Lett.*, **9**, 4, pp. 487–489, 1997.
22. R. W. Fallon, L. Zhang, A. Gloag, and I. Bennion, Identical broadband chirped grating interrogation technique for temperature and strain sensing, *Electron. Lett.*, **33**, 8, pp. 705–706, 1997.
23. A. D. Kersey, T. A. Berkoff, and W. W. Morey, Multiplexing fiber Bragg grating strain-sensor system with a fiber Fabry–Perot wavelength filter, *Opt. Lett.*, **18**, 16, pp. 1370–1372, 1993.
24. M. G. Xu, H. Geiger, J. L. Archambault, L. Reekie, and J. P. Dakin, Novel interrogating system for fiber Bragg grating sensors using an acoustic-optic tunable filter, *Electron. Lett.*, **29**, 17, pp. 1510–1511, 1993.
25. T. Gory, P. J. Ellerbrock, R. M. Measures, and J. H. Belk, Active wavelength demodulation of Bragg fiber-optic strain sensor using acoustic-optic tunable filter, *Electron. Lett.*, **31**, pp. 1602–1603, 1995.
26. A. D. Kersey, T. A. Berkoff, and W. W. Morey, High-resolution fiber-grating based strain sensor with interferometric wavelength-shift detection, *Electron. Lett.*, **28**, 3, pp. 236–238, 1992.
27. A. D. Kersey, T. A. Berkoff, and W. W. Morey, Fiber-grating based strain sensor with phase sensitive detection. *Proc. 1st European Conf. Smart Structures and Materials*, Glasgow, pp. 61–67, 1992.
28. D. A. Jackson, A. D. Kersey, and M. Corke, Pseudo-heterodyne detection scheme for optical interferometers, *Electron. Lett.*, **18**, pp. 1081–1083, 1982.
29. C. G. Askins, M. A. Putman, and E. J. Friebele, Instrumentation for interrogating many-element fiber Bragg grating arrays, *SPIE*, **2444**, pp. 257–266, 1995.
30. Y. Hu, S. Chen, L. Zhang, and I. Bennion, Multiplexing Bragg gratings using combined wavelength and spatial division techniques with digital resolution enhancement, *Electron. Lett.*, **32**, 23, pp. 1973–1975, 1997.

31. S. Chen, Y. Hu, L. Zhang, and I. Bennion, Digital spatial and wavelength domain multiplexing of fiber Bragg grating based sensors, *Proc. Int. Conf. Optical Fiber Sensors (OFS'12)*, Williamsburg, VA, pp. 448–451, Oct. 1997.
32. M. L. Dennis, M. A. Putnam, J. U. King, T.-E. Tsai, I. N. Duling III, and E. J. Friebele, Grating sensor array demodulation by use of a passively mode-locked fiber laser, *Opt. Lett.*, **22**, 17, pp. 1362–1364, 1997.
33. M. A. Davis and A. D. Kersey, Fiber Fourier transform spectrometer for decoding Bragg Grating sensors, *Proc. Int. Conf. Optical Fiber Sensors (OFS'10)*, Glasgow, pp. 167–170, 1994.
34. A. D. Kersey, A. Dandridge, A. B. Tveten, and T. G. Giallorenzi, Single-mode fiber Fourier transformer spectrometer, *Electron. Lett.*, **21**, 11, pp. 463–464, 1985.
35. D. A. Flavin, R. McBride, and J. D. C. Jones, Short optical path scan interferometric interrogation of a fiber Bragg grating embedded in a composite, *Electron. Lett.*, **33**, pp. 319–321, 1997.
36. K. B. Rochford and S. D. Dyer, Demultiplexing of interferometrically integrated fiber Bragg grating sensors using Hilbert transform processing, *J. Lightwave Technol.*, **17**, 5, pp. 831–836, 1999.
37. A. D. Kersey, T. A. Berkoff, and W. W. Morey, Fiber-optic Bragg grating strain sensor with drifting-compensated high-resolution interferometric wavelength-shift detection, *Opt. Lett.*, **18**, pp. 72–74, 1993.
38. M. G. Xu, J. L. Archambault, L. Reekie, and J. P. Dakin, Thermally compensated bending gauge using surface-mounted fiber gratings, *Int. J. Optoelectron.*, **9**, pp. 281–283, 1994.
39. G. W. Yoffe, P. A. Krug, F. Ouellette, and D. A. Thorncraft, Passive temperature-compensating package for optical fiber gratings, *Appl. Opt.*, **34**, 30, pp. 6859–6861, 1995.
40. M. G. Xu, J. L. Archambault, L. Reekie, and J. P. Dakin, Discrimination between strain and temperature effects using dual-wavelength fiber grating sensors, *Electron. Lett.*, **30**, 13, pp. 1085–1087, 1994.
41. S. E. Kanellopoulos, V. A. Handerek, and A. J. Rogers, Simultaneous strain and temperature sensing with photogenerated in-fiber gratings, *Opt. Lett.*, **20**, 3, pp. 333–335, 1995.
42. B. Guan, H. Tam, X. Tao, and X. Dong, Simultaneous strain and temperature measurement using a superstructure fiber Bragg grating, *IEEE Photon. Technol. Lett.*, **12**, 6, pp. 675–677, 2000.
43. H. J. Patrick, G. M. Williams, A. D. Kersey, J. R. Pedrazzani, and A. M. Vengsarkar, Hybrid fiber Bragg grating/long period fiber grating sensor for strain/temperature discrimination, *IEEE Photon. Technol. Lett.*, **8**, 9, pp. 1223–1225, 1996.
44. S. W. James, M. L. Dockney, and R. P. Tatam, Simultaneous independent temperature and strain measurement using in-fiber Bragg grating sensors, *Electron. Lett.*, **32**, 12, pp. 1133–1134, 1996.
45. M. G. Xu, L. Dong, L. Reekie, J. A. Tucknott, and J. L. Cruz, Temperature-independent strain sensor using a chirped Bragg grating in a tapered optical fiber, *Electron. Lett.*, **31**, 10, pp. 823–825.
46. K. Kalli, G. Brady, D. J. Webb, D. A. Jackson, L. Zhang, and I. Bennion, Possible approach for simultaneous measurement of strain and temperature with second harmonics in a fiber Bragg grating sensor, *Int. Conf. Optical Fiber Sensors (OFS'10)*, Postdeadline paper, Glasgow, 1994.

47. M. Sudo, M. Nakai, S. Suzaki, A. Wada, and R. Yamauchi, Simultaneous measurement of temperature and strain using PANDA fiber grating, *Proc. Int. Conf. Optical Fiber Sensors (OFS'12)*, Williamsburg, VA, pp. 170–173, 1997.
48. M. G. Xu, L. Reekie, Y. T. Chow, and J. P. Dakin, Optical in-fiber grating high pressure sensor, *Electron. Lett.*, **29**, 4, pp. 398–399, 1993.
49. W. W. Morey, G. Meltz, and J. M. Welss, Evaluation of a fiber Bragg grating hydrostatic pressure sensor, *Int. Conf. Optical Fiber Sensors (OFS'8), Postdeadline paper PD-4*, Monterey, CA, 1992.
50. M. LeBlanc, S. T. Vohra, T. E. Tsai, and E. J. Friebele, Transverse load sensing by use of pi-phase-shifted fiber Bragg gratings, *Opt. Lett.*, **24**, 16, pp. 1091–1093, 1999.
51. M. J. Gander, W. N. MacPheson, R. McBride, J. D. Jones, L. Zhang, I. Bennion, P. M. Blanchard, J. G. Burnett, and A. H. Greenaway, Bend measurement using Bragg gratings in multicore fiber, *Electron. Lett.*, **36**, 2, pp. 120–121, 2000.
52. M. W. Hathaway, N. E. Fisher, D. J. Webb, C. N. Pannell, D. A. Jackson, L. R. Gavrilov, J. W. Hand, L. Zhang, and I. Bennion, Combined ultrasound and temperature sensor using a fiber Bragg grating, *Opt. Comm.*, **171**, pp. 225–231, 1999.
53. T. A. Berkoff and A. D. Kersey, Experimental demonstration of a fiber Bragg grating accelerometer, *IEEE Photon. Technol. Lett.*, **8**, 12, pp. 1677–1679, 1996.
54. Y. Ogawa, J. I. Iwasaki, and K. Nakamuta, A multiplexing load monitoring system of power transmission lines using fiber Bragg grating, *Proc. Int. Conf. Optical Fiber Sensors (OFS'12)*, Williamsburg, VA, pp. 468–471, Oct. 1997.
55. P. J. Henderson, N. E. Fisher, and D. A. Jackson, Current metering using fiber-grating based interrogation of a conventional current transformer, *Proc. Int. Conf. Optical Fiber Sensors (OFS'12)*, Williamsburg, VA, pp. 186–189, Oct. 1997.
56. A. D. Kersey and M. J. Marrone, Fiber Bragg grating high-magnetic-field probe, *Proc. Int. Conf. Optical Fiber Sensors (OFS'10)*, Glasgow, pp. 53–56, Oct. 1994.
57. G. Yu, J. A. R. Williams, W. Zhang, and L. Zhang, Bragg grating based magnetic field sensor, *Proc. PREP'99*, Manchester, UK, pp. 179–186, Jan. 5–7, 1999.
58. Y. Yang, Y. T. Teng, J. S. Sirkis, B. A. Childers, J. P. Moore, and L. D. Melvin, Characterization of fiber Bragg grating based Palladium tube hydrogen sensors, *Proc. Conf. Smart Structures and Materials*, California, March 1–4, 1999.
59. G. Meltz, S. J. Hewlett, and J. D. Love, Fibre grating evanescent-wave sensors, *Proc. SPIE*, **2836**, pp. 342–350, 1996.
60. W. Erke, K. Usbeck, V. Hagemann, R. Mueller, and R. Willsch, Chemical Bragg grating sensor network basing on side-polished optical fiber, *Proc. SPIE*, **3555**, pp. 457–466, 1998.
61. M. J. Gander, G. M. H. Flockhart, R. McBride, W. N. MacPherson, J. D. C. Jones, L. Zhang, I. Bennion, P. M. Blanchard, J. G. Burnett, and A. H. Greenaway, Bend measurement using Bragg gratings in multicore fiber, *Proc. Int. Conf. Optical Fiber Sensors (OPS'14)*, Venice, pp. 720–723, Oct. 2000.
62. K. H. Wanser, K. F. Voss, and A. D. Kersey, *Proc. 10th OFS*, Glasgow, pp. 265–268, Oct. 11–13, 1994.
63. G. P. Brady, S. Hope, A. B. Lobo Ribeiro, D. J. Webb, L. Reckie, J. L. Archambaut, and D. A. Jackson, Bragg grating temperature and strain sensors, *Proc. Int. Conf. Optical Fiber Sensors (OFS'10)*, Glasgow, pp. 510–513, Oct. 1994.
64. M. A. Davis and A. D. Kersey, Matched-filter interrogation technique for fiber Bragg grating arrays, *Electron. Lett.*, **31**, 10, pp. 822–823, 1995.
65. T. A. Berkoff and A. D. Kersey, Fiber Bragg grating array sensor system using a bandpass wavelength division multiplexer and interferometric detection, *IEEE Photon. Technol. Lett.*, **8**, 11, pp. 1522–1524, 1996.

66. A. T. Alavie, S. E. Karr, A. Othonos, and R. M. Measures, A multiplexed Bragg grating fiber laser sensor system, *IEEE Photon. Technol. Lett.*, **5**, 9, pp. 1112–1114, 1993.
67. G. A. Ball, W. W. Morey, and P. K. Cheo, Single- and multiple-point fiber laser sensors, *IEEE Photon. Technol. Lett.*, **5**, pp. 263–266, 1993.
68. R. S. Weis, A. D. Kersey, and T. A. Berkoff, A four-element fiber grating sensor array with phase-sensitive detection, *IEEE Photon. Technol. Lett.*, **6**, 12, pp. 1469–1472, 1994.
69. T. A. Berkoff and A. D. Kersey, Eight element time-division multiplexing fiber grating sensor array with integrated-optic wavelength discriminator, *Proc. 2nd European Conf. Smart Structures and Materials*, Glasgow, pp. 350–353, 1994.
70. T. A. Berkoff, M. A. Davis, D. G. Bellemore, A. D. Kersey, G. M. Williams, and M. A. Putnam, Hybrid time and wavelength division multiplexed fiber grating array, *Proc. SPIE*, **2444**, pp. 288–295, 1995.
71. K. P. Koo, A. B. Tveten, and S. T. Vohra, Dense wavelength division multiplexing of fiber Bragg grating sensors using CDMA, *Electron. Lett.*, **35**, 2, pp. 165–167, 1999.
72. L. Zhang, Y. Liu, J. A. R. Williams, and I. Bennion, Enhanced FBG strain sensing multiplexing capacity using combination of intensity and wavelength dual-coding technique, *IEEE Photon. Technol. Lett.*, **11**, 12, pp. 1638–1640, 1999.
73. M. A. Davis, D. G. Bellemore, M. A. Putnam, and A. D. Kersey, Interrogation of 60 fiber Bragg grating sensors with microstrain resolution capability, *Electron. Lett.*, **32**, 15, pp. 1393–1394, 1996.
74. Y. J. Rao, K. Kalli, G. Brady, D. J. Webb, D. A. Jackson, L. Zhang, and I. Bennion, Spatial multiplexed fiber-optic Bragg grating strain and temperature sensor system based on interferometric wavelength-shift detection, *Electron. Lett.*, **31**, pp. 1009–1010, 1995.
75. Y. J. Rao, D. A. Jackson, L. Zhang, and I. Bennion, Strain sensing of modern composite materials with a spatial/wavelength-division multiplexed fiber grating network, *Opt. Lett.*, **21**, pp. 683–685, 1996.
76. M. Froggatt, Distributed measurement of the complex modulation of a photoinduced Bragg grating in an optical fiber, *Appl. Opt.*, **35**, 25, pp. 5162–5164, 1996.
77. K. Wood, T. Brown, R. Rogowski, and B. Jensen, Fiber optic sensors for health monitoring of morphing airframes. I. Bragg grating strain and temperature sensor, *Smart Mater. Struct.*, **9**, pp. 163–169, 2000.
78. D. Uttamchandani, Fiber-optic sensors and smart structures: Developments and prospects, *Electronic & Commun. Engineering J.*, pp. 237–246, Oct. 1994.
79. S. S. Roberts and R. Davidson, Mechanical properties of composite materials containing embedded fiber optic sensors, *Proc. SPIE*, **1588**, Boston, pp. 326–341, Sept. 1991.
80. J. R. Dunphy, G. Meltz, F. P. Lamm, and W. W. Morey, Multi-function, distributed optical-fiber sensor for composite cure and response monitoring, *Proc. SPIE*, **1370**, pp. 116–118, 1990.
81. R. Maaskant, T. Alavie, R. M. Measures, M. Ohn, S. Karr, D. Glennie, C. Wade, G. Tadros, and S. H. Rizkalla, Fiber-optic Bragg grating sensor network installed in a concrete road bridge, *Proc. SPIE*, **2191**, pp. 13–18, 1994.
82. R. Maaskant, T. Alavie, R. M. Measures, G. Tadros, S. H. Rizkalla, and A. Guha-Thakurta, Fiber-optic Bragg grating sensors for bridge monitoring, *Cement & Concrete Composites*, **19**, pp. 21–33, 1997.
83. J. Meissner, W. Novak, V. Slowik, and T. Klink, Strain monitoring at a prestressed concrete bridge, *Proc. Int. Conf. Optical Fiber Sensors (OFS'12)*, Williamsburg, VA, pp. 408–411, Oct. 1997.

84. C. G. Groves-Kirkby, F. J. Wilson, I. Bennion, L. Zhang, P. Henderson, D. A. Jackson, D. J. Webb, I. J. Knight, J. Latchem, and R. W. Woodward, Structural health monitoring of concrete bridges using fiber Bragg grating sensors, *IOP Meeting an Engineering Applications of Optical Diagnostic Techniques*, Coventry, UK, 1999.
85. R. S. Weis and B. D. Beadle, MWD telemetry system for coiled-tubing drill using optical fiber grating modulators downhole, *Proc. Int. Conf. Optical Fiber Sensors (OFS'12)*, Williamsburg, VA, pp. 416–419, Oct. 1997.
86. V. Dewynter-Marty, S. Rougeault, P. Ferdinand, D. Chauvel, E. Toppani, M. Leygonie, B. Jarret, and P. Fenaux, Concrete strain measurements and crack detection with surface-mounted and embedded Bragg grating extensometers, *Proc. Int. Conf. Optical Fiber Sensors (OFS'12)*, Williamsburg, VA, pp. 600–603, Oct. 1997.
87. I. J. Read, P. D. Foote, D. Roberts, L. Zhang, I. Bennion, and M. Carr, Smart carbon-fiber mast for a super-yacht, *Proc. 2nd Intl. Workshop Structural Health Monitoring*, Stanford, CA, pp. 276–286, 1999.
88. E. J. Friebele, C. G. Askins, M. A. Putnam, A. A. Fosha, Jr., J. Florio, Jr., R. P. Donti, and R. G. Blosser, Distributed strain sensing with fiber Bragg grating arrays embedded in CRTM composites, *Electron. Lett.*, **30**, pp. 1783–1784, 1994.
89. M. J. O'Dwyer, G. M. Maistros, S. W. James, R. P. Tatam, and I. K. Patridge, Relating the state of cure to the real-time internal strain development in a curing composite using in-fiber Bragg gratings and dielectric sensors, *Meas. Sci. Technol.*, **9**, pp. 1153–1158, 1998.
90. N. D. Dykes, Mechanical and sensing performance of embedded in-fiber Bragg grating devices during impact testing of carbon fiber reinforced polymer composite, *Workshop on Smart Systems Demonstrators: Concepts and Applications*, Harrogate, UK, pp. 168–175, 1998.
91. A. M. Vengsarkar, P. J. Lemaire, J. B. Judkins, V. Bhatia, T. Erdogan, and J. E. Spie, Long-period fiber gratings as band-rejection filters, *J. Lightwave Tech.*, **14**, 1, p. 5865, 1996.
92. A. M. Vengsarkar, J. R. Pedrazzani, J. B. Judkins, P. J. Lemaire, N. S. Bergano, and C. R. Davidson, Long-period fiber-grating-based gain equalizers, *Opt. Lett.*, **21**, p. 336, 1996.
93. P. F. Wysocki, J. B. Judkins, R. P. Espindola, M. Andrejco, and A. M. Vengsarkar, Broad band erbium doped fiber amplifier flattened beyond 40 nm using long-period grating filter, *IEEE Photon. Technol. Lett.*, **9**, p. 1343, 1997.
94. J. R. Qian and H. F. Chen, Gain flattening fiber filters using phase-shifted long period fiber gratings, *Electron. Lett.*, **34**, p. 1132, 1998.
95. X. J. Gu, Wavelength-division multiplexing isolation fiber filter and light source using cascaded long-period fiber gratings, *Opt. Lett.*, **23**, 7, pp. 509–510, 1998.
96. B. H. Lee and J. Nishii, Notch filters based on cascaded multiple long-period fiber gratings, *Electron. Lett.*, **34**, pp. 1872–1873, 1998.
97. B. Ortega, L. Dong, W. F. Liu, J. P. de Sandro, L. Reekie, S. I. Tsypina, V. N. Bagratashvili, and R. I. Laming, High-performance optical fiber polarizers based on long-period gratings in birefringent optical fibers, *IEEE Photon. Technol. Lett.*, **9**, 10, pp. 1370–1372, 1997.
98. A. S. Kurkov, M. Douay, O. Duhem, B. Leleu, J. F. Henninot, J. F. Bayon, and L. Rivollan, Long-period grating as a wavelength selective polarisation element, *Electron, Lett.*, **33**, 7, pp. 616–617, 1997.

99. K. S. Lee and T. Erdogan, Transmissive tilted gratings for LP_{01}-to-LP_{11} mode coupling, *IEEE Photon. Technol. Lett.*, **11**, 10, pp. 1286–1288, 1999.
100. N. H. Ky, H. G. Limberger, R. P. Salathe, and F. Cochet, Efficient broadband intracore grating LP_{01}–LP_{02} mode converters for chromatic-dispersion compensation, *Opt. Lett.*, **23**, pp. 445–447, 1998.
101. V. Bhatia and A. M. Vengsarkar, Optical fiber long-period grating sensors, *Opt. Lett.*, **21**, 9, pp. 692–694, 1996.
102. Y. G. Han, C. S. Kim, K. Oh, U. C. Pack, and Y. Chung, Performance enhancement of strain and temperature sensors using long period fiber grating, *Proc. Int. Conf. Optical Fiber Sensors (OFS'13)*, Kyongju, Korea, pp. 58–60, Apr. 1999.
103. L. Zhang, Y. Liu, L. Everall, J. A. R. Williams, and I. Bennion, Design and realisation of long-period grating devices in conventional and hi-bi fibers and their novel applications as fiber-optic load sensors, *IEEE J. Selec. Topic. Quantum Electron.* **5**, 5, p. 1373, 1999.
104. B. H. Lee, Y. Liu, S. B. Lee, and S. S. Choi, Displacement of the resonant peaks of a long-period fiber grating induced by a change of ambient refractive index, *Opt. Lett.*, **22**, p. 1769, 1997.
105. J. Meissner and W. Nowak, Strain monitoring at a prestressed concrete bridge, *Proc. Int. Conf. Optical Fiber Sensors (OFS'12)*, Williamsburg, VA, pp. 408–411, Oct. 1997.
106. H. J. Patrick, A. D. Kersey, and F. Bucholtz, Analysis of the response of long period fiber gratings to external index of refraction, *J. Lightwave Tech.*, **16**, 9, pp. 1606–1611, 1998.
107. Z. Zhang and J. S. Sirkis, Temperature-compensated long period grating chemical sensor, *Proc. Int. Conf. Optical Fiber Sensors (OFS'12)*, Williamsburg, VA, pp. 294–297, Oct. 1997.
108. T. Allsop, L. Zhang, and I. Bennion, Detection of organic aromatic compounds in paraffin by a long period fiber grating optical sensor with optimised sensitivity, *Opt. Comm.*, **191**, 3–6, pp. 181–190, 2001.
109. Y. Liu, L. Zhang, and I. Bennion, Fiber optic load sensors with high transverse strain sensitivity based on long-period grating in B/Ge co-doped fiber, *Electron. Lett.*, **35**, 8, pp. 661–663, 1999.
110. Y. Liu, L. Zhang, J. A. R. Williams, and I. Bennion, Optical bend sensor based on measurement of resonance mode splitting of long-period fiber grating, *IEEE Photon. Tech. Lett.*, **12**, 5, pp. 531–533, 2000.
111. C. C. Ye, S. W. James, and R. P. Tatam, Simultaneous temperature and bend sensing with long-period fiber gratings, *Opt. Lett.*, **25**, 14, pp. 1007–1009, 2000.
112. Z. Chen, K. S. Chiang, M. N. Ng, Y. M. Chan, and H. Ke, Bend long-period fiber gratings for sensor applications, *Proc. SPIE on Advanced Photonic Sensors and Applications*, Singapore, 1999.
113. V. Grubsky and J. Feinberg, Long-period fiber gratings with variable coupling for real-time sensing applications, *Opt. Lett.*, **25**, 4, pp. 203–205, 2000.
114. Y. Liu, J. A. R. Williams, L. Zhang, and I. Bennion, Phase shifted and cascaded long-period gratings, *Opt. Comm.*, **164**, pp. 27–31, 1999.
115. J. R. Qian and H. F. Chen, Gain flattening fiber filters using phase-shifted long-period fiber gratings, *Electron. Lett.*, **34**, p. 1132, 1998.
116. E. M. Dianov, S. A. Vasiliev, A. S. Kurkov, O. L. Medvedkov, and V. N. Protopopov, In-fiber Mach–Zehnder interferometer based on a pair of long-period gratings, *Proc. European Conf. Optical Communication (ECOC'96)*, pp. 65–68, 1996.

117. L. Tallone, L. Boschis, L. Cognolato, E. Emelli, E. Riccardi, and O. Rossotto, Narrow-band rejection filters through fabrication of in-series long-period gratings, *Proc. Int. Conf. Optical Fiber Communications (OFC'97)*, Dallas, TX, p. 175, 1997.
118. D. S. Starodubov, V. Grubsk, A. Skorucak, I. Feinberg, J. X. Cai, K. M. Feng, and A. E. Willner, Novel fiber amplitude modulators for dynamic channel power equalization in WDM systems, *Int. Conf. Optical Fiber Communications (OFC'98)*, Postdeadline paper PD8-1, PD8-4, 1998.
119. S. Ramachandran, J. Wagener, R. Espindola, and T. Strasser, Effects of chirp in long period gratings, *Technical Digest of Int. on Bragg Gratings. Photosensitivity, and Poling in Glass Waveguides*, Stuart, Florida, pp. 286–288, 1999.
120. E. M. Dianov, S. A. Vasiliev, A. S. Kurkov, O. I. Medvedkov, and V. N. Protopopov, In-fiber Mach–Zehnder interferometer based on a pair of long-period gratings, *Proc. European Conf. Optical Communication (ECOC'96)*, **1**, pp. 65–68, 1996.
121. O. Duhem, J. F. Henninot, and M. Douay, Study of in fiber Mach–Zehnder interferometer based on two spaced 3-dB long period gratings surrounded by a refractive index higher than that of silica, *Opt. Comm.*, **180**, pp. 255–262, 2000.
122. B. H. Lee and J. Nishii, Self-interference of long-period fiber grating and its application as temperature sensor, *Electron. Lett.*, **34**, 21, pp. 2059–2060, 1998.
123. L. A. Everall, R. W. Fallon, J. A. R. Williams, L. Zhang, and I. Bennion, Flexible fabrication of long period in-fiber gratings, *Proc. Conf. Lasers and Electro-Optics (CLEO'98)*, p. 513, 1998.
124. D. D. Davis, T. K. Gaylord, E. N. Glytsis, and S. C. Mettler, CO_2 laser-induced long-period fiber gratings: Special characteristics, cladding modes and polarisation independence, *Electron. Lett.*, **34**, p. 1416, 1998.
125. L. Drozin, P. Y. Fonjallaz, and L. Stensland, Long-period fiber gratings written by CO_2 exposure of H_2-loaded, standard fibers, *Electron. Lett.*, **36**, pp. 742–744, 2000.
126. S. Y. Liu, H. Y. Tam, and M. S. Demokan, Low-cost microlens array for long-period grating fabrication, *Electron. Lett.*, **35**, pp. 79–81, 1999.
127. S. G. Kosinski and A. M. Vengsarkar, Splicer-based long-period fiber gratings, *Proc. Int. Conf. on Optical Fiber Communication (OFC'98)*, pp. 278–279, 1998.
128. I. K. Hwang, S. H, Yun, and B. Y. Kim, Long-period fiber gratings based on periodic microbends, *Opt. Lett.*, **24**, pp. 1263–1265, 1999.
129. T. W. MacDougall, S. Pilevar, C. W. Haggans, and M. A. Jackson, Generalized expression for the growth of long period gratings, *IEEE Photon. Tech. Lett.*, **10**, 10, pp. 1449–1551, 1998.
130. X. Shu, T. Allsop, B. Gwandu, L. Zhang, and I. Bennion, High temperature sensitivity of long-period gratings in B/Ge co-doped fiber, accepted by *IEEE Photon. Techn. Lett.*, 13, 818–820, 2001.
131. J. B. Judkins, J. R. Pedrazzani, D. J. DiGiovanni, and A. M. Vengsarkar, Temperature-insensitive long-period gratings, *Int. Conf. Optical Fiber Communications (OFC'96)*, Postdeadline paper PDI, 1996.
132. V. Bhatia, M. K. Burford, N. Zabaronick, K. A. Murphy, and R. O. Claus, Strain and refractive index sensors using temperature-insensitive long-period gratings, *Proc. Int. Conf. Optical Fiber Sensors (OFS'11)*, Sapporo, Hokkaido, Japan, pp. 2–5, May 1997.
133. A. A. Abramov, A. Hale, R. S. Windeler, and T. A. Strasser, Widely tunable long-period fiber gratings, *Electron. Lett.*, **35**, pp. 81–82, 1999.

134. A. A. Abramov, B. J. Eggleton, J. A. Rogers, R. P. Espindola, A. Hale, R. S. Windeler, and T. A. Strasser, Electrically tunable efficient broad-band fiber filter, *IEEE Photon. Techn. Lett.*, **11**, pp. 445–447, 1999.
135. V. Bhatia, D. Campbell, R. O. Claus, and A. M. Vengsarkar, Simultaneous strain and temperature measurement with long-period gratings, *Opt. Lett.*, **22**, 9, pp. 648–650, 1997.
136. H. Patrick, G. M. Williams, A. D. Kersey, J. R. Pedrazzani, and A. M. Vengsarkar, Hybrid fiber Bragg grating/long period fiber grating sensor for strain/temperature discrimination, *IEEE Photon. Tech. Lett.*, **8**, 9, pp. 1223–1225, 1996.
137. D. B. Stegal and T. Erdogen, Long-period fiber-grating devices based on leaky cladding mode coupling, *Proc. Int. Conf. Optical Fiber Sensors (OFS'12)*, Williamsburg, VA, pp. 16–18, Oct. 1997.
138. K. S. Chiang, Y. Liu, M. N. Ng, and X. Dong, Analysis of etched long-period fiber grating and its response to external refractive index, *Electron. Lett.*, **36**, pp. 966–967, 2000.
139. V. Bhatia, D. K. Campbell, T. D. Alberto, G. A. Ten Eyck, D. Sherr, K. A. Murphy, and R. O. Claus, Standard optical fiber long-period gratings with reduced temperature sensitivity for strain and refractive index sensing, *Proc. Int. Conf. Optical Fiber Communications (OFS'97)*, Dallas, TX, pp. 346–347, 1997.
140. L. Bjerkan, K. Johannessen, and X. Guo, Measurement of Bragg grating birefringence due to transverse compressive force, *Proc. Int. Conf. Optical Fiber Sensors (OFS'12)*, Williamsburg, VA, pp. 60–63, Oct. 1997.
141. Y. Liu, Advanced fiber gratings and their applications, Ph.D. thesis, Aston University, UK, 2001.
142. H. J. Patrick, C. C. Chang, and S. T. Vohra, Long period fiber gratings for structural bend sensing, *Electron. Lett.*, **34**, 18, pp. 1773–1774, 1998.
143. H. J. Patrick and S. T. Vohra, Directional shape sensing using bend sensitivity of long period fiber gratings, *Proc. Int. Conf. Optical Fiber Sensors (OFS'13)*, Kyongju, Korea, pp. 561–564, Apr. 1999.
144. A. M. Vengsarkar, Q. Zhang, D. Inniss, W. A. Reed, P. J. Lemaire, and S. G. Kosinski, Birefringence reduction in side-written photoinduced fiber devices by a dual-exposure method, *Opt. Lett.*, **16**, pp. 1260–1262, 1994.
145. J. Rathje, M. Kristensen, and J. Hubner, Effects of core concentricity error on bend direction asymmetry for long-period fiber gratings, *Technical Digest of Int. on Bragg Gratings, Photosensitivity, and Poling in Glass Waveguides*, Stuart, FL, pp. 283–285, 1999.
146. Y. Liu, L. Zhang, J. A. R. Williams, and I. Bennion, Bend sensing by measuring the resonance splitting of long-period fiber gratings, accepted by *Opt. Comm.*, 2001.
147. X. Shu, X. Zhu. Q. Wang, S. Jiang, W. Shi, Z. Huang, and D. Huang, Dual resonant peaks of LP cladding mode in long-period gratings. *Electron. Lett.*, **35**, pp. 649–651, 1999.
148. X. Shu, X. Zhu, S. Jiang, W. Shi, and D. Huang, High sensitivity of dual resonant peaks of long-period fiber grating to surrounding refractive index changes, *Electron. Lett.*, **35**, pp. 1580–1581, 1999.
149. X. Shu and D. X. Huang, Highly sensitive chemical sensor based on the measurement of the separation of dual resonant peaks in a 100-mu m-period fiber grating, *Opt. Comm.*, **171**, pp. 65–69, 1999.

5

Femtosecond Laser-Inscribed Harsh Environment Fiber Bragg Grating Sensors

Shizhuo Yin, Chun Zhan, and Paul B. Ruffin

CONTENTS

5.1 Introduction

With the increasing need for health monitoring in structures such as bridges, tunnels, dams, highways, aircraft wings, and spacecraft fuel tanks, it is critical to develop effective sensor networks, which can measure strain, pressure, and temperature. Fiber Bragg gratings (FBGs) have the advantages of high sensitivity, high spatial resolution, and wavelength division multiplexing capability. Thus, they have been widely investigated as effective sensing elements since their first demonstration in the sensing application by Meltz et al.[1]

Typically, fiber gratings are generated by exposing the ultraviolet (UV)-photosensitive core of a germanium-doped silica core optical fiber to a spatially modulated UV laser beam, which creates permanent refractive index changes in the fiber core. A limitation of the UV-induced fiber Bragg gratings (FBGs) is that the gratings can be erased at high temperatures.

In recent years, the problems associated with the long-term thermal stability of the FBGs in a harsh environment have been largely alleviated by using an ultrashort femtosecond laser. It has been reported that an induced index modulation of 1.9×10^3 was achieved without using any fiber sensitization such as hydrogen loading.[2] The fiber gratings inscribed by femtosecond lasers have annealing characteristics similar to Type II fiber gratings and demonstrate stable operation at temperatures as high as 950°C.[2] Detailed discussions on the fabrication methods and principles can be found in many references.[2–4]

The practical challenges of real-world applications may include (1) decoupling of different sensing parameters such as strain and temperature, (2) the long-term thermal stability of the FBGs in a harsh, high-temperature environment, (3) the knowledge of the direction of the strain or bending (so-called vector sensing), and (4) development of extremely harsh environment sensors with operation temperature higher than 1500°C. Previous references mainly focused on the gratings inscribed in single mode (SM) fibers, which have only one resonant dip in the transmission spectrum. It is difficult to decouple these coexisted effects.

To overcome these limitations, recently we have investigated multiparameter sensing using asymmetric Bragg gratings fabricated in multimode silica fibers, polarization maintaining SM silica fibers and sapphire single-crystal fiber by infrared (IR) femtosecond laser irradiation. Furthermore, we also studied the ultrathin harsh environment FBGs by fabricating the gratings in etched multimode and PM fibers.

In this chapter, we briefly review our recent work on harsh environment sensing using asymmetric Bragg gratings inscribed by IR femtosecond irradiation.

The fibers used for the grating inscription include regular single mode (SM) silica fiber, graded-index multimode (MM) silica fiber, polarization-maintaining (PM) single mode silica fiber, the ultra thin multimode and polarization maintaining silica fibers, and single-crystal sapphire fiber. The outline of this chapter is as follows: The fabrication method and grating writing mechanism are described in Section 5.2. In Sections 5.3, 5.4, and 5.5, detailed discussions about their applications to harsh environment multiparameter sensing such as temperature, bending, pressure, and bio and chemical sensing are provided. In Section 5.6, a real-world test of SM gratings in the Makenna burner is performed and long-term thermal stability tests are also carried out in a high-temperature furnace. Finally, extremely high temperature sensing using higher order-mode rejected sapphire single-crystal fiber is presented. The grating is tested in a high temperature furnace up to 1600°C.

5.2 Fabrication Processes

5.2.1 Writing Mechanism

When ultrashort femtosecond laser beams hit the transmission medium, the energy deposition is initiated by multiphoton absorption rather than having to rely on impurities or defects to start an electron avalanche. The refractive index change in femtosecond-inscribed gratings is widely believed to be initiated by a nonlinear reaction through the multiphoton process, which induces the formation of the localized plasma in the bulk of the material causing densification of the latter.

Although this densification model is widely accepted, there are still other proposed mechanisms. For example, A. Streltsov and N. Borrelli suggest that densification alone cannot account for the entire effect.[4] Besides the densification model, they propose another two possible models: thermal model and color center model. Thermal model is similar to the densification model; color center model suggested that a possible source of the induced refractive-index change is that the effect of radiation is to produce color centers in sufficient numbers and strength to alter the index through a Kramers–Kronig mechanism.[5]

5.2.2 Fabrication of FBGs in Multimode Fiber by Femtosecond Laser Illumination

The fiber grating writing scheme is based on the near-field phase mask method.[2,3] Figure 5.1 illustrates the fiber grating writing scheme. A standard silica multimode fiber (infinicor 300 from Corning) with a core diameter of 62.5 μm and cladding of 125 μm is exposed to laser pulses of 150-fs duration that are generated by a Ti:sapphire amplifier at a wavelength of 800 nm with a pulse repetition rate of 1 kHz. The vertically polarized beam has a diameter of around 10 mm and is focused by a cylindrical lens with a 38-mm

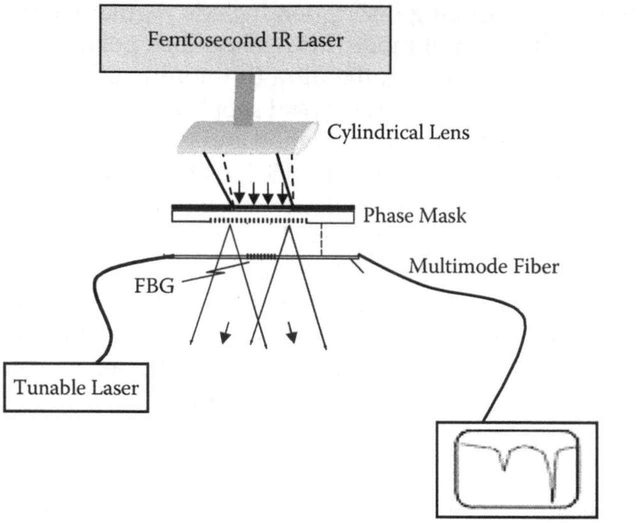

FIGURE 5.1
A schematic diagram of the experimental setup used to fabricate asymmetric Bragg gratings in multimode fiber by an IR femtosecond laser.

focal length. The multimode fiber is placed 200 μm behind a phase mask with a pitch of 2.66 μm. The phase mask is designed for 800 nm and less than 5% of the beam is diffracted into the 0th order. The pulse energy is around 1 mJ/pulse, which is strong enough to write Type II IR gratings.[6] The fiber is placed on a high-precision, three-dimensional translation stage and the beam is focused slightly off the fiber center; in this way, the gratings will be asymmetric to the fiber axis, which turns out to have brought in huge advantages in vector-sensing applications.

Since the multimode fiber has a core diameter that is wider than the focused laser beam, the beam is translated and scanned across the cross-section of the fiber in a step-and-repeat fashion by a piezo-actuator. The total exposure time is around 3 sec. Due to the involvement of multiple-beam interference, gratings with multiple pitches can be generated.[7] The pitch equals 1.33 μm—the result of the two-beam interference between the +1 and –1 diffraction orders, while the pitch equals 2.66 μm is the result of the two-beam interference between the first and the 0th diffraction orders because there is still a few percent of 0th order diffraction beam. Figure 5.2 shows the microscopic image of the grating fabricated in multimode silica fibers, which includes both the grating pitch equal to that of the mask (i.e., 2.66 μm) and half of the mask (1.33 μm). It can be clearly seen that the grating has a pitch of 1.33 μm at the edges of the pattern, while in the middle region the pitch is 2.666 μm, which corresponds to two resonant peaks in the optical spectrum analyzer.

The phase-matching condition, or Bragg condition, of an index modulation for multimode fiber with the period Λ is given mathematically by

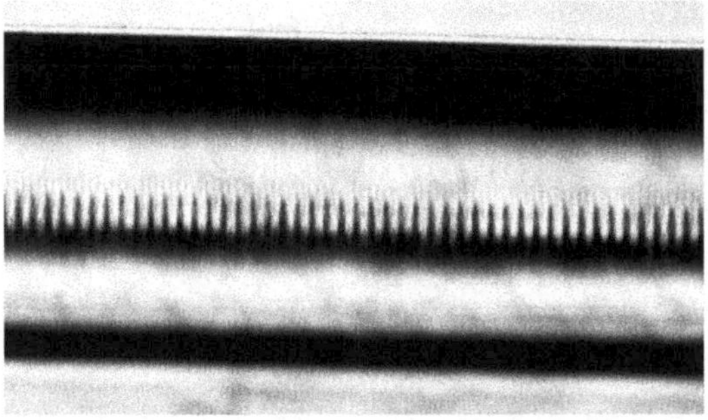

FIGURE 5.2
A microscopic image of an asymmetric FBG fabricated in multimode silica fiber by an IR femtosecond laser with 150-fs, 800-nm laser pulses, and a phase mask.

$$2n_{eff}^{q}\Lambda_{g} = m\lambda_{Bragg}^{q} \tag{5.1}$$

where m denotes the order number, n_{eff}^{q} represents the effective refractive index for the qth-order mode propagating in the multimode fiber, λ_{Bragg}^{q} is the Bragg resonant wavelength for the qth mode, and Λ_g denotes the grating pitch. It is easy to find that the transmission dip at 1320 nm is caused by the third-order Bragg grating resonance with a period of 1.333 µm and the dip at 1580 nm is formed by the fifth-order grating resonance with a period of 2.666 µm. By adjusting the distance between the phase mask and the fiber and optimizing the exposure time, we can control the depths of these two dips. By employing this method, dual wavelength grating that is capable of discriminating the temperature and strain[8] can be easily fabricated. Note that for conventionally UV Bragg grating, two phase masks with different pitches are needed, and multiple exposures are required to achieve this, but by using a femtosecond laser, it is just one exposure and one phase mask.

Because the tunable laser used for the spectral monitoring ranges from 1470 to 1570 nm, in the following, we will focus on the transmission dip resonant at 1558 nm. In addition, to ensure that the multiple dips are not due to the coupling to the cladding modes, we applied the index matching oils on the outer cladding surfaces of the Bragg gratings, which eliminated this type of coupling.

After fabricating the gratings in multimode fibers (MMFs) by femtosecond laser irradiation, the spectra of the gratings are measured by employing a tunable laser light source and an optical spectrum analyzer. It was found that the observed spectra of the Bragg gratings in MMF were very different for different excitation conditions. For example, Figure 5.3 shows a transmission spectrum obtained by a single mode tunable laser with a single mode fiber (SMF) pigtail. This corresponds to single mode exciting condition. Three or four major dips appear and the depths for these dips are around 4 dB. The bandwidth of each dip is around 0.5 nm with a separation of different dips ranging from 0.4 to 0.8 nm; the total bandwidth of all the dips is around 6 nm.

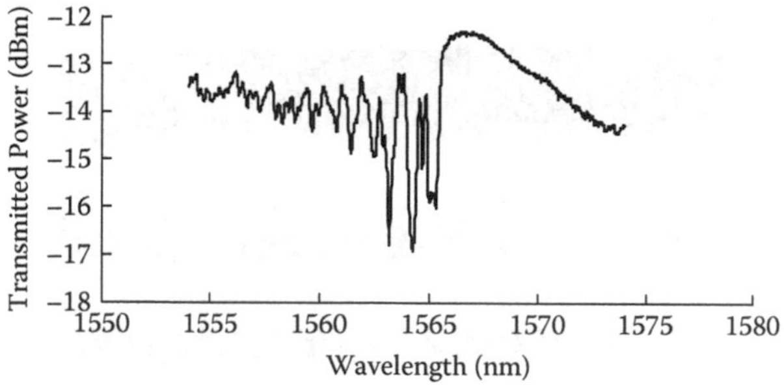

FIGURE 5.3
The measured transmission spectrum of MM FBG obtained by the single mode excitation condition: tunable laser with an SMF pigtail.

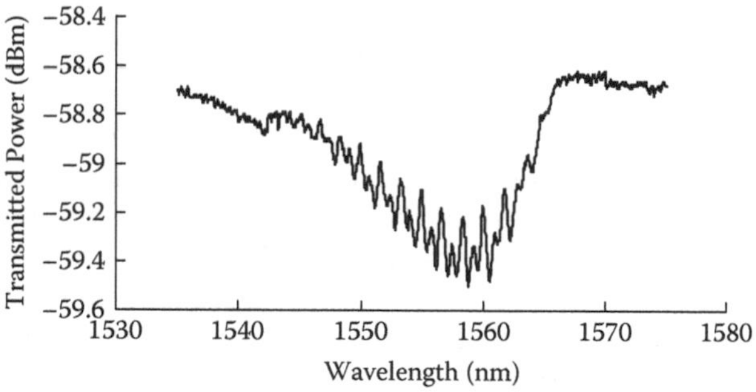

FIGURE 5.4
The measured transmission spectrum of MM FBG obtained by the multimode excitation condition: broadband light source from an optical spectra analyzer.

On the other hand, using the broadband light source from the optical spectra analyzer (OSA) as a light source, the transmission spectrum of the same grating was measured as shown in Figure 5.4. More than 10 dips appeared. The separation between these transmission dips is about 0.5 nm and the total bandwidth of these dips is 20 nm, but the average depth of each dip is less than 0.5 dB. These results show that the measured spectra have a large dependence on the conditions of excitation of modes. Measurements using a tunable laser (HP 8168E) with an SMF pigtail correspond to single mode excitation, and those using the broadband light source on OSA correspond to multimode excitation. Single mode excitation turns out to be very useful in multiparameter sensing because there are fewer dips and depth of each dip is deeper.

5.2.3 Fabrication of FBGs in PM Fiber by Femtosecond Laser Illumination

Following similar procedures to those described in Section 5.2.2, FBGs are also fabricated in a PM fiber (a bow-tie Hi-Bi (high birefringence) PM fiber

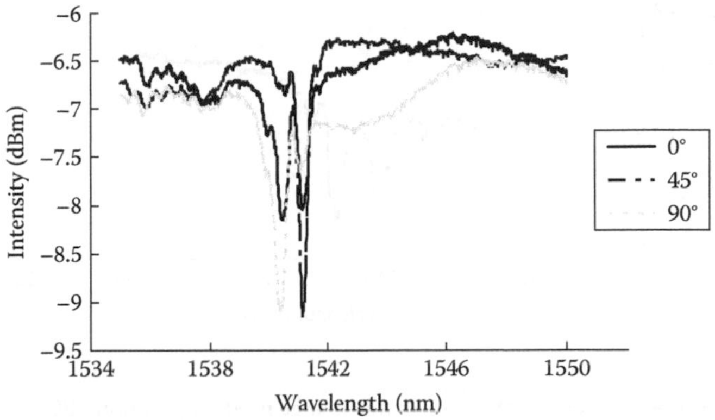

FIGURE 5.5
The evolution of the resonant transmission dips when the input light is aligned with 0, 45, and 90° to the fast axis.

made by Fibercore Ltd.). An OSA was used to measure the transmission spectra. Because the tunable laser was polarized, a Thorlabs manual paddle fiber polarization controller was used to ensure proper illumination of the birefringence axes. This allowed the proper alignment of the polarization plane with respect to the birefringence axes of the PM fiber. Figure 5.5 shows the evolution of the resonant transmission dips when the input light is aligned with 0, 45, and 90° to the fast axis.

When an FBG is formed in a PM fiber, two dips are produced in the transmission spectrum because of the slight difference of the effective refractive indices for the two orthogonal polarization modes.

$$\lambda_{Bragg,x} = 2n_{eff,x}\Lambda, \quad (5.2)$$

$$\lambda_{Bragg,y} = 2n_{eff,y}\Lambda, \quad (5.3)$$

where Λ is the period of the grating, the subscripts x and y denote the two polarization eigenmodes (slow and fast axes), $\lambda_{Bragg,i}$ ($i = x,y$) represents the Bragg resonant wavelengths in the x and y directions, respectively, and $n_{eff,i}$ ($i = x,y$) is the effective refractive indices in the x and y directions, respectively. The wavelength separation, $\Delta\lambda$, between two dips can be obtained by subtracting Eq. (5.3) from Eq. (5.2), as mathematically given by

$$\begin{aligned}\Delta\lambda &= 2\Lambda(n_{eff,x} - n_{eff,y}) \\ &= 2\Lambda \cdot \Delta n,\end{aligned} \quad (5.4)$$

where $\Delta n = n_{eff,x} - n_{eff,y}$ denotes the intrinsic birefringence of the PM fiber. Thus, the wavelength separation is directly proportional to the intrinsic birefringence of the PM fiber. To get a larger wavelength separation, a higher

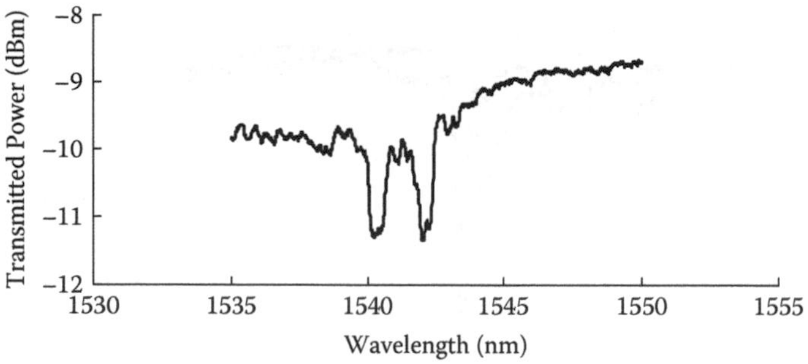

FIGURE 5.6
The measured spectrum of the photon enhanced large mode separation FBG fabricated in the polarization maintaining bow-tie fiber by femtosecond laser irradiation.

birefringence PM fiber is preferred. For example, to obtain a 0.765-nm wavelength separation, the required birefringence must be around 7.2×10^{-4}, which is considered as Hi-Bi PM fibers.[9] Further increase in the birefringence is limited by the available fiber materials and growing conditions.

Because our grating is asymmetrically written in the core, the geometric asymmetry of the side-writing process is a major cause of the induced birefringence in the grating-based fiber devices.[10] Our femtosecond laser-induced PM FBG also has a photon-enhanced birefringence. The separation between the two resonant polarized modes is as large as 1.72 nm, as shown in Figure 5.6, which solves a common problem for the conventional PM FBG fabricated by UV illumination method (i.e., when a transverse strain is applied to the grating, the separation of the two Bragg wavelengths may overlap due to the small separation of PM FBG fabricated by UV illumination).

5.2.4 Mechanism of Photo-Enhanced Birefringence

In our experiment, we believe that the photo-enhanced birefringence is caused by the strong off center femtosecond laser beam illumination over a certain time period. In general, birefringence results from the presence of asymmetries in the fiber section, which produces an anisotropic refractive-index distribution in the core region. The basic birefringence mechanisms, such as a noncircular core, asymmetrical transverse stress, bending, electric field, twist, and axial magnetic field, have been described in detail by S. Rashieigh.[11]

There are two primary types of polarization-maintaining fibers. One is the asymmetric geometric Hi-Bi fiber. The core or the cladding of this type of fiber is usually deformed from a circular shape so as to generate an asymmetry of the fiber structure—for example, the elliptical-core fiber. The other one is the stress-induced Hi-Bi fiber. This type of fiber is fabricated with stress-applying zones (SAZ) so that a particular stress status can be maintained in the fiber and its birefringence is achieved through the elasto-optic effect. The bow-tie type optical fiber is composed of three main regions: the core, the cladding, and the SAZs. Figure 5.7(a) shows the general structure of

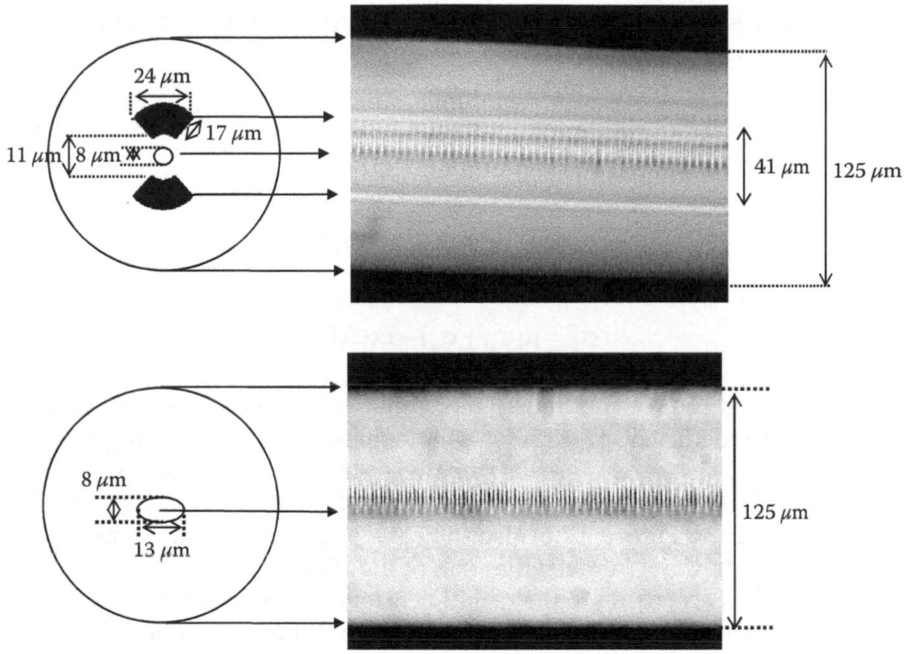

FIGURE 5.7
(a) A microscopic image and the corresponding schematic diagram of fiber cross-sections of the FBG fabricated in the bow-tie Hi Bi fiber via IR femtosecond laser illumination; (b) a microscopic image and the corresponding schematic diagram of fiber cross-sections of the FBG fabricated in the elliptical core Hi-Bi fiber via IR femtosecond laser illumination.

a bow-tie optical fiber. When a preform is heated to a temperature at which a fiber can be drawn, the preform is in a stress-free state. As the fiber cools, regimes of differential thermal expansion will induce large thermal stresses. The birefringence in bow-tie type optical fibers is mainly produced by the internal thermal stress because of the differences in the thermal-expansion coefficients at different regions. So, the thermal, mechanical, and geometric properties of different regions of the fiber determine the intrinsic birefringence. The birefringence B is defined as [12]

$$B = n_x - n_y = (C_2 - C_1)(\sigma_y - \sigma_x), \tag{5.5}$$

where σ_x and σ_y denote the axial stresses in the x and y directions, respectively, and C_1 and C_2 are the elasto-optic coefficients of the material in the x and y directions, respectively. Because the stress is distributed across the whole section of the fiber, the birefringence is evaluated by taking its average value across the core region, as given by[12]

$$\bar{B} = \frac{(C_2 - C_1)}{A} \iint_A (\sigma_y - \sigma_x) dxdy, \tag{5.6}$$

where A indicates the area of the core region. Furthermore, σ_x, σ_y, and σ_z, can be expressed as[12]

$$\sigma_x = \frac{E}{(1-\mu)(1-2\mu)}[(1-\mu)\varepsilon_x + \mu\varepsilon_y] - \frac{\alpha E \Delta T}{(1-2\mu)}, \tag{5.7}$$

$$\sigma_y = \frac{E}{(1-\mu)(1-2\mu)}[\mu\varepsilon_x + (1-\mu)\varepsilon_y] - \frac{\alpha E \Delta T}{(1-2\mu)}, \tag{5.8}$$

$$\sigma_z = \mu(\sigma_x + \sigma_y) - E\alpha\Delta T, \tag{5.9}$$

where E denotes Young's modulus, μ represents the Poisson's ratio, α is the thermal expansion coefficient, ΔT represents the temperature change during the cooling process, and ε_x, ε_y, and ε_z are the three axial strain components, respectively. Equations (5.5)–(5.9) can be used to calculate the stress-induced birefringence.

Figure 5.7(a) and Figure 5.7(b) also show the microscopic grating structures corresponding to the schematic diagram of the fiber cross-section. The drawing is magnified according to the real dimensional value. It is clearly seen that the grating structure spans the upper bow tie and the core section. The SAZ is also involved in the interaction. As we know, the working principle of the bow-tie fiber is that the asymmetrical transversal stress introduces a linear birefringence via elastic-optic index changes[13] as the result of different thermal contraction between different doped regions of the fiber during fabrication.

Similar experiments were performed on the asymmetric geometric Hi-Bi fiber. In the experiment, the elliptical core polarization maintaining fiber from IVG fiber was used. The beat length is 30 mm, and thus the intrinsic birefringence is about 0.45×10^{-4}; by using the exact same writing condition, a grating spectrum, as shown in Figure 5.8, is obtained. It was found that the femtosecond illumination also increased the birefringence of elliptical core fiber, but the amount of increase was much smaller than the bow-tie fiber case. Because the birefringence of elliptical core fiber is generated by asymmetry of the core section, while the femtosecond laser induced birefringence by grating writing process is related to the symmetric stress, these two cannot be superposed linearly, as the stress-induced Hi-Bi fiber (bow-tie fiber in our experiment) case. Also, similar to the bow-tie fiber case, for elliptical core fiber, further increase in the exposure time after saturation will only add noise and distort the grating transmission spectrum.

In summary, the key factors to achieve the large photo-enhanced birefringence are (1) the ultrashort pulse irradiation induces rapid local heating and generates large stress, (2) the photo-induced stress enhances the intrinsic birefringence via photo-elastic effect, (3) the asymmetric grating not only exists in the fiber core but also spans across part of the SAZ. Thus, there can be substantial change in the refractive index of SAZ, which results in a large photo-induced birefringence effect.

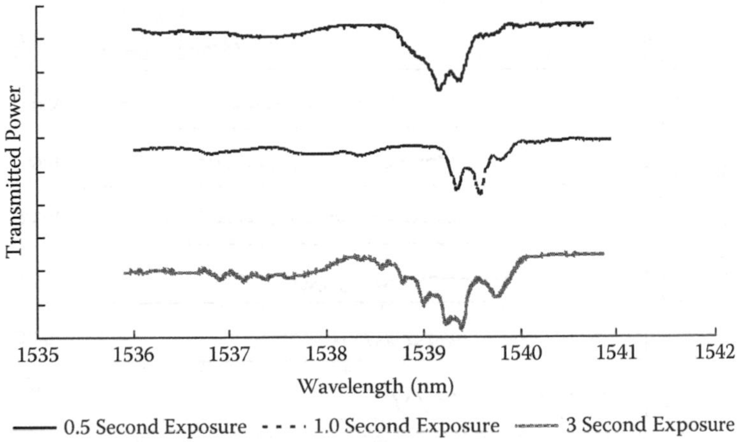

FIGURE 5.8
The measured spectra of the FBGs fabricated elliptical core Hi Bi fiber by femtosecond laser irradiation using different exposure times (0.5, 1, and 3 seconds, respectively).

5.3 Multiparameter Sensing Using MM Fiber Gratings

The unique features of this type of grating are (1) it can withstand temperatures as high as 1000°C, which makes it suitable for harsh environment sensing; (2) for MM Bragg grating, bending (or pressure, strain) and ambient temperature fluctuations affect the shapes of multiple transmission dips in different ways, which enables multiparameter sensing capability. The principle of multiparameter sensing of the multimode fiber gratings is based on the fact that multimode fiber gratings have multiple resonant dips in spectrum and the change of spectral shape is different for different types of perturbations (e.g., temperature and pressure).

5.3.1 Temperature Sensing

In the temperature sensing experiment, the FBGs were put in a high-temperature furnace. One end of the fiber was connected to a tunable laser and the other end was connected to an OSA. The spectrum of FBG was measured at different temperatures. Figure 5.9(a) shows the measured spectra of an MM FBG at a temperature range from room temperature to 700°C. As the temperature increased, the spectrum shifted, but the spectral profile remained unchanged. Figure 5.9(b) depicts the spectra of dip locations as a function of temperature for different dips. Four spectral curves corresponding to four dips are parallel to each other.

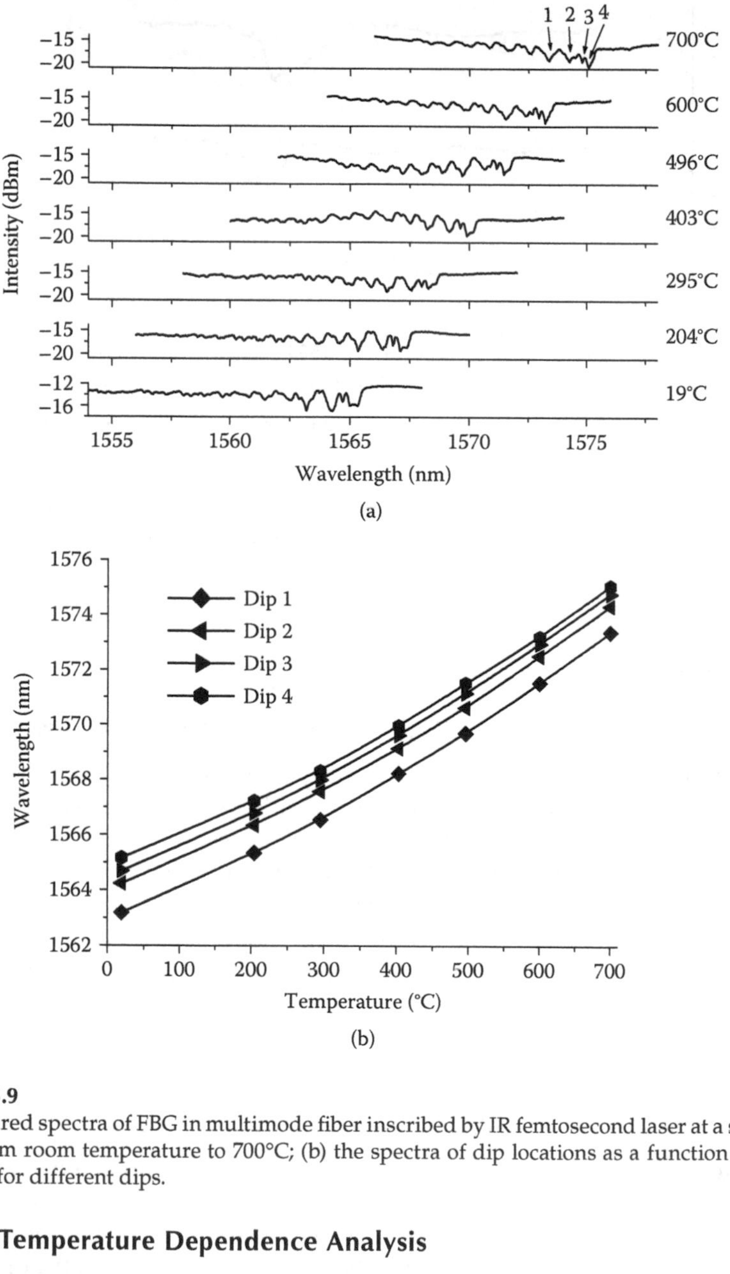

FIGURE 5.9
(a) Measured spectra of FBG in multimode fiber inscribed by IR femtosecond laser at a spectral range from room temperature to 700°C; (b) the spectra of dip locations as a function of temperature for different dips.

5.3.2 Temperature Dependence Analysis

To obtain the temperature dependence of the grating, a method similar to that used in single mode Bragg grating analysis is employed. By differentiating the grating equation (i.e., Eq. 5.1) with temperature, we obtain:

$$\frac{d\lambda_r}{dT} = 2\left(\frac{dn_{co}^{eff}}{dT}\bullet\Lambda + n_{co}^{eff}\bullet\frac{d\Lambda}{dT}\right). \tag{5.10}$$

By further simplification, Eq. (5.10) becomes

$$\frac{d\lambda_r}{dT} = 2\left(\frac{dn_{co}^{eff}}{dT}\bullet\frac{\lambda_r}{2n_{co}^{eff}} + \frac{\lambda_r}{2\Lambda}\bullet\frac{d\Lambda}{dT}\right) = \lambda_r\left(\frac{dn_{co}^{eff}}{n_{co}^{eff}dT} + \frac{d\Lambda}{\Lambda dT}\right) \tag{5.11}$$

Equation (5.11) predicts that the thermally induced change in the Bragg wavelength is due to two major contributors: the changes in the effective index of the fiber core and in the grating period, where

$$\frac{d\Lambda}{\Lambda dT}$$

is the thermal expansion coefficient for the fiber (approximately 0.55×10^{-6} for silica).[14] The quantity

$$\frac{dn_{co}^{eff}}{n_{co}^{eff}dT}$$

represents the thermo-optic coefficient and typically equals 8.6×10^{-6} for the silica core fiber.[14] According to Eq. (5.11), the theoretical temperature sensitivity at the 1560-nm Bragg grating wavelength is estimated to be 0.014 nm/°C. On the other hand, the experimental Bragg grating resonant wavelength shift as a function of temperature is shown in Figure 5.9(b), which gives the temperature sensitivity around 0.013 nm/°C. Thus, in terms of temperature sensing, the experimental and theoretical results agree very well.

5.3.3 Bending Sensing

In the bending sensing experiment, the FBG was wrapped on a cylindrical surface with a 25-mm radius. Again, one end of the fiber was connected to a tunable laser light source and the other end of the fiber to OSA. The spectrum was measured for the cases with and without bending. To determine the vectorial- (or directional-) sensing capability for this asymmetric FBG, the FBG was bent in two orthogonal directions and the spectra for both bending directions were measured. Figure 5.10(a) and Figure 5.10(b) show the measured spectra and dip locations of MM FBG for all three cases: (1) without bending, (2) bending in x direction, and (3) bending in y direction. These experimental results clearly demonstrated that there were significant changes in spectral profiles when the FBG was bent, which was totally different from the case of temperature sensing, as described in Sections 5.3.1 and 5.3.2. Different dips have different changing amounts in terms of dip locations and depths when bending is applied. In addition, the change in the spectral profile was also sensitive to the bending direction due to the use of an asymmetric FBG.

5.3.4 Bending Dependence Analysis

As aforementioned, our experimental results clearly demonstrated that there were significant changes in spectral profiles when the FBG was bent, which

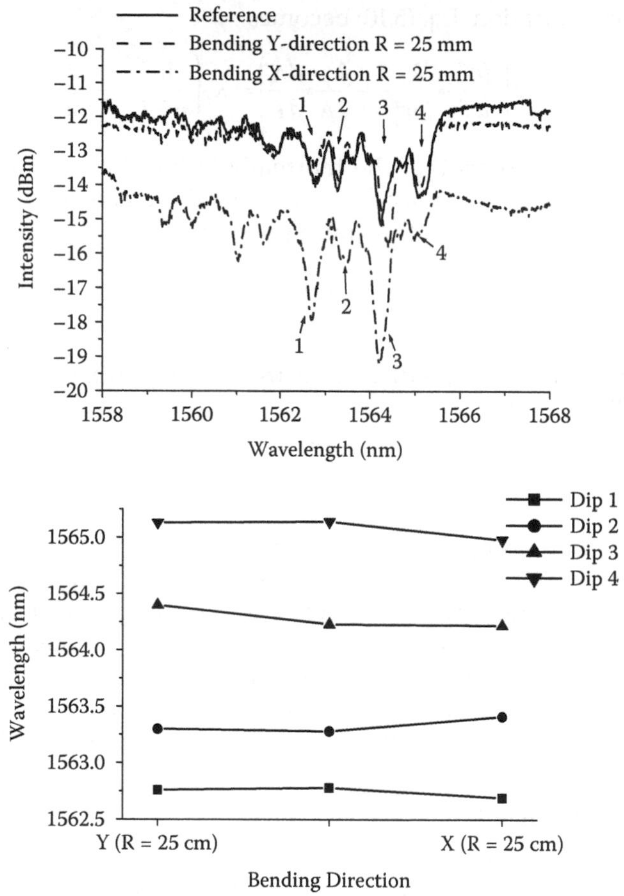

FIGURE 5.10
(a) Measured spectra of FBG in multimode fiber inscribed by IR femtosecond laser at different bending conditions: (i) without bending; (ii) bending in x-direction; and (iii) bending in y-direction; (b) the MM FBG spectra of dip locations as a function of bending direction for different dips.

was totally different from the case of temperature sensing. In terms of bending; different dips have difference changes in both dip locations and depths. Furthermore, the change in the spectral profile was also sensitive to the bending direction due to the use of an asymmetric FBG. Since the dominant effect caused by the bending is the bending-induced birefringence, our theoretical analysis is based on the bending-induced birefringence. The modal birefringence produced by bending is approximately given by[15]

$$B_B = \frac{b^2}{2R^2} EC \ , \tag{5.12}$$

where E is the Young's modulus, C is the photo-elastic constant, b is the radius of effective mode field, and R is the bending radius. For silica glass, $EC/(1 + v) = 0.146$, where v is the Poisson's ratio. Note that most practical engineering materials have ν between 0.0 and 0.5. In our calculation, $v = 0.18$ for silica glass. For multimode fiber, different orders of modes have different mode field radii. In general, higher order modes have larger mode field radius, and thus are more sensitive to bending. In our analysis, to match our experiments, we chose the parameters $b = 60$ μm and $R = 25$ mm.

Based on Eq. (5.12), the calculated induced birefringence by bending is around 5×10^{-7}. By measuring the resonant wavelength shift, the birefringence can be experimentally determined from the Bragg condition. As shown in Figure 5.10, the experimentally measured wavelength shift is about 0.22 nm, which corresponds to a birefringence around 2×10^{-4}. Thus, experimental results are much larger (1000 times) than the calculated result. This is because we are using asymmetric FBGs in our experiment. In other words, the bending sensitivity comes from the asymmetric nature of the FBGs.

5.4 Multiparameter Sensing Using PM Fiber Gratings

The main challenge of sensing using multimode fiber is that the shape of the spectrum also depends on the excitation condition, which makes the analysis of the sensing data complicated. On the other hand, PM fiber is a relatively convenient approach. The basic working principle of the PM fiber approach may be summarized as follows. The PM fibers support two linearly polarized modes. When a Bragg grating is written into this type of fiber, the physical periodicity of the grating is identical for both polarization modes, but the effective refractive indices are different. Thus, the Bragg resonant wavelength is different for each polarization mode, which results in two resonant dips in the transmission spectrum of the FBG. Since measured parameters such as the temperature and strain affect the transmission spectrum in different ways, one can distinguish two effects by monitoring the wavelength shift and the evolution of the depth of the two resonance peaks. Note that it is much easier to analyze the data of two dips (in the case of PM fibers) instead of multiple dips (in the case of multimode fibers).

However, the main challenge of this approach is that the separation between the two Bragg wavelengths is typically very small (in the range of 0.5 nm),[11–17] which makes it very difficult to resolve two dips. Furthermore, the separation can be further reduced or eliminated by the external perturbations (e.g., strain). Although efforts have been made to increase the wavelength separation (e.g., using Hi-Bi PM fiber), the increase in the separation is still limited (i.e., <1 nm). Fortunately, this limitation is largely alleviated by employing the photo-enhanced birefringence.

5.4.1 Temperature Sensing

To illustrate the multiparameter sensing capability of this large polarization mode separated FBG, we conducted the following experiments for sensing both the temperature changes over a large temperature range (from room temperature to 800°C) and transversal bending. These external perturbations induce both the refractive index changes (via elasto-optic and thermal-optic effects) and periodicity changes.[18] The corresponding Bragg peak shifts for the two polarization directions, $\Delta\lambda_{Bragg,i}$ ($i = x,y$), can be expressed as[16,19,20]

$$\Delta\lambda_{Bragg,x} = 2\Delta n_{eff,x}\Lambda + 2n_{eff,x}\Delta\Lambda, \tag{5.13a}$$

$$\Delta\lambda_{Bragg,y} = 2\Delta n_{eff,y}\Lambda + 2_{neff,y}\Delta\Lambda, \tag{5.13b}$$

where $\Delta\Lambda$ is the periodicity change of the grating and $\Delta n_{eff,i}$ ($i = x,y$) denotes the effective refractive index changes in the x and y polarization directions, respectively. The amount of change induced by temperature is similar for both x and y polarization directions. In other words, $\Delta n_{Bragg,x}(T) = \Delta n_{Bragg,y}(T)$, where T represents temperature; thus, the temperature change simply causes the two dips in the transmission spectrum to shift uniformly while the spacing between them remains the same. However, the transverse bending will induce different amounts of change in x and y polarization directions. In other words, the transverse bending will not only cause resonant dips to shift but will also change the separation between two dips due to the bending-induced birefringence. The following experimental results verify the preceding theoretical predictions.

In the temperature sensing experiment, this polarization mode separated FBG was put in a furnace. One end of the fiber was connected to a tunable laser source and the other end of the fiber was connected to an OSA. Also, a polarization controlling paddle was connected to the tunable laser source for rotating the polarization direction of the light. The spectra of the FBG were measured at different temperatures. Figure 5.11(a) shows the measured spectra of the FBG at different temperatures ranging from room temperature to 800°C. One can clearly see that, as the temperature increases, the spectrum shifts but the spacing between two dips remains the same. Figure 5.11(b) is a plot of the dip location as a function of temperature, which shows that the curves for two dips are nearly parallel with each other. This again confirms that the influence from the temperature is the same for the two dips, and is consistent with the theoretical modeling. The experimental result also shows that the large polarization mode separated FBG not only increases the separation between the two dips but also can survive in harsh, high-temperature environments.

To analyze the temperature dependence of the grating, a method similar to that used in single mode Bragg grating analysis is employed. Equation (5.13) predicts that the thermally induced change in the Bragg wavelength is due to two major factors: (1) the change in the effective index of the fiber core,

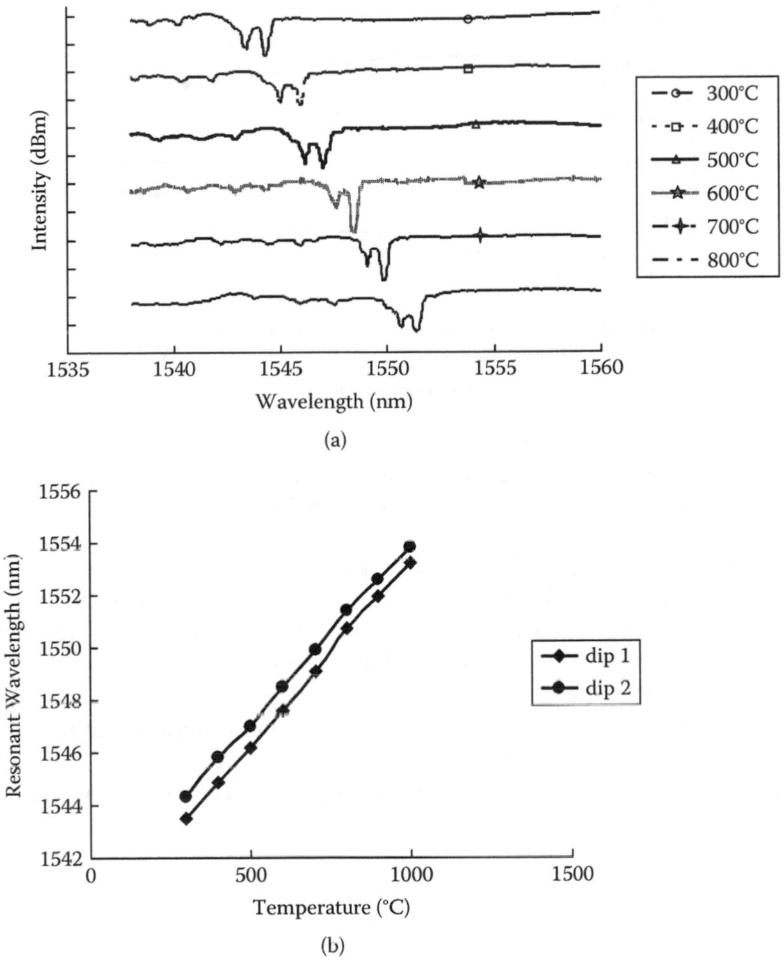

FIGURE 5.11
(a) The measured spectra of the FBG fabricated in bow-tie fiber by femtosecond laser irradiation at different temperatures ranging from room temperature to 800°C; (b) dip locations of two polarization modes as a function of temperature.

and (2) the change in the grating period. According to Eq. (5.13), the temperature sensitivity at the 1540-nm Bragg grating wavelength is estimated to be 0.014 nm/°C. Our experimental result is around 0.013 nm/°C, which is consistent with the theoretical value.[14]

5.4.2 Bending Sensing

In the bending sensing experiment, the PM FBG was freely bent along a cylindrical surface. The bending radii for the x and y directions are 10 and 15 mm, respectively. To measure the transmission spectrum, again, one end of the fiber was connected to a tunable laser and the other end to an OSA.

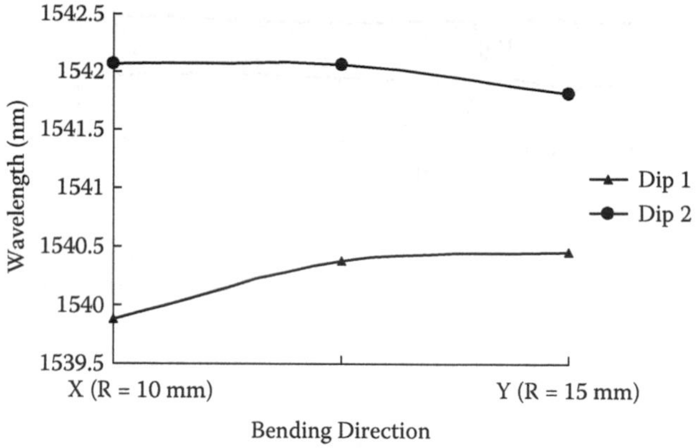

FIGURE 5.12
The measured spectral locations of two dips of FBG in bow-tie fiber for three cases: without bending, with x-direction bending, and with y-direction bending.

The spectra were measured for three cases: with *x*-direction bending, with *y*-direction bending, and without bending. Figure 5.12 shows the measured spectra and the spectral locations of the dips for all three cases. These experimental results clearly demonstrate that, when the transversal bending is applied, there are not only changes in the spectral locations of the dips but also in the spacing between the two dips. This phenomenon is mainly due to the bending-induced birefringence.

Mathematically, the bending-induced birefringence can be approximately given by[15]

$$\beta_B = \frac{r^2}{2R^2} C_s \tag{5.14}$$

$$C_s = 0.5 k_0 n_0^3 (p_{11} - p_{12})(1 + v_p), \tag{5.15}$$

where r is the radius of the fiber, R is the radius of the bending, n_0 is the average refractive index of the fiber, p_{11} and p_{12} are the components of the strain-optical tensor of the fiber material, v_p is Poisson's ratio, and k_0 is the wavevector. For fused silica, $n = 1.46$, $p_{11} = 0.12$, $p_{12} = 0.27$, and $v_p = 0.17$ at $\lambda = 1.5$ µm. Substituting these parameters into Eq. (5.15), we get $Cs = 1.14 \times 10^6$. These data suggest that for large diameter coils, the birefringence is small, but for small diameter coils, it can be large. Furthermore, the beat length, L_p, can be calculated based on the birefringence

$$L_p = \lambda / B = 2\pi / \beta_B \tag{5.16}$$

In our bending experiment, the theoretical bending-induced birefringence, which can be calculated by using Eqs. (5.14)–(5.16), is around 6×10^{-6}, which corresponds to a 6-pm wavelength shift. However, as shown in Figure 5.12, the experimentally measured wavelength shift can be as large as 400 pm. This difference is because the theoretical calculation is based on conventional symmetric grating. Since the grating used in our experiment is asymmetric grating, it is more sensitive to the bending. Thus, asymmetric grating can also increase the bending sensitivity when it is used as a bending sensor. Furthermore, the spectral changes are also sensitive to the direction of the bending. Thus, we can sense temperature change and vector bending simultaneously.

5.5 Ambient Refractive Index Sensing Using FBGs Fabricated in Thickness-Reduced MM and PM Fibers

Fiber Bragg grating couples the forward mode to the backward propagation mode; because no cladding modes are involved, fiber Bragg gratings are intrinsically insensitive to the surrounding medium for the standard (e.g., 125 μ cladding) fibers. However, if FBGs are fabricated in thickness-reduced (ultrathin) fibers with cladding layer partially or totally removed, the effective refractive index is significantly affected by the surrounding medium as a result of the geometrical modification. This leads to the shifts in the resonant wavelength combined with a modulation of the amplitude along with the change of the surrounding refractive index (SRI).[21,22]

5.5.1 MM Fibers

In our experiment, to reduce the thickness of the fibers, the multimode fiber was immersed in 20% buffered hydrofluoric acid (HF) solution, and etched down to 60 μm. In this case, almost all the cladding layer is gone. FBGs are fabricated in these thickness-reduced MM fibers by following the similar procedure used for fabricating FBGs using femtosecond lasers in the standard unetched fibers, except that the exposure time is reduced to 1 sec. Since there is no silica cladding (and air becomes the cladding), even for the single mode excitation condition, the observed spectra of the FBGs were highly multimode due to the large amount of existing modes in this case. The number of propagation modes can be estimated by[23]

$$M = \frac{1}{2} a^2 k^2 n_1^2 \Delta \tag{5.17}$$

where a is the core radius, k is wave number, n_1 is the refractive index of the core and Δ is the maximum relative index difference. Apparently, increasing the relative index difference will increase the number of modes that can be transmitted. Even for the single mode excitation, the resonant dips are dramatically increased. Fortunately, the depth of each dip is still about 2 or 3 dB, which is big enough in terms of sensing applications.

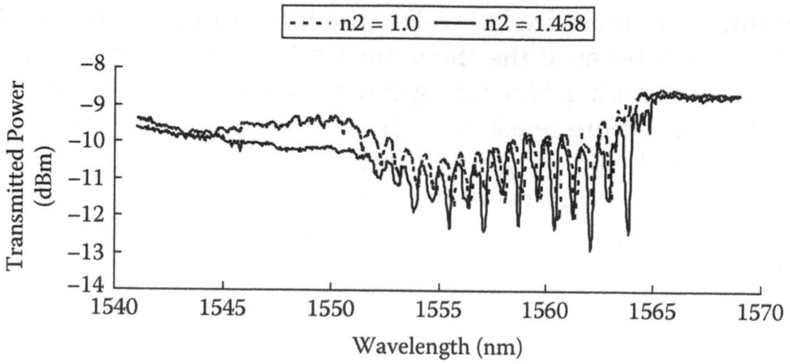

FIGURE 5.13
The measured spectra of FBG in thickness-reduced MM fibers with different surrounding refractive indices.

To check whether this type of FBG (fabricated in thickness reduced fibers by femtosecond-laser illumination) was sensitive to the ambient refraction index or not, an FBG was immersed in the index-matching oil with a refractive index 1.458, and the spectrum change was monitored by a tunable laser and OSA. Figure 5.13 shows the measured spectra of FBG corresponding to the cases with and without index matching oil, respectively. A big shift of the resonant wavelength was observed.

5.5.2 PM Fibers

The most commonly used PM fiber is bow-tie fiber, which is a stress-induced birefringence fiber. Because of the complex structure in the cladding, this type of fiber is not suitable for etching to the thinner diameters. In the experiment, we used the elliptical core PM fiber from IVG fiber. The beat length is 30 mm; thus, the intrinsic birefringence is about 0.45×10^{-4}, and the fiber is etched to 38-μm diameter. Bragg grating is written in the fiber using a procedure similar to that described in Section 5.2.2.

5.5.3 Liquid Pressure/Depth Sensing Using FBGs Fabricated in Thickness-Reduced Fibers

Pressure sensors have been widely used in a large number of applications from automotive and aerospace industries to healthcare. To implement liquid pressure sensing, an FBG was fabricated in a thickness-reduced MM fiber. This FBG was attached to a long rod and immersed at different depths of the water, and the reflection spectrum was measured. The MM fiber with the FBG was butted coupled to a four-port coupler made of single mode silica core fiber as shown in Figure 5.14. The light from the tunable laser was launched through a single mode fiber into the input port of the coupler and the OSA was connected to the output port. The end of the fiber that contained the Bragg grating along with the supporting rod was then immersed

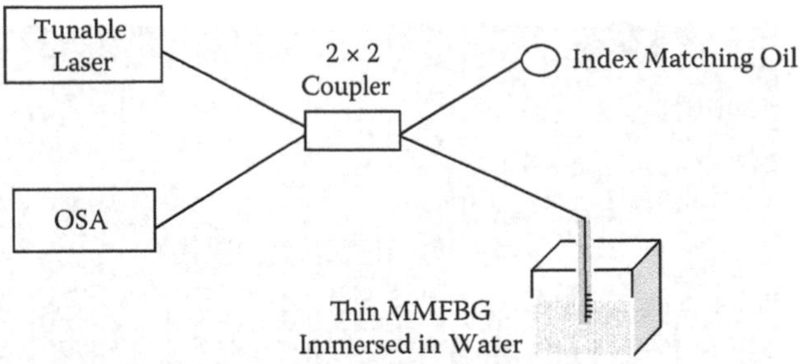

FIGURE 5.14
A schematic diagram of the liquid pressure sensor using the FBGs fabricated in the thickness-reduced MM and PM fibers.

into the water and the reflection signals were measured at different immersion depths.

Experimental results are shown in Figure 5.15(a) and Figure 5.15(b), respectively. It is clearly shown that spectra are different for different immersion depths (12 and 18 cm, respectively; to be discussed in the following). First, the whole spectrum shifts to the longer wavelength when the grating is immersed into the water. Second, more modes are excited for the deeper immersion. This phenomenon may be explained in the following way. The deeper the FBG goes into the water, the higher the pressure on the FBG will be, which results in a higher refractive index due to the increased density by the liquid pressure squeezing effect. According to the Bragg condition, the resonant peak wavelength will shift to the longer wavelength as the effective refractive index increases. Also, the increase in the fiber refractive index increases the refractive index difference between the fiber and the surrounding medium (i.e., liquid). This in turn increases the number of propagation modes. Thus, more modes can be excited for the deeper immersion.

Besides liquid pressure sensing using the FBG fabricated in the thickness-reduced MM fiber, we also conducted the experiments using FBGs fabricated in the thickness-reduced PM fiber (the one fabricated in Section 5.5.2). The sensing principle is that the transverse pressure will induce the birefringence and change the separation between the different polarization modes. Previous investigations by other groups indicate that the induced birefringence from the water pressure on the fiber can be estimated by[24]

$$\Delta n_{WP} = 4C_s \frac{P}{\pi r E}, \tag{5.18}$$

where $C_s = 1.14 \times 10^{-6}\ \text{m}^{-1}$ is *a strain-optical coefficient related parameter*, P is the water pressure (i.e., $P = \rho g h$), r is the outer radius of the fiber (in our experiment, the diameter of the PM fiber was etched down to 40 μm, so $r = 20$ μm), and E is Young's modulus of the fiber material (for fused silica, it is around $7.6 \times 10^{10}\ \text{N m}^{-2}$).

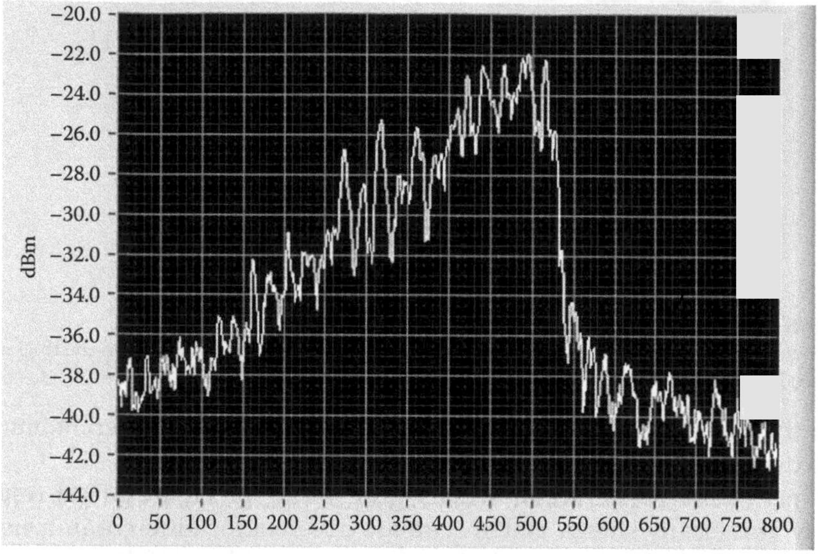

(a)

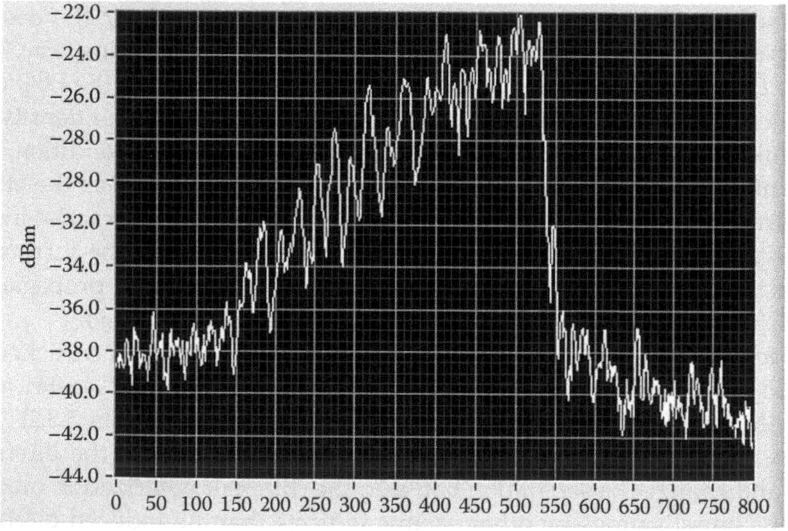

(b)

FIGURE 5.15
(a) The measured reflection spectrum of the FBG fabricated in the thickness-reduced MM fiber at an immersion depth of 12 cm. The figure is captured by the Labview program from the optical spectrum analyzer; (b) the measured reflection spectrum of the FBG fabricated in the thickness-reduced MM fiber at an immersion depth of 18 cm. The figure is captured by the Labview program from the OSA. (Note: the horizontal axis in these two figures represents the sampling points of the OSA. These 800 points are sampled in the wavelength range of 1545–1575 nm.)

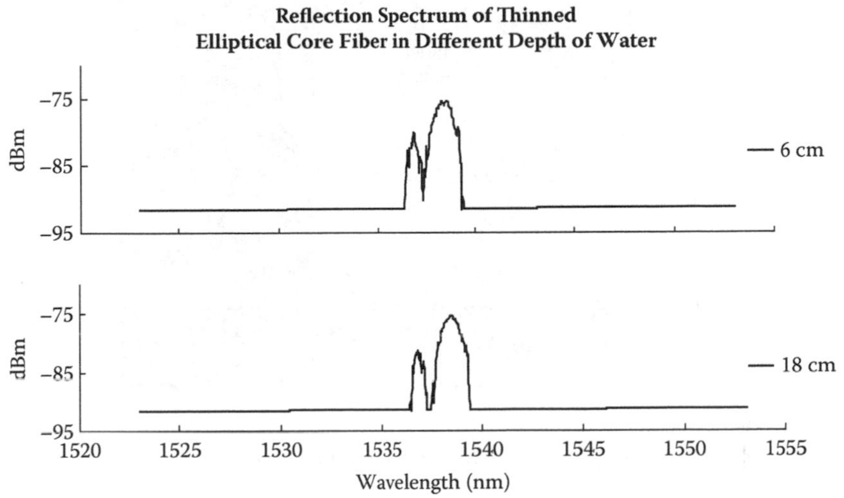

FIGURE 5.16
Measured reflection spectra of FBG fabricated in the thickness-reduced PM fiber at immersion depths 6 and 18 cm, respectively. The deeper immersion induces a larger spectral separation between two polarization modes.

Using an experimental setup[21] similar to that illustrated in Figure 5.14, Figure 5.16 shows the measured spectra as a function of immersion depths. The separation between two polarization modes is 1.69 nm (corresponding to 6 cm immersion depth), and 1.88 nm (corresponding to 18 cm immersion depth). Note that this is a very rough estimation, since our fiber grating is asymmetric and tiled in the core, and that it is elliptical core polarization maintaining fiber, so the real situation is much more complicated than this formula can described. But the experimental results and simple calculation clearly show that when the grating is immersed into the deeper water, the pressure applied on the grating is also larger. Thus, the induced birefringence is also larger, which results in a larger separation between two polarization modes. Compared with the multimode fiber sensing case, the spectrum is much simpler and it has only two peaks, which makes the practical implementation of the sensing easier.

5.6 Distributed Thermal Sensing and Long-Term Stability of Femtosecond Laser-Inscribed Bragg Grating Sensors

5.6.1 Distributed Thermal Sensing

A distributed thermal sensing experiment was investigated using the McKenna burner system (at Dr. Andre L. Boehman's lab at Penn State Energy Institute), as shown in Figure 5.17. In the experiment, an unconfined flat

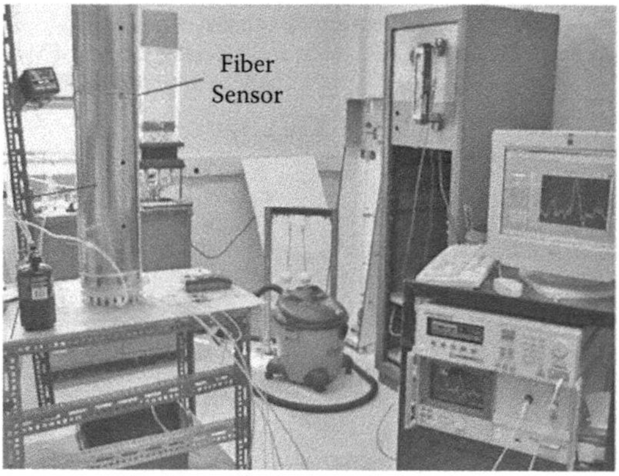

FIGURE 5.17
The picture of the experimental setup of McKenna burner system.

laminar premixed flame was established using a McKenna burner, which consisted of a 60.2-mm diameter water-cooled porous plate. The burner is covered by a chimney with holes in it, and the ceramic rod with fiber grating attached is inserted in the holes to measure the temperature profile of the burner. In the horizontal direction, a measurement is taken at 1.7 cm above the burner surface plate, as shown in Figure 5.18(a). Five positions are chosen. The distance between each position is 1 cm. In the vertical direction, the measurement is conducted at four different heights above the burner surface, as shown in Figure 5.18(b) and Figure 5.18(c). To calibrate the measurement, the temperatures were also measured by a thermal coupler, as tabulated in Table 5.1. Figure 5.19(a) and Figure 5.19(b) show the reflection spectra at different locations. The peak locations of the reflection spectra are different for different locations due to the temperature difference among these positions. It clearly shows the distributed sensing capability of our proposed system.

5.6.2 Long-Term Stability of the Femtosecond Laser-Inscribed FBG Sensors

In order to test long-term thermal stability, the femtosecond laser-induced Bragg grating was inserted in a stainless tube and put in a high-temperature furnace for about 10 days, as shown in Figure 5.20 (the facility at GE Global Research Center). The grating spectrum was monitored by a micron optics sensing interrogator. Figure 5.21(a) and Figure 5.21(b) show the measured temperature as a function of time. In Figure 5.21(a), the upper curve and lower curve are the measured results from fiber grating and the thermal couple, respectively, while in Figue 5.21(b) the upper and lower curve denote the measured results from thermocouple and fiber grating, respectively. From this experimental result, one can conclude that (1) the fiber grating has very good long-term stability, and (2) the fiber grating also has a lower noise.

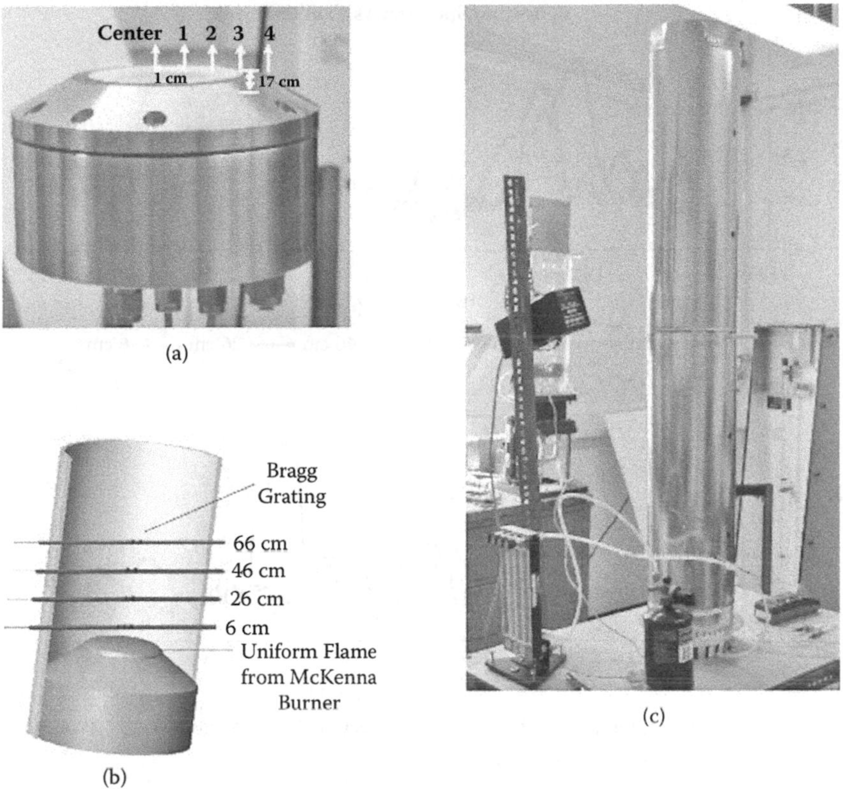

FIGURE 5.18
(a) Horizontal direction along burner; (b) vertical direction along burner; (c) hole distributions in the chimney along vertical direction.

TABLE 5.1
Temperature Calibrated by Thermal Coupler

Vertical Direction		Horizontal Direction	
Height above the Flame (cm)	**Temperature (°C)**	**Position**	**Temperature (°C)**
		Center	980
6	730	1	940
26	180	2	900
46	138	3	690
66	127	4	400

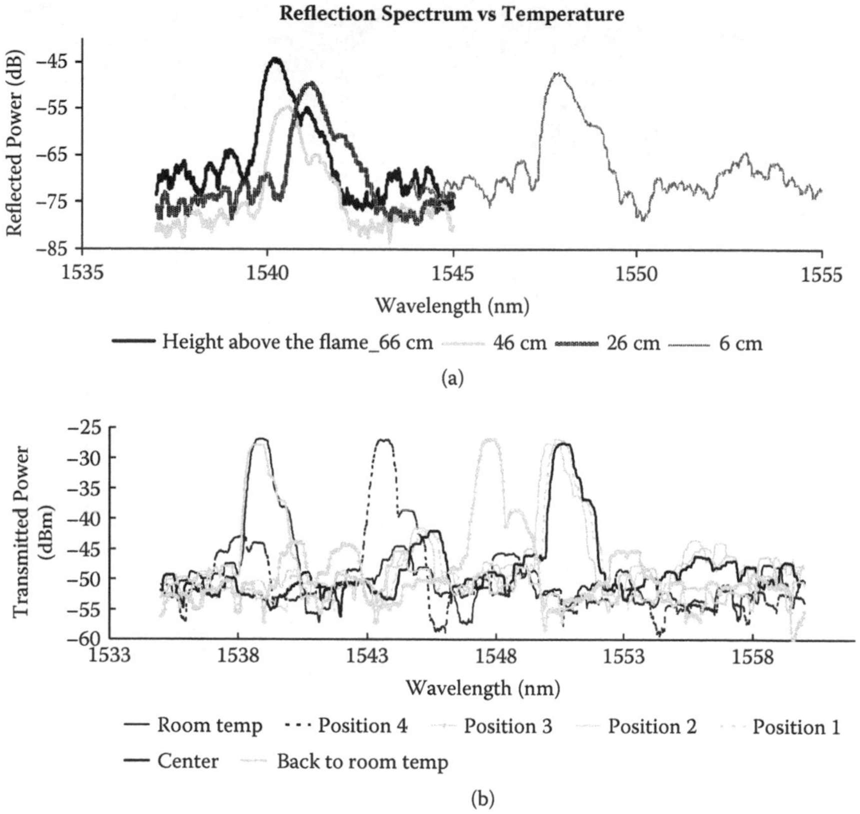

FIGURE 5.19
(a) Grating spectrum measured with different locations along the vertical direction; (b) grating spectrum measured with different locations along the horizontal direction.

5.7 High-Temperature Sensing Using Higher Order-Mode Rejected Sapphire Crystal Fiber

In this section, the fabrication of higher order mode rejected fiber Bragg gratings (FBGs) in sapphire crystal fiber using IR femtosecond laser illumination is presented. The grating is tested in a high-temperature furnace up to 1600°C. By using an ultrathin sapphire crystal fiber (60 μ in diameter), the fiber is able to be bent to a certain radius, which is carefully chosen to provide low loss for the fundamental mode and high loss for the other high-order modes. After bending, less than 2-nm resonant peak bandwidth is achieved. The grating spectrum is dramatically improved, and higher resolution sensing measurement can be achieved. This mode filtering method is very easy to implement. Furthermore, the sapphire fiber is sealed with high-purity alumina ceramic cement inside a flexible, high-temperature titanium tube,

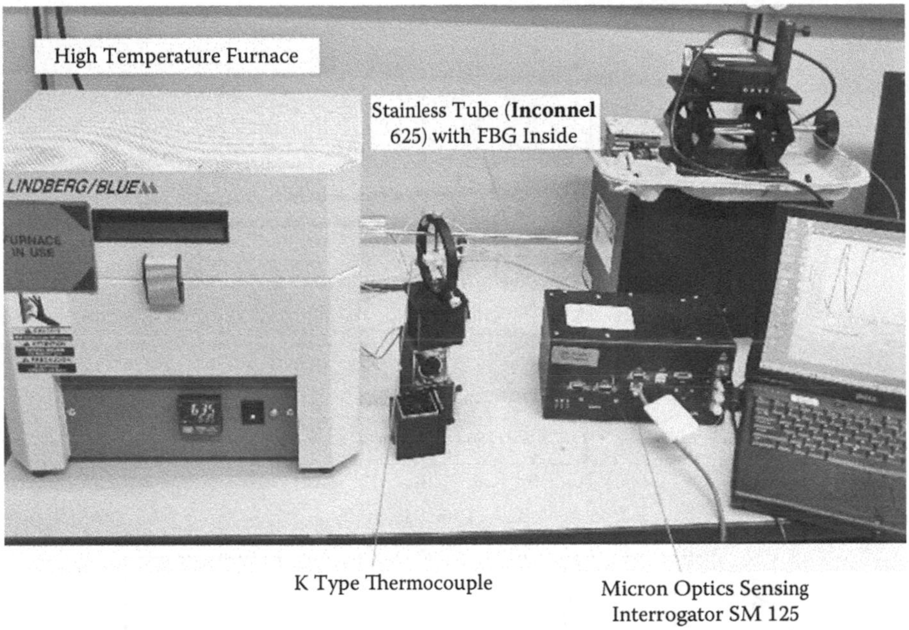

FIGURE 5.20
Experimental setup of thermal durability test.

and the highly flexible titanium tube offers a robust packaging to the fragile sapphire fiber. Thus, the sapphire grating sensor is a promising technology for extremely high temperature sensing applications.

5.7.1 Background on Sapphire Fiber Grating Sensors

To develop extremely harsh environment sensors, in recent years, optical sensors based on sapphire fibers were investigated.[25–27] Sapphire ($A1_2O_3$) crystal possesses a melting point of over 2000°C, excellent transparency (200–5000 nm), and well-documented resistance to corrosion and reasonable optical propagation properties, which make it an ideal candidate for high-temperature fiber-based environmental sensing applications. However, the growth of optical-grade sapphire crystal fibers is still in the research phase. Growth methods do not permit sapphire fiber to be drawn as a single mode waveguide or with a cladding. Without a cladding, the entire sapphire fiber may be considered to be a core, with the surrounding lower index air as the fiber cladding.[28] In this case, the large diameter sapphire core supports the propagation of thousands of different modes. Thus, the spectrum from this extremely multimode fiber is not good because it has too many peaks. To alleviate this problem, D. Grobnic et al proposed using tapered fiber for exciting the mode.[27] But the procedure of matching the mode fields is a challenging job. In this section, we report another method that uses the higher order mode rejected ultrathin sapphire crystal fiber via proper bending.[29]

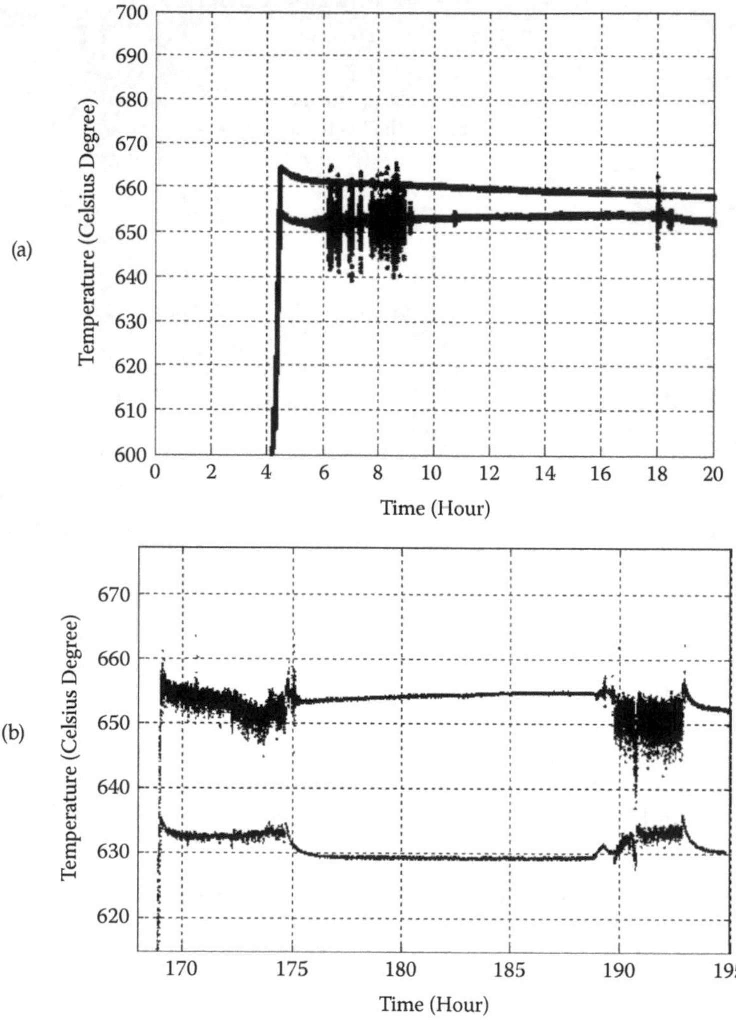

FIGURE 5.21
A comparison between FBG measurement results and thermocouple results. (a) Upper curve measured the result from fiber grating; lower curve measured the result from thermalcouple. (b) Upper curve measured the result of the grating; lower curve measured the result from the grating.

5.7.2 Experimental Procedure

The Quantronix Titan IR femtosecond laser system was used for the sapphire FBG fabrication.[30] The laser system includes the seed laser, femtosecond pulse regenerative, and multipass Ti:Sapphire amplifier. The laser has a 150-fs pulse duration, an output wavelength of 800 nm, a maximum output pulse energy around 2.5 mJ/pulse, and a pulse repetition rate up to 1 kHz. The vertically polarized beam from this laser is focused onto the sapphire fiber through a silica phase mask by a 38-mm focal length cylindrical lens. The fiber is placed on a high precision three-dimensional translation stage. The pitch of the phase mask is 2.66 μm. The phase mask is designed for 800 nm and less than 5% of the beam is diffracted into the 0th order. Since the multimode sapphire fiber has a core diameter that is wider than the focused laser beam, the beam is translated and scanned across the cross-section of the fiber in a step-and-repeat fashion by a piezo-actuator.[27] The total exposure time is around 2 sec.

The phase-matching condition, or Bragg condition, is given mathematically by

$$2n^{q}_{eff-s}\Lambda_{g-s} = m\lambda^{q}_{Bragg}, \tag{5.19}$$

where m denotes the diffraction order number, n^{q}_{eff-s} represents the effective refractive index for the qth-order mode propagating in the multimode sapphire fiber, λ^{q}_{Bragg} is the Bragg resonant wavelength for the qth mode, and Λ_{g-s} denotes the grating pitch in sapphire fiber. In our experiment, the grating pitch is 2.666 μm; the refractive index for sapphire fiber at 1550 nm is around 1.74. Thus, the third-order grating is estimated to have a resonant peak in the 1545-nm region. Figure 5.22 shows the microscopic image of the grating fabricated in sapphire fibers.

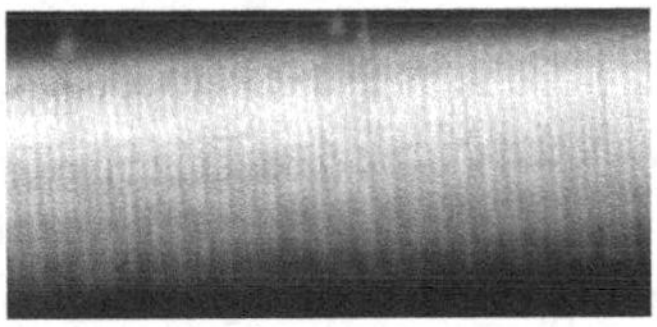

FIGURE 5.22
A microscopic image of an FBG fabricated in sapphire fiber by an IR femtosecond laser with 150-fs, 800-nm laser pulses and a phase mask.

To measure the spectrum, one end of the grating was buttcoupled to a four-port coupler made of multimode silica fiber. The signal from the tunable laser was launched into the input port of the coupler and an OSA at the output port collected the reflected signal from the grating. The grating spectrum before bending is shown in Figure 5.23. It can be seen that there are

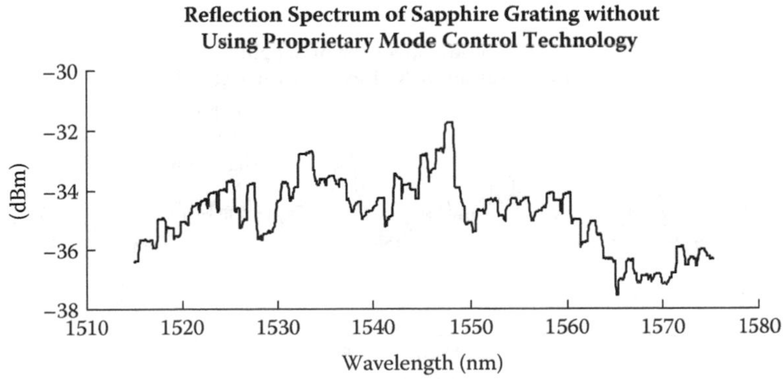

FIGURE 5.23
The measured reflection spectrum of sapphire FBG before bending obtained by tunable laser and optical spectrum analyzer.

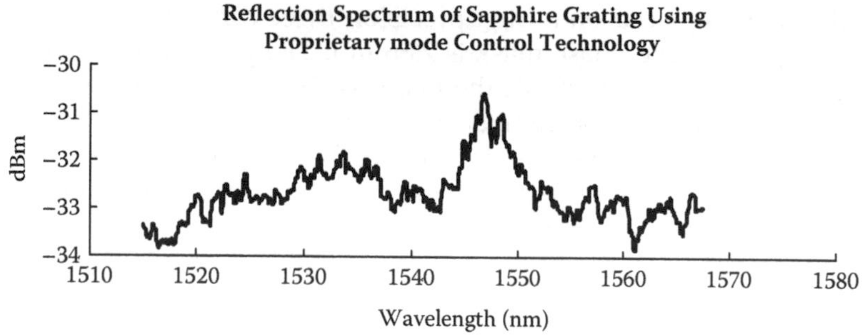

FIGURE 5.24
The measured reflection spectrum of sapphire FBG after bending.

higher order modes in the shorter wavelength portion of the grating spectrum. Since higher order modes "see" more cladding during prorogation, their effective refractive index is smaller than that of lower order modes.

The fiber is inserted into a titanium tube from New England Small Tube Corp. (0.050 OD × 0.033 ID, melting point ±1660°C). The fiber and the tube are bent together to filter out the parts of the higher modes and achieve a narrow and stable reflection spectrum. Then, the fiber is fixed inside the titanium tube by blowing in high-purity alumina ceramic paste from Cotronics Corp. After the paste cures, a robust sensor is ready for test. The grating spectrum is measured again after bending, as shown in Figure 5.24. It is very clear that some high-order modes disappear and the complex peaks at shorter wavelength range are filtered out.

To test the high-temperature sensing capability, the sensor was put into a high-temperature muffle furnace from MTI Corporation (KSL-1700X) and the reflection signal was measured from 200 to 1600°C. The measured experimental results are shown in Figure 5.25 and Figure 5.26. Similar to the silica fiber gratings, as temperature increases, the resonant peak shifts to the

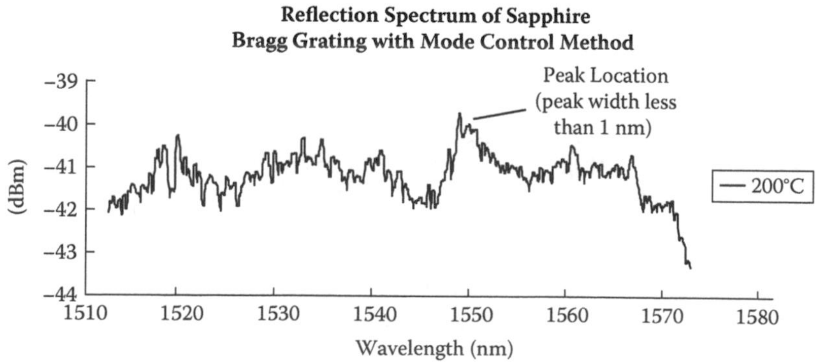

FIGURE 5.25
The measured spectra of the FBG fabricated in sapphire fiber at 200°C.

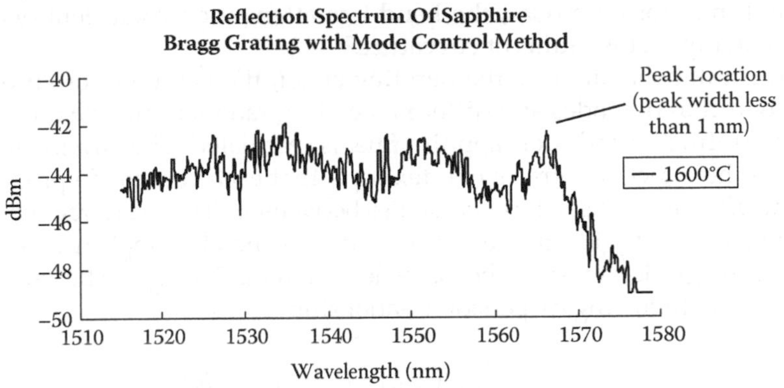

FIGURE 5.26
The measured spectra of the FBG fabricated in sapphire fiber at 1600°C.

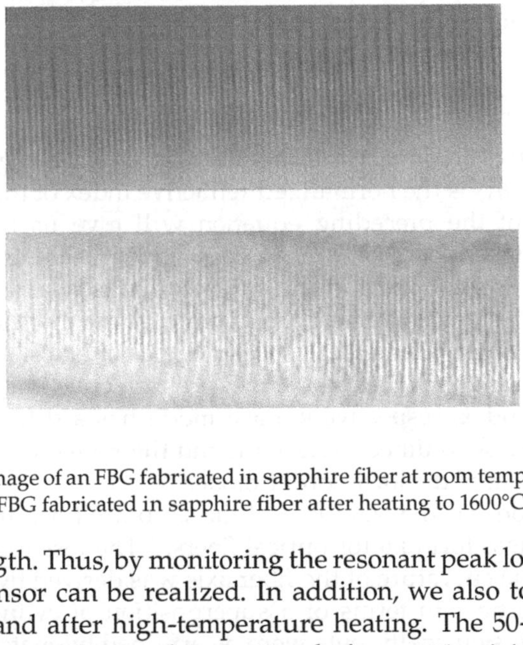

FIGURE 5.27
(a) A microscopic image of an FBG fabricated in sapphire fiber at room temperature; (b) a microscopic image of an FBG fabricated in sapphire fiber after heating to 1600°C for over 5 hours.

longer wavelength. Thus, by monitoring the resonant peak location, the high-temperature sensor can be realized. In addition, we also took microscopic images before and after high-temperature heating. The 50-time magnified microscopic image of the sapphire grating before putting it into a 1600°C furnace is shown in Figure 5.27(a) and the image of the grating, after heating for over 10 hours at 1600°C is shown in Figure 5.27(b). It is clear that the grating remains in good condition after being heated to this very high temperature.

5.7.3 Results and Discussions

Bending the fiber grating to certain radius provides low loss for the low-order mode and high loss for other high-order modes. This technique is also based on the fact that the fundamental mode is the least sensitive to bend loss and that, for all modes, the bend-loss attenuation coefficient depends exponentially on the radius of curvature.[31]

To quantitatively analyze the bending effect, the bend loss of an optical fiber for different modes and different bending radius is modeled. First, the effective refractive index of sapphire fiber is calculated. The physical significance of the effective refractive index n_{eff} is that each core mode propagates with an effective index whose value lies between n_1 (i.e., the refractive index of the core) and n_2 (i.e., the refractive index of the cladding). A core mode ceases to be guided when n_{eff} becomes less than n_2. The n_{eff} can be calculated based on the following dispersion relationship[23]:

$$V\sqrt{1-b}\,\frac{J_1\left(V\sqrt{1-b}\right)}{J_0\left(V\sqrt{1-b}\right)} = V\sqrt{b}\,\frac{K_1\left(V\sqrt{b}\right)}{K_0\left(V\sqrt{b}\right)}, \tag{5.20}$$

where J_m ($m = 0, 1$) is a Bessel function of the first kind, K_m ($m = 0, 1$) is a modified Bessel function of the second kind,

$$V = 2\pi a_1/\lambda\sqrt{n_1^2 - n_2^2}$$

is the normalized frequency of the fiber at a wavelength λ, and $b = (n_{eff\_co}^2 - n_2^2)/(n_1^2 - n_2^2)$ is the normalized refractive index of the core.

The solution of the preceding equation will give us universal curves describing the dependence of b on V. For a given value of l, there will be a finite number of solutions and the mth solution is referred to as the LP_{lm} mode. Figure 5.28 shows that there are 25 propagating modes within the sapphire fiber. The fiber parameters used for the calculation are 75 μm, 1.746, and 1.0, representing the core radius, the core refractive index, and the cladding refractive index, respectively. Each mode has a different n_{eff}. Higher order modes are less confined in the core, and their effective refractive index is smaller than that of the lower order mode.

Second, we modeled the bend loss of the sapphire fiber. In 1976, Marcuse derived a bend loss formula for optical fibers.[31] The loss formula with constant radius of the curvature of the fiber axis was derived by expressing the field outside the fiber in terms of a superposition of cylindrical outgoing waves. The expansion coefficients were determined by matching the superposition field to the field of the fiber along a cylindrical surface that was tangential to the outer perimeter of the curved fiber. For an optical fiber with a core radius a, refractive index of the core n_{co}, refractive index of the cladding n_{cl}, and a bend radius R, the ratio of the output power P_{out} to the input power P_{in} is given by

$$\frac{P_{out}}{P_{in}} = \exp(-2\alpha l_b), \tag{5.21}$$

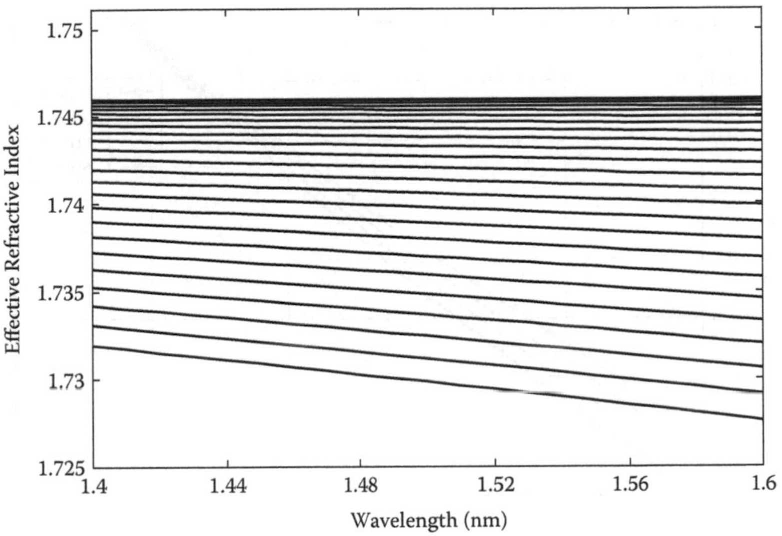

FIGURE 5.28
Effective refractive indices of different modes in sapphire fiber.

where l_b is the length of bend and 2α is the pure bend loss coefficient. When an infinite cladding thickness of the fiber is assumed, the bend loss coefficient is given by[32]

$$2\alpha = \frac{1}{2}\left(\frac{\pi}{\gamma^3 R}\right)\frac{\kappa^2}{V^2 K_1(\gamma a)}\exp\left(-\frac{2\gamma^3 R}{3\beta^2}\right) \tag{5.22}$$

where $k = 2\pi/\lambda$, $\kappa = (k^2nco^2 - \beta^2)^{1/2}$, $V = ak(n_{co}{}^2 - n_{cl}{}^2)^{1/2}$, $\gamma = (\beta^2 - k^2n_{cl}{}^2)^{1/2}$, and β is the propagation constant of the fundamental mode. Following the analysis of Marcuse, we calculated the attenuation for each of the LP_{lm} that can prorogate within the sapphire fiber.

Figure 5.29 shows the bending loss curves of the sapphire fiber as a function of the effective refractive index for different wavelengths. Higher order modes have lower effective refractive indices, and the attenuation coefficients for higher order modes are larger than lower order modes; these modes will therefore be more effectively suppressed. As shown in Figure 5.29, the bending loss curves are monotonic decreasing curves with the wavelength since the guided mode is less confined at the longer wavelength. Figure 5.30 shows the bending loss against bending radius for different wavelengths. As expected, the bending loss is bigger with smaller bending radius. Figure 5.29 and Figure 5.30 indicate that >3 dB suppression for the higher order modes is possible. The corresponding bending radii are calculated to be 10 mm, which is compatible with the compact packaging of the sensor, described previously in this section.

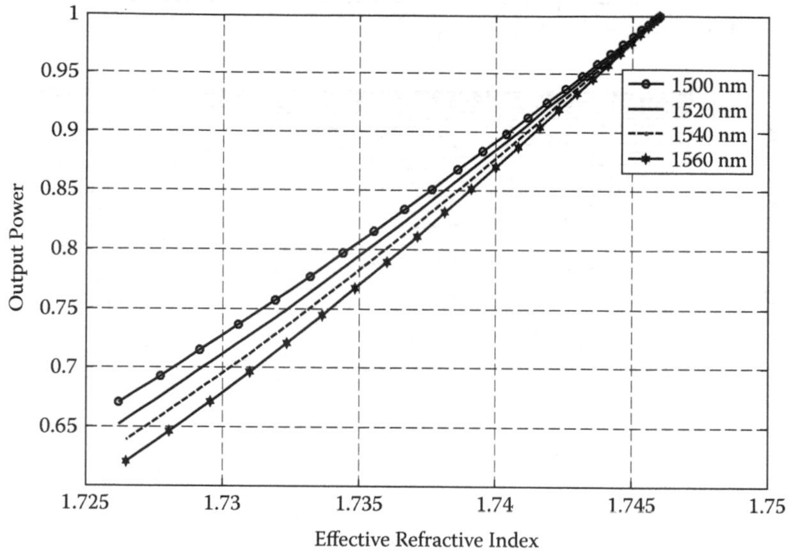

FIGURE 5.29
Calculated bend loss curves of the sapphire fiber as a function of effective refractive index for different wavelengths.

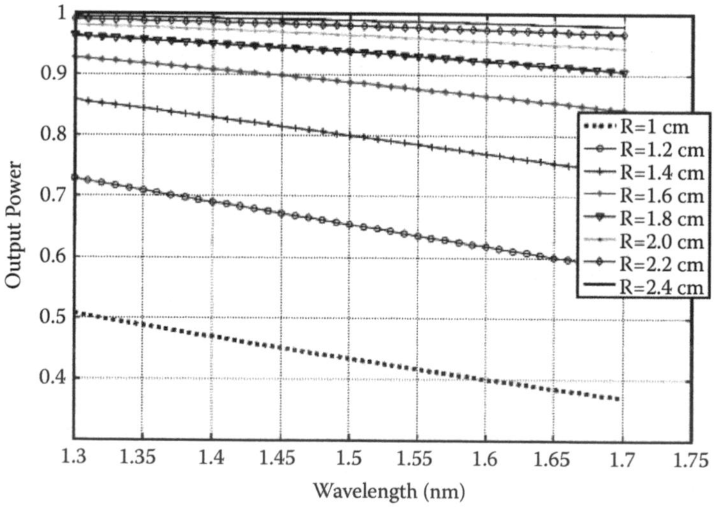

FIGURE 5.30
Calculated bend loss curve as a function of bend radius for different wavelengths.

Third, we also analyzed the temperature dependence of the grating based on Eq. (5.11). To conduct this analysis, we differentiate the grating equation (Eq. 5.1) with temperature and obtain

$$\frac{d\lambda_r}{dT} = \lambda_r \left(\frac{dn_{co-s}^{eff}}{n_{co-s}^{eff} dT} + \frac{d\Lambda_s}{\Lambda_s dT} \right), \tag{5.23}$$

where

$$\frac{d\Lambda_s}{\Lambda_s dT}$$

is the thermal expansion coefficient for the sapphire fiber (approximately $5.4 \times 10^{-6}/\mathrm{K}$ for c-axis sapphire), and

$$\frac{dn_{co-s}^{eff}}{n_{co-s}^{eff} dT}$$

represents the thermo-optic coefficient that is typically around $12 \times 10^{-6}/\mathrm{K}$[33] for sapphire at wavelength of 633 nm.

At 633 nm, the wavelength thermal sensitivity of the sapphire grating was evaluated to be around 20 pm/°C. As shown in Figure 5.31, our experimental results show that the thermal sensitivity of our ultrathin sapphire grating is 12.4 pm/°C.

The difference between the theoretical calculation and experimental results may be due to the following facts. First, the thermo-optic coefficient is a function of wavelength, and it decreases at longer wavelength.[34] So, at 1550 nm, the thermo-optic coefficient should be smaller than $12 \times 10^{-6}/\mathrm{K}$. Second, notice in Eq. (5.23) that it is the effective thermo-optic coefficient of sapphire instead of the value of bulk material. The effective thermo-optic coefficient of the sapphire fiber is composed of the thermo-optic coefficient of the sapphire bulk material (fiber core) and the air (fiber cladding). Since the refractive index of air does not change much with the temperature, the effective thermo-optic coefficient of the sapphire should be much smaller

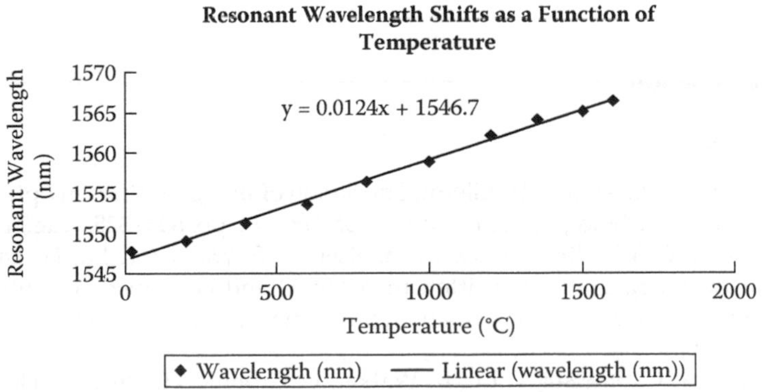

FIGURE 5.31
The resonant spectral peak locations as a function of temperature.

than the bulk material. If we estimate the thermo-optic coefficient of the sapphire fiber to be half of the bulk material—that is, 6×10^{-6}/K, then the calculated thermal sensitivity of our sapphire grating is around 13 pm/°C, which is close to our experimental results.

5.8 Conclusion

In this chapter, we have briefly reviewed our recent work on harsh environmental multiparameter sensing based on the FBGs fabricated by IR femtosecond laser irradiation in MM fibers, PM fibers, and sapphire fibers. First, we presented the fabrication methods and experimental procedures of the FBG fabrications in the MM and PM fibers. Second, we discussed the unique features of these FBGs, including (1) good for harsh environment sensing (gratings can survive up to 1000°C), and (2) multiparameter sensing capability (the changes of the spectral profiles of multidips are different under different types of perturbations such as temperature and bending. Third, the applications of these FBGs to sense different parameters, including (1) temperature, (2) bending, (3) refractive index, and (4) liquid pressure, were addressed. Fourth, we also discussed the fabrications and applications of FBGs in the thickness-reduced MM and PM fibers. These experimental results clearly demonstrated that FBGs fabricated by IR femtosecond laser irradiations in the MM and PM fibers were very suitable for the harsh environment multiparameter sensing. Finally, we proposed and developed a high-temperature sensor by fabricating higher order mode rejected fiber Bragg gratings in ultrathin sapphire crystal fiber. The grating performance is tested in a high-temperature furnace up to 1600°C. This unique in-fiber grating in single crystal sapphire fiber could result in a new generation of high-temperature, harsh environment distributed sensors.

References

1. G. Meltz, W. Morey, and W. Glenn, Formation of Bragg gratings in optical fibers by a transverse holographic method, *Opt. Lett.*, **14**, pp. 823–825, Aug., 1989.
2. J. Stephen, C. W. Mihailov, D. G. Smelser, R. B. Walker, P. Lu, H. Ding, and J. Unruh, Bragg gratings written in All-SiO2 and Ge-doped core fibers with 800-nm femtosecond radiation and a phase mask, *J. Lightwave Technol.*, **22**, 1, 94, January 2004.
3. S. Mihailov, C. Smelser, P. Lu, R. Walker, D. Grobnic, H. Ding, G. Henderson, and J. Unruh, Fiber Bragg gratings made with a phase mask and 800-nm femtosecond radiation, *Opt. Lett.*, **28**, pp. 995–997, 2003.

4. D. Grobnic, C. Smelser, S. Mihailov, and R. Walker, Long-term thermal stability tests at 1000°C of silica fiber Bragg grating made with ultrafast laser radiation, SPIE, **5855**, pp. 106–109, 2005.
5. A. M. Strelsov and N. F. Borrelli, Study of femtosecond laser written waveguides in glasses, *J. Opt. Soc. Am. B. Opt. Phys.*, **19**, pp. 2496–2504, 2002.
6. C. Smelser, S. Mihailov, and D. Grobnic, Formation of Type I-IR and Type II-IR gratings with an ultrafast IR laser and a phase mask, *Optics Express*, **13**, pp. 5377–5386, 2005.
7. C. Christopher, W. Smelser, S. Mihailov, D. Grobnic, P. Lu, R. Walker, H. Ding, and X. Dai, Multiple-beam interference patterns in optical fiber generated with ultrafast pulses and a phase mask, *Optics Letters*, **29**, pp. 1458–1460, 2004.
8. M. G. Xu, J.-L. Archambault, L. Reekie, and J P. Dakin, Discrimination between strain and temperature effects using dual-wavelength fibre grating sensors, *Electronics Letters*, **30**, 13, 1994.
9. G. Chen, L. Liu, H. Jia, J. Yu, L. Xu, and W. Wang, Simultaneous strain and temperature measurements with fiber Bragg grating written in novel Hi-Bi optical fiber, *IEEE Photonics Technology Letters*, **16**, 221, 2004.
10. A. M. Vengsarkar, Q. Zhong, D. Inniss, W. A. Reed, P. J. Lemaire, and S. G. Kosinski, Birefringence reduction in side-written photoinduced fiber devices by a dual-exposure method, *Optics Letters*, **19**, 16, 1994.
11. S. C. Rashleigh, Origins and control of polarization effects in single-mode fibers, *Journal of Lightwave Technology*, **LT-1**, 2, 1983.
12. Y. Liu, B. M. A. Rahman, and K. T. V. Grattan, Thermal-stress-induced birefringence in bow-tie optical fibers, *Applied Optics*, **33**, 1994.
13. K-H. Tsai, K-S. Kim, and T. F. Morse, General solutions for stress-induced polarization in optical fibers, *Journal of Lightwave Technology*, **9**, 1, 1991.
14. M. Bass Ed., *Handbook of Optics*, Vol. II, McGraw–Hill, Inc., New York, pp. 33–60, 1995.
15. J. Noda, K. Okamoto, and Y. Sasaki Polarization-maintaining fibers and their applications, *Journal of Lightwave Technology*, **LT-4**, 8, p. 1071, 1986.
16. C-C. Ye, S. E. Staines, S. W. James, and R. P. Tatam, A polarisation maintaining fibre Bragg grating interrogation system for multi-axis strain sensing, in *Smart Structures and Materials 2002: Smart Sensor Technology and Measurement Systems*, D. Inaudi and E. Udd, eds., Proc. SPIE, **4694**, 195, 2002.
17. C. Caucheteura, H. Ottevaereb, T. Nasilowskib, K. Chaha, G. Statkiewiczc, W. Urbanczykc, F. Berghmansb, H. Thienpontb, and P. M'egreta, Superimposed Bragg gratings written into polarization maintaining fiber for monitoring micro-strains, *SPIE*, **5952**, p. 59520M-1, 2005.
18. V. R. Bhardwaj, P. B. Corkum, and D. M. Rayner, C. Hnatovsky, E. Simova, and R. S. Taylor, Stress in femtosecond-laser-written waveguides in fused silica, *Optics Letters*, **29**, 12, 2004.
19. W. Zhao and R. O Claus, Optical fiber grating sensors in multimode fibers, *Smart Materials and Structures*, **9**, pp. 212–214, 2000. Printed in the UK.
20. T. Mizunami, T. V. Djambova, T. Niiho, and S. Gupta, Bragg gratings in multimode and few-mode optical fibers, *Journal of Lightwave Technology*, **18**, 2, 2000.
21. A. Iadicicco, Cusano, A. Cutolo, R. Bernini, and M. Giordano, Thinned fiber Bragg gratings as high sensitivity refractive index sensor, *IEEE Photonics Technology Letters*, **16**, 4, p. 1149, 2004.

22. K. Zhou, X. Chen, L. Zhang, and I. Bennion, Optical chemosensors based on etched fibre Bragg gratings in D-shape and multimode fibres, *Proceedings of SPIE*, **5855**, p. 158, 2005.
23. A. Ghatak and K. Thyagarajan, *Introduction to Fiber Optics*, Cambridge University Press, 1998.
24. Y. Namihira, M. Kudo, and Y. Mushiaka, Effect of mechanical stress on the transmission characteristics of optical fibers, *Transactions of the Institute of Electronics Communications Engineering*, Japan, **60-C**, pp. 107–115, 1977.
25. S. Yin, S.-H. Nam, J. Chavez, Z. Chun, and C. Luo, Innovative long period gratings: Principles and applications, *Proceedings of SPIE*, **5206**, pp. 30–44, 2003.
26. D. Grobnic, S. J. Mihailov, C. W. Smelser, and H. Ding, Sapphire fiber bragg grating sensor made using femtosecond laser radiation for ultrahigh temperature applications, *IEEE Photonics Technology Letters*, **16**, 11, pp. 2505–2507, 2004.
27. D. Grobnic, S. J. Mihailov, H. Ding, F. Bilodeau, and C. W. Smelser, Single and low order mode interrogation of a multimode sapphire fiber Bragg grating sensor with tapered fibers, *Proceedings of SPIE*, **5855**.
28. D. H. Jundt, M. M. Fejer, and R. L. Byer, Characterization of single-crystal sapphire fibers for optical power delivery systems, *Applied Physics Letters*, **55**, pp. 2170–2172, 1989.
29. J. P. Koplow, D. A. V. Kliner, and L. Goldberg, Single-mode operation of a coiled multimode fiber amplifier, *Optics Letters*, **25**, 7, 2000.
30. C. Zhan, Y. Zhu and S. Yin, Asymmetric Bragg gratings inscribed by IR femtosecond irradiation for harsh environment multiparameter sensing applications, *Optical Fiber Technology*, **13**, pp. 98–107 (invited paper), 2007.
31. D. Marcuse, Curvature loss formula for optical fibers, *Journal of the Optics Society of America*, **66**, 3, 1976.
32. A. Snyder and J. Love, *Optical Waveguide Theory*, Chapman and Hall, New York, p. 479, 1983.
33. M. Bass, ed., *Handbook of Optics*, 2nd ed., Vol. II, McGraw–Hill Inc., New York, pp. 33–56, 1995.
34. J. Tapping and M. L. Reilly, Index of refraction of sapphire between 24 and 1060°C for wavelengths of 633 and 799 nm, *Journal of the Optics Society of America A*, **3**, 5, 1986.

6

Fiber Specklegram Sensors

Francis T. S. Yu

CONTENTS

6.1 Introduction

Optical fiber is one of the most efficient media for transmitting temporal information to date. However, so far it has not been possible to use it for transmitting spatial information efficiently. Although, in principle, multimode fiber offers high space-bandwidth products (roughly equal to the number of the modes), transmitting spatial information is still beyond the reach of practical reality. This is mainly due to the modal coupling and modal dispersion for which the transmitted spatial information will be severely scrambled. With present techniques, it is difficult, or impossible, to unwrap the scrambled signal.

Methods [1] have been proposed to unscramble the spatial information—for example, by means of a phase conjugate mirror (PCM). However, two strictly identical fibers are needed: one for the information transmission and the other for the information unscrambling. It is obvious that this method is not practically realistic, since it is impossible to have two physically identical fibers. Nevertheless, image transmission through a multimode fiber can be sent back to the transmission fiber using a PCM [2]. The recovered image can be observed as the transmitting end. This is obviously not a useful technique for information transmission, since the receiver is still unable to receive the information. So far, little work has been done in exploiting the spatial-information transmission from a multimode fiber for practical applications.

In recent years, sensing with optical fibers has grown to become a viable technique for a variety of applications [3–5]. For example, the embedded fiber sensor and the fiber optic actuator can be used for smart monitoring of the mechanical aspects of composite structures [6]. The increasing usage of composite materials for vehicle structural applications (e.g., aircraft, submarines, missiles, etc.) has given the impetus to develop smart structure components. However, to develop such kinds of structural components, a reliable and sensitive fiber sensor is needed. Currently, a variety of fiber sensors has been suggested for this purpose. Among them are the *interferometric fiber sensor* and the *intensity detection fiber sensor.* For the interferometric-type sensor, the sensing is primarily based on either the measurand-induced change of optical path length or the polarization properties of the fiber. Although the interferometric sensors are very sensitive, they usually require a reference-arm fiber that is vulnerable to environmental factors. To compensate for the environmental influence, complicated devices are used with the sensing system. On the other hand, the intensity detection sensor depends on the measurand-induced alternation of the sensing fiber. Although this type of fiber sensor is less vulnerable to environmental perturbations, the sensitivity is rather low and that limits its application. In addition, these two types of fiber sensors generally use a single-mode fiber, which is very brittle, so that they may not be easy to implement or even strong enough in some practical applications.

In this chapter, we discuss a new type of optical fiber sensor [7,8] called the *fiber specklegram sensor* (FSS). One of the advantages of the FSS is its capability of exploiting the complex modal interference of a multimode fiber for sensing, using a speckle hologram, which is called a *specklegram.* This type of fiber sensor is highly sensitive to modal phase changes. Since FSS is a common arm interferometric sensor, it is less vulnerable to environmental factors. One of the important aspects of FSS must be the multiplexing capability, by which multiple sensing channels can be integrated within a single sensing fiber.

6.2 Fiber Specklegram Formation

When a laser beam is launched into a multimode fiber, the output field from the fiber produces a complex speckle pattern, which is known as *modal noise* in fiber optic communication. However, the speckle pattern, in some sense,

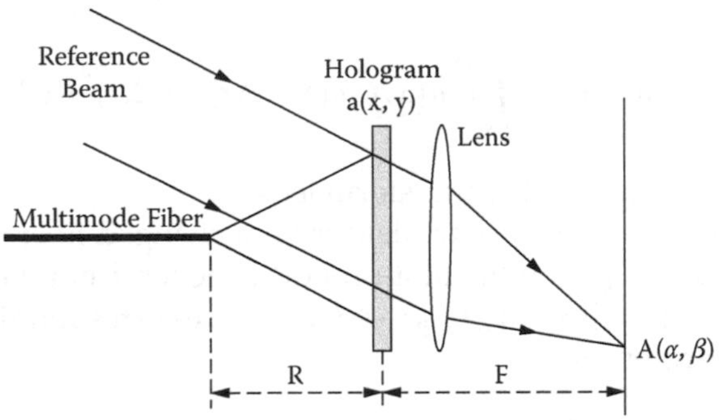

FIGURE 6.1
Fiber specklegram formation.

contains the input *spatial information* of the fiber's status. In fact, the output speckle field can be represented in terms of relative phase and amplitude quantities of the fiber modes. Although it is difficult to recover the entire spatial-information content [1], it can be used for correlation detection by means of a holographic method [7,8]. In other words, if a hologram is formed with the output fiber speckle field, the fiber status information can be recorded in a hologram, which we have called the *fiber specklegram* in Figure 6.1.

Suppose the output fiber speckle field impinging on the hologram plane is denoted by $a(x, y)$ with a phase distribution of $\phi_a(x, y)$. The recorded specklegram can be written as

$$\left|1+\left|a(x,y)\right|\exp\left[i\phi_a(x,y)\right]\right|^2 = 1+\left|a(x,y)\right|^2 + \left|a(x,y)\right|\exp\left[i\phi_a(x,y)\right] + a(x,y)\exp\left[-i\phi_a(x,y)\right] \tag{6.1}$$

where $a(x, y) = |a(x, y)|\exp[i\phi_a(x, y)]$ is the complex speckle-field distribution. If the specklegram is read out by the same speckle field (i.e., assuming an unchanged fiber status), the first-order diffraction term would be equal to $|a(x, y)|^2$, which is the intensity speckle pattern at the holographic plane. It is trivial that the *complex light field* $A(\alpha, \beta)$ at the back focal plane of the transform lens can be written as

$$A(\alpha,\beta) = \iint \left|a(x,y)\right|^2 \exp\left(i2\pi\frac{ax+\beta y}{\lambda f}\right) dx\,dy \tag{6.2}$$

and its corresponding *spectral distribution*, $S(\alpha, \beta)$, is given by

$$S(\alpha,\beta) = \iint C_0(\Delta x, \Delta y)\exp\left(i2\pi\frac{\alpha\Delta x+\beta\Delta y}{\lambda f}\right) d\Delta x\, d\Delta y \tag{6.3}$$

where

$$C_0(\Delta x, \Delta y) = \iint \left| a(x,y)a(x+\Delta x, y+\Delta y) \right|^2 dx\,dy$$

and f is the focal length of the transform lens.

In contrast, if a readout speckle field from a changed fiber status is given as $b(x, y) \exp[i\phi_b(x, y)]$, then the first-order diffraction from the specklegram would be $|a(x, y)b(x, y)| \exp\{i[\phi_a(x, y) - \phi_b(x, y)]\}$. The corresponding $A(\alpha, \beta)$ and $S(\alpha, \beta)$ can be written as

$$A(\alpha,\beta) = \iint \left| a(x,y)b(x,y) \right| \exp\left\{ i\left[\phi_a(x,y) - \phi_b(x,y) \right] \right\} \times \exp\left(i2\pi \frac{\alpha x + \beta y}{\lambda f} \right) dx\,dy \tag{6.4}$$

$$S(\alpha,\beta) = \iint C(\Delta x, \Delta y) \exp\left[i2\pi \frac{\alpha \Delta x + \beta \Delta y}{\lambda f} \right] d\Delta x\, d\Delta y \tag{6.5}$$

where

$$C(\Delta x, \Delta y) = \iint \left| a(x,y)b(x,y)a(x+\Delta x, y+\Delta y) \cdot b(x+\Delta x, y+\Delta y) \right| \exp\left\{ i\left[\Delta\phi(x,y) - \Delta\phi(x+\Delta x, y+\Delta y) \right] \right\} dx\,dy$$

and $\Delta\phi(x, y) = \phi_a(x, y) - \phi_b(x, y)$. We stress that Eqs. (6.2)–(6.5) can be regarded as the correlation output of the specklegram under unchanged and changed fiber status, respectively.

For simplicity, to analyze the correlation aspects of the fiber specklegram, we will present a one-dimensional analysis, as shown in Figure 6.2. Because the cross-reading between even and odd modes does not exist in the fiber specklegram, we would only consider the even modes for analyses. The output light field due to the mth mode is given by

$$g(\xi) = \sum_{m=0}^{M=1} \left\{ \exp\left[i\frac{(2m+1)\pi}{d} \xi \right] + \exp\left[-i\frac{(2m+1)\pi}{d} \xi \right] \right\} \mathrm{rect}\left(\frac{\xi}{d} \right) \exp(i\Psi_m) \tag{6.6}$$

where d is the thickness of the fiber and Ψ_m represents the mth mode phase factor.

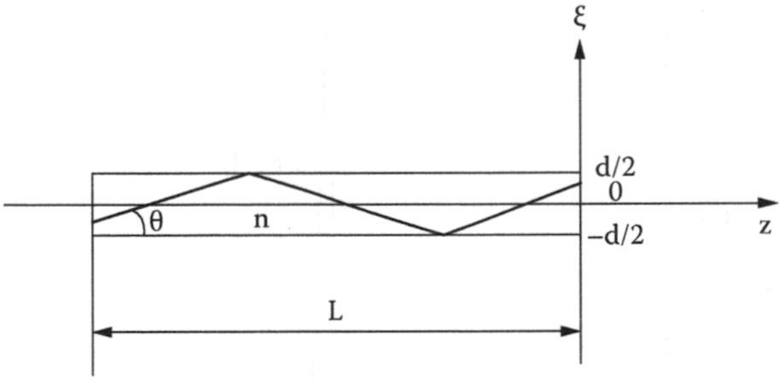

FIGURE 6.2
One-dimensional step-index waveguide.

If we assume that the holographic aperture is located at the far-field condition from the fiber, the overall speckle-field distribution at the hologram aperture would be

$$a(x) = \sum_{m=M}^{M} \text{sinc}\left(\frac{d}{\lambda R}x - \frac{m}{2}\right)\exp\left(-i\frac{\pi x^2}{\lambda R} + i\Psi_m\right) \tag{6.7}$$

where R is the distance between the fiber and the hologram aperture, M is the number of existing modes in the fiber, and $\psi_{-m} = \psi_m$.

Let us now assume that M is sufficiently large and ψ_m is a random variable over $(-\pi,\pi)$. The intensity distribution at the back focal plane for the unchanged fiber status is given by

$$S(\alpha) = \text{tri}^2\left(\frac{R}{df}\alpha\right)\left\{M^2\,\text{sinc}^2\left(\frac{R}{df}M\alpha\right) + \sum_{n=1}^{M-1}(2M-2n+1)\,\text{sinc}^2\left[n\left(1-\frac{R}{df}|\alpha|\right)\right]\right\} \tag{6.8}$$

where "tri" represents a triangular function. We note that the first term represents the readout light fields from each mode (i.e., reads its own hologram), which are coherently added, called the *coherent term*. The corresponding transform distribution has a correlation peak intensity propositional to M^2, and a spread width equals $2df/RM$. However, the second term (summation) represents the *incoherent term*, which is due to cross-modal readout of the specklegrams formed by other modes.

Figure 6.3 shows the reconstruction as a function of $R\alpha/df$ for $M = 15$ and $\mu = 100$, respectively. We see that the correlation distribution becomes sharper as M increases. Nevertheless, the spreadwidth of the correlation distribution is $2df/RM$, which is inversely proportional to M. Furthermore, the surrounding background as shown in Figure 6.3 represents the incoherent term. By integrating the coherent term with respect to α, we found that the overall

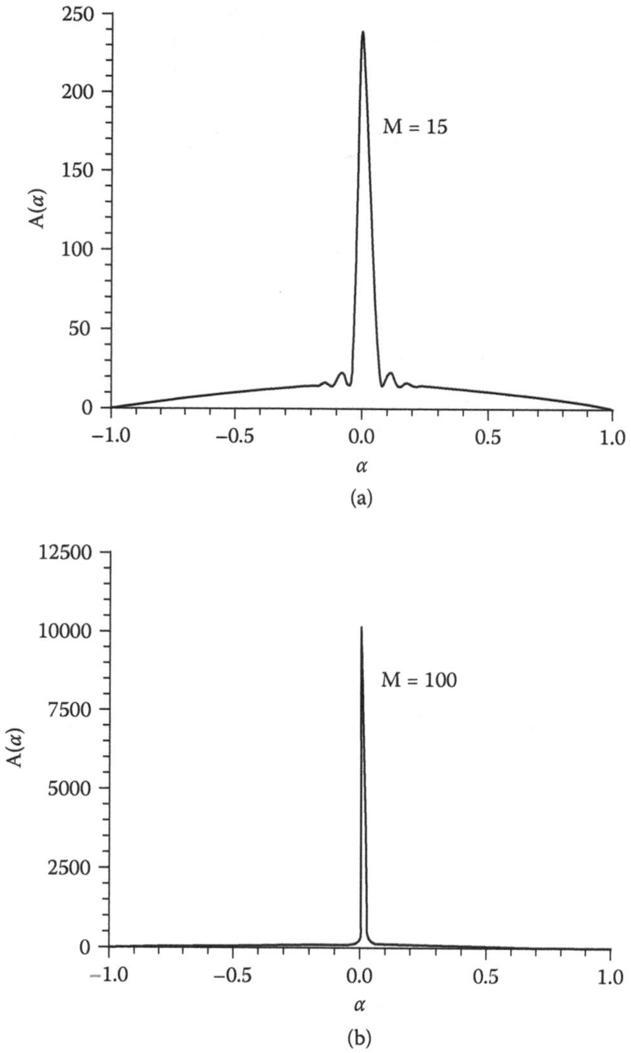

FIGURE 6.3
Correlation distributions as obtained from the fiber specklegrams: (a) for $M = 15$; (b) for $M = 100$.

energy of the correlation term would be equal to 0.43, which represents 43% of the energy that can be recovered from the speckle field with the specklegram. Notice the percentage of readout recovered from the speckle field with the specklegram. Notice that the percentage of readout energy would remain constant as the number of modes M increases. In fact, the incoherent readout term for a large modal index difference ΔM would not have a significant effect.

If the status of the fiber is changed, then the intensity output distribution would be

$$S(\alpha) = \text{tri}^2\left(\frac{R}{df}\right)\left(\left|\sum_{m=-M}^{M-1} \exp\left\{i\left[\pi\frac{R}{df}(2M+1)\alpha + \Delta\Psi_m\right]\right\}\right|^2 + \sum_{n=1}^{M-1}(2M-2n+1)\,\text{sinc}^2\left[n\left(1-\frac{R}{df}|\alpha|\right)\right]\right) \quad (6.9)$$

where we assume every mode within the fiber is equally excited, $\Delta\psi_m$ represents the corresponding phase deviation caused by the fiber status change, and $\Delta\psi_{-m} = \Delta\psi_m$.

As compared with the result of Eq. (6.8), the change of the fiber status affects only the coherent term. Let us assume that $\Delta\psi_m$ is randomly distributed over an internal $(-\delta,\delta)$; the correlation peak intensities as a function of δ for $M = 100$ and $M = 10^4$ are respectively plotted in Figure 6.4. From this figure we see that peak intensity decreases rapidly as δ increases. There are, however, two cases of δ that are discussed separately:

For $\delta \ll 1$, the peak intensity linearly decreases as the square of the standard deviation σ^2 caused by the modal phase change of the fiber, as given by

$$S_0(0) = M^2(1-\sigma^2) \quad (6.10)$$

where

$$\sigma^2 = \frac{1}{M}\sum_{m=0}^{M}\Delta\Psi_m^2$$

Thus, the detection sensitivity of the speckle hologram can be determined, such as $\partial S(0)/\partial(\sigma^2)$, which can be shown to be proportional to M^2.

For $\delta > \pi$, the coherent term (i.e., the first term) becomes completely incoherent. In this case the average output intensity would be proportional to M. Thus, the *extinction ratio* of the fiber specklegram would be on the order of $1/M$. Notice that if the mode number is large, the extinction ratio can be made very small.

For the experimental demonstration, the fiber specklegram is recorded in a $LiNbO_3$ photorefractive (PR) crystal, so that it is possible to read out the speckle hologram in situ. We have used a step-index multimode fiber 50 cm long with a numerical aperture equal to 1.3. The light source used is a 100-mW argon laser tuned at a wavelength of 514 nm, as illustrated in the experimental setup of Figure 6.5. A microbending device was used to induce the fiber status change, and the specklegram is made with no pressure applied on the microbending device. For the readout process, with no

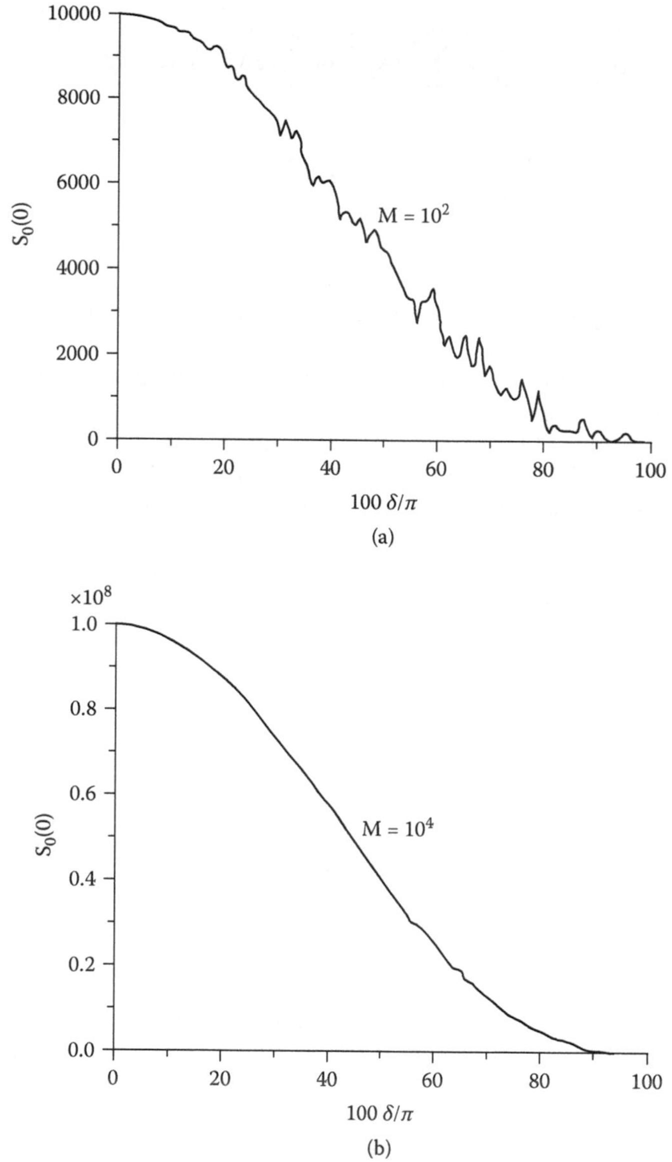

FIGURE 6.4
Correlation peak intensity as obtained from the coherent term: (a) $M = 10^2$; (b) $M = 10^4$.

fiber status change and the reference beam blocked, the correlation intensity distribution at the output plane is shown in Figure 6.6(a). The bright spot at the top of this figure shows the correlation distribution with some noise background. We have seen that the reconstruction process is rather stable, which demonstrated one of the advantages of using fiber specklegram. When

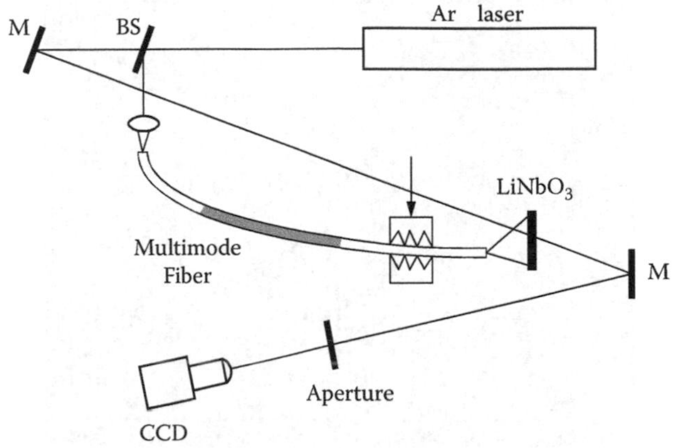

FIGURE 6.5
Experimental setup for specklegram formation: BS, beamsplitter; *M*, mirrors.

a small amount of pressure is applied on the microbending device, the correlation spot observed at the output plane immediately disappears and the noise background remains as shown in Figure 6.6(b). Note that the amount of intensity reduction from the fiber is about 1%, which indicates that the FSS is highly sensitive to a status change of the fiber. When the pressure on the microbending device is removed, we observe that the same correlation intensity profile reappears at the output end, which indicates that the sensing fiber returns to its original physical status.

6.2.1 Some Remarks

Fiber sensors, like any other sensing device, are designed to detect certain status changes and at the same time avoid the influence of other parameter changes such as environmental factors. For example, in structural health monitoring, the strain or acoustic waves are the signals we are interested in detecting. Although the FSS is a sensitive device used to detect the acoustic wave, the fiber speckle pattern also changes as the environmental temperature changes. Because the FSS is a common path interferometric sensor, a few centimeters in length is usually needed. The sensor is less affected by environmental factors than those of the two-arm interferometric technique. For instance, if one uses a fiber about 50 cm long, the temperature-tolerance (e.g., the correlation peak) decreases to half-height can be greater than 1°C. As compared with the two-arm interferometric technique, the temperature tolerance would be in the order smaller than 0.1°C.

To alleviate the influence of the excitation condition, one can use a single-mode fiber to couple the light beam into the sensing fiber. As long as the coupling between the single-mode and the sensing fiber is firmly fixed, the excitation condition remains unchanged. Furthermore, the intensity variation of the light source will not alter the relative distribution of the speckle

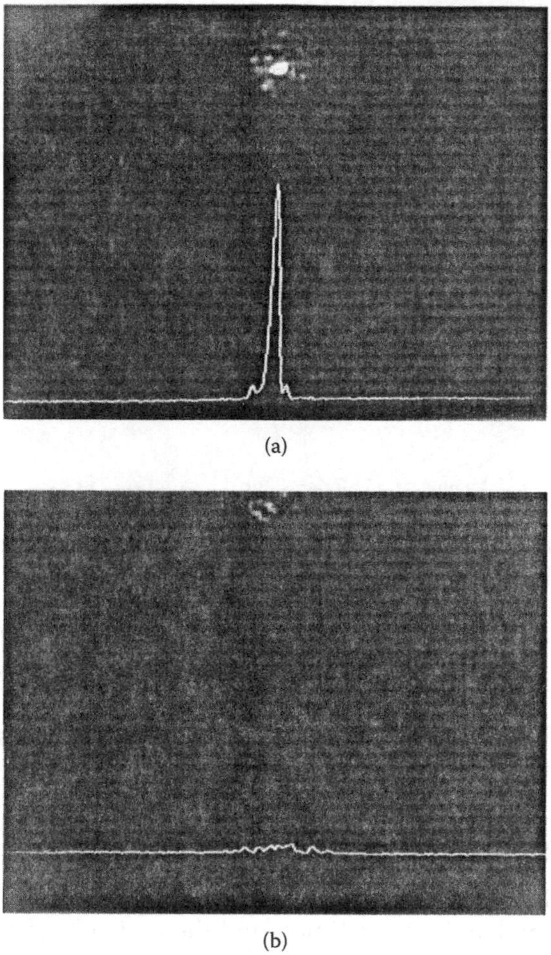

FIGURE 6.6
Correlation detection as obtained from (a) unchanged fiber status and (b) changed fiber status.

field; the perturbations due to the light source would not significantly affect the result.

To mitigate the *flexing effect* of the sensing fiber, we found that the sensing section should be as short as possible. In other words, the sensing fiber must be fixed in the sensor system so that the flexing effect can be minimized.

To improve the temperature tolerance in the FSS, we propose using two kinds of multimode sensing fibers: One has a *positive* thermal-expansion coefficient, and the other has a *negative* thermal-expansion coefficient [9]. If these fibers are connected as depicted in Figure 6.7(a), the overall thermal effect can be reduced to a minimum (i.e., $d\phi/dT = 0$), by adjusting the length ratio of fiber 1 and fiber 2. Here we have assumed a uniform temperature distribution over the entire sensing fiber. However, if the temperature variation is not uniform over the sensing fiber, then the latter can be subdivided

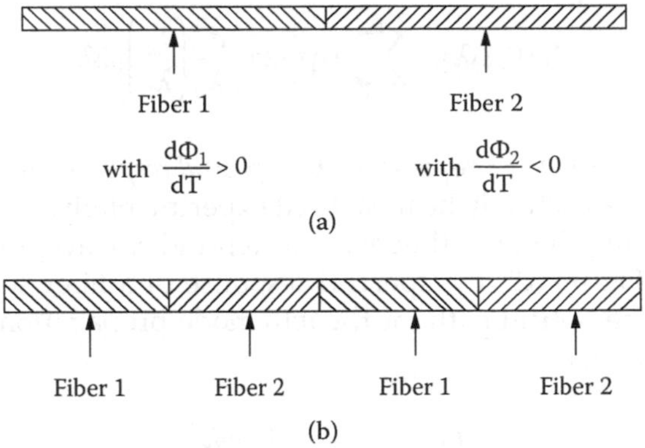

FIGURE 6.7
(a) Two cascaded fibers and (b) cascaded fibers.

into several sections. For example, two kinds of fibers will be connected in small sections as illustrated in Figure 6.7(b), where we assume the temperature distribution in each section is assumed to be uniform.

6.3 Spectral Response

Since the fiber specklegram is recorded using a specific wavelength, the reconstructed reference beam would be severely degraded if the readout wavelength deviated from the writing wavelength. This degraded reconstruction is primarily due to intermodal dispersion. Since modal-propagation velocities are different within the multimode fiber, the intermodal dispersion introduces phase differences (PDs) among the modes at the output end of the fiber. Because the PD is produced by the deviation of the relative phasings, the compensation can be achieved by using the same writing wavelength for the readout process. However, as the readout wavelength deviates, the relative phasings among the modes are different from the recorded phasings, meaning that the reconstruction of all the modes is not in phase. Suppose that the PD for the mth mode with respect to the fundamental mode is l_m, the recording wavelength is λ_0, and the recorded relative phasings are $2\pi l_m/l_0$; then the recorded mth-mode hologram, due to the wavelength deviation $\Delta\lambda$, produces a phase difference:

$$\frac{d}{d\lambda_0}\left(\frac{2\pi l_m}{\lambda_0}\right)\Delta\lambda \tag{6.11}$$

If we assume that all modes are equally excited, the peak value from the coherent term can be shown as

$$A_0(0, \Delta\lambda) = \sum_{m=0}^{M} \exp\left[i2\pi \frac{d}{d\lambda_0}\left(\frac{l_m}{\lambda_0}\right)\right]\Delta\lambda \tag{6.12}$$

for $\alpha = 0$. Since $|A_0(0, \Delta\lambda)|^2$ represents the *spectral response function* of the fiber specklegram, it can actually be measured experimentally.

Again, by taking the one-dimensional step-index waveguide approximation shown in Figure 6.2, the spectral response of a fiber specklegram can be simulated. The optical path for the *m*th mode propagation in the sensing fiber can be written as

$$\frac{nL}{\cos\theta} = nL\left[1 + \frac{1}{2}\left(\frac{m\lambda}{d}\right)^2\right] \tag{6.13}$$

where θ is the propagation angle, n is the refractive index, d is the thickness of the waveguide, and L represents the fiber length. Thus, we have

$$l_m = \frac{nl}{2}\left(\frac{ml}{d}\right)^2 \tag{6.14}$$

By substituting Eq. (6.14) into Eq. (6.12), we have

$$A_0(0, \Delta\lambda) = \sum_{m=-M}^{M} \exp\left(i\frac{\pi m^2 \Delta\lambda nL}{d^2}\right) \tag{6.15}$$

If the preceding summation is approximated by an integration over m, the correlation peak of the reconstructed light field can be expressed by a *Fresnel integral* as given by

$$A_0(0, \Delta\lambda) = 2M \frac{d}{(2\lambda L\Delta\lambda M)^{1/2}} \int_0^{p_0} \cos\frac{\pi}{2} P^2\, dI \tag{6.16}$$

where $P_0 = [(2nL\Delta\lambda_0)^{1/2}/d]M$.

If the *first minimum value* of the Fresnel integral is used to define the bandwidth of the spectral response—that is,

$$P_0 = \frac{(2nL\Delta\lambda_0)^{1/2}}{d} M = 1.74 \tag{6.17}$$

then the *spectral bandwidth* of the fiber specklegram can be shown as

$$(\Delta\lambda)_0 = 1.5 \frac{d^2}{nLM^2} \tag{6.18}$$

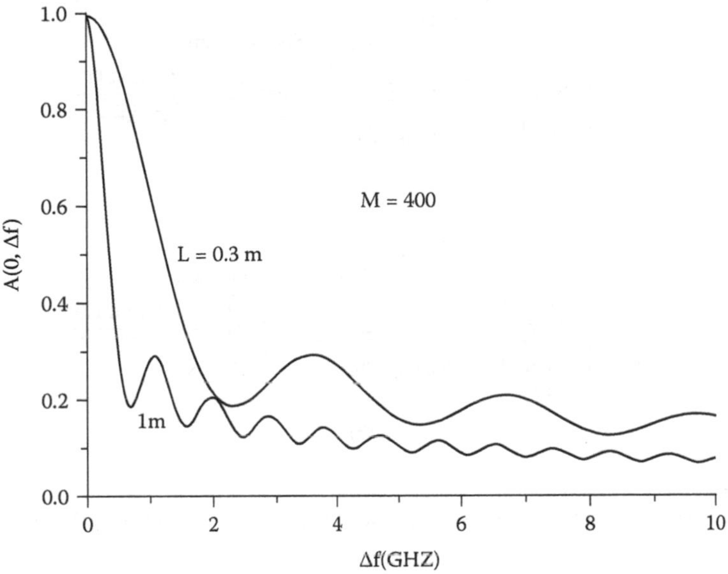

FIGURE 6.8
Spectral-response function of fiber specklegrams: $l = 0.3$ m and $L = 1.0$ m.

and the corresponding *temporal frequency bandwidth* is given by

$$(\Delta f)_0 = 1.5nc \frac{1}{(NA)^2 L} \tag{6.19}$$

where $NA = M\lambda n/d$ is the numerical aperture and c is the velocity of the light in free space. From this equation we see that the bandwidth of the fiber specklegram is inversely proportional to L, the length of the sensing fiber. We further note that $(\Delta f)_0$ is inversely proportional to $(NA)^2$. Thus, by increasing the numerical aperture of the fiber, one can reduce the bandwidth of the FSS.

A plot of the reconstructed light field $A_0(0, \Delta\lambda)$ against Δf is given in Figure 6.8. We see that the bandwidth of the FSS is indeed very narrow, which is in the order of a few gigahertz. The high-wavelength selectivity of the speckle hologram is one of the advantages of using wavelength division multiplexing (WDM) in an FSS system.

6.4 Multiplexing and Demultiplexing

The FSS can be angular division multiplexing (ADM) and WDM, as can be seen in Figure 6.9(a) and Figure 6.9(b), respectively [8]. When the phase

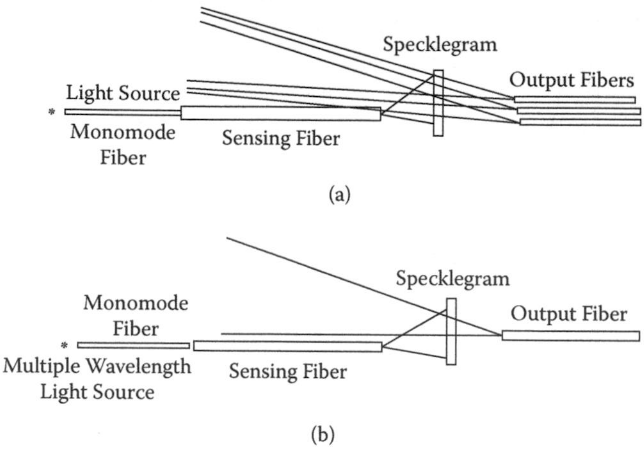

FIGURE 6.9
FSS multiplexings: (a) angular division multiplexing (ADM); (b) wavelength division multiplexing (WDM).

variation δ is greater than π, the reconstructed light field from the specklegram is extremely weak for a large M. The changes of fiber status in different aspects can be treated as mutually orthogonal statuses. Thus, by simply exploiting these orthogonal fiber statuses, a speckle hologram can be angular multiplexed, wavelength multiplexed, or both. However, arrays of the detectors are needed for ADM FSS. On the other hand, if WDM is used, only one detector is needed, but time-sharing or spectroscopic analysis to separate the channels is required. Even by using a thin-emulsion specklegram, we have found that a reasonably good ADM and WDM specklegram can be used. Needless to say, if thicker specklegram is used, higher ADM and WDM are expected. We stress that multiplexing capability in the fiber specklegram is one of the major features for the application of the FSS.

We note that for ADM the angular separation should be greater than $2df/R$, which is determined by the diameter d of the fiber. For WDM the wavelength separation should be larger than $10(\Delta\lambda)_0$, where $(\Delta\lambda)_0$ is the spectral bandwidth of the fiber specklegram.

For an experimental demonstration, we use a photorefractive crystal for the recording specklegram. By using different angular reference beams, we have constructed an ADM specklegram with respect to different pressures as applied on the microbending device. In other words, a specific reference-beam angle is used for the construction of a specific subhologram by applying a specific pressure on the microbending device. Figure 6.10(a) and Figure 6.10(b) show the correlation results obtained from the ADM specklegram with and without applied pressure, respectively. It is obvious that the correlation results can be switched back and forth simply by applying the "on" and "off" pressure on the microbending device.

A WDM FSS can be made by using different illuminating wavelengths for different applied pressures. Figure 6.11 shows one of the typical examples

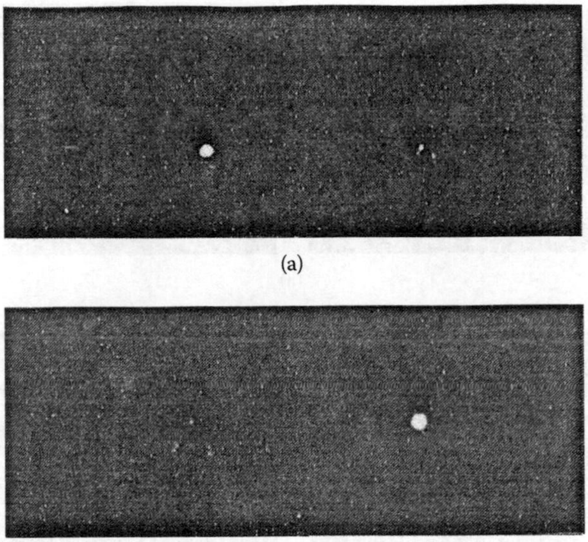

(a)

(b)

FIGURE 6.10
Detection from ADM specklegram: (a) and (b) obtained from two fiber statuses.

where correlation detections were obtained from a WDM specklegram. The orthogonal fiber statuses were multiplexed in a WDM specklegram with red, green, and blue lights. The spectral contents were obtained by applying the same respective pressures with the same polychromatic (red, blue, and green) readout.

We see that N sensors (e.g., microbending devices) can in fact be incorporated in a single-sensing fiber. The output modal-phase distribution would be the combination of the modal-phase changes, induced by the sensors. However, we have to determine those changes induced by the specific sensor. Nevertheless, this problem can be solved by using a demultiplexing technique, to be discussed next.

The simplest demultiplexing method is using a binary decoding scheme, in which (on–off) binary sensors are used. It is trivial to note that if N sensors are used, we would have 2^N orthogonal fiber statuses recorded in an ADM specklegram. Thus, by reading the angular reconstructed light fields, the status from the N sensors can be detected. Although this technique is a simple method for the FSS implementation, it requires 2^N recordings, which would be cumbersome for practical implementation.

Another approach for demultiplexing is using the so-called harmonic decoding techniques for which the harmonic signals can be detected. Suppose that two vibrational sensors are integrated into the sensing fiber, in which these two holograms are multiplexed in the specklegram. Let us assume that the first hologram is recorded when a bias pressure P_0 is applied to the first transducer and that no bias is exerted on the second transducer.

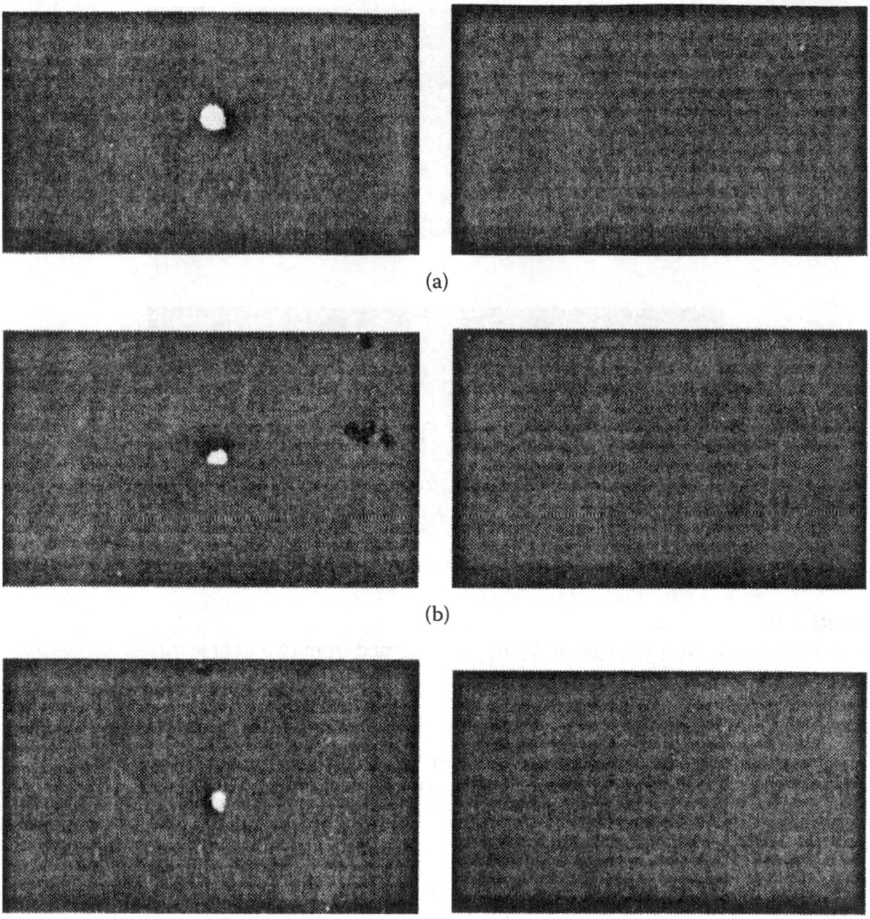

(a)

(b)

(c)

FIGURE 6.11
Detection from WDM specklegram; (a), (b), and (c): red, blue, and green correlation peaks as obtained from three different fiber statuses.

Let us further assume that the second hologram is made with no bias pressure applied on both transducers, as shown in Figure 6.12. Thus, in the reconstruction process, the output-reconstructed light field from the first hologram would be operating in the linear region in the P_1 direction. Because the second hologram is operating in a quadratic region in the P_2 direction, it will generate harmonic signals in the quadratic region, as illustrated in Figure 6.12. Thus, if a small variation of P_1 is applied at the first sensor, it would generate a fundamental signal at the correlation plane, whereas the variation of P_2 applied at the second sensor would generate a harmonic signal. Therefore, it is apparent that one can decode the time signal from the different transducers. Thus, we see that N sensing channels can indeed be integrated into a single-sensing fiber by using the FSS system.

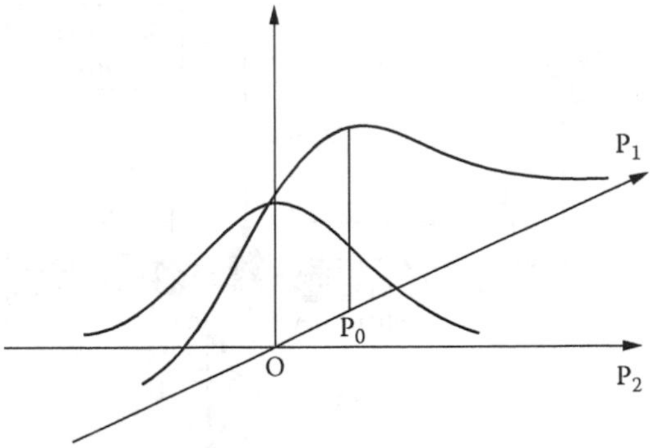

FIGURE 6.12
Harmonic decoding.

6.5 Coupled-Mode Analysis

It is essential to understand the effect of fiber perturbation on the output speckle field. We shall now investigate the speckle field as affected by modal coupling due to fiber status changes [10]. Since the perturbation of a multimode fiber can be detected by simply measuring the changes in the fiber speckle field, the changes can be measured either by subtracting the preceding with the current speckle field or by correlating between two speckle fields. These methods primarily use the spatial intensity content, while the complex amplitude of the speckle fields is not utilized. To actually use the complex speckle-field sensing, it can be coupled with a photorefractive fiber, as shown in Figure 6.13. The complex speckle field at point *A*, before transmitting through the bending part of the fiber [11], can be written as

$$E_1 = \sum_n b_n g_n(x, y) \tag{6.20}$$

where b_n and $g_n(x, y)$ represent the coefficient and the spatial complex field of the *n*th mode. If the fiber is, in fact, deformed at point *P*, there would be a modal coupling at this point. Thus, the complex light field at the deformation end (i.e., point *B*) can be written as

$$E_2(x, y) = \sum_n c_m g_m(x, y) \tag{6.21}$$

where

$$c_m = \sum_n b_n T_{nm} \tag{6.22}$$

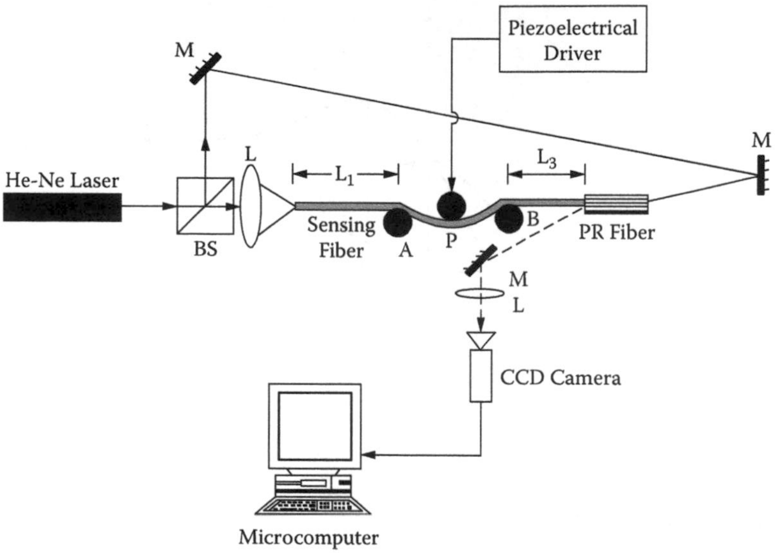

FIGURE 6.13
Experimental setup of a PR FSS: M, mirrors; BS, beamsplitter; L, lenses; L_1, L_3, fiber length.

and T_{nm} represents the *modal transfer factor* from the nth to mth mode. By combining the preceding equations, we have

$$E_2(x,y) = \sum_{n,m} b_n T_{nm} g_m(x,y) \tag{6.23}$$

If we assume that the section between point B and the exit end is short and straight, the modal coupling within this section can be disregarded. Thus, the complex speckle field emerging from that fiber would be

$$E_3(x,y) = \sum_{n,m} b_n T_{nm} g_m(x,y) \exp(i\beta_m L_3) \tag{6.24}$$

where β_m denotes the mth-mode wave constant. By assuming the fiber has been deformed (i.e., bent), the exit field should have a modal transfer factor of T_{nm}. As the fiber is further perturbed, a new complex speckle field is denoted by $E'_3(x,y)$, and the new modal transfer factor is given by T'_{nm}. Thus, the correlation peak value (CPV) between these two fiber speckle fields can be obtained by

$$\text{CPV} = \iint E'_3(x,y) E_3^+(x,y)\,dx\,dy, \tag{6.25}$$

which can be shown as

$$\begin{aligned} \text{CPV} &= \iint \sum_{s,t} b_s T'_{st} g_t(x,y) \exp(i\beta_t, L_3) \sum_{n,m} b_n^* T_{nm}^* g_m^*(x,y) \\ &\quad \times \exp(-i\beta_m, L_3)\, dx\, dy \\ &= \sum_{s,t} b_s T'_{st} \exp(i\beta_t, L_3) \sum_{n,m} b_b^* T_{nm}^* \exp(-i\beta_m L_3) \\ &\quad \times \iint g_t(x,y) g_b^*(x,y)\, dx\, dy \\ &= \sum_{s,t} b_s T'_{st} \exp(i\beta_t, L_3) \sum_{n,m} b_n^* T_{nm}^* \exp(-i\beta_m L_3) \delta_{t,m} \\ &= \sum_{n,m,s} b_s b_n^* T'_{sm} T_{nm}^* \end{aligned} \tag{6.26}$$

where

$$b_n = a_n \exp(i\phi_n) \tag{6.27}$$

The asterisk denotes the complex conjugation, and a_n and ϕ_n are the corresponding amplitude and phase distributions, respectively. For simplicity, a_n is assumed Gaussian distributed. By substituting Eq. (6.27) into Eq. (6.26), we have

$$\text{CPV} = \sum_{n,m,s} a_s a_n^* T'_{sm} T_{mn}^* \exp(i\phi_s - \phi_n) \tag{6.28}$$

Since ϕ_s and ϕ_n are random phase distributions, the ensemble average of the preceding equation equals zero, for $s \neq n$. Thus, Eq. (6.28) reduces to

$$\text{CPV} = \sum_{n,m} |a_n|^2 T'_{nm} T_{nm}^* \tag{6.29}$$

which, depending on the modal transfer factor, is determined by the geometry of the fiber deformation.

Let us assume that the sensing fiber is a step-index fiber and the transversal electric field satisfies the scalar *Helmholtz equation*, such as

$$\nabla^2 E(r) + k^2 n^2(r) E(r) = 0 \tag{6.30}$$

where $n(r)$ represents the radial refractive index distribution of the fiber. When the sensing fiber is deformed, as depicted in Figure 6.14(a), the refractive index of the fiber remains the same, but the fiber geometry is changed. However, obtaining a solution based on fiber-bending geometry is rather forbidden. We shall analyze this problem using an equivalent approach [12], as

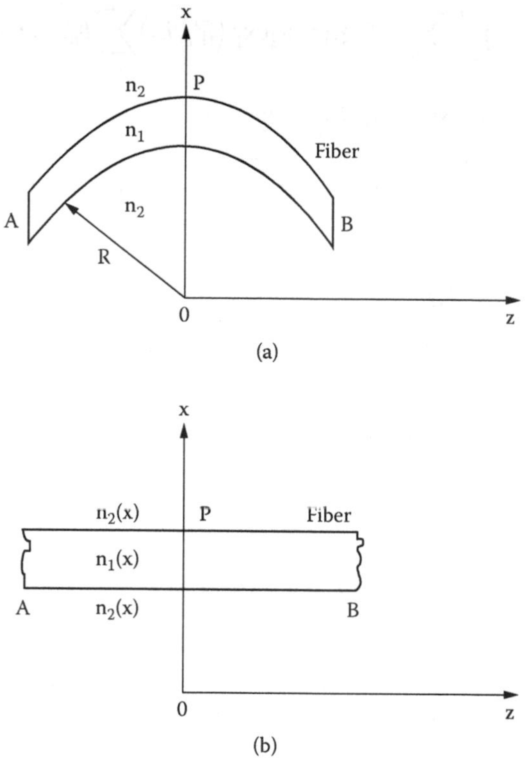

FIGURE 6.14
(a) Geometry of a bending fiber and (b) equivalent geometry. *A*, input end; *B*, output end; *R*, bending radius; n_1, n_2, refractive indices.

depicted in Figure 6.14(b), in which the equivalent refractive index variation represents the deformed geometry. In other words, the bending geometry of the fiber is converted into an equivalent straight-line (nonstep-index) fiber, so that the refractive index varies along the x direction and can be written as

$$n^2(r) - n_0^2 = \frac{2x}{R} n_0^2 \tag{6.31}$$

where n_0 is the refractive index of the fiber (without bending) and R is the radius of bending curvature, which can be approximated by

$$R \approx \frac{D^2}{8\delta} \tag{6.32}$$

in which δ is the transversal displacement and D is the distance between two supporting points on the fiber, as illustrated in Figure 6.15. If we assume that the fiber is slowly curved, then the scalar electric field shown in Figure 6.14(b) can be expressed as the combination of the local modal fields within a step-index fiber, such as

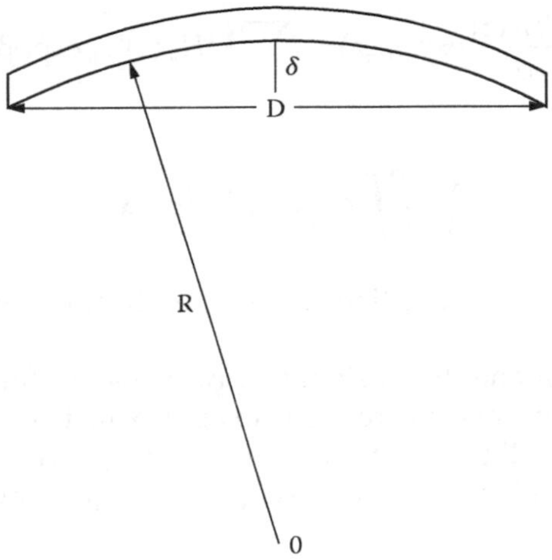

FIGURE 6.15
Curvature of fiber bending.

$$E(x,y,z)=\sum_{m} H_m(z) g_m(x,y) \exp(i\beta_m z) \tag{6.33}$$

where $g_m(x, y)$ is the *mode function* for a step-index fiber, which satisfies the following modal equation:

$$\left(\frac{\partial^2}{\partial x^2}+\frac{\partial^2}{\partial y^2}\right) g_m(x,y)+\left(k^2 n_0^2-\beta_m^2\right) g_m(x,y)=0 \tag{6.34}$$

and $H_m(z)$ is a slowly varying function given by

$$\left|\frac{\partial^2 \mathrm{H}_m(z)}{\partial z^2}\right| \ll \left|\beta_m \frac{\partial \mathrm{H}_m(z)}{\partial z}\right| \tag{6.35}$$

Thus, by substituting Eqs. (6.31) and (6.33) into Eq. (6.30) and subjecting them to the constraints of Eqs. (6.34) and (6.35), we have

$$\sum_{m} i\beta_m \frac{\partial H_m(z)}{\partial z} g_m(x,y) \exp(i\beta_m z) = -\sum_{m} k^2 \frac{x}{R} n_0^2 H_m(z) g_m(x,y) \exp(i\beta_m z) \tag{6.36}$$

By considering the orthogonality of the eigenfunction $g_m(x, y)$, we show that the preceding equation can be written as

$$\beta_n \frac{\partial H_n(z)}{\partial z} \exp(i\beta_n z) = i \sum_m k^2 H_m(z) k_{mn} \exp(i\beta_m z) \tag{6.37}$$

where

$$K_{mn} = \frac{k^2}{\beta_n} \iint g_n^*(x,y) \frac{x}{R} n_0^2 g_m(x,y)\, dx\, dy \tag{6.38}$$

Thus, we see that when the fiber is bent, each mode is coupled with all other modes.

Let us consider a one-dimensional *dielectric slab* of finite thickness with infinite extension in y and z directions to approximate the fiber geometry of Figure 6.15. There will be two types of existent eigenfunctions: namely, the *symmetric* $g_m(x, y)$ and the *antisymmetric* $g_s(x, y)$, as given by

$$g_m(x) = \begin{cases} A_m \cos\gamma_m x = \left(\dfrac{\alpha_m}{1+\alpha_m d}\right)^{1/2} \cos\gamma_m & -d \le x \le d \\ C_m \exp(-\alpha_m x) = \dfrac{\alpha_m^{1/2} \exp(\alpha_m d) \cos\gamma_m d}{(1+\alpha_m d)^{1/2}} \exp(-\alpha_m x) & x > d \\ C_m \exp(-\alpha_m x) = \dfrac{\alpha_m^{1/2} \exp(\alpha_m d) \cos\gamma_m d}{(1+\alpha_m d)^{1/2}} \exp(\alpha_m x) & x < -d \end{cases} \tag{6.39}$$

where α_m and γ_m are the eigenvalues, which can be determined from the following equations:

$$\tan\gamma_m d = \frac{\alpha_m}{\gamma_m}$$

$$\alpha_m^2 + \gamma_m^2 = k^2\left(n_1^2 - n_2^2\right)$$

$$g_s(x) = \begin{cases} A_s \cos\gamma_s x = \left(\dfrac{\alpha_s}{1+\alpha_s d}\right)^{1/2} \sin\gamma_s x & -d \le x \le d \\ C_s \exp(-\alpha_s x) = \dfrac{\alpha_s^{1/2} \exp(\alpha_s d) \sin\gamma_s d}{(1+\alpha_m d)^{1/2}} \exp(-\alpha_s x) & x > d \\ C_s \exp(-\alpha_s x) = \dfrac{\alpha_s^{1/2} \exp(\alpha_s d) \sin\gamma_s d}{(1+\alpha_s d)^{1/2}} \exp(\alpha_s x) & x < -d \end{cases} \tag{6.40}$$

where

$$\cot(\gamma_s d) = -\frac{\alpha_s}{\gamma_s}$$

$$\alpha_s + \gamma_s^2 = k^2(n_1^2 - n_2^2)$$

We note that if $g_m(x)$ and $g_n(x)$ are symmetric eigenfunctions of the same kind, the kernel of the integration in Eq. (6.39) would be an odd function, by which the integration over the (x, y)-coordinates would be zero. Thus, the coupling between two different symmetric modes would be significant, in which the *coupling coefficient* can be shown as

$$K_{sm} = \frac{k}{\beta_s}\left[\int_{-x}^{-d} C_s \exp(\alpha_s x)\frac{x}{R} n_2^2 C_m \exp(\alpha_m x)\,dx \right. \tag{6.41}$$

$$+ \int_{d}^{x} C_s \exp(-\alpha_s x)\frac{x}{R} n_2^2 C_m \exp(-\alpha_m x)\,dx$$

$$\left. + \int_{-d}^{d} A_s \sin(\gamma_s x)\frac{x}{R} n_1^2 A_m \cos(\gamma_m x)\,dx \right]$$

$$= \frac{k n_1^2}{\beta_s R\left[\left(\frac{1}{\alpha_s + d}\right)\left(\frac{1}{\alpha_m} + d\right)\right]^{1/2}} \left[\frac{\sin(\gamma_s + \gamma_m)d}{(\gamma_s + \gamma_m)^2} + \frac{\sin(\gamma_s - \gamma_m)d}{(\gamma_s - \gamma_m)^2}\right.$$

$$\left. - \frac{\cos(\gamma_s + \gamma_m)d}{(\gamma_s + \gamma_m)^2} - \frac{\cos(\gamma_s - \gamma_m)d}{(\gamma_s - \gamma_m)^2}\right]$$

To have a feeling of magnitude, we let the diameter of the fiber $2d = 50\ \mu m$, the refractive index of the fiber core $n_1 = 1.5$, and the numerical aperture of the fiber $NA = 0.2$. The coupling coefficients between the seventh and other modes are plotted in Figure 6.16, in which we see that the strongest coupling strengths exist between the seventh and eighth modes. It is also true that strong coupling existed for all other adjacent modes. In other words, the strongest couplings exist only between the adjacent modes, and the coupling between the nonadjacent modes can be neglected in the analysis.

By referring to Eq. (6.37), we note that the number of coupling equations is equal to the number of modes. Since the mode number in a multimode fiber is very large (~10^3), solving N coupling equations simultaneously is practically forbidden. However, if one retains only the two strongest adjacent modes

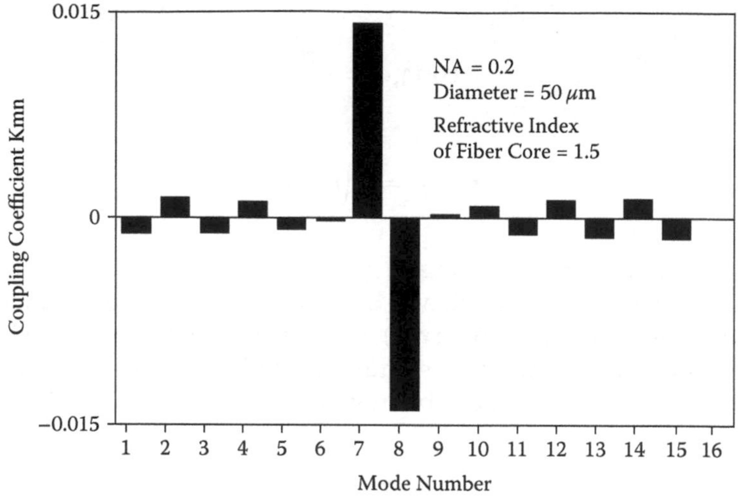

FIGURE 6.16
Coupling coefficient k_{mn} as a function of mode number.

and neglects the other weak couplings, the coupling equations between the *m*th and *s*th adjacent modes can be written as

$$\begin{cases} \dfrac{\partial H_s(z)}{\partial z} \approx -ikH_m(z)K_{sm}\exp\left[-i(\beta_s-\beta_m)z\right] \\ \dfrac{\partial H_m(z)}{\partial z} \approx -ikH_s(z)K_{ms}\exp\left[i(\beta_s-\beta_m)z\right] \end{cases} \tag{6.42}$$

The solution can be shown to be

$$H_m(z)=\exp\left[i(\beta_s-\beta_m)z/2\right]\left\{H_m(0)\left[\cos(w_{sm}z)+\frac{i(\beta_s-\beta_m)}{2w_{sm}}\sin(w_{sm}z)\right]\right. \tag{6.43}$$

$$\left. -i\frac{k}{2}\frac{K_{ms}}{w_{ms}}\sin\left(w_{sm}zH_s(0)\right)\right\}$$

$$T_{sm}=\left.\frac{H_m(L_2)\exp(i\beta_mL_2)}{H_s(0)}\right|_{H_m(0)=0} \tag{6.44}$$

$$=\exp\left[i(\beta_s+\beta_m)L_2/2\right]\left[-ik\frac{K_{ms}}{2w_{sm}}\sin(w_{sm}L_2)\right]$$

where

$$W_{sm} = \left[\left(\frac{\beta_s - \beta_m}{2}\right)^2 + k^2 L_{sm} K_{ms}\right]^{1/2}$$

By referring to Eqs. (6.29), (6.41), and (6.44), the normalized correlation peak value (NCPV) as a function of the transversal displacement for different bending period *D* can be plotted as shown in Figure 6.17. We see that the dynamic range is very much dependent on the bending radius *R*. In other words, the smaller the bending radius is, the higher the sensitivity is, but it offers a narrower dynamics range. For example, given that *D* = 20 mm, the sensitivity measure is about one order higher than that for *D* = 2.4 mm. Therefore, by using a different bending radius (i.e., the period of a microbending device), a highly sensitive FSS with an acceptable dynamic range can be fabricated.

Besides the bending curvature, the sensitivity of the FSS is also affected by the core size of the fiber. Figure 6.18 shows the NCPV as a function of the transversal displacement for different fiber-core diameters. We see that the larger the fiber core used, the higher the sensitivity of the FSS. Plots of the NCPV as a function of the transversal displacement for different *NA*s are also shown in Figure 6.19, in which we see that *NA* of the sensing fiber does not appreciably affect the sensitivity of the FSS. For comparison, the intensity NCPV as a function of transversal displacement is also plotted (by the dashed curve) in the same figure. We see that the complex amplitude sensing is indeed far more sensitive than the intensity speckle sensing [13,14].

Let us provide a result as obtained by a step-index silica of 50-μm core diameter and about 50 cm long. The PR fiber used for specklegram construction is a specially doped Ce:Fe:LiNbO$_3$ single-crystal fiber about 7 mm in

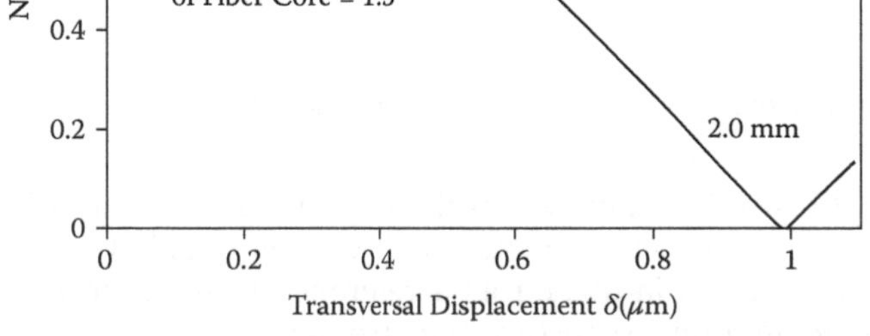

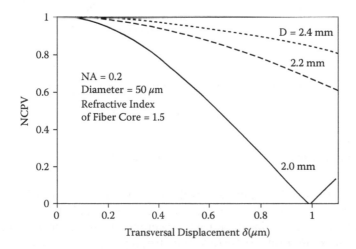

FIGURE 6.17
NCPV as a function of transversal displacement for various bending curvatures.

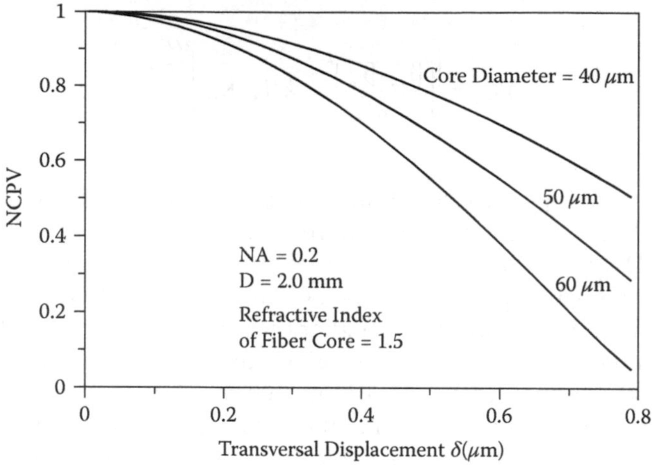

FIGURE 6.18
NCPV as a function of transversal displacement for various core diameters.

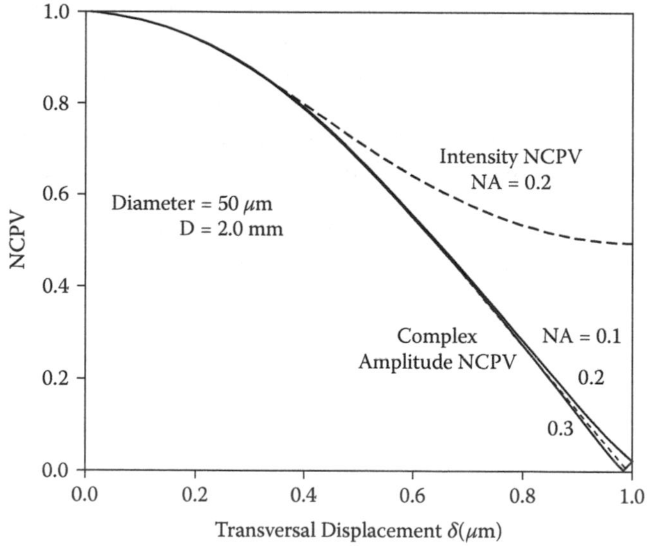

FIGURE 6.19
NCPV as a function of transversal displacement for various NAs.

length and 0.7 mm in diameter. In the experiments the laser beam is split into two paths as shown in Figure 6.13: One is coupled into the sensing fiber, and the other is launched as a reference beam to form a reflection-type fiber specklegram. To perturb the fiber, a piezoelectric driver is applied on the microbending device. The bending radius (i.e., the distance between bending points A and B) is measured to be about 2 mm. The corresponding

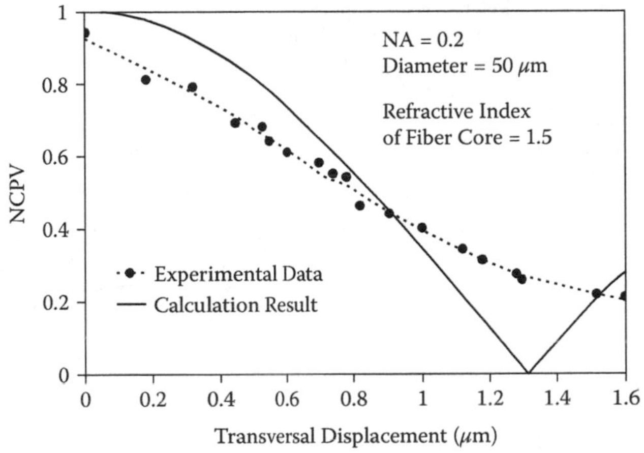

FIGURE 6.20
NCPV as a function of transversal displacement for calculated and experimental results.

NCPV as a function of transversal displacement is plotted in Figure 6.20; it is fairly linear from 0 to 1.4 μm transversal displacement. The dynamic range is calibrated to about 1.3 m, and the sensitivity is measured as high as 0.05 μm. Since the FSS is a common-path sensor, it is less vulnerable to the environmental factors than the two-arm interferometer-type fiber sensors. As compared with the analytical result shown by the continuous curve, our theoretical analysis (at least for the first-degree approximation) is fairly compatible with the experimental result. To conclude this section we have that the dynamic range and sensitivity of the FSS are both affected by the curvature of the fiber bending and its core size. We have also demonstrated that our analysis is fairly compatible with the experimental reality.

6.6 Specklegram Signal Detection

We now present a quantitative analysis of the relationship between the correlation peak intensity of the FSS system and the fiber status changes caused by environmental factors. Let us refer to Figure 6.21 in which W_0 represents the speckle wave front emerged from the sensing fiber status S_0. If the record specklegram is illustrated by W_0, a correlation peak will appear at point 0 at the observation plane *P*. However, if the fiber status slightly deviated from status S_0, presumably due to environmental changes, the speckle field emerging from the fiber would be slightly altered. The complex light field a point *Q* away from 0 at the observation plane would be a coherent addition caused by every modal phasing of the fiber. If the complex amplitude at point *Q* caused by the *m*th-mode scattering under status S_0 is given by

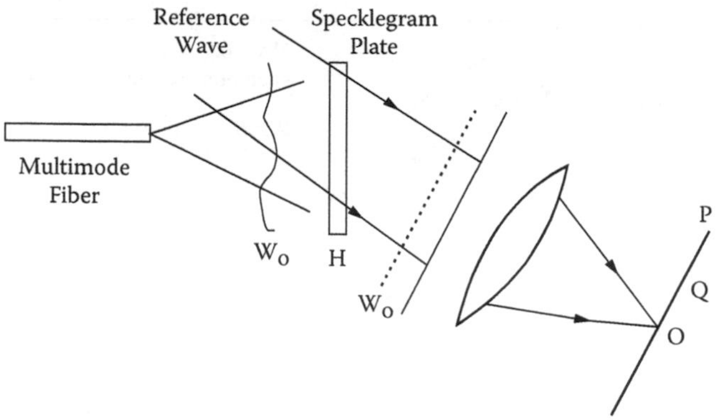

FIGURE 6.21
Specklegram reconstruction process.

$$u_{om} = a_{om} \exp(i\phi_{om}) \tag{6.45}$$

then the overall complex wave field at Q would be

$$u_0 = \sum_{m=1}^{M} u_{om} = \sum_{m=1}^{M} a_{om} \exp\left(i\phi_{om}\right) \tag{6.46}$$

where M is the number of modes.

Now let us assume that the fiber status is deviated to a new status S; the complex amplitude at point Q caused by the mth mode is then given by

$$u_m = (a_{om} + \Delta a_m) \exp[i(\phi_{om} + \Delta\phi_m)] \tag{6.47}$$

The overall complex wave field contributed by every mode is

$$u = \sum_{m=1}^{M} u_m = \sum_{m=1}^{M} \left(a_{om} + \Delta a_m\right) \exp\left[i\left(\phi_{om} + \Delta\phi_m\right)\right] \tag{6.48}$$

where Δa_m and $\Delta\phi_m$ are the random amplitude and the phase variables, respectively. If we assume that Δa_m and $\Delta\phi_m$ are Gaussian distributed, then their corresponding distributions are given by

$$P_{\phi m}\left(\phi_m\right) = \frac{1}{\sqrt{2\pi}\sigma_{\phi m}} \exp\left(-\frac{\phi_{\mathrm{m}}^2}{2\sigma_{\phi m}^2}\right) \tag{6.49}$$

$$P_{am}\left(a_m\right) = \frac{1}{\sqrt{2\pi}\sigma_{am}} \exp\left(-\frac{a_{\mathrm{m}}^2}{2\sigma_{\phi m}^2}\right) \tag{6.50}$$

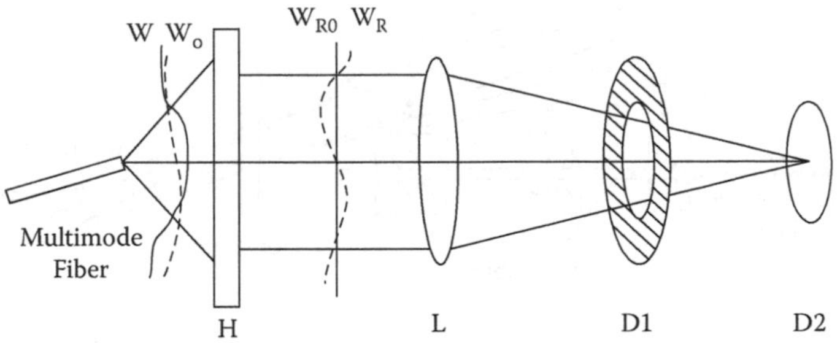

FIGURE 6.22
Detection method of the core-ring ratio: *L*, lens; *H*, specklegram.

where $\sigma\phi_m$ and σ_{am} are the standard deviations of $\Delta\phi_m$ and Δa_m, respectively.

Now let us consider the case for $\sigma\phi_m > 2\pi$, which is similar to the case of the scattering of a laser beam from a strong diffuser. The complex amplitude of the scattered light field would be more or less uniformly distributed on the observation plane.

However, for the case of $0 < \sigma\phi_m < 2\pi$, the standard deviation $\sigma\phi_m$ can be quantitatively measured. Instead of observing at the receiving plane, two concentric circular photocells, D_1 and D_2, are used (Figure 6.22). An opened-aperture D_1 permits the correlation peak to arrive at D_2, for which the detected intensity is the zero-order diffraction plus some speckle distribution that passes the center core of D_1. The intensity ratio between D_2 and D_1 (i.e., the core-ring ratio) can be written as [15]

$$R = \frac{2\pi \exp\left(-\sigma_{\phi m}^2\right) + \Omega_1\left[1 - \exp\left(\sigma_{\phi m}^2\right)\right]}{\left(\Omega_1 - \Omega_2\right)\left[1 - \exp\left(-\sigma_{\phi m}^2\right)\right]} \tag{6.51}$$

where Ω_1 and Ω_2 are the solid angles subtended by the inner and outer peripheries of D_1 and D_2, respectively. Thus, we see that the standard deviation $\sigma\phi_m$ can be determined by

$$\sigma_{\phi m} = \left[\ell n \frac{2\pi + R\left(\Omega_2 - \Omega_1\right) - \Omega_1}{R\left(\Omega_2 - \Omega_1\right) - \Omega_1}\right]^{1/2}$$

Because $\sigma\phi_m$ represents the status changes of the sensing fiber, a quantitative relationship between the correlation peak intensity and the fiber status can be obtained. In other words, by using Eq. (6.52), we can experimentally measure the standard variation of the random phase variation that emerges from the sensing fiber. To improve the dynamic ranges of this fiber specklegram sensing, we can use ADM and WDM multiplexing techniques. For instance, ADM can be used to detect different pressure ranges applied on a

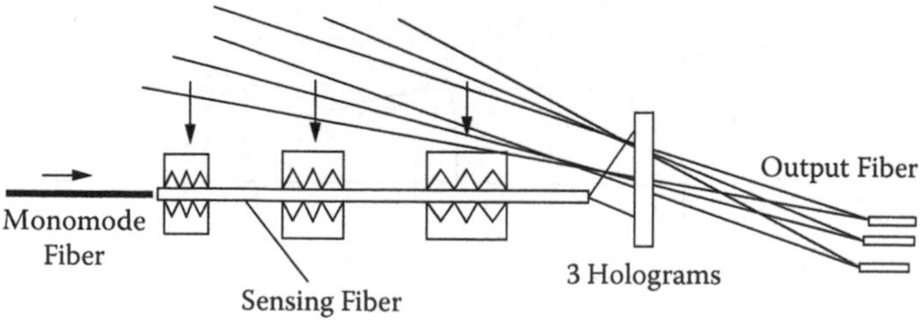

FIGURE 6.23
Application of ADM to different pressure ranges (e.g., F_1, F_2, and F_3) corresponding to different pressure ranges (e.g., P_1, P_2, and P_3).

specific microbending device. Since the backgrounds from other multiplexed specklegrams would also be read out as illustrated in Figure 6.23, the ADM angular detections for different pressure ranges can be obtained (one at a time) as illustrated in this figure.

6.7 Potential Applications

In the preceding sections we have shown that the FSS is highly sensitive for detection, is less vulnerable to environmental factors, and can be integrated in a single fiber for multisensing aspects. These unique features are difficult to accomplish with a single-mode fiber without complicated instrumental support. We shall now discuss a few of the possible applications using the FSS.

6.7.1 Acoustic-Sensing Array

An acoustic sensor using the FSS technique is shown in Figure 6.24, in which a single-mode fiber is used to launch the light beam into the sensing fiber. The emerging speckle field combined with a converging reference beam forms a specklegram. When an acoustic wave exerts pressure on the sensing

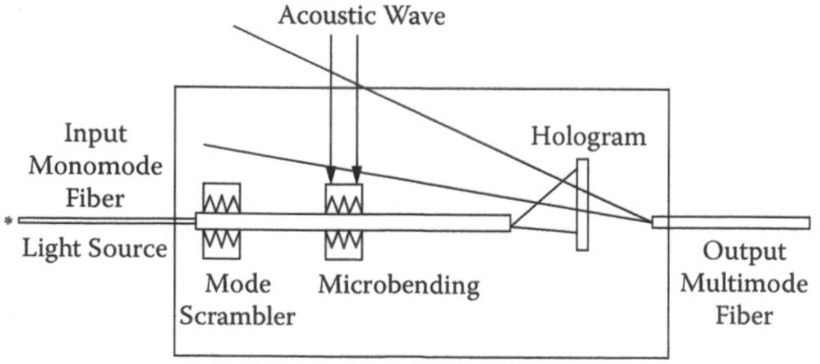

FIGURE 6.24
Fiber specklegram acoustic sensor.

fiber, the microbending device induces modal-phasing changes so that an output signal can be detected. We stress that as long as all the components inside the sensing channel are well implemented, outside disturbances on the single-mode fiber and the fluctuation of the light source would not produce significant changes of the modal phasings.

To induce modal phase variations of the sensing fiber with acoustic pressure, we can employ a specially designed microbending transducer. For example, if the bending period of the transducer matches the different wave propagations between adjacent modes, the transducer will generate a greater modal phase deviation. Thus, the sensitivity of the fiber sensor would be proportional to $d\sigma^2/dP$, which is approximately the same order of the two-arm interferometric fiber sensors, where σ is the standard deviation of the modal phase variation and dP is the applied pressure on the transducers. As compared with the intensity fiber sensor, its sensitivity is proportional to dT/dP, where T is the intensity transmittance of the fiber. As T changes, it causes the guided modes to couple into the radiation modes. Once again we see that FSS is far more sensitive than the intensity fiber sensor, since dT/dP is much smaller than $d\sigma^2/dP$.

It is trivial to construct an array of the acoustic sensors, by using a series of microbending transducers as applied on the same sensing fiber. To ensure that the modal phase variations induced by these transducers are mutually independent, one can either use different bending periods for each transducer or apply a modal scrambler between the neighboring transducers. For example, ADM with harmonic decoding method can be applied to the array of the acoustic sensors (e.g., for multiparameter sensings).

6.7.2 Structural-Fatigue Monitoring

It is important that certain structural components such as for aircraft, submarines, missiles, etc. can be monitoring while they are in operation. For example, monitoring the strain (or the vibration) may provide viable information concerning the structural fatigue and degradation that may prevent sudden disastrous destruction.

Optical fiber sensors have been identified as attractive candidates for nondestructive assessment [12,16]. One approach to the problem relies on the breakage of the sensing fiber that is embedded or attached within the composite structures. However, this method requires a significant amount of destruction to the fibers before the damage is registered. There are, however, many types of destructions, such as cracking and delamination, that could go completely unnoticed. This type of fiber-sensing method cannot detect these damages. Another drawback of the destructive-type fiber sensing is that it is an *irreversible* sensing process.

An alternative fiber-sensing technique is the use of an array of interferometers to detect the energy released from the matrix cracking, structural fiber breakage, or delamination within the composite materials. Although this method offers highly sensitive detection, the system is too complicated

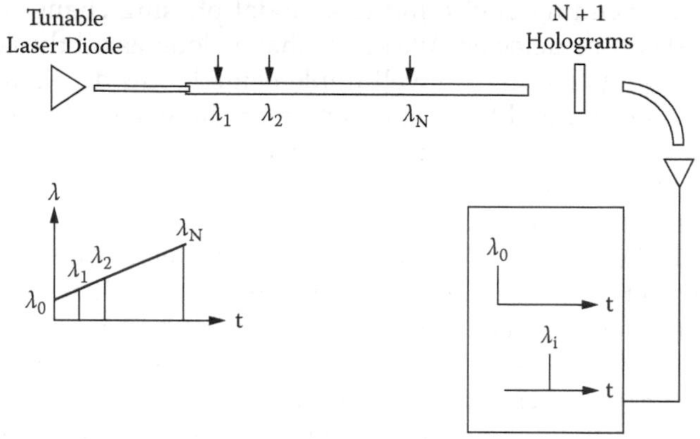

FIGURE 6.25
Structural-fatigue monitoring.

to offer practical implementation [17]. On the other hand, the multiplexing capability of the proposed FSS would provide a simple sensor array system for this purpose. For example, if N strain locations are required to be monitored, then $N + 1$ specklegrams are needed for WDM, as illustrated in Figure 6.25. In other words, if the first specklegram is taken with no strain applied by using wavelength λ_0, then the next specklegram will be taken with strain applied at the first location by using wavelength λ_1, and so forth. For reconstruction of the specklegram, a tunable laser diode can be used for scanning the wavelengths. It is trivial that if there is no strain applied on the sensing fiber, a correlation peak read out by λ_0 can be detected. Similarly, if a correlation peak detection is caused by v_i, then pressure must be applied on the ith transducer, as illustrated in the figure. We note that the WDM with a binary decoding scheme is, in fact, a mixture of time division with wavelength division multiplexing (TDM–WDM) techniques.

We must also mention the commonly used optical time-domain reflectometry (OTDR) [3]. The OTDR relies on the measurement of the back-reflection from the strain locations, which requires large strain or even breakage of the fiber for the signal to be detected. However, to locate the strain positions, OTDR requires an expensive time-delay electronic device to measure the reflected light pulses.

6.7.3 Smart Acoustic-Emission Detection

The potential importance of acoustic-emission detection and analysis for the inspection of composite structures has been reported in the literature [18]. Time- and frequency-domain analyses yield important information concerning the types of defects, the geometries, the locations, and the material fatigues. If a measurement of the acoustic emission of the composite material is required, fiber yields the major advantage of being easily embedded in the structural components.

6.8 Inner-Product FSS

We have shown that complex speckle-field sensing is far more sensitive than intensity speckle sensing. In this section we show that submicrometer displacement sensing can be achieved by utilizing the *intensity inner product* of the speckle fields. In other words, by taking the intensity speckle patterns before and after the perturbation of the fiber, the intensity inner product between the two speckle patterns can be calculated. Since the fiber speckle field is caused by the modal phasing of the fiber, the intensity inner product would be highly sensitive to the fiber status changes, which can be used for submicrometer displacement sensing as described in the following discussion.

Let us assume that a laser beam is launched into the sensing fiber and all the modes in the fiber are equally excited. With reference to the optical system illustrated in Figure 6.26, the excited complex speckle field is the coherent superposition of all the modal complex fields, as given by

$$A_0(x,y)=\sum_{m=0}^{M} a_{om}(x,y)\exp\{i[\phi_{om}(x,y)]\} \tag{6.53}$$

where M is the modal number, and $a_{om}(x, y)$ and $\phi_{om}(x, y)$ represent the amplitude and the phase distributions resulting from the mth mode, respectively. The intensity speckle field captured by the charge coupled device (CCD) camera can be written as

$$I_0(x,y)=|A_0(x,y)|^2=\sum_{m=0}^{M}\sum_{n=0}^{M} a_{om}a_{on}\exp[i(\phi_{om}-\phi_{on})] \tag{6.54}$$

If the fiber status changes, the output complex speckle field can be written as

$$A_0(x,y)=\sum_{m=0}^{M}[a_{om}(x,y)+\Delta a_m]\exp\{i[\phi_{om}(x,y)+\Delta\phi_m]\} \tag{6.55}$$

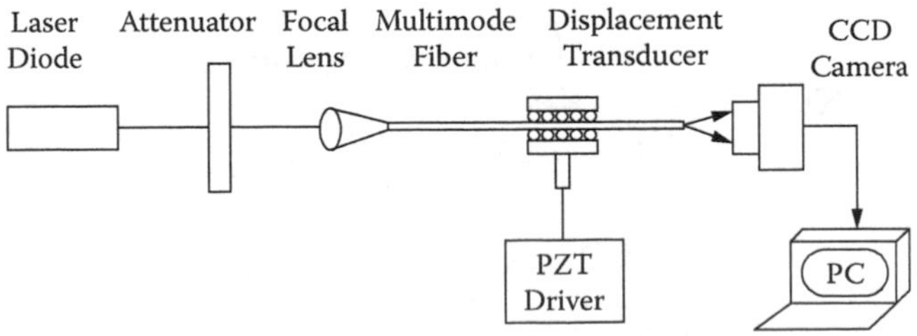

FIGURE 6.26
Multimode fiber sensor for submicrometer displacement; PZT, piezoelectric transducer.

where Δa_m and $\Delta\phi_m$ represent the amplitude and the phase deviation due to fiber status changes, respectively. The intensity distribution of the speckle field can be shown as

$$I(x,y) = \left|A(x,y)\right|^2 \tag{6.56}$$

$$= \sum_{m=0}^{M}\sum_{n=0}^{M}\left(a_{om} + \Delta a_m\right)\left(a_{on} + \Delta a_n\right)\exp\left[i\left(\phi_{omn} + \Delta\phi_{mn}\right)\right]$$

in which we have dropped the (*x*, *y*) notation for convenience. The intensity inner product of the speckle fields can be written as

$$\text{IPC} = \iint I_0(x,y)I(x,y)\,dx\,dy \tag{6.57}$$

$$= \iint \sum_{i=0}^{M}\sum_{j=0}^{M}\sum_{m=0}^{M}\sum_{n=0}^{M} a_{om}a_{oi}\left(a_{oi} + \Delta a_{oi}\right)\left(a_{oj} + \Delta a_{oj}\right)$$

$$\times \exp\left[i\left(\phi_{omn} + \phi_{oij} + \Delta\phi_{ij}\right)\right]dx\,dy$$

which can be shown as

$$IPC = \sum_{i=0}^{M}\sum_{j=0}^{M} B_{ij}\exp\left[i\left(\Delta\phi_{ij}\right)\right] \tag{6.58}$$

where

$$B_{ij} = \iint \sum_{m=0}^{M}\sum_{n=0}^{M} a_{om}a_{on}\left(a_{oi} + \Delta a_{oi}\right)\left(a_{oj} + \Delta a_{oj}\right)$$

$$\times \exp\left[i\left(\phi_{omn} + \phi_{oij}\right)\right]dx\,dy$$

The normalized intensity inner product can be calculated by

$$\text{NIPC} = \frac{\iint I_0(x,y)I(x,y)\,dx\,dy}{\left[\iint I_0^2(x,y)\,dx\,dy \iint I^2(x,y)\,dx\,dy\right]^{1/2}} \tag{6.59}$$

$$= \frac{\sum_{i=0}^{M}\sum_{j=0}^{M} B_{ij}\exp\left[i\left(\Delta\phi_{ij}\right)\right]}{\left(\sum_{i=0}^{M}\sum_{j=0}^{M} B'_{ij}\sum_{i=0}^{M}\sum_{j=0}^{M} B''_{ij}\right)^{1/2}}$$

where

$$B'_{ij} = \iint \sum_{m=0}^{M} \sum_{n=0}^{M} a_{om} a_{on} a_{oi} a_{oj} \exp\left[j\left(\phi_{onm} + \phi_{oij}\right)\right] dx\,dy$$

$$B''_{ij} = \iint \sum_{m=0}^{M} \sum_{n=0}^{M} \left(a_{om} + \Delta a_m\right)\left(a_{on} + \Delta a_n\right)\left(a_{oi} + \Delta a_m\right)\left(a_{oj} + \Delta a_j\right)$$

$$\times \exp\left[i\left(\phi_{omn} + \phi_{oij} + \Delta\phi_{mn} + \Delta\phi_{ij}\right) dx\,dy\right.$$

We see that if the sensing fiber is not perturbed, the two speckle patterns are virtually identical, for which we have normalized inner product (NIP) = 1. However, if the fiber is perturbed, the modal phase deviates. It produces a different speckle field that causes the NIP to reduce. If the perturbation of the fiber becomes extremely severe, one would expect the NIP to level off to a constant value, as we show later. Suppose that the fiber status is moderately changed; the NIP can be approximated by

$$\text{NIPC} \approx \frac{1}{\left(M+1\right)^2} \sum_{i=0}^{M} \sum_{j=0}^{M} \exp\left[i\left(\Delta\phi_{ij}\right)\right] \tag{6.60}$$

where the amplitude deviations Δa_{om} have been disregarded for which we see that $B_{ij} = B'_{ij} = B''_{ij} = \text{constant}$.

Let the maximum phase deviation between the Mth mode and the 0th mode be denoted by

$$\delta = \Delta\phi_{M0} = \Delta\phi_M - \Delta\phi_0 \tag{6.61}$$

and the phase deviation $\Delta\phi_{ij}$ given by

$$\Delta\phi_{ij} = \frac{i-j}{M}\delta \tag{6.62}$$

Then the variation of NIP as a function of δ can be plotted in Figure 6.27. We see that NIP monotonically decreases as a function of δ, within the phase variation between 0 and π. We note that this analysis is quite compatible with the results obtained in Figure 6.18.

By referring to the optical setup in Figure 6.26, we assume that a 50-cm-long multimode fiber, which has a numerical aperture of 0.2, is used as the sensing fiber. A laser diode of δ = 0.67 μm is used as the light source, and a piezoelectric transducer is applied on the microbending device. Notice that a small transverse displacement Δx would cause a small elongation of the sensing fiber. Thus, by capturing the fiber speckle intensity distributions with a

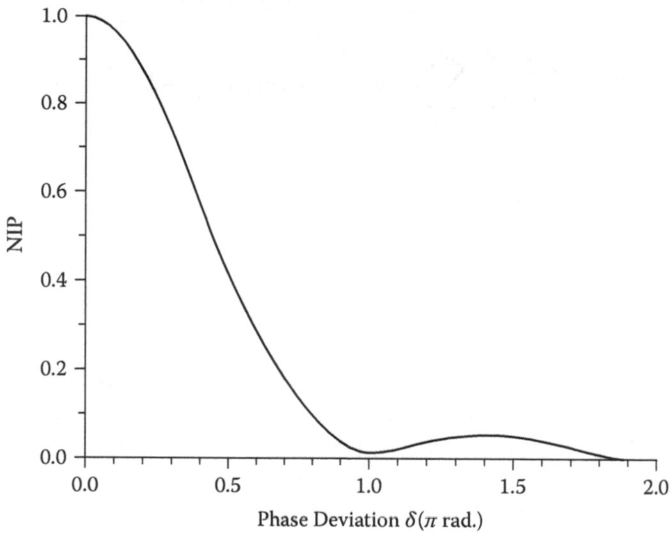

FIGURE 6.27
Variation of the NIP as a function of modal phase deviation δ.

CCD camera, the intensity inner-product evaluation can be performed by a microcomputer. Let us assume that the modal phase change caused by fiber elongation due to transversal displacement vx is written by

$$\Delta\phi_k = \frac{2\pi}{\lambda} C_0 n \Delta L \frac{1}{\cos\theta_k} \tag{6.63}$$

where λ is the wavelength of the light source, C_0 is the strain optic factor, n is the refractive index of the fiber, ΔL represents the fiber elongation, and θ_k the propagation angle in the fiber, resulting from the kth mode, with respect to the fiber axis. It is trivial that the *relative* and *maximum* modal phase deviations can be shown as

$$\Delta\phi_k = \frac{2\pi}{\lambda} C_0 n \Delta L \left(\frac{1}{\cos\theta_i} - \frac{1}{\cos\theta_j} \right) \tag{6.64}$$

$$\delta = \frac{2\pi}{\lambda} C_0 n \Delta L \left(\frac{1}{\cos\theta_M} - \frac{1}{\cos\theta_0} \right) < \pi \tag{6.65}$$

where $\delta = \Delta\phi_M - \Delta\phi_0$.

We stress that if M is in the order of several hundreds of modes, the propagation angle within the fiber can be approximated by

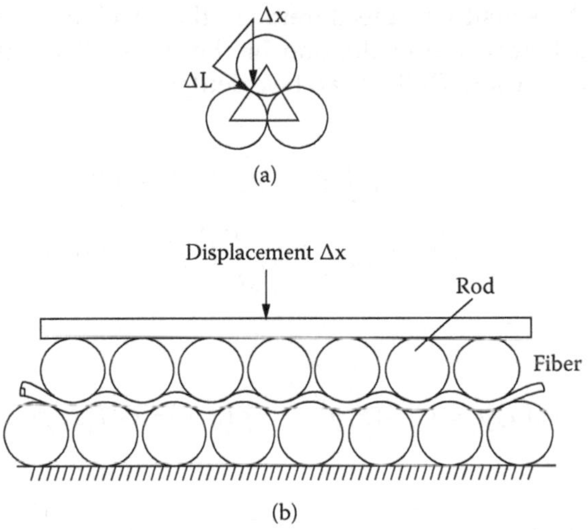

FIGURE 6.28
(a) Relationship between the transverse displacement and the fiber elongation; (b) microbending device.

$$\theta_0 \approx 0^\circ \quad \theta_M \approx \arcsin\left(\frac{\sin\theta_c}{n}\right) \tag{6.66}$$

In Figure 6.28(a), the displacement Δx can be shown related to the elongation of the fiber, such as

$$\Delta x = \frac{\Delta L}{2\alpha \sin 30^\circ} \tag{6.67}$$

where α is equal to the number of rods imposed on the sensing fiber, as shown in Figure 6.28(b).

As an example, given $\lambda = 0.67$ μm; $C_0 = 0.78$; $\alpha = 7$; $\sin\theta_c = 0.2$, where θ_c is the critical angle; and $n = 1.5$, the propagation angle in the fiber is $\theta_m = 7.7°$. Thus, the *maximum allowable displacement* of the fiber with respect to the first zero point of the NIP (shown in Figure 6.27) can be estimated by Eqs. (6.65) and (6.67). We see that the maximum allowable displacement should be limited to $\Delta x_{max} < 4.5$ μm. If the measurable deviation of the NIP is in the order of 0.01, the minimum allowable displacement of the fiber is $\Delta x_{min} > 0.05$ μm. In other words, the measurable displacement of the fiber should be limited within the range as given by

$$0.05 \ \mu m < \Delta x < 4.5 \ \mu m$$

To improve the sensitivity measurement, the speckle patterns have been edge enhanced before the evaluation of the NIPs. Thus, the normalized inner-product coefficient (NIPC) can be calculated as

$$\mathrm{NIPC} = \frac{\iint g_0(x,y)g(x,y)dx\,dy}{\left[\iint g_0^2(x,y)dx\,dy \iint g^2(x,y)dx\,dy\right]^{1/2}} \tag{6.68}$$

where

$$g_0(x,y) = \mathrm{G}\left[I_0(x,y)\right] \qquad g(x,y) = \mathrm{G}\left[I(x,y)\right] \tag{6.69}$$

$$\mathrm{G}\left[f(x,y)\right] = \left\{\left[df(x,y)/dx\right]^2 + \left[df(x,y)/dy\right]^2\right\}^{1/2} \tag{6.70}$$

where $I_0(x,y)$ and $I(x,y)$ are the intensity's speckle patterns before and after the perturbation, respectively. We note that by clipping the correlation of the speckle patterns, we reduce the calculating time and also increase the sensitivity of measurement. As shown in Figure 6.29, we see that the sensitivity improves as the thresholding increases. However, the maximum sensing displacement (i.e., the *dynamic range*) is limited to 6 μm. Nonetheless, the results indicate that an optimum thresholding level of about 25% can be used, which will provide a reasonably good linearity and *sensitivity* measurement.

One may note that from the analysis of Figure 6.27, the smallest NIP would be approaching zero. However, the NIP will never go to zero, since

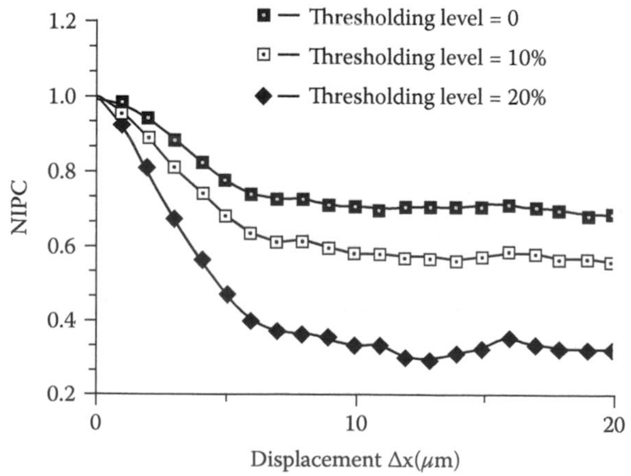

FIGURE 6.29
NIPC as a function of transversal displacement Δx for various thresholding levels.

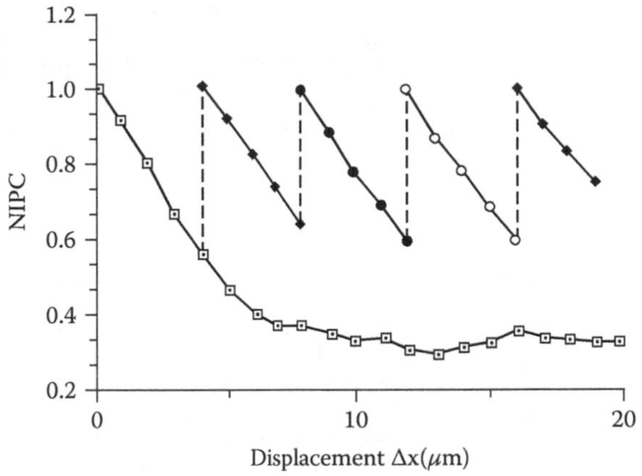

FIGURE 6.30
Extended dynamic range by an autonomous processing technique.

the intensity speckle distribution is a positive, real quality. Furthermore, as shown in Figure 6.29, the NIP is fairly linear within the range $0 < \Delta x < 6$ μm, and it fluctuates slowly as the displacement Δx increases beyond 6 μm. This is primarily due to the modal-phasing deviation δ being larger than π.

We note that the dynamic sensing range can be extended if one updates the reference speckle pattern. For example, if the last speckle pattern (e.g., at $\Delta x = 6$ μm) is used as the reference pattern for $\Delta x > 6$ μm sensing, then an expanded dynamic range can be initiated. As illustrated in Figure 6.30, the dynamic range of the FSS can indeed be extended beyond the 20-μm range.

In summing up this section, we would stress that *complex speckle field* detection would be far more sensitive than this *intensity speckle field.* By comparing the result obtained in Figure 6.20 with Figure 6.29, it is evident that the sensitivity obtained by using the complex speckle field is about an order *higher* than the sensitivity obtained with intensity patterns. In other words, the sensitivity obtained by complex fields is about 0.1 μm, while the one obtained by intensity patterns is about 1 μm.

6.9 Sensing with Joint-Transform Correlation

Because of the simplicity and adaptive nature of a joint-transform correlator (JTC), it has been used for pattern recognition [19,20], target tracking [21,22], and other applications. For instance, a JTC system using LCTVs (liquid crystal television) as spatial light modulators (SLMs) has been reported to detect laser speckle patterns deflected from a ground glass for displacement measurement [23]. We shall now illustrate a JTC system that can be used to process the intensity speckle fields for sensing [24], as depicted in Figure 6.31. A

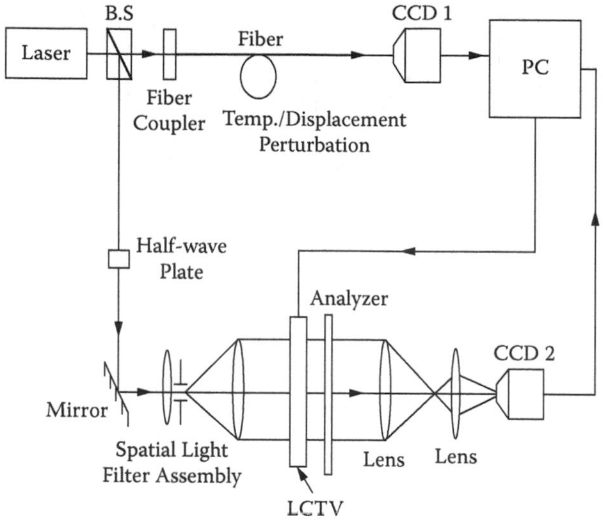

FIGURE 6.31
JTC-FSS sensing system.

laser beam is split into two paths, one coupled into a multimode fiber and the other used as a coherent source for the joint-transform operation.

For simplicity of illustration, we assume the sensing fiber is subjected to temperature perturbation for which a section of the fiber is embedded in a temperature-varying chamber. We note that fiber speckle-field variation is mainly affected by the modal phase deviation due to perturbation. The phase deviation between the *m*th and *n*th modes, as shown by Eq. (6.65), is given by

$$\delta_{mn} = \Delta\phi_m - \Delta\phi_n = k\xi n\Delta L\left(\frac{1}{\cos\theta_m} - \frac{1}{\cos\theta_n}\right) \tag{6.71}$$

where k is the wavenumber; ξ represents the strain optics correction factor, which is ~0.78 for a silica fiber; n is the refractive index of the fiber; ΔL is the change in fiber length caused by applied strain; and θ_m and θ_n are the incident angles of the modal wave fields with respect to the fiber axis. Thus, the induced phase difference due to the *temperature* gradient can be shown as

$$\delta'_{mn} = kL\left(\frac{1}{\cos\theta_m} - \frac{1}{\cos\theta_n}\right)\left(\frac{n}{L}\frac{dL}{dT} + \frac{dn}{dT}\right)\Delta T \tag{6.72}$$

where L is the fiber length, and $dL/(L\,dT)$ and dn/dT represent the temperature-induced strain and the temperature-induced refractive index, respectively.

As an example, a fused-silica fiber would have the induced strain and induced refractive index given by

$$\frac{dL}{LdT} = 5 \times 10^{-7}\ \mathrm{K}^{-1} \tag{6.73}$$

$$\frac{dn}{dT} = 10^{-5}\ \mathrm{K}^{-1} \tag{6.74}$$

We further assume that the sensing fiber has a numerical aperture (NA) equal to 0.2, $n = 1.5$, the wavelength of illumination is $\lambda = 632.8$ nm; assuming $\delta'_{M0} = \pi$, the measurable temperature range would be about ~3.51°C. From this illustration we see that the temperature-induced strain can be deduced from the phase deviation δ'_{M0}. Thus, by equating Eqs. (6.71) and (6.72), the temperature-induced strain is calculated to be about ~8.56 μ strain/°C.

By referring to the JTC setup of Figure 6.31, we see that the sensing operation takes place with two cycles of transform operations: namely the joint-transform and correlation operations, respectively. In the first cycle operation, the speckle patterns at the fiber output can be collected by the charge coupled device (CCD)I camera and then transferred to the LCTV panel for the joint-transform operation such that the joint-transform power spectrum (JTPS) is captured by CCD2. In the second-half cycle operation, the JTPS is transferred back to the LCTV panel for the correlation performance. The output correlation distribution is displayed on the PC monitor as shown in Figure 6.32, in which a pair of correlation peaks can be observed. It must be mentioned that the duty cycle of the operation is primarily limited by the PC, which is estimated to about 2.5 s, if a 486 IBM-compatible PC is used. Notice that once the initial state of the fiber is established, a real-time temperature measurement can be performed, and the temperature variation relative to the reference speckle intensity pattern can be measured. The input speckle

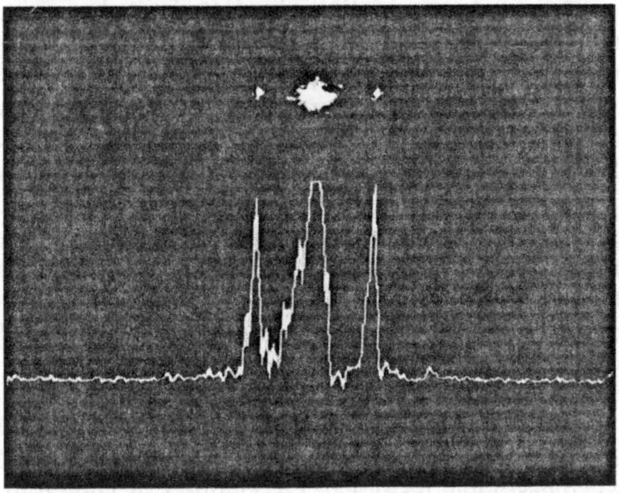

FIGURE 6.32
Output correlation distribution.

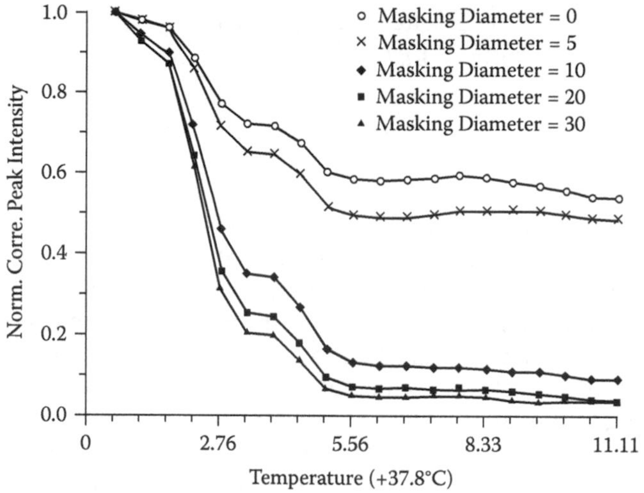

FIGURE 6.33
NIPC as a function of fiber temperature with 38°C used as the reference temperature.

pattern contains about 80 speckles that occupy about 27 by 42 pixels in the LCTV panel.

To obtain a higher correlation performance, the center part of the JTPS is blocked by a circular disk. To illustrate this effect, computer simulation is performed in which we assume that a 12-cm-long probing fiber is embedded in a temperature chamber. Correlation peak intensities are taken by correlating the reference speckle pattern with respect to the perturbed fiber speckle patterns. Thus, by slowly increasing or decreasing the temperature with respect to the reference speckle field, normalized correlation peak intensity (NCPI) versus temperature variations is plotted, as shown in Figure 6.33, where temperature of the reference speckle field is assumed at T = 38°C. We have used various sizes of low spatial frequency maskings to improve the correlation performance, as shown in the figure, in which we see that by masking the low-frequency spectrum with a mask diameter equivalent to 30 pixels, a broader dynamic range can be obtained. From these simulated results we have shown that a high-sensitivity measurement can be obtained if an appropriate mask size is used. Since the speckle size affects the correlation performance, we have found that the speckle size should be made larger than the pixel size of the LCTV panel. However, if the speckle size is too large compared to the pixel size, the changes of the speckle fields would provide an arbitrary result, which prevents us from obtaining a quantitative measurable result, as shown in Figure 6.34. This is primarily caused by losing the statistical property imposed by a few speckles. Nevertheless, if the speckle size is adequately small but larger than the pixel size of the LCTV, a linear measurable result can be obtained, as shown in Figure 6.35. In this figure, only the linear region is plotted, where the reference pattern in assumed at T = 21°C. We have seen that the sensitivity can be as high as

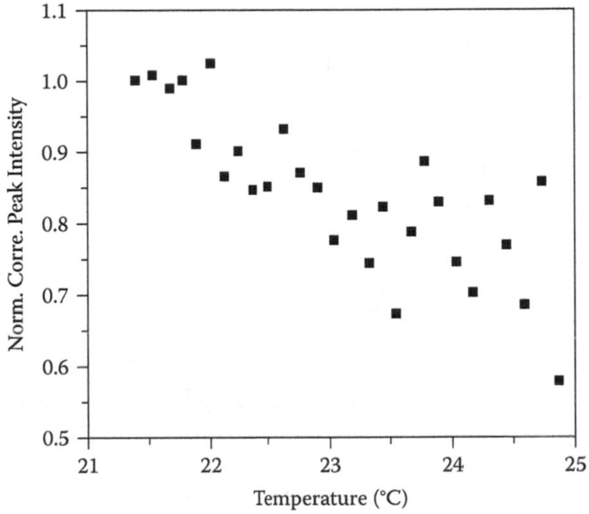

FIGURE 6.34
NIPC as a function temperature variation for larger speckle size (e.g., 25 speckles in a 27 × 42 pixel frame). The reference temperature is 21°C.

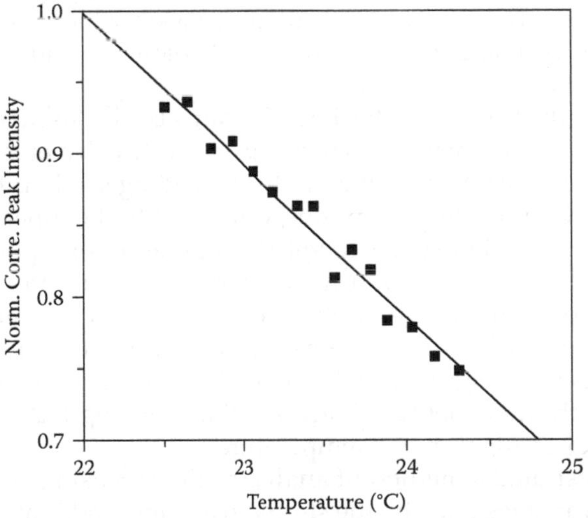

FIGURE 6.35
Normalized correlation peak intensity as a function of temperature variation for smaller speckles size (e.g., 80 speckles in a 27 × 42 pixel frame). The reference temperature is 21°C.

0.1°C, with a small degree of error. We have also found that at the lower end of the normalized intensity curves, the peak intensities become rather low (as can be seen in Figure 6.33), which is primarily due to the relative phase difference of the correlation speckle fields that exceed a phase shift greater than π (i.e., $\delta_{mn} > \pi$).

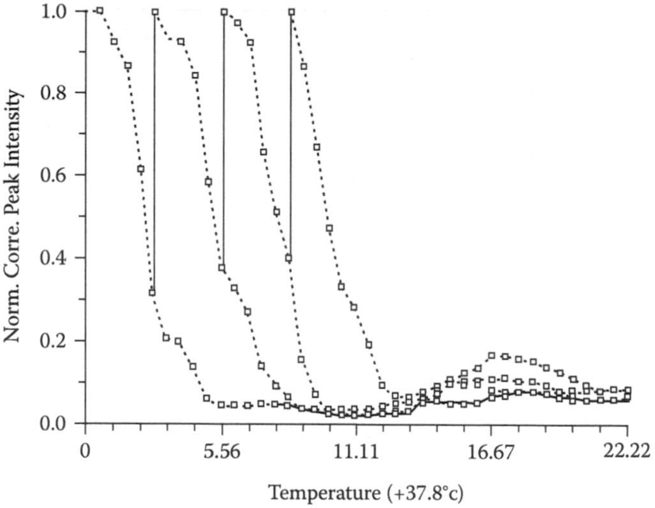

FIGURE 6.36
Extended dynamic range for the sensing measurement.

Again, the dynamic range can be extended by adopting a new reference speckle pattern at the lower end of the dynamic range, and so forth, as illustrated in Figure 6.36. There is an apparent drawback to extending the dynamic range by using the updating speckle patterns: namely, the measurable error would be accumulated.

We further illustrate that the JTC-FSS can also be used for submicron displacement measurement, for which a piezoelectric driver is used as a displacement transducer, as described in the preceding section. The advantage of using the JTC processing technique, compared to the inner-product technique, must be the high-speed correlation operation by optics. Figure 6.37 shows the NCPI variation as a function of displacement. We see that it has a measurable dynamic range of about 6 μm with a displacement sensitivity of about 1 μm, which is the same result obtained by the inner-product technique shown in Figure 6.29. We note that the displacement sensing is dependent on the incremented changes of the fiber length ΔL, instead of the refractive index changes due to temperature.

We have illustrated a method of analyzing the FSS using an adaptive JTC. The major advantages must be the simple, real-time, and low-cost operation, by which it may offer a wide variety of applications.

6.10 Dynamic Sensing

We shall now extend the adaptive JTC-FSS to dynamic sensing [25], by which we can determine the rate change (i.e., change per cycle) and the upward

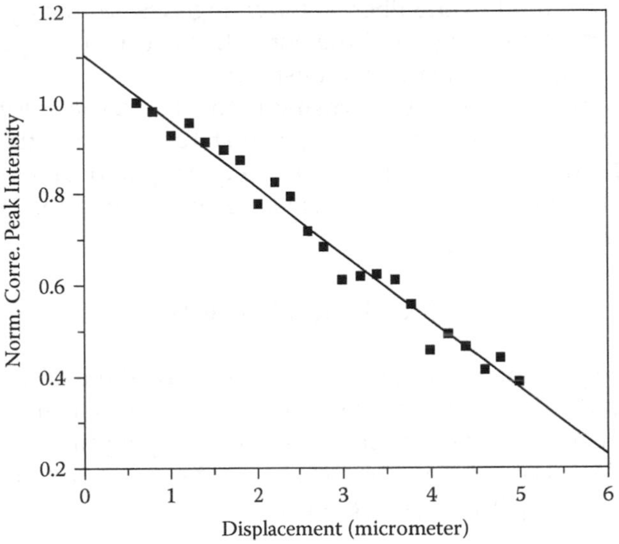

FIGURE 6.37
Normalized correlation peak intensity as a function displacement. The reference displacement is zero.

or downward trend change of perturbation. Notice that the rate and trend changes of the fiber status can provide significant and sometimes vital information, which is important in some types of high-risk environmental sensing (e.g., mechanical fatigue monitoring, seismic monitoring, etc.).

What we mean by "dynamic sensing" is a means by which the rate change of the sensing environment can be continuously monitored. In other words, by correlating a preceding speckle field with an updated speckle field, we can determine the dynamic status of the sensing fiber. If one autonomously displays multiple speckle patterns on an electronically addressable SLM in a JTC (i.e., one uses a set of continuously updating speckle patterns), both the trend and the rate change of the fiber status can be detected. The major difference between dynamic sensing and conventional sensing is that *autonomous updating sensing* can be continuously tracked.

For the autonomous target-tracking JTC [21,22], the reference pattern of the joint-transform speckle fields can be continuously updated so that the rate change of the fiber status can be detected. For the JTC-FSS, the dynamic range of the sensing is limited by the linearity of the normalized correlation peak intensity. However, if we use the autonomous updating technique, the dynamic range of the FSS can be extended. Using autonomous (dynamic) sensing, as we describe in a moment, we can determine the dynamic status of the fiber (i.e., the rate change), but the increasing or decreasing trend of the fiber perturbation is unknown. This limitation can easily be alleviated using a multispeckle pattern technique. In other words, by autonomously displaying three sequential speckle patterns on the SLM, we can determine the rate change and the trend of the fiber status changes. Since the dynamic range is limited by the linear region of the normalized correlation peak intensity, dynamic sensing can alleviate this constraint.

An interesting dynamic fiber sensing is the *heterodyne sensing* of a periodically or quasi-periodically varying fiber status. By using the heterodyne sensing technique, we show that both frequency and amplitude perturbations can be simultaneously detected. Let us assume that the intensity speckle pattern is given by

$$I(x,T)=I(x,\ T_1\cos\omega_1 t) \tag{6.75}$$

where **x** is the position vector of the speckle pattern at the output coordinate of the fiber and T represents an external perturbation measurand (e.g., stress or strain), which is assumed to be a sinusoidal varying function, given by $T = T_1 \cos \omega_1 t$. We now correlate the time-varying speckle patterns $I(\mathbf{x}, T)$ with a reference speckle pattern $I(\mathbf{x}, T_0 \cos \omega_0 t)$, given by

$$R(t)=I(x,\ T+\Delta T)\otimes I(x,\ T) \tag{6.76}$$

where $R(t)$ denotes a time-varying correlation output, $\otimes$ represents the correlation operator, and $\Delta T = T_1 \cos \omega_1 t - T_0 \cos \omega_0 t$, in which we assume that T_0 and ω_0 are given (a priori) and $|\omega_1 - \omega_0| \ll \omega_1$. Since the correlation operation is assumed to be a memoryless nonlinear process, Eq. (6.76) can be expanded into a McLaurin's series, such as

$$R(t)=A+B\Delta T+C(\Delta T)^2+\cdots \tag{6.77}$$

where $A, B, C, \ldots$ are arbitrary constants. By substituting ΔT into Eq. (6.77), we can show that

$$R(t)=A+B'(T_1\cos\omega_1 t-T_0\cos\omega_0 t) \tag{6.78}$$

$$+\ C'\,T_1T_0\big[\cos(\omega_1+\omega_0 t)+\cos(\omega_1-\omega_0 t)+\cdots$$

where B′ and C′ are time-independent constants.

By further restricting the time response of the sensor, we see that a dc component with a modulation term cos $[\omega_1 - \omega_0 t)]$ can be detected. Since ω_0 is assumed given, the selected data give rise to ω_1 and T_1, for which the dc component can be used as a normalizing factor. We note that heterodyne detection can be used to detect the frequency and the amplitude fiber perturbations simultaneously.

In most fiber sensors the fiber status caused by different aspects of perturbations may induce the same output detection, for which the sensing parameters may not be identified. However, if a multimode fiber is used, different sensing parameters would produce different speckle patterns. Thus, by exploiting the fiber speckle content with a dynamic search technique, we can determine the parameters of perturbation (e.g., temperature, displacement, strain, and stress). In other words, by assuming that fiber speckle patterns caused by different sensing parameters are different and then by correlating the updated speckle pattern with the previously recorded speckle patterns (for different parameters), one can detect the sensing parameters for each event per duty cycle. Note that the rate change and the trend of the fiber status can also be determined by using the autonomous sensing (multispeckle-pattern) technique.

As we noted in the preceding sections, the fiber speckle field changes due to modal phase changes (as long as the phase deviation $\delta_{mn} < \pi$), and the correlation peak intensity is relatively linearly proportional to the fiber elongation ΔL. However, for dynamic fiber speckle sensing, the sequential speckle patterns are used for joint-transform correlation, by which the *rate* and the *trend* changes of the fiber status can be determined. For example, a set of three sequential speckle patterns is displayed on the SLM shown in Figure 6.38(a). The locations of the output correlation distribution are shown

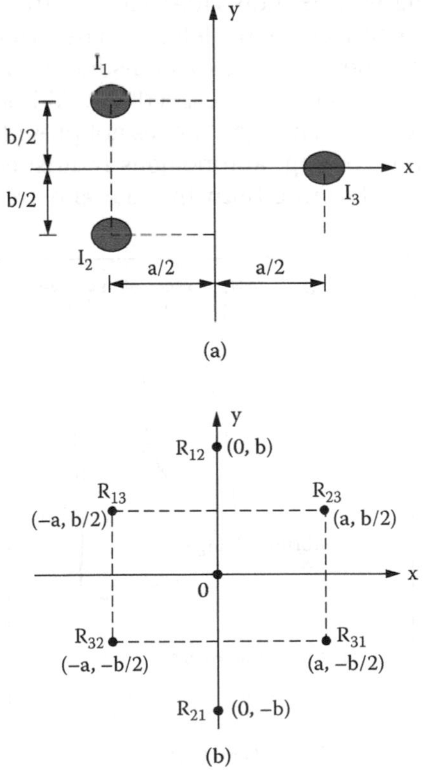

FIGURE 6.38
Autonomous multispeckle-pattern sensing: (a) input patterns; (b) output correlation distribution.

in Figure 6.38(b). Since $R_{ij} = R_{ji}$ $(i, j = 1, 2, 3)$, the intensity of the correlation spot should be the same. By normalizing the correlation peak intensities, fiber perturbation can be conducted. For example, if the peak intensities are lower than 1, the fiber has been perturbed. Thus, by autonomously addressing the speckle patterns, the dynamic sensing of the fiber status can be detected. One of the advantages of using LCTV as the SLM is the electronic addressability so that the input signal can be continuously updated.

Referring to the JTC-FSS of Figure 6.31, a sensing fiber that has a 53-μm diameter with an NA of 0.2 and is about 1 m long is used. The fiber speckle pattern is captured by CCDI and then addressed onto the LCTV with an IBM 486 compatible PC, which has a clock speed of 33 MHz. Displacement perturbation is induced when a piezoelectric driver is used as a transducer on a microbending device of ~8 mm in length. Notice that each speckle pattern displays about 80 speckles within a 27 × 42 pixel area of the LCTV panel. By displaying the speckle patterns (one represents the preceding speckle pattern, and the other is the updated one), a JTPS is obtained. Sending the JTPS back to the LCTV, a set of correlation peaks can be observed at the output plane. Thus, by continuously updating the speckle patterns for each duty-cycle operation, the change per cycle (i.e., the rate change) of the fiber status can be determined. Figure 6.39 shows the NCPI as a function of JTC cycles, in which the fiber status changes can be detected. The abrupt changes represent the transversal displacement of the piezotransducer, which are measured to about 0.9, 1.3, and 2.3 μm per cycle, respectively. Although the autonomous JTC operation offers the rate changes, it does not provide the trend changes.

To alleviate this shortcoming, autonomous multispeckle-pattern sensing can be employed. To determine both the rate and trend changes, we can

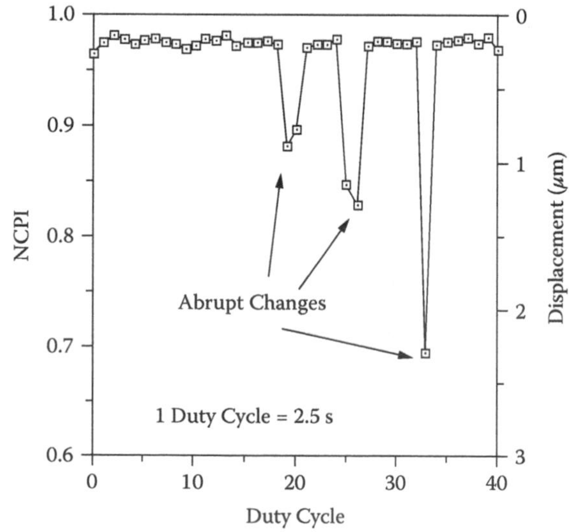

FIGURE 6.39
Output NCPI as a function of duty cycle.

use an autonomous multispeckle-pattern sensing scheme, as illustrated in Figure 6.38(a). For example, I_1, I_2, and I_3 are the sequential speckle patterns derived from the initial fiber status, a known perturbation, and the current updated speckle field, respectively. The corresponding cross-correlation distributions R_{13}, R_{12}, and R_{23} are shown in Figure 6.38(b), in which R_{ij} represents the correlation peak intensity between I_i and I_j $(i, j = 1, 2, 3)$. Note that this set of peak intensities can be detected by a CCD camera displaying on a PC monitor, as shown in Figure 6.40(a) and Figure 6.40(b). In Figure 6.40(a) we see that the peak intensity R_{13} is the smallest one; that means the current fiber status change is an increasing trend (providing that the previous one has an increasing trend). On the other hand, if R_{13} is not the smallest, as shown in Figure 6.40(b), the trend of the fiber status change must be decreasing. Figure 6.41 shows a multispeckle-pattern dynamic sensing result. The lower curve shows the dynamic displacement as a function of the sensing duty cycle, in which we see the rate and trend changes of the dynamic displacement.

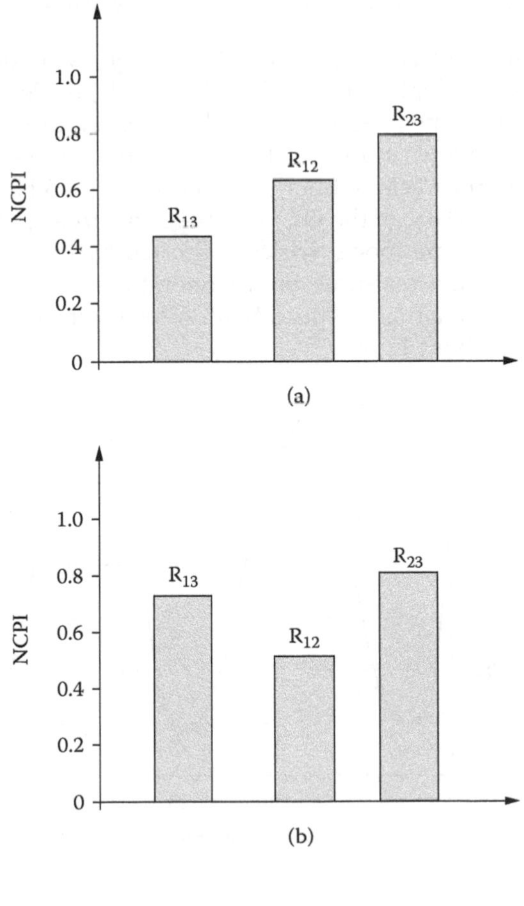

FIGURE 6.40
Examples of correlation peak intensities: (a) $R_{13} < R_{12} < R_{23}$; (b) $R_{12} < R_{13} < R_{23}$.

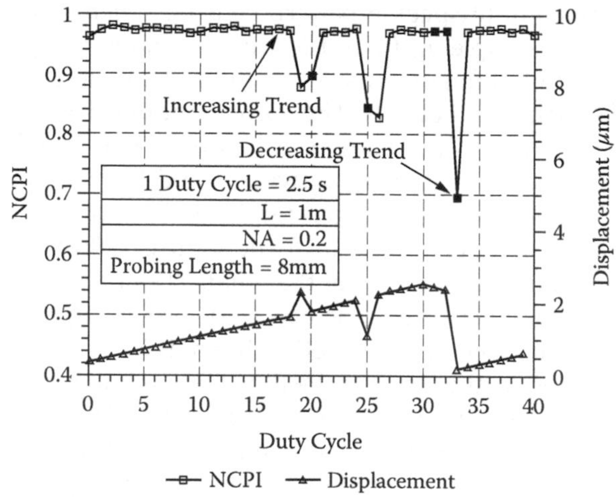

FIGURE 6.41
Displacement as a function of JTC duty cycle, using the autonomous sensing technique.

Furthermore, if the sensing fiber is perturbed by a vibrational signal (e.g., a sinusoidal varying displacement), then the multispeckle-pattern dynamic sensing technique can be used [26,27] as shown in Figure 6.42, in which we see that the NCPI follows closely with the vibration cycles of perturbation. In this experiment the cycle of fiber perturbation is ~1.5 Hz, which is faster than the duty cycle (~0.6 s) of the JTC. We stress that this operation is primarily run in a nonreal-time mode, which is limited by the software program (C language). To have a real-time device operation, we should use a faster software program and a higher speed SLM. A ferroelectric device [28] has an operating speed of about 12 μs.

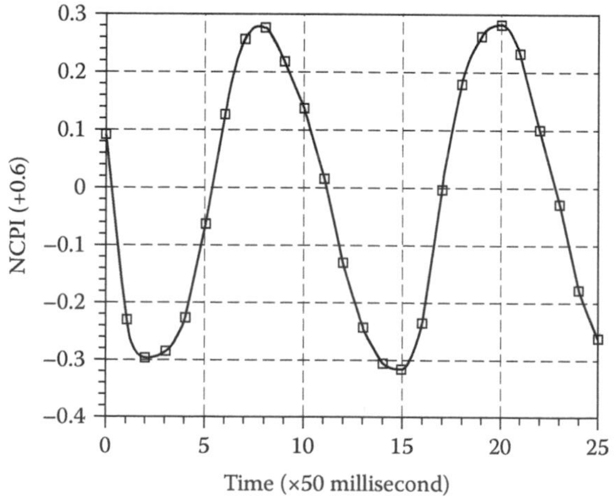

FIGURE 6.42
Dynamic sensing for vibration perturbation.

6.11 Concluding Remarks

We have introduced a different type of fiber sensor, called the fiber specklegram sensor (FSS), that exploits the spatial content instead of the temporal content for sensing. The major advantages of using the FSS can be summarized in the following:

High sensitivity: We have shown that the sensitivity of the FSS can be as high as the two-arm Mach–Zehnder interferometric fiber sensor.

Low cost: Since the FSS uses multimode fiber, it is less expensive to draw and more materials can be drawn into fiber forms.

Single-path sensor: The FSS is a single-path sensor, which is less vulnerable to environmental factors.

Multiplexing: We have shown that the FSS can be easily multiplexed with angular division multiplexing (ADM) and wavelength division multiplexing (WDM) schemes for multiparameter or multichannel sensing.

Materials: Since more material can be drawn in (multimode) fiber forms, the FSS offers a wider range of applications—for example, under a high-temperature environment or high-tensile condition.

Doping and implanting: Impurities can be doped or implemented in a multimode fiber to enhance the sensing aspects. Small microchips may be implanted within a multimode filter to improve the sensing algorithms.

In short, we believe that the FSS technology will offer a broader range of practical applications. If the FSS system is carefully designed (e.g., by using the doping, implantation, ADM, and WDM capabilities), it can be applied to smart fiber optic sensing, such as smart skin detection, fatigue monitoring, and other areas.

References

1. A. Yariv, On transmission and recovery of 3-D image information in optical waveguides, *J. Opt. Soc. Amer.*, **66**, p. 301, 1976.
2. G. J. Dunning and R. C. Lind, Demonstration of image transmission through fibers by optical phase conjugation, *Opt. Lett.*, **7**, p. 558, 1982.
3. B. Culshaw and J. Dakin, *Optical Fiber Sensor: Systems and Applications*, Artech House, Boston, 1989.
4. A. Dandridge, Fiber-optic sensors make waves in acoustic control and navigation, *IEEE Circuits and Devices*, **6**, p. 13, 1990.
5. E. Udd, *Fiber Optic Sensors*, John Wiley & Sons, New York, 1991.

6. E. Udd and R. O. Claus, Fiber optic smart structures and skins III, *SPIE,* **1370**, 1990.
7. S. Wu, S. Yin, and F. T. S. Yu, Sensing with fiber specklegrams, *Appl. Opt.,* **30**, p. 4468, 1991.
8. S. Wu, S. Yin, S. Rajan, and F. T. S. Yu, Multichannel sensing with fiber specklegrams, *Appl. Opt.,* **31**, p. 5975, 1992.
9. S. Musikant, *Optical Materials,* Marcel Dekker, New York, 1990.
10. F. T. S. Yu, J. Zhang, S. Yin, and P. B. Ruffin, Analysis of a fiber specklegram sensor by using coupled-mode theory, *Appl. Opt.,* **34**, p. 3018, 1995.
11. H. F. Taylor, Bending effects in optical fibers, *IEEE J. Lightwave Technol,* **LT-2**, p. 617, 1984.
12. R. A. Pappert, E. E. Gossard, and I. J. Rothmuller, An investigation of classical approximation used in VLF propagation, *Radio Sci.,* **2**, p. 387, 1967.
13. F. T. S. Yu, M. Wen, S. Yin, and C. M. Uang, Submicrometer displacement sensing using inner-product mulitmode fiber speckle fields. *Appl. Opt.,* **32**, p. 4685, 1993.
14. F. T. S. Yu, S. Yin, J. Zhang, and R. Guo, Application of fiber speckle hologram to fiber sensing, *Appl. Opt.,* **33**, p. 5202, 1994.
15. L. Cheng and G. G. Siu, Measurement of surface roughness with core-ring-ratio method using incoherent light, *Meas. Sci. Technol.,* **1**, p. 1149, 1990.
16. A. Yariv, *Optical Electronics,* Saunders College Publishing, Orlando, FL, 1991.
17. R. Kirst, Point sensor multiplexing principles, *Photon. Spectra,* **17**, pp. 511–515, 1989.
18. K. D. Bennett and R. O. Claus, Internal monitoring of acoustic emission in graphite epoxy composites using embedded optical fiber sensors, in *Review of Progress in Quantitative Nondestructive Evaluation,* D. O. Thompson and D. E. Chimenti, eds., p. 331, Plenum, New York, 1990.
19. F. T. S. Yu and S. Jutamulia, *Optical Signal Processing. Computing and Neural Networks,* Ch. 5, John Wiley & Sons, New York, 1992.
20. F. T. S. Yu and X. J. Lu, A real-time programmable joint transform correlator, *Opt. Commun,* **52**(1), p. 10, 1984.
21. E. C. Tam, F. T. S. Yu, D. A. Gregory, and R. Juday, Autonomous real-time object tracking with an adaptive joint transform correlator, *Opt. Eng.,* **29**(4), p. 314, 1990.
22. E. C. Tam, F. T. S. Yu, D. A. Gregory, and R. Juday, Data association multiple target tracing using a phase-mostly liquid crystal television, *Opt. Eng.,* **29**(9), p. 1114, 1990.
23. T. Okamato, Y. Egawa, and T. Asakura, Liquid crystal television applied to a speckle correlation method: Real-time measurement of the object displacement, *Opt. Commun.,* **88**, p. 17, 1992.
24. F. T. S. Yu, K. Pan, C. Uang, and P. B. Ruffin, Fiber specklegram sensing by means of an adaptive joint transorm correlator, *Opt. Eng.,* **32**, p. 2884, 1993.
25. F. T. S. Yu, K. Pan, D. Zhao, and P. B. Ruffin, Dynamic fiber specklegram sensing, *Appl. Opt.,* **34**, p. 622, 1995.
26. S. Yin, P. Purwosumarto, and F. T. S. Yu, Application of fiber specklegram sensor to fine angular alignment. *Opt. Commun.,* **170**, p. 15, 1999.
27. B. Yang, H. Lee, and B. Lee. Optical pattern recognition by using speckle-multiplexed holograms, *SPIE Proc.,* **3801**, p. 190, 1999.
28. K. M. Johnson, M. A. Handschy, and L. A. Pagano-Stauffer, Optical computing and image processing with ferroelectric liquids crystals, *Opt. Eng.,* **26**, p. 385, 1987.

7

Interrogation Techniques for Fiber Grating Sensors and the Theory of Fiber Gratings

Byoungho Lee and Yoonchan Jeong

CONTENTS

7.1 Introduction

That refractive index variation patterns (i.e., gratings) can be formed in optical fibers was discovered and reported by Hill et al. in 1978 [1]. After Meltz et al. devised a controllable and effective method for fabricating the fiber gratings by side-illuminating optical fibers with a UV laser [2], intensive studies on the fabrication and application of the devices for optical communications and optical fiber sensors began. Today, fiber gratings are generally made by the side-illumination method using KrF lasers or frequency-doubled argon ion lasers. The intensity variation patterns, which are required to write gratings in fibers, are made by interference using a phase mask method or a holographic method or by scanning the laser beam over the fibers with intensity modulation. Detailed discussions on the fabrication methods and principles can be found in many references, including that by Othonos and Kalli [3].

Fiber gratings can be categorized into periodic gratings and aperiodic gratings. *Periodic* gratings include fiber Bragg gratings (FBGs), long-period fiber gratings (LPFGs), and others such as tilted (or blazed) gratings. *Aperiodic* gratings include chirped fiber gratings and others. These gratings are used in sensor heads or data-extracting systems (interrogators). Most of the sensor heads that adopt fiber gratings use FBGs. Figure 7.1 shows the principle of the FBG sensor. A light that has a broadband spectrum is launched to the FBG sensor. At the FBG the optical wave is partially reflected from each part of

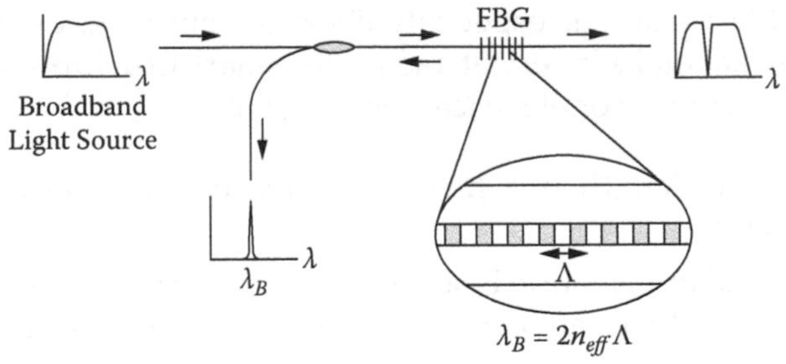

FIGURE 7.1
Principle of the fiber Bragg grating sensor.

the grating. However, the optical waves that are partially reflected from each part constructively interfere with each other only for a specific wavelength of light, which is called the *Bragg wavelength*. Hence, for the given broadband light, only a narrow spectrum at the Bragg wavelength is reflected, while other wavelength components are transmitted through the FBG. The Bragg wavelength is given by

$$\lambda_B = 2n_{\text{eff}}\Lambda \tag{7.1}$$

where n_{eff} is the effective refractive index of the fiber core and Λ is the grating period. A detailed theory of fiber gratings can be found in Section 7.5.

In Eq. (7.1) it can be seen that the Bragg wavelength is changed with a change in the effective refractive index or grating period. Strain applied to an FBG elongates it (or compresses it for negative strain); hence, the grating period is increased (or decreased), which results in a shift of the Bragg wavelength to longer (or shorter) wavelengths. The strain applied to the fiber is usually expressed in the unit of strains (ε). In fact, ε is "unitless" because it is a relative comparison concept; that is, if a 1-m-long fiber is elongated by 1 μm, the strain is 1 μm/1 m = 1 με (microstrain). Typically, the Bragg wavelength shift with strain is ~0.64 pm/με near the Bragg wavelength of 830 nm, ~1 pm/με near 1300 nm, and ~1.2 pm/με near 1550 nm [4]. The difference arises from the difference in the effective refractive indices at the wavelengths. With temperature change, the grating period also changes due to thermal expansion (or compression) of the fiber, but the effect of the change in the refractive index is about one order of magnitude larger than that of the thermal expansion (or compression). Hence, with the temperature change, the Bragg wavelength shifts mainly due to the change of n_{eff}. The overall change is ~6.8 pm/°C near the Bragg wavelength of 830 nm, ~10 pm/°C near 1300 nm, and ~13 pm/°C near 1550 nm [4].

As discussed, the FBGs can be used as strain or temperature sensor heads. They can also be used for probing a variety of other types of measurands such as pressure, erosion, liquid or chemicals, bending, or even magnetic fields. (Many references are available, including Grattan and Meggitt [5].)

The use of fiber gratings, especially FBGs, as sensor heads has a number of advantages that make it very attractive for smart structures over the other conventional fiber optic or electrical sensors [4,6]:

The Bragg wavelength is a linear function of the measurands over large ranges.

The measurand information is spectrally encoded; hence, the sensor signals are basically unaffected by environmental noise or power loss.

FBGs have inherent advantages over fiber devices such as signal transmission capability with small loss over fiber channels.

FBGs can be low in price and are easily available.

FBGs have high reflectivity for the Bragg wavelength light, while their sizes are small (~1 cm, typically), and they can be quasi-point sensors.

FBGs are lightweight and because of their small diameters can be inserted into composite materials without disturbance.

Various types of sensor multiplexings such as spatial division multiplexing (SDM), wavelength division multiplexing (WDM), time division multiplexing (TDM), and code division multiple access (CDMA), and their combinations, can be implemented to form quasi-distributed or quasi-point sensor array systems.

Interrogators or *demodulators* in fiber grating sensor systems are the measurand-reading units that extract measurand information from the light signals coming from the sensor heads. As mentioned, the measurand is typically encoded spectrally, and hence the interrogators are usually meant to measure the Bragg wavelength shifts and convert the results to measurand data. In the laboratory, when one is developing fiber grating sensor heads, optical spectrum analyzers are indispensable in monitoring grating reflection or transmission spectra. However, optical spectrum analyzers are not appropriate for real sensor systems, not only because of their high prices but also because their slow scanning speed limits dynamic sensing.

Several important topics (for example, discrimination between strain and temperature effects, and multiplexing of FBG sensors) can be addressed related to fiber grating sensors, and many review articles and books have already been published. In this chapter we focus on the interrogation techniques of fiber grating sensors. Excellent reviews on similar topics are available [3,7], but here we try to review the interrogators, including some of the most recent techniques, under the assumption that the readers are not intimately familiar with fiber grating sensors. In the last part of this chapter, we briefly explain the theory of fiber gratings. Although a detailed knowledge of the theory is not necessarily needed in understanding fiber grating sensor systems, such knowledge might enrich the design and developing capabilities of various sensor heads and interrogators adopting various kinds of fiber gratings.

7.2 Passive Detection Schemes

Passive detection scheme interrogators refer to those that do not use any electrical, mechanical, or optical active devices. Interrogators using linearly wavelength-dependent devices, performing measurand monitoring by detecting optical power, using identical chirped grating pairs, and known as the charge-coupled device (CCD) spectrometer are discussed in this section.

7.2.1 The Use of Linearly Wavelength-Dependent Devices

Linearly Wavelength-Dependent Optical Filters

The simplest way to think about measuring the wavelength change to light reflected from an FBG is to use a linearly wavelength-dependent optical filter. Indeed, this method was one of the first proposed for the practical wavelength change interrogation system of FBG sensors [8].

Figure 7.2(a) shows the concept of the wavelength demodulator. The light transmittance of the filter is linearly dependent on wavelength. According to the linear response range, this type of filter is sometimes called an *edge filter* (which has a narrow linear range with a sharp slope, as a sharp edge of a bandpass filter) or a *broadband filter* (which has a wide range with a less sharp slope, as a boundary of a broadband filter). There is a trade-off between the measurable range and the sensitivity.

This wavelength-change interrogator is based on intensity measurement; that is, information relative to wavelength change is obtained by the intensity monitoring of the light at the detector. A number of interrogators discussed in the upcoming text are based on the intensity measurement. For the intensity-based demodulators, the use of intensity referencing is necessary because the light intensity might be changed due to not only the reflection wavelength (Bragg wavelength) change of the FBG but also the power fluctuation of the light source, the disturbance in the light-guiding path, or the dependency of light source intensity on the wavelength. In a sense, although the intensity-based measurement has the advantage of being a simple structure, it does not use a key advantage of an FBG sensor—the fact that the information of the measurand is contained in the reflection light wavelength, and not in its intensity.

Figure 7.2(b) shows the schematic diagram of the FBG sensor system adopting the wavelength-dependent optical fitter demodulator, where the light reflected from the FBG is split into two; one of them passes through the wavelength-dependent filter, while the other is used as a reference. The intensity ratio at the two detectors is given by

$$\frac{I_S}{I_R} = A(\lambda_B - \lambda_0 + B) \tag{7.2}$$

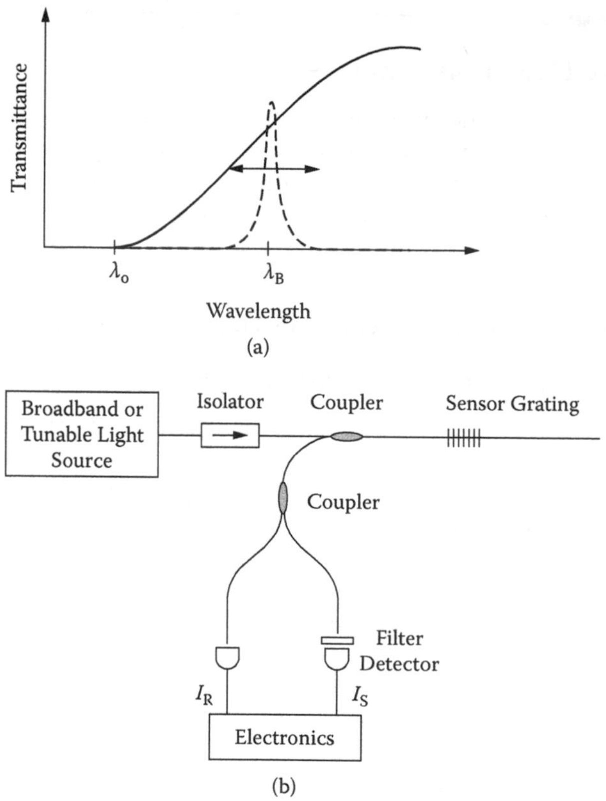

FIGURE 7.2
(a) Transmittance of the linearly wavelength-dependent optical filter interrogator. The dashed peak shows the spectrum of the light reflected by a sensor Bragg grating. (b) Sensor system schematic adopting the interrogator. (From S. M. Melle et al., *IEEE Photonics Technol. Lett.*, **4**, 5, pp. 516–518, 1992.)

where A is a constant determined by the slope of the filter and B is a constant arising from the nonzero reflection bandwidth of the FBG. Equation (7.2) is linearly dependent on the Bragg wavelength change but independent of light intensity variation due to the source fluctuation, etc. That is because the intensity variations are cancelled out by comparing the signal I_s with the reference I_r.

The first experiment of this method was done by using a commercial infrared high-pass filter (RG830), which has a linearly wavelength-dependent edge of 815–838 nm [8]. The use of a biconical fiber filter was also proposed in the wavelength region around 1520 nm, with an unambiguous wavelength interrogation range of ~20 nm [9]. Static and dynamic resolutions of ~ ±3.5 $\mu\varepsilon$ and 1.5 $\mu\varepsilon\sqrt{\text{Hz}}$, respectively, were obtained.

It is necessary here to explain the dynamic resolution unit. When the dynamic strain signal is measured using a spectrum analyzer, the minimum detectable strain is determined by the background-noise level in the spectrum. The magnitude of the noise, however, changes with the frequency span because the noise power in the detector and the amplifying circuit depends on the frequency span. Therefore, to allow measurements with different spans to be compared, it is necessary to normalize all measurements to a 1-Hz bandwidth. If the noise is Gaussian in nature, then the amount of noise magnitude in other bandwidths may be approximated by scaling the power spectrum by the square root of the bandwidth. Thus, this normalized minimum detectable strain is displayed in units of $\varepsilon/\sqrt{\text{Hz}}$. For example, when the signal-to-noise ratio (SNR; the difference of signal component and background noise, when expressed in log scale) normalized to a 1-Hz bandwidth is 20 dB with a 1-με rms input strain, the minimum detectable strain is calculated as

$$\begin{aligned}\varepsilon_m &= \frac{1\ \mu\varepsilon/\sqrt{\text{Hz}}}{10^2} \\ &= 10\ n\varepsilon/\sqrt{\text{Hz}}\left(\because 20\ \text{dB} = 10\log\frac{V_{pd}}{V_m} = 10\ \log\frac{1\ \mu\varepsilon/\sqrt{\text{Hz}}}{\varepsilon_m}\right)\end{aligned} \tag{7.3}$$

where V_{pd} and V_m are photodetector voltage outputs corresponding to strain signal and noise, respectively.

An LPFG has a broad rejection band and has transmission spectrum regions that are linearly dependent on wavelength, as described in Section 7.5. The linear region of the LPFG was also tested as an optical filter to interrogate a fiber laser sensor with an FBG mirror [10].

In general, a linearly wavelength-dependent optical filter demodulator system has the advantage of low cost. Hence, sensor systems that adopt this type of demodulator have been commercialized.

Linearly Wavelength-Dependent Couplers

The linearly wavelength-dependent optical filter interrogator just discussed deteriorates the SNR because the filter decreases optical power. An alternative interrogator has also been proposed by Melle et al. [8] and demonstrated by Davis and Kersey [11]. The scheme uses a wavelength division multiplexer coupler (usually called *WDM coupler*), which has a linear and opposite change in the coupling ratios between the input and two output ports (see Figure 7.3). The power loss is reduced, and a static strain resolution of ~ ±3 με for the range of 1050 με was obtained. The minimum detectable dynamic strain was 0.5 $\mu\varepsilon/\sqrt{\text{Hz}}$. A highly overcoupled coupler was also used to try to increase the steepness of the slope, and hence the sensitivity [12]. The use of a coupler made of a dichroic mirror sandwiched between two graded-index (GRIN) lenses has also been proposed for a similar purpose as well as to reduce polarization dependency [13].

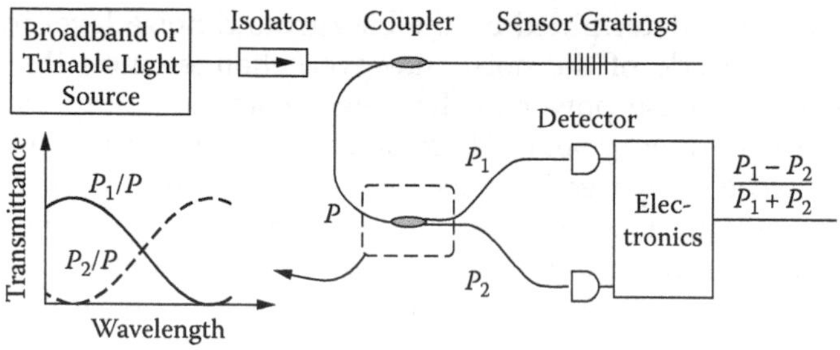

FIGURE 7.3
Schematic diagram of the sensor system adopting the linearly wavelength-dependent coupler interrogator. (From M. A. Davis and A. D. Kersey, *Electronics Lett.*, **30**, 1, pp. 75–77, 1994.) P, P_1, and P_2 indicate optical powers at each port ($P = P_1 + P_2$) if the insertion loss is neglected.

7.2.2 Power Detection

In some applications of fiber grating sensors, a simple detection of reflected or transmitted power is sufficient for the measurand interrogation.

Instead of using a linearly wavelength-dependent optical filter, we can use a light source that has intensity linearly dependent on wavelength. An example is to use the amplified spontaneous emission (ASE) profile of an erbium-doped fiber amplifier (EDFA) [14]. As shown in Figure 7.4, the Bragg wavelength of a sensor grating is located in the linear region of the ASE spectrum. The change in the Bragg wavelength results in a power change at the photodiode. Primitive dynamic tests up to 1 kHz for a strain range up to 2700 με with 50-με resolution were performed.

Kim et al. [15] recently proposed a chirped fiber grating strain sensor that is immune to temperature change. As we discuss in Section 7.5, a chirped fiber grating has a position-dependent grating period (or pitch) and a wide reflection bandwidth. Half of a linearly chirped grating (the longer period part) is fixed to a glass tube, as shown in Figure 7.5. With temperature increase, the overall reflection band moves to a longer wavelength, but the bandwidth

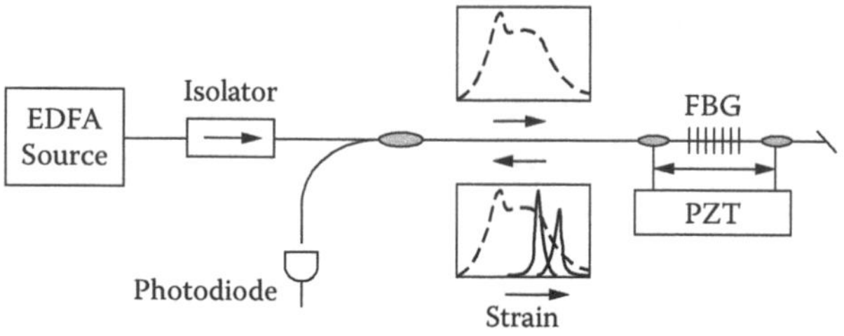

FIGURE 7.4
Schematic diagram of a sensor system using the linear region of the ASE of EDFA [14]. The dashed lines indicate the ASE spectrum and the solid peaks show the spectra of lights reflected by an FBG. The piezoelectric transducer (PZT) is used to apply strain in the experimental test, which is the typical way of testing.

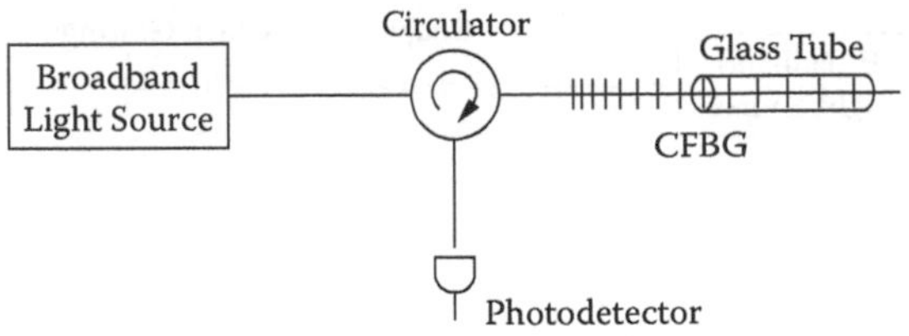

FIGURE 7.5
Schematic diagram of a sensor system adopting a chirped fiber Bragg grating (CFBG). Half of the grating is fixed to a glass tube in order not to respond to strain. (From S. Kim et al., *IEEE Photonics Technology Lett.*, **12**, 6, pp. 678–680, 2000.)

remains unchanged because the entire chirped grating experiences the same thermo-optic effect. However, with strain, the bandwidth becomes narrower because the reflection band due to the shorter period grating part moves to a longer wavelength with an elongation in the grating period, and the band becomes partially overlapped with the reflection band of the fixed grating part. Therefore, if a broadband light is launched to the fiber sensor system, the strain can be measured independently of temperature by monitoring the light power reflected by the fiber grating. Note that in Figure 7.5, a circulator is used instead of a coupler. Either one can be used in most of the sensor systems monitoring reflected lights from fiber gratings. The circulator has the advantage of less loss but is more expensive than a 3-dB coupler.

Another example of using a chirped fiber grating sensor and monitoring its reflected power is the sensor proposed by Kersey et al. [16]. They use a strongly apodized chirped grating sensor that has an asymmetric broadband spectral response. One side of the reflection spectrum has a ramp profile that produces a gradual change in reflectivity. The reflection spectrum shifts to a longer wavelength with strain. Therefore, for a narrow band light with the wavelength located in the linear ramp region of the sensor grating reflection spectrum, the reflected light power is linearly dependent on the strain.

Another recent example of the case in which power detection is enough for interrogating measurand is strain measurement using an LPFG [17]. By using a special case of a quadratic-dispersion resonance, the dip wavelength in the transmission spectrum is made fixed, while the transmittance changes with strain. Therefore simple monitoring of transmitted optical power gives the information of strain applied to the LPFG.

Although the preceding examples do not require any complex interrogator, the reference power monitoring is necessary in order to avoid errors arising from effects such as an optical source power fluctuation or a disturbance in a guiding path.

7.2.3 Identical Chirped Grating Pair Interrogator

A passive sensor system using identical chirped grating pairs was proposed by Fallon et al. [18]. This is similar to the matched fiber Bragg grating

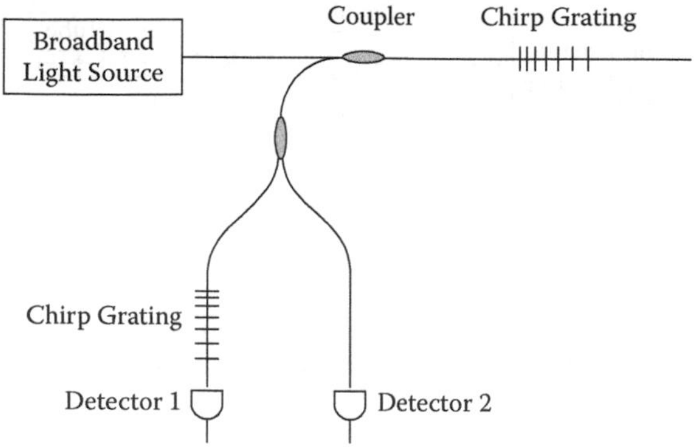

FIGURE 7.6
Schematic diagram of the identical chirped grating pair interrogator sensor system. (From R. W. Fallon et al., *Electronics Lett.*, **33**, pp. 705–706, 1997.)

pair detection, discussed in Section 7.3.4, but the present method does not involve the use of scanning. Two identical chirped gratings with a quasi-square reflection spectrum are used as a sensor head and an interrogating filter, as shown in Figure 7.6. If no strain is applied to the sensor grating, the received power at Detector 1 is minimal. With strain applied to the sensor grating, the received power increases due to the mismatch of the reflection spectra of the two gratings. The linear power increase with strain is ceased if the overlap between the two spectra diminishes to zero. Hence, the grating bandwidth determines the measurement range. For example, a 10-nm bandwidth grating (around 1300-nm wavelength) gives a sensing range of 10 mε. Detector 2 in Figure 7.6 is necessary to subtract power reflected from other gratings, such as those in a WDM sensor system. This system can be extended to a multiplexing scheme [19] where multiple sensors are arranged in a serial or parallel or a combination of both. A configuration disadvantage of this sensor system is that each sensor grating occupies a broadband in spectral domain, and hence the amount of multiplexing within source light bandwidth is more limited.

7.2.4 CCD Spectrometer Interrogator

One of the wavelength-change interrogators suitable for multipoint fiber grating sensors is to use parallel detection using a detector array such as a CCD. Lights reflected from FBGs are given to a fixed diffractive element such as finely ruled diffraction gratings and then focused to a CCD. For a light incident to the diffraction grating, the diffraction angle is dependent on the wavelength of the light. Therefore, as shown in Figure 7.7, lights with different wavelengths illuminate different areas of pixels [20]. The change in the light wavelength results in the shift of the light at the detector array of the CCD. Therefore, this system can be used as a wavelength-change interrogator for multipoint fiber grating sensors. The CCD spectrometer interrogator

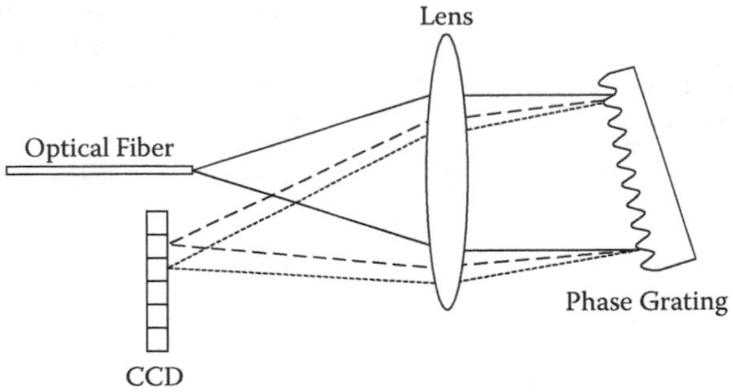

FIGURE 7.7
Schematic diagram of the wavelength interrogation system using a CCD and a plane phase grating. The dashed and dotted lines indicate lights with different wavelengths. (From A. D. Kersey et al., *J. Lightwave Technol.*, **15**, 8, pp. 1442–1463, 1997.)

was developed for instrumenting wavelength-stepped FBG sensor arrays fabricated online as part of the fiber draw process [21].

This approach collects all the light returned by each FBG over the entire scan period of the CCD. Therefore, this system is able to detect weaker reflected light, compared to the usual scanning interrogators such as a scanning Fabry–Perot interferometer, which is discussed in Section 7.3.2. In the experiment by Askins et al. [21], the reflectivities of FBGs were ≤3%.

By dispersing a 24-nm bandwidth over a 256-pixel CCD, as many as 22 FBGs spaced by 1-nm intervals may be resolved, with more than a 0.4-nm overlap-free range [20]. The center-to-center pixel spacing often corresponds to ~0.1 nm. Therefore, a strain resolution of 1 $\mu\varepsilon$, which corresponds to a wavelength resolution of about 0.7 pm (near 830-nm wavelength), requires a resolution of less than 1/100 of a pixel. As the image of each FBG is spread over several adjacent pixels, a weighted average of those illuminated pixel positions scaled by the detected signals from each pixel gives a computed wavelength of the light. A strain sensitivity below 1 $\mu\varepsilon$ (without averaging) at repetition rates above 3.5 kHz has been reported with 20 FBGs of 1 ~ 3% reflectivity, illuminated by several hundred microwatts of broadband light [20].

Two recent examples of the sensor systems that adopted this interrogation technique are the FBG refractometer sensor system aimed for online quality control of petroleum products [22] and the FBG temperature profiling system for a biomedical study [23].

Chen et al. [24,25] fully extended this method for use in two-dimensional CCD. Although their experiment was done as a primitive test, their goal was to use a CCD as an interrogator for multiline FBG sensors. With the configuration shown in Figure 7.8, lights from different optical fibers can be spatially multiplexed and focused on different columns of the CCD. Here, the curved grating acts as both a diffractor and a focusing device. Each optical fiber line has multiple FBGs with different reflection wavelengths, whose lights are focused on different positions in each column of the CCD. Therefore, this demodulator can interrogate spatial and wavelength multiplexed sensors.

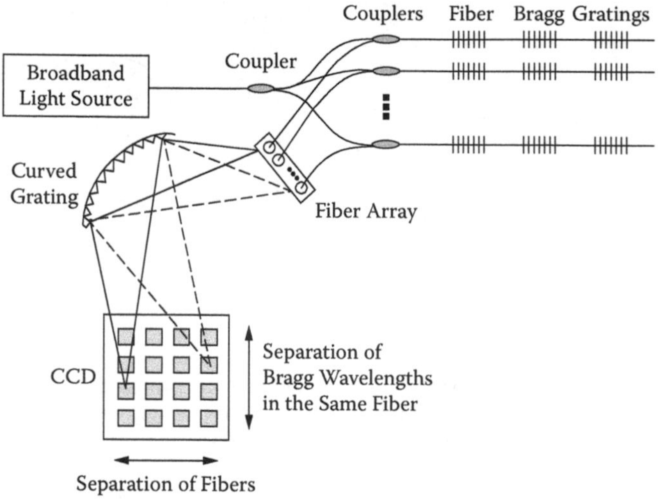

FIGURE 7.8
Schematic diagram of a sensor system with spatial and wavelength multiplexing using a two-dimensional CCD. (From Y. Hu et al., *Electronics Lett.*, **33**, 23, pp. 1973–1975, 1997.)

7.3 Active Detection Schemes

In this section we overview active detection scheme interrogators that usually involve tracking, scanning, or modulating mechanisms to monitor Bragg wavelength shifts from single or multiple FBGs. In general, although the active detection schemes require more complex systems compared to the passive detection schemes, the active schemes show better resolution.

7.3.1 Fiber Fourier Transform Spectrometer Interrogator

One method of direct spectroscopic analysis of the Bragg wavelength is to apply Fourier transform spectroscopy [26,27]. Figure 7.9 shows its schematic diagram, which is based on a fiber Michelson interferometer. This approach resembles low-coherent light interferometry. A light reflected from an FBG with a reflection bandwidth Δv has a coherence time $\tau = 1/\Delta v$ and a coherence length $\Delta L = \tau c/n_{eff} = c/(n_{eff}\Delta v)$ in optical fiber, where c is the speed of light in free space and n_{eff} is the effective refractive index of the optical fiber core. In many cases, we express the bandwidth in free-space wavelength $\Delta\lambda$, which is given by $\Delta\lambda = \Delta(c/v) = -c\Delta v/nu^2 = -\lambda\Delta v/v = -\lambda^2\Delta v/c$. Here we usually neglect the minus sign because we are considering the absolute bandwidth. Therefore, the coherence length is given by

$$\Delta L = \frac{\lambda^2}{n_{eff}\Delta\lambda} \tag{7.4}$$

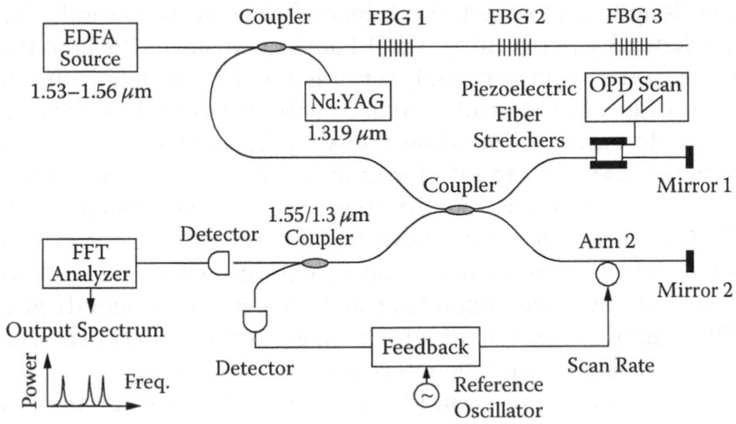

FIGURE 7.9
Schematic diagram of a sensor system using a fiber Fourier transform spectroscopy interrogator (FFT: fast Fourier transform). (From M. A. Davis and A. D. Kersey, *J. Lightwave Technol.*, **13**, 7, pp. 1289–1295, 1995.)

For an FBG at a 1550-nm reflection wavelength with $\Delta\lambda$ of 0.2 nm, the coherence length in optical fiber (typically, $n_{eff} \approx 1.45$) is about 8 mm. The interference pattern (i.e., the *interferogram*) appears only when the optical path length difference (OPD) between the two arms of the interferometer falls within the coherence length. Hence, if one arm of the interferometer is shorter than the other, and if the shorter arm is linearly lengthened with time by a sawtooth-like signal applied to a piezoelectric transducer (PZT) tube on which a portion of the shorter arm is wound, an interferogram is obtained as shown in Figure 7.10 for a light reflected from an FBG. The period of the fringe patterns corresponds to a 2π phase change; that is, it corresponds to the round optical path length change the same as the wavelength in fiber. (The temporal coherence characteristics of the light reflected from the FBG give the slowly varying envelope shape in Figure 7.10. That is, for the two lights that make round trips of the two arms and combine at the detector, their coherence is reduced as their relative time delay is increased. If the relative time delay becomes larger than the coherence time, then no interference pattern appears.) If the

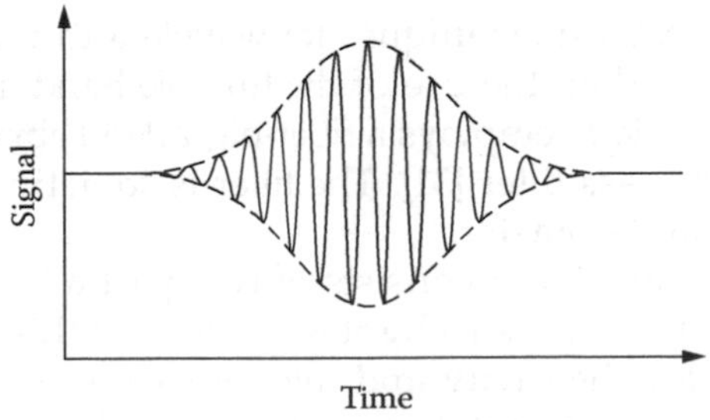

FIGURE 7.10
Example of an interferogram using a modulated Michelson interferometer for a single FBG.

Bragg wavelength is changed, the interval between the periodic fringes is changed. Hence, by performing a fast Fourier transform (FFT) of the signal with an electronic spectrum analyzer, it is possible to determine the measurand. If multiple FBG sensors are used with different Bragg wavelengths, the total interferogram in the time domain will show a complex shape due to the superposition of the interferograms corresponding to each grating. However, in the frequency domain, they are clearly separated, as shown in Figure 7.9. Note that here "frequency" does not refer to the optical frequency but, rather, to the low frequency basically determined by the arm scanning speed. The Nd:YAG laser operating at a 1319-nm wavelength is used in Figure 7.9 to monitor the rate at which fringes are produced, correcting the scan using a feedback loop when necessary.

Davis and Kersey [27] used an FFT analyzer with a resolution of ~6 mHz, or a fractional change in frequency of ~$1{:}10^5$ that represents an equivalent wavelength shift resolution of 15 pm, which translates to a strain resolution of ~12 $\mu\varepsilon$ at 1550 nm. This scheme overcomes the 2π measurement range limitation of the typical interferometric wavelength-shift detection that is explained later.

Flavin et al. [28] reported a modified method in which only a short section of the interferogram is scanned. While the method by Davis and Kersey scans about 10 cm to obtain the full interferogram, Flavin et al. obtained a 5-pm resolution with an optical path scan of 1.2 mm. In the method, by using an electronic Hilbert transform and software process, the effect of the envelope shape in the interferogram is removed and only the phase change with the optical path change is unwrapped from the sinusoidally varying function. The Bragg grating reflection wavelength is measured by comparing the phase change with that of a high-coherence inteferogram derived from a He-Ne laser. There have been studies to extend this method to interrogate multiple grating sensors [29,30]. In particular, Rochford and Dyer [30] showed that closely spaced (1.4-nm separation) Bragg wavelengths from different gratings can be demultiplexed. Therefore, this interrogation system might be applied to a dense, multiple-FBG sensor system.

7.3.2 Fabry–Perot Filter Interrogator

One of the most successful techniques for wavelength-change interrogators of FBG sensors is based on the use of the tunable bandpass filter. The most commonly used technique employs a fiber-pigtailed Fabry–Perot tunable filter as a narrow bandpass filter [31]. The filter is sometimes referred to as a fiber Fabry–Perot interferometer.

The Fabry–Perot filter (FPF) consists of two partially reflecting surfaces with a spatial separation. When a light is incident on this cavity, due to multiple reflections inside the cavity and interference of the multiply reflected lights, the transmittance of light through this cavity has a periodic characteristic with the variation of optical frequency or the spacing between the two

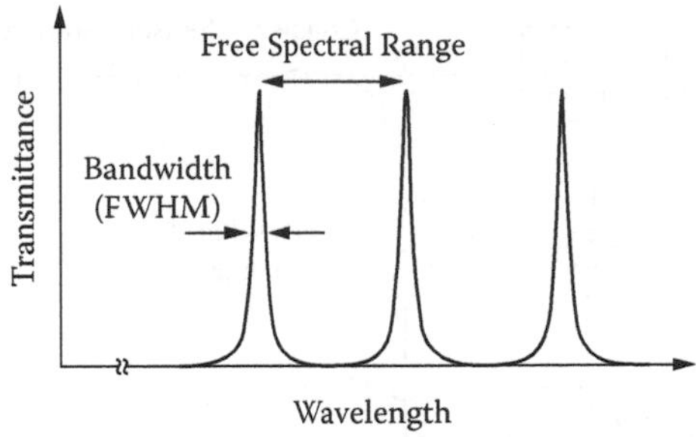

FIGURE 7.11
Typical transmittance versus wavelength of the Fabry–Perot filter (FWHM: full width at half-maximum).

reflecting surfaces. For an optical wave, which is partially reflected by the surfaces, the extra phase it experiences during each round trip is given by

$$\varphi = 2\pi\frac{2nl}{\lambda} + \varphi_0 \tag{7.5}$$

where n is the refractive index of the cavity material, l is the cavity length, λ is the wavelength in free space, and ϕ_0 is the phase that might come from the reflections at both ends of the cavity (0 or 2π). If the phase difference (i.e., the extra phase) is a multiple of 2π radians, then the transmittance becomes maximum due to the constructive interference. Figure 7.11 shows a typical transmittance versus wavelength plot for a regular Fabry–Perot interferometer. The transmittance curve is usually plotted as a function of the cavity length or optical frequency. However, the curve plotted as a function of the wavelength is more useful in understanding the interrogation principle because we usually specify a light with a wavelength rather than a frequency. Although the phase difference in Eq. (7.5) is inversely proportional to the wavelength, the small change in the phase difference $\Delta\phi$ is proportional to the small wavelength change because $\Delta(1/\lambda) = -(\Delta\lambda)/\lambda^2$. Equation (7.5) also shows that the phase difference is proportional to the cavity length. Hence, if the cavity length is changed with PZT, the transmission band is tuned.

Typically, tunable fiber FPFs have bandwidths of about 0.2 to 0.6 nm, a free spectral range (FSR; see Figure 7.11) of 40 to 60 nm, and a finesse factor (which is the ratio of the FSR to the bandwidth) of 100 to 200. Filter tuning is achieved by accurately displacing the reflection surface separation using a piezoelectric element. Currently available FPFs can have scan rates close to 1 kHz or higher.

Figure 7.12 shows the schematic diagram of the tunable fiber FPF for demodulating the wavelength shift from a single FBG. In this case the demodulator is working in a tracking or closed-loop mode. Typically, the FPF bandwidth is comparable to the grating bandwidth. The FSR should be larger than the operating range of the grating to avoid measurement ambiguity. Kersey

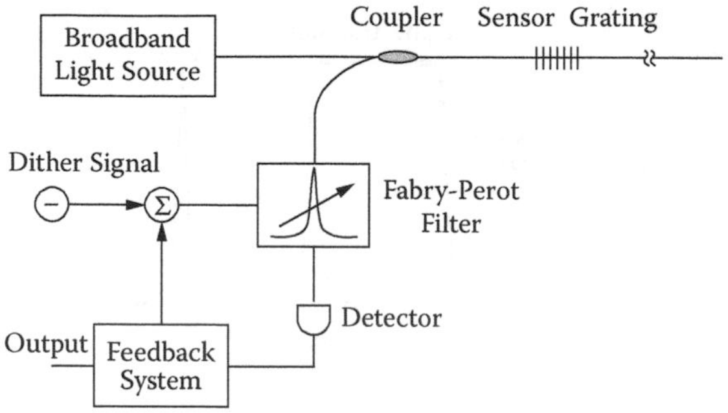

FIGURE 7.12
Schematic diagram of the Fabry–Perot filter interrogator sensor system working in a lock-in mode. (From A. D. Kersey et al., *Optics Lett.*, **18**, pp. 1370–1372, 1993.)

et al. [31] reported a resolution of ~1 pm over a working range over 40 nm. The closed-loop arrangement locks the Fabry–Perot passband to the grating reflection signal. By the sinusoidal dithering of the cavity length of the filter, the filter transmission wavelength (i.e., resonance wavelength) is periodically changed by a fraction (~0.01 nm) of its passband (~0.3 nm). If the resonance wavelength of the filter matches the reflection wavelength of the FBG, the detected power reaches a maximum due to the maximal overlap (in the frequency space) of the filter passband and the grating reflection spectrum. (The multiplication in frequency domain corresponds to convolution in the time domain.) If the passband shifts to longer or shorter wavelength by dithering, the detected power becomes smaller. Therefore, if the passband maximally overlaps the FBG reflection spectrum, even with dithering the received power does not have a varying component with the same frequency of the dithering. Rather, it has the second harmonic component. The first harmonic component, however, exists when the filter resonance wavelength and grating Bragg wavelength do not precisely match, although the two bands partially overlap. The first harmonic component of the detected power is monitored and used as an error signal to lock the filter. The voltage applied to the piezoelectric material is used as measurement data.

Although the tracking method is applicable to a single grating sensor interrogation, multiple grating sensor signals can also be interrogated using the FPF by scanning the resonance wavelength. If the grating Bragg wavelengths and their ranges of change due to measurands do not overlap and yet fall within the spectral bandwidth of the light source and the FSR of the FPF, a number of gratings along the same optical fiber can be interrogated. Figure 7.13 shows a typical example of the sawtooth scanning signal and detected signals. In real systems, the dithering is also used for fine measurement, and the zero-crossing of the time derivative of the received power is monitored.

The U.S. Naval Research Laboratory group has field-tested multipoint sensor systems for civil structure monitoring adopting the scanning fiber FPF interrogation approach [32]. Figure 7.14 shows the schematic of their

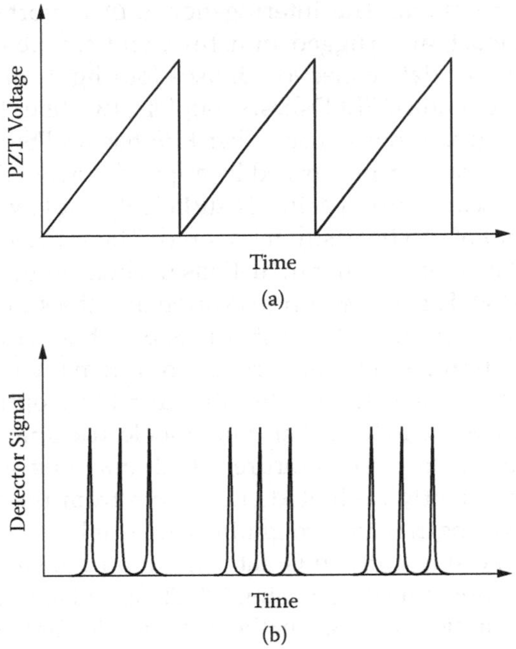

FIGURE 7.13
Typical scanning of the Fabry–Perot filter for wavelength-multiplexed sensors. (a) PZT driving signal; (b) detector signal for three wavelength-multiplexed sensors.

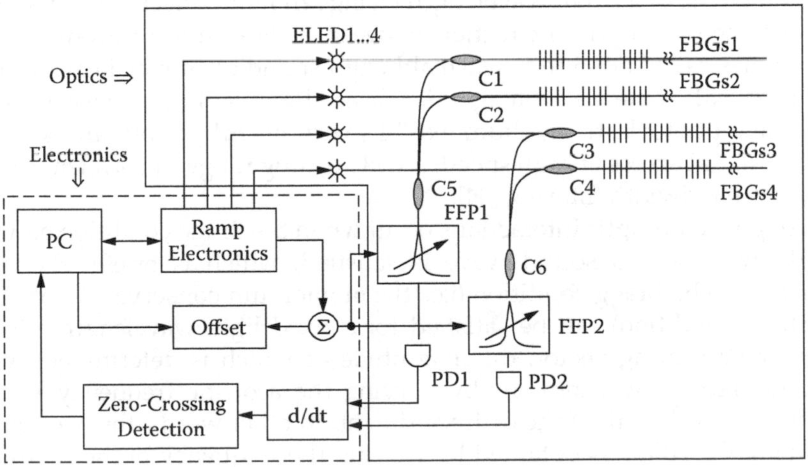

FIGURE 7.14
Schematic diagram of a 64-channel sensor fiber Bragg grating sensor system adopting Fabry–Perot filters. (From S. T. Vohra et al., *Proc. 13th Int. Conf. Optical Fiber Sensors (OFS-13)*, Kyongju, Korea, *SPIE*, **3746**, pp. 32–37, 1999.)

64 grating sensor system. The interrogation unit is portable, lightweight (<5 pounds), compact, and rugged in nature, and can be easily used with a laptop computer for data collection. It uses four light sources (edge-emitting, light-emitting diodes [ELEDs]), six couplers, two fiber FPFs, two optical detectors, and electronic parts. Each fiber FPF has an FSR of about 45 nm, allowing 16 individual sensors spaced by approximately 2.7 nm to be interrogated per filter scan. This spacing is sufficient to allow strains of about ±1300 με for each FBG to be easily monitored. The intrinsic fluctuations in the system translated into strain fluctuations of about ±2 με with an effective scan rate of about 45 Hz. The strain resolution and the scan rate are considered adequate for monitoring large structures, such as bridges. In applications where the interrogation unit is required to remain at the field site for a long period of time (over 12 months), the fiber FPFs, optical sources, and reference gratings can be mounted on a thermoelectric unit. This permits the maintenance of a constant temperature, which maximizes system stability and provides consistently calibrated strain data from the arrays. Three or four absolute wavelength reference gratings are used.

There also has been a study in Great Britain to implement a 64-point (or more) strain measurement system using 128 FBGs and adopting Fabry–Perot wavelength demultiplexing [33]. In the system, 16 fibers with eight gratings each were spatially multiplexed. At each sensing point, two gratings in neighbor fiber lines were used as a pair to distinguish the strain and temperature effects.

7.3.3 Acousto-Optic Tunable Filter Interrogator

The acousto-optic tunable filter (AOTF) is a device for detection, over a wide spectral range, of a small wavelength change in light reflected from an FBG [34]. The acousto-optic interaction between a light and an acoustic wave (sometimes called a "sound" wave although sound cannot be heard from it, just as we call some electromagnetic waves "light" although we cannot see them) in a photoelastic medium exhibits very useful performances, which include signal access at high speed, a wide tuning range in wavelength, and a narrow bandwidth filtering [34].

Strong acousto-optic interactions occur when the Bragg condition between a light wave and a sound wave is satisfied, which is briefly shown in Figure 7.15. The Bragg condition (i.e., the momentum conservation, or phase matching condition) can be satisfied for a fixed light wavelength with an appropriate grating vector, the magnitude of which is determined by the acoustic frequency. Therefore, by varying the acoustic frequency, we can select the wavelength of the diffracted light. In other words, the wavelength tuning of the AOTF is obtained by varying the frequency of the RF (radio-frequency) signal that generates the sound wave via the transducer. As a result, by changing the RF of the AOTF, it is possible to interrogate a sensor grating. It can also be extended to multipoint grating sensors. It is possible to

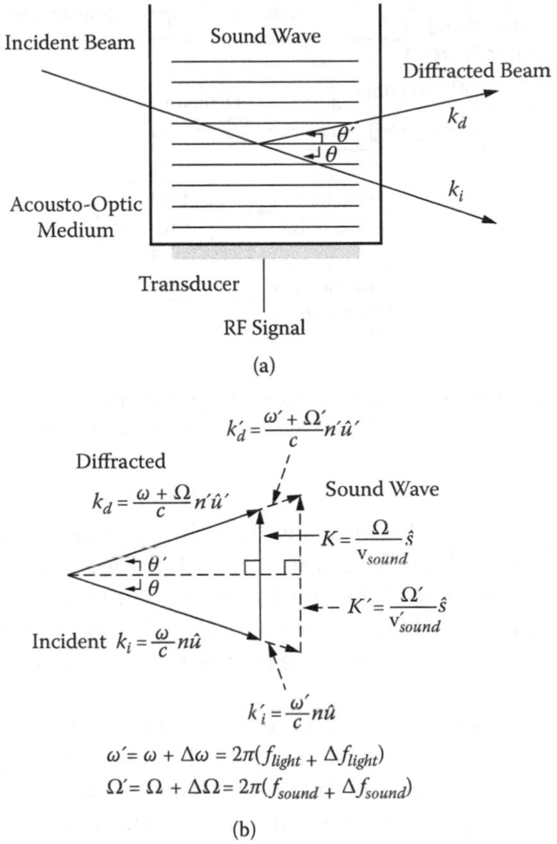

FIGURE 7.15
Bragg diffraction in AOTF: (a) diffraction of a light beam from a sound wave; (b) conservation of momentum. (From A. Yariv and P. Yeh, *Optical Waves in Crystals*, John Wiley & Sons, New York, 1984.)

change the acoustic wave frequency from tens of hertz to hundreds of megahertz, thereby extending spectral tuning range. Thus, it is suitable for use in both dynamic and quasi-static detection of wavelength change in a broadband spectral range that may include multiplexed sensors. Furthermore, the AOTF technique has several advantages. It can be accessed at multiple wavelengths simultaneously as well as at random wavelengths [35]. This is obtainable by applying multiple RF signals of different frequencies. Hence, the AOTF can offer a parallel interrogation and a reduction of interrogation time in a multiplexed sensor array system.

In general, the AOTF is operated in two principal modes: the *scanning mode* and the *lock-in mode* [36]. In the scanning mode, the light wavelength is gradually scanned over the range of interest by means of varying the applied RF signal frequency. The diffracted light power as a function of the RF signal frequency (which determines the diffracted light wavelength) simply

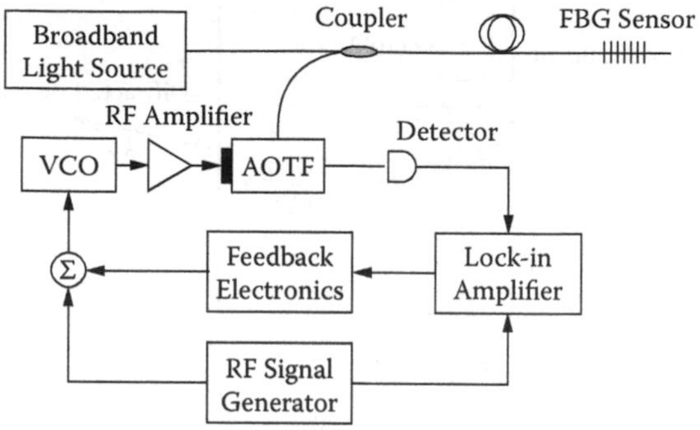

FIGURE 7.16
Schematic diagram of AOTF interrogation (VCO: voltage-controlled oscillator). (From M. G. Xu et al., *J. Lightwave Technol.*, **14**, pp. 391–396, 1996; M. G. Xu et al., *Electronics Lett.*, **29**, pp. 1510–1511, 1993.)

represents the power spectrum of the light. In the lock-in mode, the AOTF is dithered with a feedback loop, and the lock-in signal with the dithering frequency is detected. The dithering frequency is much lower than that of the RF signal, say hundreds of kilohertz for the RF of tens of megahertz. The lock-in regime is relatively immune to the intensity noise compared with the scanning mode.

Several demonstrations of AOTF detection in FBG interrogation are as follows. An electronic tracking system for multiplexed fiber grating sensors using an AOTF has been demonstrated by Geiger et al. [36] in which a standard deviation of 0.4 με was achieved at a measurement period of 100 ms. Both the analysis and experiment for an FBG interrogation system using an AOTF have been presented by Xu et al. [37,38] (see Figure 7.16 for the schematic diagram). The measured strain response factor was –96.6 Hz/με, with the measurement period of 100 ms. The measurement resolution could be improved by measuring the AOTF mean frequency over a longer period. However, in this case, the interrogation system requires a longer response time. In addition, it was pointed out that the reflective configuration and the transmissive configuration show, respectively, better noise performance for low-reflectivity fiber gratings and for high-reflectivity gratings. The temperature sensitivity of the AOTF was also studied—the temperature-tuning coefficient was 2.68 kHz/°C, or 0.03 nm/°C. That is, a temperature change of 1°C in the AOTF would be equivalent to a strain change of 25 με or a temperature change of 2.3°C in a typical fiber grating sensor around 1550-nm wavelength. Thus, for the best performance of the AOTF, it must be temperature stabilized or temperature compensated.

A multiple-wavelength interrogation using multiple RF signals of different frequencies has been demonstrated by Volanthen et al. [35]. Two FBGs written at 1300 and 1550 nm were successfully monitored simultaneously by means of a single AOTF. An electronic demultiplexing as well as an optical demultiplexing were verified (see Figure 7.17). In other words, the optical signals

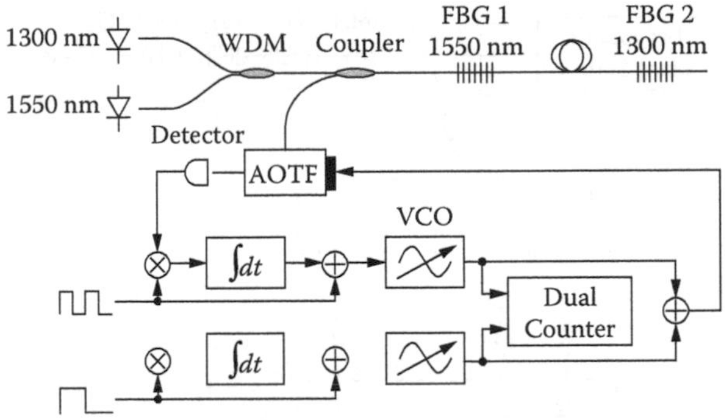

FIGURE 7.17
Schematic diagram of AOTF interrogation with electronic demultiplexing technique. (From M. Volanthen et al., *Electronics Lett.*, **32**, pp. 1228–1229, 1996.)

of different wavelengths were simultaneously dithered with a combination of different RF signals, and both lock-in signals could be detected with a single receiver. The rms errors with the electronic demultiplexing technique were 2 με at both wavelengths. The cross-talk levels between the different wavelengths were as low as −52 and −56 dB. However, the maximum number of arrayed sensor gratings that can be interrogated simultaneously is limited by the available number of frequencies with low cross-talk, within the receiver bandwidth. A combination of the electronic demultiplexing and the optical demultiplexing in the AOTF would increase the maximum number of multiplexed sensor gratings.

An AOTF interrogation system with an ELED source has recently been used in a series of routine tests for monitoring destructive cracks in composite materials, as part of an evaluation for aerospace applications [39]. Five lines of fiber, each with eight FBGs, were embedded within a sandwiched structure of a composite material. Using a single AOTF, each of the individual fiber sensing lines was interrogated in turn by a switch. Hence, this sensor system is a combination of WDM and SDM (or TDM).

7.3.4 Matched Fiber Bragg Grating Pair Interrogator

The interrogation technique by the matched FBG pairs, proposed by Jackson et al. [40], is based on matching a receiving grating to a corresponding sensor grating. The basic concept of the sensor–receiver grating pair is that the Bragg wavelength of the sensor grating is monitored by filtering the light reflected from the sensor grating with a receiver grating that is nearly identical to the sensor grating at rest. When a strain or a temperature variation is applied to the sensor grating, the matched condition of the Bragg wavelength between the grating pair is not satisfied. But the condition can be easily recovered by tuning the receiver grating with a suitable method. It is possible to utilize a piezoelectric stretcher, to tune the receiver grating [40]. This technique can be readily applied to the simultaneous interrogation of

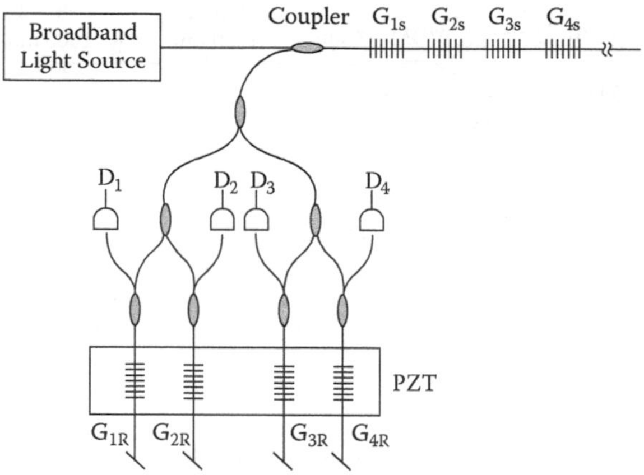

FIGURE 7.18
Schematic diagram of matched fiber Bragg grating pair interrogator in parallel configuration. [The FBGs G_{iR} and G_{iS} (i = 1, 2, 3, 4) are matched pairs.] (From D. A. Jackson et al., *Optics Lett.*, **18**, pp. 1192–1194, 1993.)

a multiplexed arrayed sensor system with a large number of gratings if all the matched gratings can be provided at the receiver end. Current fabrication techniques for fiber gratings can afford to manufacture grating arrays with good reproducibility, hence making this a simple, useful, and low-cost interrogation technique. The resultant resolution is largely dependent on the bandwidth of the gratings and the resolution of piezoelectric stretcher.

The matched FBGs can be positioned both in parallel and in series at the receiver end. In parallel configuration as shown in Figure 7.18, the light filtered by each matched grating is monitored with a separate detector, which has been demonstrated by Jackson et al. [40]. This has the advantage of simple detection (all matched gratings can be simultaneously tuned with a piezoelectric stretcher) with low cross-talk, but the number of detectors should be increased with the number of sensors in an array. The minimum resolution of the measured quasi-static strain was 4.16 με [40].

In series configurations as shown in Figure 7.19, each matched grating is tracked by a different piezoelectric stretcher with a different dithering frequency [41,42]. The signal is detected with feedback electronics, and hence the lock-in signals can be detected separately with a single receiver. Therefore, the single receiver can afford to interrogate the multiplexed sensors simultaneously. Both the transmissive and reflective configurations can be applied to this technique. Demonstrated by Brady et al. [41], the reflective configuration had two sensor–receiver grating pairs, and the resultant strain and temperature resolutions were 3.0 με and 0.2°C, respectively. The transmissive configuration, demonstrated by Davis et al. [42], had six sensor–receiver grating pairs and two FBGs were independently tracked. The resultant dynamic strain resolution was 0.01 $\mu\varepsilon/\sqrt{\text{Hz}}$ with a linear response over ±100 με.

Kang et al. [43] proposed a tilted grating matched filtering. A tilted grating can have two dips in a transmission spectrum, one coming from the

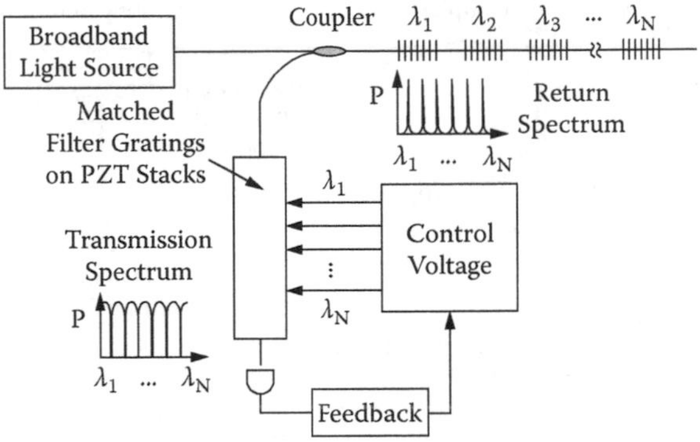

FIGURE 7.19
Schematic diagram of matched FBG pair interrogator in series configuration. (From M. A. Davis and A. D. Kersey, *Electronics Lett.*, **31**, pp. 822–823, 1995.)

coupling of the forward-propagating core mode to the backward-propagating core mode, and the other from the coupling of the forward-propagation core mode to the backward-propagating cladding mode. The spacing between the two dips in the transmission spectrum is unaffected by temperature change, and the dips can be scanned by applying a sawtooth signal to a PZT that stretches the tilted grating. Using this temperature-immune demodulator to a temperature-discriminating dual-grating sensor, Kang and coworkers were able to obtain a strain sensor system that allows stable measurements independent of temperature perturbation at both sides of the sensor and demodulator without any additional temperature-isolation or temperature-referencing process.

7.3.5 Unbalanced Mach–Zehnder Interferometer Interrogator

A very high resolution interrogator for an FBG sensor is an unbalanced Mach–Zehnder interferometer (MZI) demodulator [44]. By using the unbalanced MZI, the Bragg wavelength change of the fiber grating sensor is converted to the phase variation of the interference signal at the detector. As shown in Figure 7.20, the two arms of the fiber Mach–Zehnder interferometer have different lengths. If the physical length difference of the two arms is d, the optical path length difference (OPD) is $n_{\text{eff}}d$, where n_{eff} is the effective refractive index of fiber core. This result stems from the fact that inside an optical fiber the wavelength is smaller than that in the air by a factor of $1/n_{\text{eff}}$. Hence, the effective length difference that the light feels is n_{eff} times the difference is physical length. If the OPD falls within the coherence length of the light reflected by the FBGs, then an interference signal is obtained; that is, the light intensity at the detector in Figure 7.20 is given by

$$I = I_0\left[1 + a\cos\left(2\pi\frac{n_{\text{eff}}d}{\lambda}\right)\right] \tag{7.6}$$

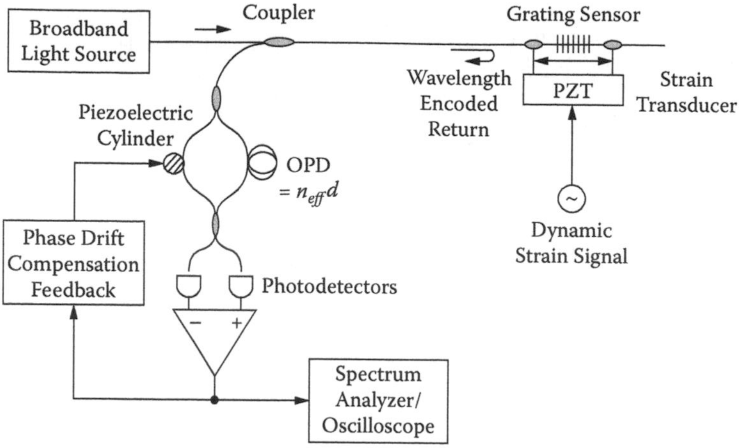

FIGURE 7.20
Unbalanced Mach–Zehnder interferometer interrogator (basic structure). (From A. D. Kersey et al., *Electronics Lett.*, **28**, 3, pp. 236–238, 1992.)

where I_0 is a constant, a is related to the temporal coherence of the light reflected by the FBG, and λ is the free-space wavelength of the light. If the Bragg wavelength is changed due to the measurand at the FBG, then the phase in Eq. (7.6) is changed; by analyzing the phase change, the applied measurand information can be obtained. In MZI systems that use broadband sources such as ELEDs, superluminescent diodes, or EDFA ASE, the signal level at the photodetector is very low due to the small spectral power density of the light sources. Using the balanced detection technique as shown in Figure 7.20, the SNR can be substantially improved. At the 3-dB coupler in front of the detectors in Figure 7.20, the lights that cross-coupled the coupler experience a 90° phase change. Therefore, there is a 180° relative phase difference between the two signals at the photodetectors. By subtracting one signal from the other, the signal level is doubled and the phase-independent noises are cancelled out.

The advantage of using MZI over the linearly wavelength-dependent optical filter [8] is the ease of customizing the filter function. By controlling the path length difference d, we can easily increase or decrease the system sensitivity, as we discuss later. However, the longer the path length difference is, the smaller the interference signal becomes because of the low coherence of the light reflected from an FBG. (The parameter a in Eq. 7.6 decreases with the deviation of d from 0.) Therefore, the maximum sensitivity is related with the interferometer's OPD and the coherence of reflected light (which is inversely dependent on the FBG reflection bandwidth). Weis et al. [45] calculated the condition for the maximum sensitivity to be

$$n_{\text{eff}}\, d\Delta k = 2.355 \tag{7.7}$$

where $n_{\text{eff}}\, d$ is the OPD and Δk the bandwidth of FBG reflection spectrum expressed in wavenumber ($k = 2\pi/\lambda$) units. From this relation, when a fiber grating that has $\Delta\lambda = 0.2$ nm at a 1550-nm center wavelength is used, the OPD for the maximum sensitivity is calculated as 4.5 mm (~3 mm is free space). When a fiber grating with a strain sensitivity of 1.2 pm/με and a reflection wavelength of 1550 nm is used with the 4.5-mm optical path unbalanced MZI, the phase change response is ~12 rad/nm, which corresponds to ~0.014 rad/με. By using a phase meter with a 0.1° resolution and a dynamic phase shift detector with capability of $\sim \mu\text{rad}/\sqrt{\text{Hz}}$—both of which are currently available—we can obtain the strain resolution of ~0.13 με and $\sim 70\,\text{p}\varepsilon/\sqrt{\text{Hz}}$ for the quasi-static and dynamic strain, respectively.

The measurement sensitivity can be dramatically increased by using a fiber laser. The sensor head consists of two FBGs inscribed on erbium-doped fiber, and they work as end mirrors of the laser cavity as well as sensors. The laser spectrum has a much narrower wavelength bandwidth (i.e., longer coherence length). Therefore, the sensitivity can be increased by enlarging the OPD of the MZI. Koo et al. [46] demonstrated a dynamic strain resolution of $\sim 7\times10^{-15}/\sqrt{\text{Hz}}$ with a 100-m OPD interferometer.

Pseudo-Heterodyne Method

One of the practical ways to analyze the phase variation is pseudo-heterodyne processing, which involves the application of optical path length modulation to one of the interferometer arms [47], shown in Figure 7.21. A length of optical fiber is wound around a piezoelectric tube. A ramp signal applied to the tube expands the tube and the fiber, altering the optical path length of the arm periodically. If we denote the physical path difference by $d = d_0 + d_1 t$ as a function of time during one period, then the phase term in Eq. (7.6) is given by $2\pi n_{\text{eff}} d_0/\lambda + 2\pi n_{\text{eff}} d_1 t/\lambda$. If there is a change in light wavelength by $\Delta\lambda$ (i.e., if $\lambda' = \lambda + \Delta\lambda$), then the phase term becomes

$$2\pi n_{\text{eff}} d_0/\lambda' + 2\pi n_{\text{eff}} d_1 t/\lambda' \approx$$

$$2\pi n_{\text{eff}} d_0/\lambda - 2\pi n_{\text{eff}} d_0 \Delta\lambda/\lambda^2 + 2\pi n_{\text{eff}} d_1 t/\lambda - 2\pi n_{\text{eff}} d_1 t\Delta\lambda/\lambda^2$$

where the last term is negligible. The ramp signal is given in such a way that its period T is equal to $\lambda/n_{\text{eff}} d_1$. If we define the angular frequency ω by $\omega = 2\pi/T$, then Eq. (7.6) becomes

$$I \approx I_0\left[1 + a\cos\left(\omega t + 2\pi\frac{n_{\text{eff}} d_0}{\lambda} - 2\pi\frac{n_{\text{eff}} d_0}{\lambda^2}\Delta\lambda\right)\right] \tag{7.8}$$

Therefore, we see that the detector signal has a sinusoidal form whose period, $2\pi/\omega$, is equal to that of the ramp signal T. In practice, the amplitude

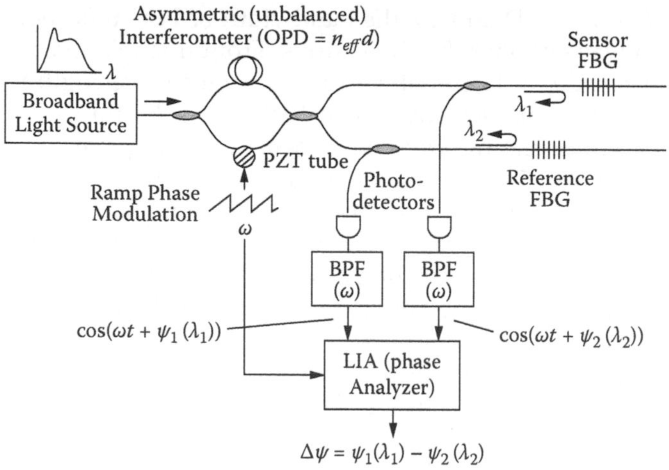

FIGURE 7.21
Pseudo-heterodyne detection adopting unbalanced Mach–Zehnder interferometer with reference grating (LIA: lock-in amplifier). (From A. D. Kersey and T. A. Berkoff, *IEEE Photonics Technol. Lett.*, **4**, 10, pp. 1183–1185, 1992; A. D. Kersey et al., *Optics Lett.*, **18**, 1, pp. 72–74, 1993.) In general, the reference grating is isolated from environmental change such as temperature or strain and can be located near the interrogator system. This reference grating should not be confused with another kind of reference grating that is used to separate temperature effect from strain effect at the sensor head.

of the ramp signal should be carefully controlled to make the waveform sinusoidal (i.e., 2π modulation). The detector output is then bandpass filtered at the modulation frequency to eliminate the other frequency components (spikes made by discontinuities in the ramp signal). By using the bandpass-filtered signal and ramp signal as the input and reference signals of the lock-in amplifier, respectively, we can measure the change in light wavelength in phase form, which can be seen in Figure 7.22. The trade-off between the resolution and measurement range is apparent. That is, if we reduce the

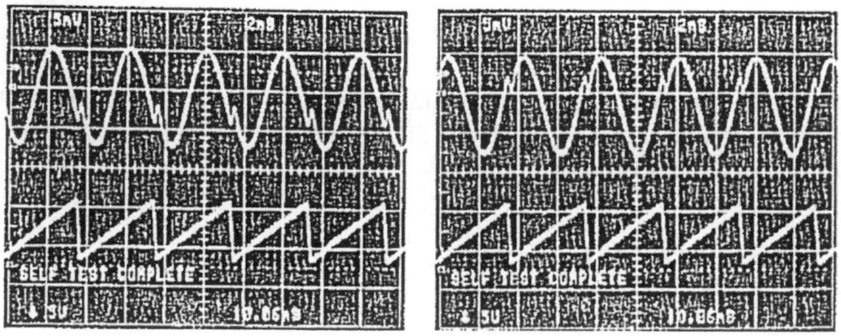

FIGURE 7.22
Input signals (lower) to piezoelectric tube and output intensity signals (upper) of photodetector for different reflection wavelengths of a sensor grating.

unbalanced path length d_0, the unambiguous measurement range within 2π radians is increased, while the phase response to the wavelength change becomes less sensitive, and vice versa.

A serious problem with the unbalanced MZI interrogator is the fact that the interference signal is very sensitive and drifts over time due to fluctuations in environmental factors such as temperature. The OPD changes as a result of irregular temperature distributions. Although it can be reduced to some degree by shielding the interferometer part, a completely drift-free system is nearly impossible. To solve this problem, Kersey et al. [48] proposed the reference grating method, shown in Figure 7.21. Because the fluctuation arises from the interferometer, both the sensor and reference grating signals experience the same amount of thermal phase drift. Therefore, the effect can be cancelled by subtracting the reference grating output from that of the sensor grating. Note that in Figure 7.21, the interferometer is inserted between the light source and the gratings, while it was applied after the sensor grating in Figure 7.20. Both cases are basically the same. Sometimes the method described in Figure 7.21 is more applicable.

Quadrature Signal Processing Techniques

Although pseudo-heterodyne processing is very effective, it is not suitable for high-bandwidth applications because of the complex phase measurement mechanism. The simplest way to use the MZI interrogator is to limit the input wavelength range within the linear portions of the sinusoidal filter function. The input–output relationship is then assumed to be linear, and the wavelength shift is analyzed by simple observation of the detector output. In this case, however, it is not easy to sustain this linear relation because the filter function of the interferometer drifts by temperature variation (signal fading). In addition, as explained earlier, a trade-off exists between the measurement resolution and the input wavelength range. To overcome these problems, quadrature signal processing techniques were proposed [49–52]. The techniques consist of two parts: generation of two quadrature signals and phase extraction. From simple trigonometric relations, when two signals have a 90° phase difference, the phase information (i.e., measurand applied to the fiber grating sensor) can be extracted using differentiate-and-cross-multiplication (DCM) [49] or an arctangent demodulation algorithm [50].

The phase-generated carrier (PGC) demodulation technique proposed by Dandridge et al. [49] has been widely used. As shown in Figure 7.23, the path length difference of the MZI is modulated by a sinusoidal wave with a modulation angular frequency ω_0. The intensity variation then becomes

$$I = A + B\cos(C\cos\omega_0 t + \phi(t)) \tag{7.9}$$

where A is the DC component, B is the constant related to the coherence and mixing efficiency, C is the modulation amplitude, and $\phi(t)$ is the phase signal containing the measurand information. Expanding Eq. (7.9) in terms

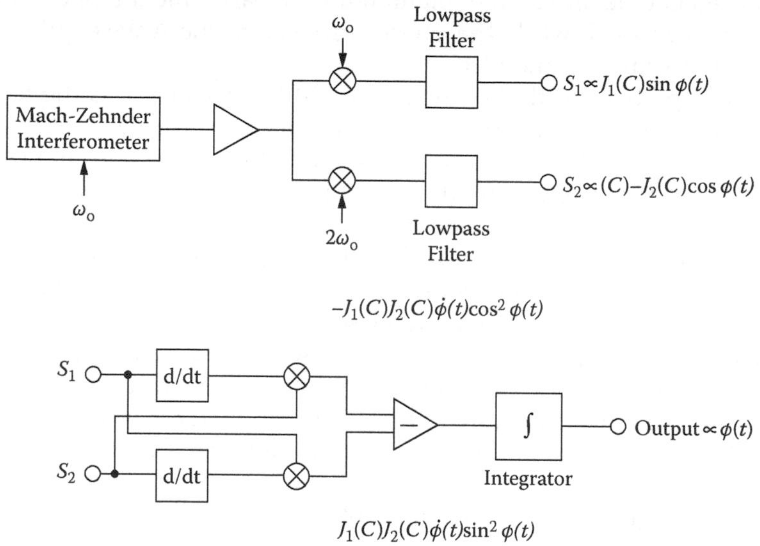

FIGURE 7.23
Schematic diagram of the phase-generated carrier demodulator system. (From A. Dandridge et al., *IEEE J. Quantum Electronics*, **18**, pp. 1647–1653, 1982.)

of Bessel functions (this is basically Fourier series representation for angular frequency ω_0 and its harmonics) produces

$$I = A + B\left\{\left[J_0(C) + 2\sum_{k=1}^{\infty}(-1)^k J_{2k}(C)\cos 2k\omega_0 t\right]\cos\phi(t) - \left[2\sum_{k=0}^{\infty}(-1)^k J_{2k+1}(C)\cos\left[(2k+1)\omega_0 t\right]\right]\sin\phi(t)\right\} \tag{7.10}$$

This interference output is electronically mixed with the first (ω_0) and second ($2\omega_0$) harmonics of the modulation frequency and low-pass filtered, generating the $BJ_1(C)\sin\phi(t)$ and $-BJ_2(C)\cos\phi(t)$ signals, respectively. Hence, two signals that have a 90° phase difference are obtained. The DCM process is utilized to extract the phase from the signals. The time derivatives of the output signals are given by $BJ_1(C)\,\phi(t)\cos\phi(t)$ and $BJ_2(C)\,\phi(t)\sin\phi(t)$, respectively. Cross-multiplying these results with the signals before the differentiation and then subtracting them give

$$B^2 J_1(C)J_2(C)\dot{\phi}(t)(\sin^2\phi(t) + \cos^2\phi(t)) = B^2 J_1(C)J_2(C)\dot{\phi}(t) \tag{7.11}$$

By integrating Eq. (7.11), we obtain the signal $\phi(t)$. An alternative phase-extracting technique involves arctangent demodulation. The sine signal is divided by the cosine signal. An arctangent function is then applied to the dividend, thus extracting the phase $\phi(t)$. This arctangent demodulation is more suitable for the digital signal processing scheme, while the DCM is preferred in analog circuit form. The PGC demodulation technique of the MZI shows various advantages over the basic MZI system, such as an unlimited input range, no signal fading, and the identification of input signal direction, etc.

Song et al. [51] recently proposed an interesting approach that uses timed digital sampling of the interference signal. The modulated interference output is digitally sampled twice in one modulation period, and the two sampled data streams become the quadrature signals when the time delay between the samplings is carefully controlled to achieve a 90° phase difference. With this technique it is much easier to control the system parameters over the PGC. Furthermore, because the quadrature signals are obtained from the same output port, any intensity variations caused by external perturbation are cancelled in the dividing process of arctangent demodulation, making the final output very strong against unwanted perturbation effects. They demonstrated both quasi-static and dynamic strain measurements up to a 2-kHz system bandwidth with resolution of 3.5 με and $6\,n\varepsilon/\sqrt{\text{Hz}}$, respectively.

Another approach is to use a dual-grating sensor and an OPD-controlled MZI, as shown in Figure 7.24 [52]. An MZI with a tunable optical delay line in one arm is adjusted so that the two Bragg wavelengths of the dual-grating sensor correspond to the 90° (or multiples of 360° plus 90°) spacing of the MZI interference pattern in wavelength domain, as shown in Figure 7.25. A chirped FBG is used in the interrogator as a band reflection filter to separate the two quadrature signals.

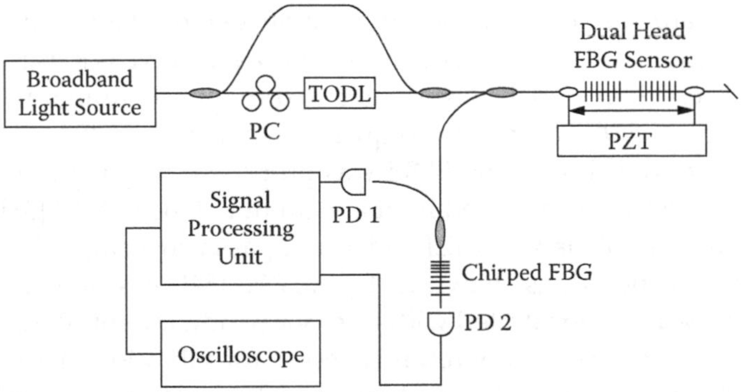

FIGURE 7.24
Schematic diagram of the quadrature processing sensor system using a dual-grating sensor and an OPD-controlled MZI (PC: polarization controller; TODL: tunable optical delay line; PD photodiode. (From S. C. Kang et al., *Proc. Conf. Lasers and Electro-Optics—Pacific Rim (CLEO/Pacific Rim '99)*, Seoul, Korea, pp. 135–136, Sept. 1999.)

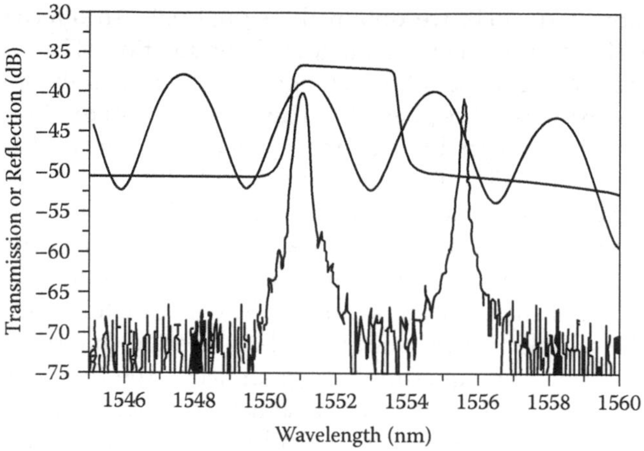

FIGURE 7.25
MZI interference pattern in wavelength domain (sinusoidal curve), dual-grating sensor reflection signals (two peaks), and chirped FBG reflection spectrum (square curve). (From S. C. Kang et al., *Proc. Conf. Lasers and Electro-Optics—Pacific Rim (CLEO/Pacific Rim '99)*, Seoul, Korea, pp. 135–136, Sept. 1999.)

Interrogation for Multiplexed Sensors

It has been shown that wavelength-shift detection using the unbalanced MZI offers extremely high resolution. To achieve the practicality level required in the real fields, however, it is highly desirable that sensors be multiplexed, enabling distributed sensing at a lower cost. Therefore, many studies have been initiated to implement multipoint FBG sensors adopting the unbalanced MZI interrogators. A TDM method using a pulse modulator and FBGs with different Bragg wavelengths (hence, it is not purely TDM as it uses the wavelength division characteristic also) was tested to give a minimum detectable dynamic strain of $2\text{n}\varepsilon/\sqrt{\text{Hz}}$ with frequencies greater than 10 Hz [45].

A combination of WDM and SDM was proposed by using a tunable FPF to spectrally slice the broadband light and an unbalanced MZI [53]. A WDM scheme using an unbalanced MZI and a bandpass wavelength division multiplexer was proposed, as shown in Figure 7.26 [54]. A strain resolution of $1.52\,\text{n}\varepsilon/\sqrt{\text{Hz}}$ was demonstrated with a sensor bandwidth of 10 ~ 2000 Hz for four sensors. The system was recently tested for vibration monitoring of a ship waterjet [55]. Arrays of this system were also tested for a vessel to monitor higher harmonics in the turbine and rapid rise times during slamming events on the wet deck [56]. The interrogator system had a bandwidth of 2.5 kHz, and the noise floor of the system was less than $10\,\text{n}\varepsilon/\sqrt{\text{Hz}}$ at 1 kHz. The 5.5-nm channel width in the WDM led to a ±2500-με dynamic range.

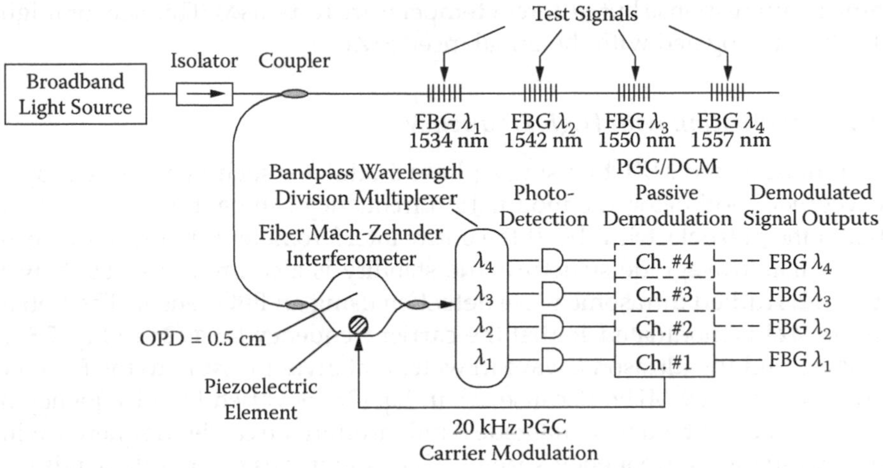

FIGURE 7.26
A wavelength division multiplexing scheme adopting unbalanced MZI. (From T. A. Berkoff and A. D. Kersey, *IEEE Photonics Technol. Lett.*, **8**, 11, pp. 1522–1524, 1996.)

Interrogation for Two-Grating Sensors

The Bragg wavelength shift is subject to both strain and temperature, and the discrimination of the two physical values is not possible by one wavelength-shift measurement from a single grating sensor. In quasi-static strain measurement, wavelength shift by temperature variation can seriously affect the accuracy of strain measurement. Of the techniques presented to solve the problem, a dual-grating sensor system composed of two different-diameter fiber gratings employs the unbalanced MZI [57]. Here the involved dual gratings reveal the same wavelength shifts with temperature but different wavelength shifts with strain because of the different cladding diameters [58,59]. Two interference signals (formed by the unbalanced MZI) for the two gratings are added at the detector, generating a total intensity variation according to the phase difference of the two signals. When the phases are in phase, the intensity variation reaches a maximum, while it becomes a minimum out of phase. Because the amounts of phase variation as a result of temperature are the same for the two grating signals, the total intensity value is not affected by temperature, enabling temperature-insensitive strain measurements to be made. By monitoring the amplitude (not the phase) of a detector signal, the strain can be measured. A dynamic strain resolution of $0.1\,\mu\varepsilon(\mathrm{rms})/\sqrt{\mathrm{Hz}}$ has been obtained [60].

Cavaleiro et al. [61] recently proposed a different type of dual-grating sensor head. One grating is formed on a germanosilicate fiber, while the other is formed on a boron-codoped germanosilicate fiber. The two gratings have the same strain response but different temperature responses. This sensor might also be interrogated with the unbalanced MZI.

Phase Modulation with High Frequency

In an unbalanced MZI, by using a phase modulator such as an electro-optic waveguide modulator, the modulation frequency can be much higher than that of the piezoelectric tube. If the entire interferometer is made in the form of a planar waveguide structure, the stability is also improved. Hathaway et al. [62] studied ultrasonic wave detection using an FBG sensor. The unbalanced MZI is modulated so that the carrier frequency $f = \omega/2\pi$ in Eq. (7.8) is 10 MHz, and the ultrasonic wave in water that gives pressure to the FBG has a frequency of 1.9 MHz. Because $\Delta\lambda$ in Eq. (7.8) is varied by a frequency of 1.9 MHz, the light intensity has side bands around the carrier frequency with the acoustic wave frequency spacing. The amplitudes of the side bands can be monitored by an RF spectrum analyzer to ascertain the strength of the ultrasonic wave. Hathaway et al. simultaneously measured the temperature of water by using a CCD spectrometer demodulator to interrogate the dc wavelength shift of the FBG. They obtained a noise-limited pressure resolution of $\sim 4.5\times10^{-4}\ \text{atm}/\sqrt{\text{Hz}}$ and a temperature resolution of 0.2°C.

7.3.6 Michelson Interferometer Interrogator

Techniques involving the use of the MZI can also be applied to the Michelson interferometer. The fiber Fourier transform spectrometer interrogator discussed in Section 7.3.1 is also based on the fiber Michelson interferometer.

Sometimes bulk Michelson interferometers are used for FBG sensor interrogators. The resolution of the interrogator is proportional to the OPD between the two arms of the interferometer, while the unambiguous range of measurement (FSR) is inversely proportional to it, as discussed in Section 7.3.5. In the case of the bulk Michelson interferometer, it is easy to lengthen the OPD, while the OPD change in a fiber interferometer is limited by the breaking strength of the fiber. However, the use of bulk devices suffers from optical loss and alignment preservation.

A recent example of using the bulk Michelson interferometer for a time division multiplexed fiber grating sensor system can be found in Ashoori et al. [63].

Rao et al. [64] proposed a stepped Michelson wavelength scanner, shown in Figure 7.27. One mirror is modulated by a PZT with a sawtooth signal. The other mirror can be stepped to two positions, shown in Figure 7.27. For the mirror location at Position 1, the resolution is high, because the OPD between the two arms is large, while the FSR is narrow, as can be seen in Figure 7.28. For the mirror location at Position 2, the resolution is low, while the FSR is wide. Therefore, the dual-cavity (with a stepped mirror) Michelson interferometer can take advantage of both cases: high resolution and wide FSR. This method can also be applied to the MZI interrogator.

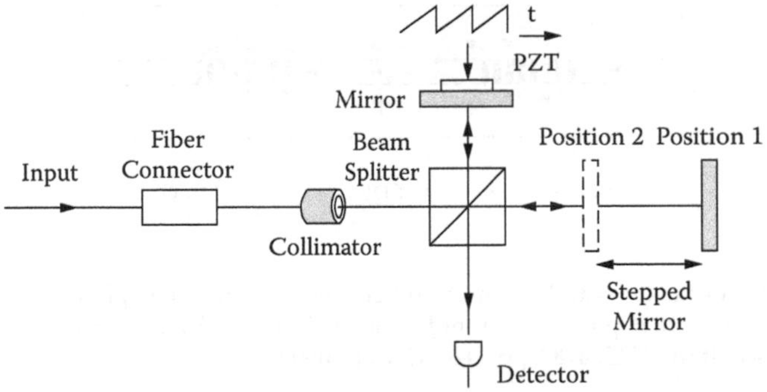

FIGURE 7.27
Schematic diagram of the stepped Michelson interferometer interrogator. (From Y. J. Rao et al., *Optics Lett.*, **21**, 19, pp. 1556–1558, 1996.)

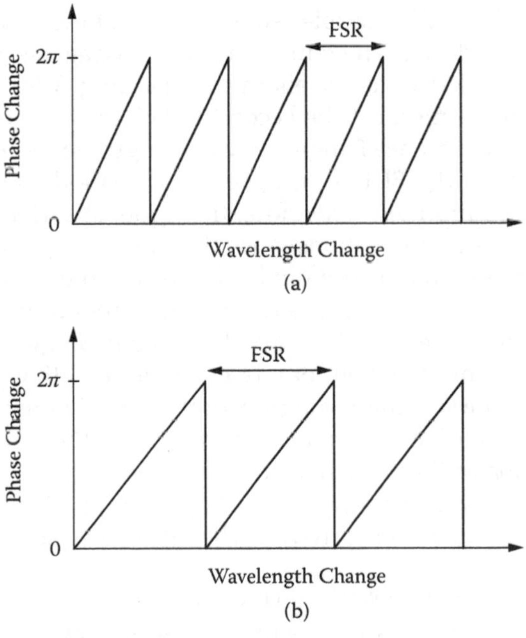

FIGURE 7.28
The phase change in the output interference signals versus the wavelength change of light: (a) for the mirror in Figure 7.27 at Position 1; (b) for the mirror in Figure 7.27 at Position 2. (From Y. J. Rao et al., *Optics Lett.*, **21**, 19, pp. 1556–1558, 1996.)

7.3.7 Long-Period Fiber Grating Pair Interferometer Interrogator

Dianov et al. [65] proposed a new type of MZI based on two cascaded LPFGs. The cladding mode coupled by the front LPFG is recoupled into the core by the following LPFG, which produces a series of interference fringes in the

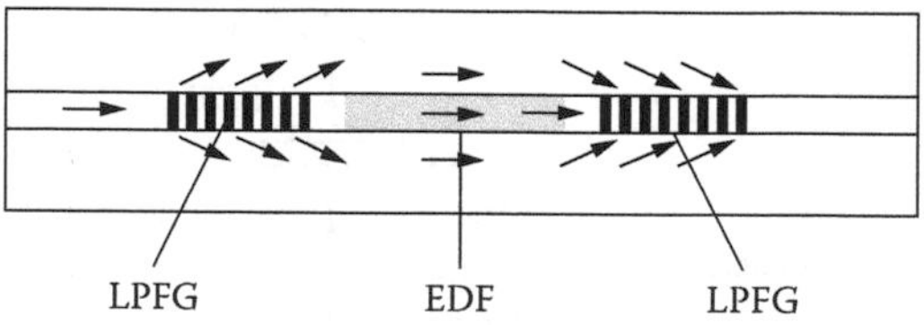

FIGURE 7.29
LPFG pair interferometer with EDF inserted between the pair. The light in the core is amplified by EDF if the EDF is pumped. (From J. Jung et al., *The 14th Int. Conf. Optical Fiber Sensors (OFS-14)*, Venice, Italy, *SPIE*, **4185**, pp. 114–117, Oct. 2000.)

corresponding stopband (the stopband that appears if only a single LPFG is used). The interference can be used for many practical purposes.

Jung et al. [66] proposed a device that consists of an LPFG pair and an EDF inserted between the two gratings, shown in Figure 7.29. The device is applicable to high-resolution interrogation of an FBG sensor. For the resonant wavelengths, the first LPFG couples some portion of incident light out of the core and into the cladding. The remaining core propagates in the core of the EDF, while the coupled light travels in the cladding. When the two lights in the core and cladding meet the second LPFG, they are recoupled by the grating, and the interference fringes (in wavelength domain) appear behind the second grating. If the EDF is pumped, the core mode propagating in the core of the EDF is amplified. Therefore, the lights at different wavelengths experience different amplifications since the amplifying amount differs by the intensity of the remnant core mode; that is, the more light that is left in the core, the larger is the amplification with sufficient pump conditions. Jung et al. used this device to monitor the FBG sensor Bragg wavelength shift by modulating the input current of a laser diode (LD) that pumps the EDF. Figure 7.30 shows the schematic diagram of the experimental setup. The harmonic current signal of the pump LD is also given to the lock-in amplifier for low-noise detection of the same harmonic component.

The transmission spectrum (without LD pumping) of the device used in the experiment is shown in Figure 7.31. The figure also shows the difference

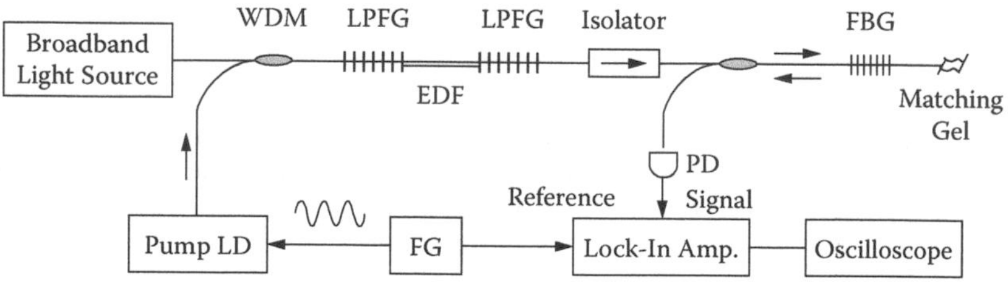

FIGURE 7.30
Schematic diagram of a sensor system adopting the LPFG-EDF interferometer interrogator (FG: function generator). (From J. Jung et al., *The 14th Int. Conf. Optical Fiber Sensors (OFS-14)*, Venice, Italy, *SPIE*, **4185**, pp. 114–117, Oct. 2000.)

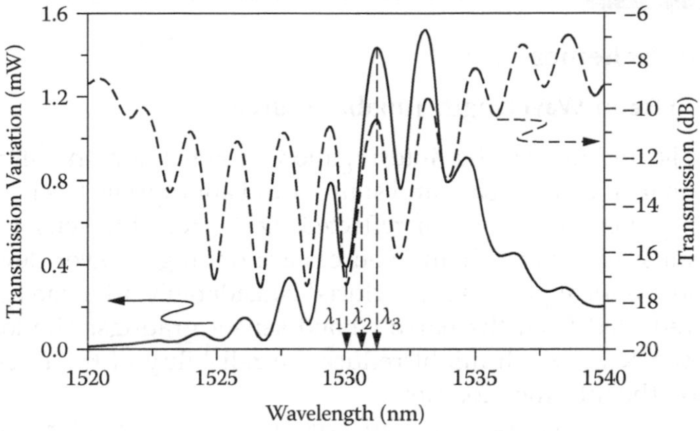

FIGURE 7.31
Transmission spectrum without LD pump (dashed line), and difference (solid line) in optical transmission powers with and without the 90-mW pump. (From J. Jung et al., *The 14th Int. Conf. Optical Fiber Sensors (OFS-14)*, Venice, Italy, *SPIE*, **4185**, pp. 114–117, Oct. 2000.)

in transmitted optical powers with and without the 90-mW pump. Therefore, for a given strain or temperature at the sensor FBG, if the EDF is harmonically (sinusoidally) pumped with time, the detector signal has the same harmonic component, the amplitude of which is dependent on the strain or temperature at the sensor grating. The fundamental harmonic component is expected to have minimum and maximum amplitudes at λ_1 and λ_3, respectively, in the quasi-linear range between λ_1 and λ_3. The average wavelength resolution in this system was evaluated to be 0.05 pm (which corresponds to nearly 40 nε and 4×10^{-3} °C around 1550 nm) by a lock-in amplifier with detection scale of 10 μV at 100-Hz EDF pumping. Experiments show that the rms voltage decreases at higher modulation frequencies because the excited erbium ions do not fully respond to the modulated pumping at high frequencies. Hence, a trade-off exists between the bandwidth and the wavelength resolution of this interrogation system. The resolution can be increased by a longer EDF, but it decreases the measurement range since the interference fringe spacing (in wavelength domain) decreases with larger length.

One distinction of this device from the standpoint of the conventional unbalanced MZI is the thermal stability. In this device the interference fringes originate from the phase difference produced by the mismatch in effective refractive indices of core and cladding in the same fiber with tens of centimeters, while in the conventional unbalanced MZI, the light is split by a 3-dB coupler and propagates in two different fibers, in which physical environmental conditions such as temperature may not be the same and the output signal drifts with time. The LPFG pair interferometer device features almost no thermal drift in the transmission spectrum.

7.4 Other Schemes

7.4.1 The Use of Wavelength Tunable Sources

Many configurations of the interrogation systems for the multiplexed arrayed sensors include broadband sources and appropriate filtering devices for detecting spectral change in reflected lights from the sensor gratings. However, most of the broadband sources are not high-power devices, and the receivable optical powers are reduced considerably after the broadband lights are reflected from the narrow-band sensor gratings. The low power leads to low SNRs, which might reduce the reliability of the interrogation and increase the interrogation time.

Thus, in order to increase the SNR, active high-power interrogation has been proposed [67,68]. Rather than a superfluorescent broadband source, this technique utilizes a wavelength tunable source that has a relatively high power and a narrow linewidth. If the wavelength of the line source is known and is launched to a sensor grating, the detected optical power reflected from the sensor grating indicates the spectral response of the sensor at the given wavelength. As a result, it is possible to fully interrogate the spectral change in the sensor grating by tuning the wavelength of the laser source over a spectral range of interest. Both the tunable fiber laser and tunable semiconductor laser are readily utilized.

The wavelength tunable EDF laser has been demonstrated for the interrogation of a three-FBG sensor by Ball et al. [67]. In the configuration shown in Figure 7.32, the LD pump power at 980 nm was adjusted to yield 100 μW of laser power, and the measured lasing linewidth was less than 20 kHz. The lasing wavelength was tuned by varying the fiber cavity length using a PZT, and the resultant tuning range was limited to 2.3 nm around 1553 nm. The resolution of the fiber laser source/analyzer was measured to approximately 2.3 pm, which corresponds to a temperature resolution of 2°C.

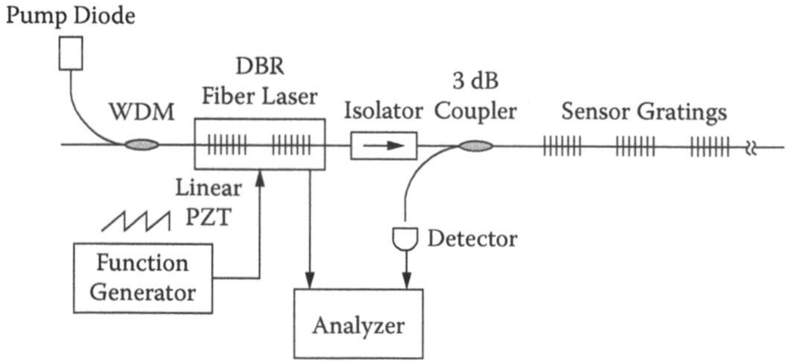

FIGURE 7.32
Schematic diagram of an interrogation by wavelength tunable source (WDM: wavelength division multiplexer; DBR: distributed Bragg reflector). (From G. A. Ball et al., *J. Lightwave Technol.*, **12**, pp. 700–703, 1994.)

The peak detection demodulation of a Bragg fiber optic sensor using a gain-coupled distributed feedback (DFB) tunable laser has been demonstrated by Coroy et al. [68]; it provides a continuous tuning range from 1536.5 to 1544.5 nm. The measured resolution was $\pm 0.076\,\mu\varepsilon/\sqrt{\text{Hz}}$ over 2000 με.

Yun et al. [69] demonstrated a 0.1-nm linewidth, wavelength-swept EDF laser scanned over a 28-nm range and used it to interrogate an FBG sensor array with a strain resolution of 0.47 με rms at 250 Hz ($42\,\text{n}\varepsilon/\sqrt{\text{Hz}}$).

7.4.2 The Use of Mode-Locked Fiber Lasers with Wavelength-Time Conversion

The mode-locked fiber laser interrogation scheme to be discussed is based on the wavelength-time conversion by means of highly dispersed optical pulses for the fast interrogation of a large number of FBG sensor arrays [70–72]. The short pulses generated exhibit high-energy and broadband characteristics. The broad spectrum of the short pulse is potentially useful for the simultaneous interrogation of a large number of wavelength multiplexed sensor arrays. In other words, it is possible to reflect optical pulses from the sensor gratings that are spread in a wide spectral range. In an experiment by Dennis et al. [70] shown in Figure 7.33, the optical bandwidth was in excess of 85 nm. The pulses are launched to the sensor gratings through a highly dispersive fiber, and the reflected lights also traverse back through the dispersive fiber. The carrier wavelength of the reflected pulse is determined by the Bragg wavelength of the sensor grating. That is, only the spectral components that rely on the bandwidth of the FBG are reflected among the broad spectral components of the short pulse. The reflected pulse with a different carrier wavelength endures a different traversal time through the dispersive fiber. As a result, the wavelength change in the sensor grating by strain or temperature variation causes the change of the pulse arrival time at the receiver end. Temporally, the reflected light consists of a sequence of pulses separated by the time of flight between the arrayed gratings, and each pulse has

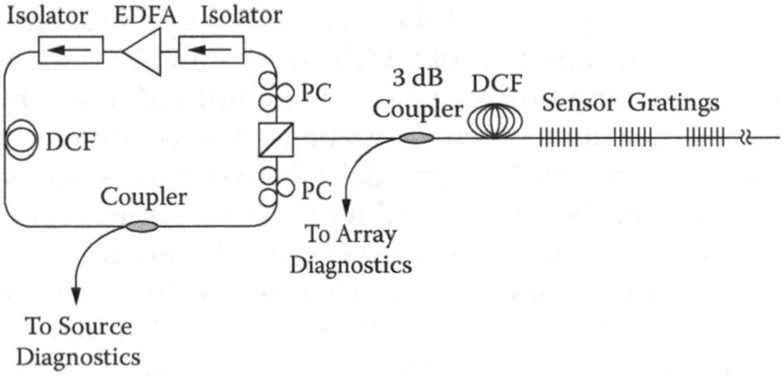

FIGURE 7.33
Schematic diagram of interrogation by passively mode-locked fiber lasers with wavelength-time conversion. (DCF: dispersion-compensation fiber.) (From M. L. Dennis et al., *Optics Lett.*, **22**, pp. 1362–1364, 1997.)

a self-time delay caused by the dispersive propagation through the dispersive fiber. Thus, a fast time-domain interrogation is possible for multiplexed array sensor gratings. The integrated signal change can be analyzed by a hybrid approach by combining optical time-domain reflectometry (OTDR) with the dispersive wavelength-to-time interrogation [70].

The sensor grating array demodulation using a passively mode-locked fiber laser has been demonstrated by Putnam et al. [71,72], and a similar result can be found in Putnam et al. [71,72]. In the experiment of Putnam et al. [71], the mode-locked output power was in excess of 50 mW, and the bandwidth and the repetition rate were 80 nm and ~7 MHz, respectively. A standard dispersion-compensation fiber (DCF, $D = -83$ ps/nm/km at $\lambda = 1550$ nm) with a length of 3.25 km was used. The 3.67-nm shift of the peak wavelength yields a 1.96-ns advance in the arrival time. The sensitivity was determined to be approximately ±20 με over 3500 με. Another demonstration of the broadband square-pulse operation of a passively mode-locked fiber laser for FBG interrogation has also been presented by Putnam et al. [72], in which the laser output pulse had a 4-W peak power and a 10-ns width, and the spectral bandwidth was greater than 60 nm. If a hybrid interrogation system were utilized, this configuration could support more than 30 WDM channels and 20 TDM channels, or 600 arrayed sensors [72].

7.4.3 Interrogation for Optical CDMA Fiber Grating Sensors

Code division multiple access (CDMA) is one of the widely used multiplexing techniques in communication systems. It was investigated for the interrogation of multiplexed array sensor gratings [73,74]. Briefly, the CDMA is based on correlation techniques for coded signals. Let us examine a simple example of the CDMA. Assuming m shift registers, a pseudorandom bit sequences (PRBS) would be $2^m - 1$ in length. We can then encode the sequence so that the autocorrelation value should vanish for any shifted code (i.e., for any asynchronously aligned code), so that it has a peak value only for the synchronous code. It is noteworthy that the shift should be discrete or an integer multiple of 1 bit. This is possible if the sequence is appropriately coded with bipolar state (+1, −1), the numbers of which are $2^{(m-1)}$ and $2^{(m-1)} - 1$, respectively [73]. As a result, synchronous detection can be utilized as an interrogation technique for the distributed sensor gratings. The conditions required to realize this scheme are that the time delay between the sensors should be the integer multiple of the 1-bit period for the code sequence, and the code sequence length $2^m - 1$ should be greater than the number of distributed sensors. The delay time from the individual sensor indicates its positional information; hence, it is possible to interrogate the detected signals from the distributed sensors by correlating them with reference code sequences generated with a variety of time shifts. Consequently, the correlation value with an appropriate time shift in the reference code readily shows the sensor information at the specified position.

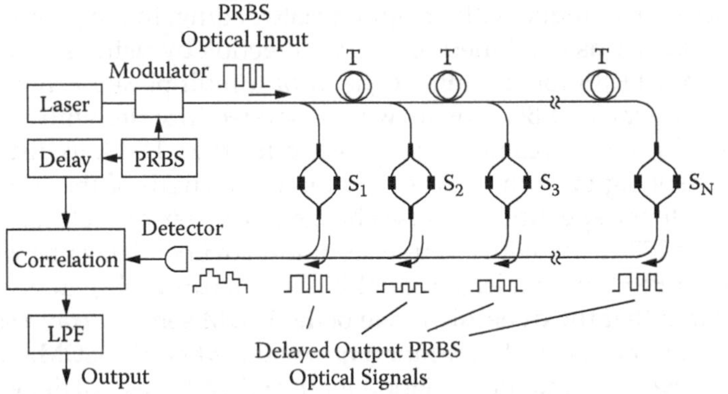

FIGURE 7.34
Schematic diagram of interferometric (not fiber grating) sensor array using CDMA (LPF: low-pass filter; T: time-delay amount). (From A. D. Kersey et al., *Electronics Lett.*, **28**, pp. 351–352, 1992.)

In the scheme demonstrated by Kersey et al. [73], shown in Figure 7.34, the arrayed eight interferometric sensors were investigated in a ladder topology with 30-m fiber delay lines for separate sensors. Using a continuous wave (CW) 830-nm diode laser and an acousto-optic modulator, the PRBS was produced to be 31 bits ($m = 5$) in length with a single bit length of ~145 ns that matched the fiber delay line of 30 m. The measured sensitivity was $\sim 100\ \mu\text{rad}/\sqrt{\text{Hz}}$, and the cross-talk levels were determined to be ~60 dB.

Koo et al. [74] recently demonstrated dense WDM of FBG sensors using CDMA by combining code division and wavelength division multiplexing schemes, shown in Figure 7.35. The WDM scheme using FBGs requires the spectral separation of the sensor arrays to be greater than the maximum spectral change of the individual sensor. For example, 3 nm of sensor spectral separation is required for a sensor dynamic range of ±1000 με [74]. However, the CDMA is capable of reducing the spectral separation of the fiber grating bandwidth, since it is based not on the spectral detection but rather on

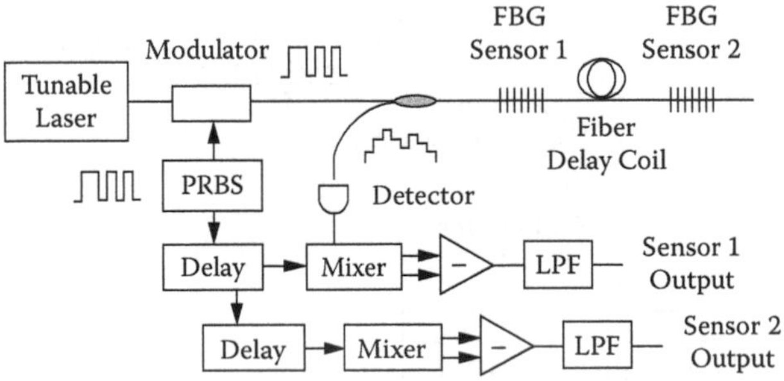

FIGURE 7.35
Schematic diagram of interrogating fiber Bragg grating sensor array using WDM and CDMA (LPF: low-pass filter). (From K. P. Koo et al., *Electronics Lett.*, **35**, pp. 165–167, 1999.)

the correlation technique with an appropriate coding. In the preceding work [74], the CDMA was combined to WDM detection by using a tunable laser source with a PRBS for the interrogation of a multiplexed array of sensor gratings. The arrayed FBG sensors were separated in series with appropriate fiber delay lines. As a result, it was possible for the CDMA to distinguish a specific sensor signal from others, and the wavelength of the tunable laser could indicate the spectral response change of the sensor grating. Two FBGs at 1535.2 and 1535.5 nm with a 100-m separation of fiber delay line were interrogated by a PRBS with a length of 31 bits and a 0.5-μs chip interval. It was demonstrated that the dynamic range of the WDM sensor array was no longer limited by the spectral separation of sensors when the CDMA was combined. However, the channel isolation for the CDMA was ~20 dB, which was relatively not very good.

7.4.4 Frequency Modulation Techniques

Chan et al. [75,76] proposed frequency modulation techniques for FBG sensor multiplexing. One of these is the subcarrier frequency division multiplexing shown in Figure 7.36 [75]. The light intensity in each optical fiber line is modulated with a subcarrier angular frequency ω_i(i = 1, 2 in Figure 7.36). The light reflected by each FBG passes through the intensity modulator again, and hence it has dc, ω_i, and $2\omega_i$ angular frequency components. After passing through a tunable optical filter such as a tunable Fabry–Perot interferometer, the light is detected and the electrical signal passes through bandpass filters centered at ω_i. With this method, even if the two gratings in different fiber lines have the same Bragg wavelength, they can be separated electronically.

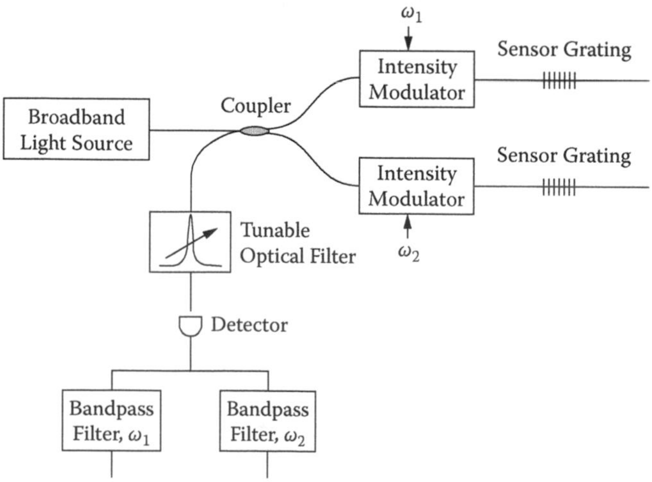

FIGURE 7.36
Schematic diagram of the subcarrier frequency division multiplexing. (From P. K. C. Chan et al., *Optics Laser Technol.*, **31**, pp. 345–350, 1999.)

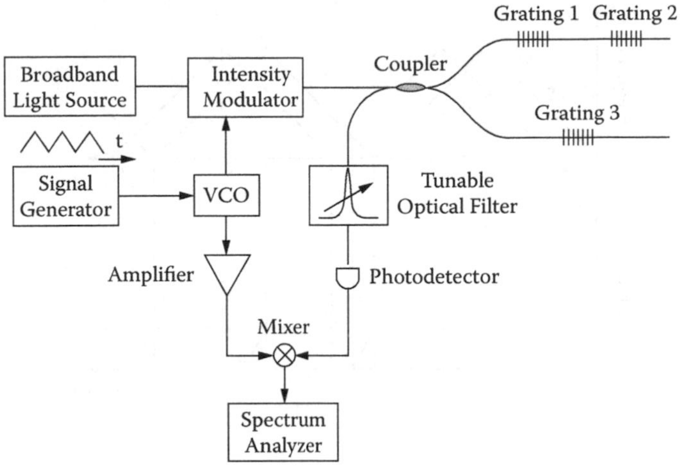

FIGURE 7.37
Schematic diagram of a sensor system using the frequency-modulated, continuous-wave technique. (From P. K. C. Chan et al., *IEEE Photonics Technol. Lett.*, **11**, 11, pp. 1470–1472, 1999.)

This method can be extended via a combination with WDM, for example. However, for cost effectiveness, the number of subcarrier modulation frequencies is limited and dynamic sensing is also limited. Instead of assigning different subcarrier frequencies, the same subcarrier frequency can be used for different fiber lines [75]. Only one bandpass filter is then needed, and the light intensity modulation is done in a time division manner: The electrical signal of the subcarrier frequency is either applied (on) or not applied (off) to each light intensity modulator with time division. Hence, this scheme is a TDM based on an electronic switch.

The FBG sensor multiplexing using a frequency-modulated continuous wave was also proposed [76]. As Figure 7.37 shows, light from a broadband source is modulated with a triangular swept frequency carrier generated from a voltage-controlled oscillator (VCO) and launched into the sensor array. The reflected signals from the FBGs are guided back to a tunable optical filter, then to a photodetector, and finally mixed with a reference signal from the VCO. The system output will consist of a number of beat signals, with their beat frequencies determined by the delay differences between the sensor signals and the reference signal (see Figure 7.38). The Bragg wavelength of the individual FBG can be interrogated by scanning the tunable optical filter and recording the control voltage of the tunable filter that corresponds to the peak of the corresponding frequency components.

7.4.5 Intragrating Sensing

Intragrating sensing is a technique that interrogates an internal grating distribution (and hence the measurand distribution) within a fiber grating (see Figure 7.39). The method of detecting a change in the peak reflection wave-

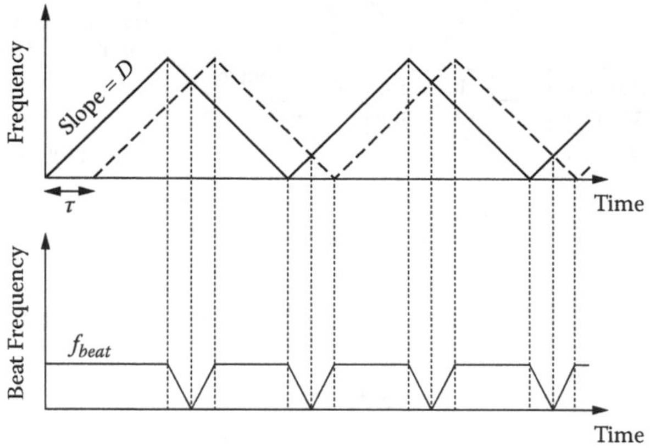

FIGURE 7.38
The reference signal (solid line in upper graph) and its delayed version (reflected signal from a sensor grating: dashed line in upper graph) give a beat signal with the beat frequency f_{beat} equal to the slope D times delay τ. (From P. K. C. Chan et al., *IEEE Photonics Technol. Lett.*, **11**, 11, pp. 1470–1472, 1999.)

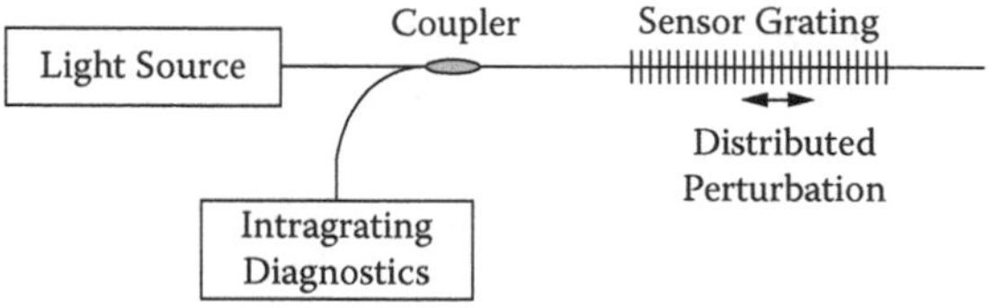

FIGURE 7.39
Basic schematic diagram of intragrating sensing.

length of an FBG is limited, to provide only the averaged information over the FBG length. Therefore, the FBG is a quasi-point sensor. However, the intragrating sensing technique can be used to resolve the distributed profile of the grating structure and its change at high resolution. This is based on the local Bragg wavelength-detecting technique. The local Bragg wavelength is determined by $\lambda_B = 2n_{\text{eff}}(z)\Lambda(z)$, where $n_{\text{eff}}(z)$ is the effective refractive index of the fiber core as a function of the position parameter z, and $\Lambda(z)$ is the local grating period. Hence, it exhibits the information of the local grating structure. Thus, the distributed profile or the local change of fiber gratings (which can initially be uniform or nonuniform) can be interrogated. The interrogation of the local Bragg wavelength can be realized by means of three categorized techniques [3]: reflection spectrum analysis, group-delay measurement, and a hybrid measurement of the reflection spectrum and the group delay. To detect the sensor grating change, the initially distributed profile of the sensor grating should be known. The induced changes are readily resolved by comparing measurement results with the original profile.

Reflection Spectrum Analysis Method

Reflection spectrum analysis is based on analyzing the relation between the reflection spectrum and the grating profile. The relation can be described by [77]

$$\int_{\lambda_B(z-0)}^{\lambda_B(z)} \ln(1-R(\lambda'))\,d\lambda' = \mp\frac{\pi^2}{2}\int_0^2 \frac{[\Delta n_{\text{eff}}(z')]^2}{n_{\text{eff}}(z')}dz' \tag{7.12}$$

with

$$\lambda_B(z) = 2n_{\text{eff}}(z)\Lambda(z) = 2n_{\text{eff},0}\Lambda_0(1+\delta(z)) \tag{7.13}$$

where $R(\lambda)$ and $\delta(z)$ denote, respectively, the reflection spectrum and the local variation function for optical quantities, and $n_{\text{eff/0}}$ and Λ_0 are the average effective refractive index of core and the average grating period, respectively. Note that $\delta(z)$ comes from position dependency of either n_{eff} (due to temperature profile) or Λ (due to strain profile). Prior to the analysis for interrogating the local Bragg wavelength $\lambda_B(z)$, the reflection spectrum $R(\lambda)$ of the sensor grating must be known. Considering that $\lambda_B(z)$ is related to $n_{\text{eff}}(z)$ by Eq. (7.13), the local Bragg wavelength $\lambda_B(z)$ can be obtained numerically by equating the integrals on both sides of Eq. (7.12). When strain or temperature variation is induced on the sensor grating, the reflection spectrum is changed. As a result, the perturbation can be analyzed by measuring the new reflection spectrum. In LeBlanc et al. [77], the measured spatial resolution was 0.8 mm for a 5-mm grating with a gradient of 250 με/mm, and the strain resolution was 80 με. In fact, the effective interaction length around $z = z_i$ (i.e., the position-dependent resolution) is given by [77]

$$L_{\text{eff}} = \sqrt{\Lambda_0 \Big/ \left|\frac{d\delta(z)}{dz}\right|_{z=z_1}}, \tag{7.14}$$

Note that the spatial resolution of the measurement is dependent on the spatial change rate of the effective index or the grating period. A larger spatial gradient of the index change is required for better resolution. Hence, better resolution is possible with prechirped gratings.

Group-Delay Measurement Method

The group-delay measurement method is based on analyzing the relation between the phase response of the reflected light and the grating profile. The group delay of the light reflected by the fiber grating can be described by [78]

$$\tau_d \equiv \frac{d\phi_d}{d\omega} = -\frac{\lambda^2}{2\pi c}\frac{d\phi_d}{d\lambda} \tag{7.15}$$

where ϕ_d is the phase of the reflected light. As in the case of reflection spectrum, the phase response is also dependent on the distributed profile of the grating structure. Thus, if the phase response is analyzed, collecting information on the distributed profile of the grating structure can be achieved.

For example, let us assume a chirped grating as the distributed profile. The group delay for the local Bragg wavelength can be approximated as the double-penetration time through the grating. Thus, the penetration depth—that is, the position for the local grating with local Bragg wavelength λ_B—can be described by [3]

$$z_B = -\frac{\lambda_B^2}{4\pi n_{\mathrm{eff}}} \frac{d\phi_d}{d\lambda}\bigg|_{\lambda=\lambda_B} \tag{7.16}$$

Thus, the local change in the grating structure can be interrogated by directly measuring the group delay for various wavelengths. In other words, the change of the local Bragg wavelength due to strain or temperature variation gives rise to the group-delay change for that wavelength.

The distributed profile of a chirped grating, which has a 33-nm bandwidth of approximately 1535 nm and is 19 mm in length, has been interrogated by measuring the phase response using a Michelson interferometer [79].

Hybrid Measurement Method

In general, the relationship between the complex reflectance and the grating structure can be described in the low-reflectivity approximation as follows [80]:

$$|\kappa(z)|\exp[-i\psi(z)] = -\frac{2n_{\mathrm{eff}}}{\lambda_0}\int_0^{\infty} r(\lambda)\exp\left(-4\pi n_{\mathrm{eff}}\frac{\lambda-\lambda_0}{\lambda_0^2}z\right)d\lambda \tag{7.17}$$

with

$$r(\lambda) = \sqrt{R(\lambda)}\exp[-i\phi_d(\lambda)] \tag{7.18}$$

where λ_0 and $\kappa(z)$ are the nominal center wavelength and the distributed coupling constant, respectively. It is noteworthy that the complex reflectance includes both the amplitude and phase for the reflectance. The coupling constant, which includes the information of the local grating structure, is related to the response $r(\lambda)$. Thus, if both the reflected power and phase response are measured, it is possible to interrogate the distributed profile of the grating structure. A detailed approach to this aspect can be found in Duck and Ohn [80] (see Figure 7.40).

Experiments

Huang et al. [81] exploited the equivalence between the intensity/phase response and the Fourier transform of the structure of a low-reflectivity

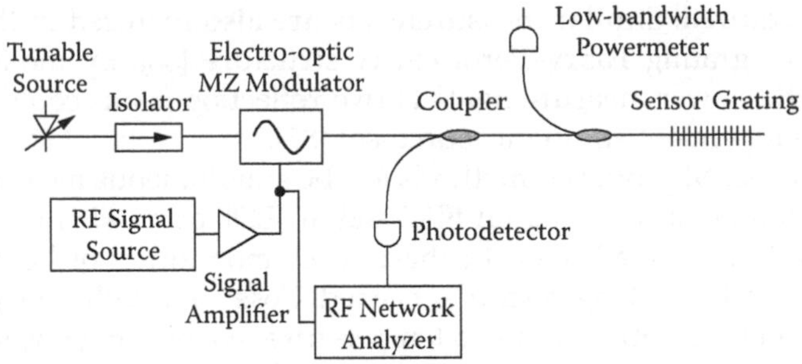

FIGURE 7.40
Schematic diagram of intragrating sensing using a hybrid measurement technique (MZ: Mach–Zehnder). (From G. Duck and M. M. Ohn, *Optics Lett.*, **25**, pp. 90–92, 2000.)

grating, which could be utilized as a technique for determining an arbitrary measurand profile. An arbitrary strain profile measurement within fiber gratings using an interferometric Fourier transform technique has been presented by Ohn et al. [82] using an ~86-mm-long grating sensor, in which a theoretical strain accuracy of ~25 με was achieved with spatial accuracy of 2.58 mm. Recently, a distributed Bragg grating sensing with a direct group-delay measurement technique was presented by Duck et al. [80]. The interrogation of two loaded samples was demonstrated by measuring the reflectivity of the sensor and the group-delay characteristics. The strain resolution was found to be ±24 με, with a spatial resolution defined by a minimum spatial wavelength component of the coupling distribution of 1.65 mm.

A technique for reconstructing arbitrary strain distributions within fiber gratings by time-frequency signal analysis has been presented based on the hybrid detection of the reflectivity and phase response [83]. This method could be used to permit accurate strain reconstruction for an arbitrary strain profile and for an arbitrary grating length. By computing the instantaneous average frequency, it is possible to determine the arrival time for each reflected frequency (local Bragg frequency), which is directly related to the local position of the grating. A detailed numerical approach to this aspect can be found in Azana and Muriel [83].

7.4.6 Other Techniques

Although some of the sensors discussed already require measurements of more than one parameter, there are many other sensor heads for which two or more parameters need to be measured for practical purposes. A simple example that requires the measurements of two Bragg wavelengths (near 850 and 1300 nm, for example) is the dual-wavelength, superimposed FBG sensor, which is capable of discriminating strain and temperature effects [84].

Another example of a sensor head for the simultaneous measurement of strain and temperature is the hybrid FBG and LPFG sensor [85]. Measurements of the shift in Bragg wavelength and optical powers for two reflected lights are required. Similar measurements are also required in the scheme using a fiber grating Fabry–Perot cavity structure [86]. By modifying the structure, the power measurements of two reflection peaks could give temperature-independent strain measurement [87].

Jung et al. [88,89] proposed methods for the simultaneous measurement of strain and temperature using an FBG and an EDF or an FBG written in an erbium:ytterbium-doped fiber. In these cases monitoring of both reflected light power and wavelength shift is needed. Posey and Vohra [90] tested an eight-channel fiber optic strain and temperature sensor using FBGs and stimulated Brillouin scattering. In this method, measurements of Brillouin frequency shifts as well as Bragg wavelength shifts are needed.

Zhang et al. [91] proposed the concept of intensity and wavelength dual-code multiplexing. Dual-peak, low-reflection gratings and single-peak, high-reflection gratings are alternately placed along an optical fiber. The measurand-induced reflection wavelength shifts of dual-peak, low-reflection gratings can cross over the reflection wavelength of the single-peak, high-reflection gratings. This technique for enhancing the measurement range requires the monitoring of reflection powers, wavelength shifts, and a decision scheme when the reflection wavelengths cross.

In addition to the fiber grating sensors discussed thus far, other varieties of fiber sensors using fiber gratings also exist. A few examples are a macro-bending sensor (which is immune to temperature and strain) [92] and a chirped grating Fabry–Perot sensor with multiple wavelength-addressable FSRs [93]. Many modified and novel structures are continually introduced as well.

7.5 The Theory of Fiber Gratings

Theoretical aspects of fiber gratings are discussed in this section. Although this rather detailed (but not fully detailed) theory is not necessarily required for understanding most fiber grating sensors and interrogators, a good understanding of the theory might provide novel ideas in the area and enrich the design capabilities of the fiber grating sensor systems. We begin with the coupled-mode theory—the most widely used method for the analysis of periodic media [34,94]. The conventional coupled-mode theory is limited to the analysis of uniform gratings. However, several numerical techniques based on the coupled-mode theory have been developed as an extension to the analysis of aperiodic grating structures. Kogelnik [95] utilized the numerical solution of two coupled-mode equations for aperiodic structures. The transfer matrix method has also been developed for almost-periodic gratings [96]. In this approach the waveguides are divided into short segments, and in each segment the gratings are assumed to be periodic. The characteristics of almost-periodic grating can then be obtained by multiplying each transfer matrix of a short segment.

Furthermore, the discretized coupled-mode theory approach has been developed as a generalized seminumerical approach for describing multi-mode couplings in highly nonuniform gratings [97–99]. In this approach the waveguides are divided into short segments in a manner similar to the transfer matrix approach; however, the segment can be shorter than the grating period and is assumed to have a constant perturbation rather than a sinusoidal one. This method is capable of analyzing arbitrarily structured gratings, even if they have no periodicity, and therefore represents an efficient path to a nonuniform grating analysis. In addition to the seminumerical techniques based on coupled-mode theory, several other useful methods are available, including the Bloch wave approach [34,100], the effective index method [101], the extension of Rouard's method by Weller-Brophy and Hall [102], and the elegant discrete-time approach by Frolik and Yagle [103].

Here we focus on the general coupled-mode theory approach to understand the fundamental characteristics of fiber gratings in sensor applications, and the transfer matrix approach for nonuniform gratings is briefly discussed. We discuss guided modes in optical fibers and their interactions with a periodic perturbation of index change and derive the coupled-mode approach, which is described to be applicable for general cases of gratings. In addition, we briefly discuss a methodological technique for a nonuniform grating analysis. Next we discuss uniform gratings for both the contradirectional and codirectional couplings, which include FBGs and LPFGs. The property of a uniform and periodic index change is primary and fundamental to understanding the fiber grating action in most sensor applications. Finally, we discuss the spectral response of several nonuniform gratings, including chirped FBGs and phase-shifted and cascaded LPFGs.

7.5.1 Guided Modes in Optical Fibers and Resonant Couplings in Fiber Gratings

Guided optical fiber modes can be classified in two types. One is for core modes that are totally reflected at the core–cladding boundary (in terms of ray optics) and are bounded in the core region. The other is for cladding modes that are totally reflected at the cladding–air (surrounding medium) boundary and are bounded in cladding and core regions. The ray-optic illustration of the guided modes, which include the core and cladding modes, is shown in Figure 7.41.

In the case of a common single-mode fiber, the fundamental core mode is described by HE_{11} [104]. However, the linearly polarized approach is often utilized due to its simple and easy approach, which is sufficiently quantitative for the core-mode evaluation [105]. LP_{01} can be found by solving the following equation [105–107]:

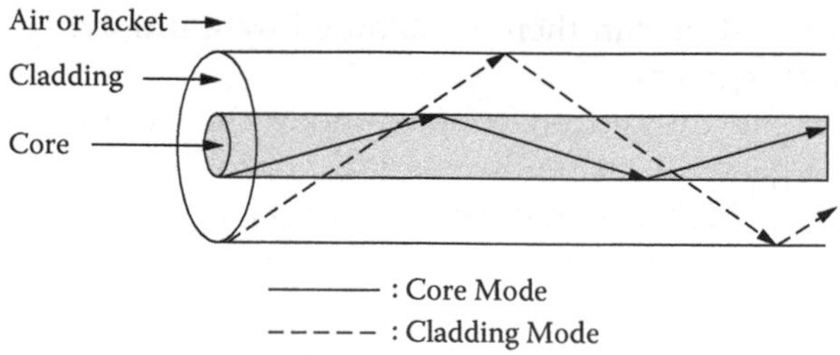

FIGURE 7.41
Ray-optic illustration of the guided modes.

$$V\sqrt{1-b}\frac{J_1\left(V\sqrt{1-b}\right)}{J_0\left(V\sqrt{1-b}\right)} = V\sqrt{b}\frac{K_1\left(V\sqrt{b}\right)}{K_0\left(V\sqrt{b}\right)} \quad (7.19)$$

where

$$V = \frac{2\pi r_{co}}{\lambda}\sqrt{n_{co}^2 - n_{cl}^2} \quad (7.20)$$

$$b = \frac{n_{eff}^2 - n_{cl}^2}{n_{co}^2 - n_{cl}^2} \quad (7.21)$$

where J_n is a Bessel function of the first kind; K_n a modified Bessel function of the second kind; r_{co} the core radius; λ the wavelength in vacuum; and n_{co}, n_{cl} and n_{eff} the core refractive index, the cladding refractive index, and the effective refractive index of the core, respectively. With respect to the cladding modes, their derivation is more complicated, since the waveguide geometry includes two boundaries. The details of their derivations can be found in Tsao [104] and Erdogan [107].

In fact, if there are some perturbations in the fiber, the modes can be coupled to other modes. The main coupling directions can be determined as contradirectional or codirectional based on whether the traveling directions of modes coupled to each other are opposite or the same. Based on the direction of the mode coupling, fiber gratings can be classified in two types. One type is a short-period or reflection grating, where coupling occurs between modes traveling in opposite directions. The FBGs, chirped gratings, and tilted short-period gratings belong to this category. The other type is a transmission grating, represented by LPFGs, where coupling occurs between modes traveling in the same direction [106].

The intuitive illustration of mode couplings is shown in Figure 7.42: (a) illustrates a contradirectional coupling for reflection-type gratings and (b) a codirectional coupling for transmission-type gratings. By means of the refractive index perturbation, diffracted modes can be excited, as shown in Figure 7.42. For the diffracted mode to be accumulated constructively, each

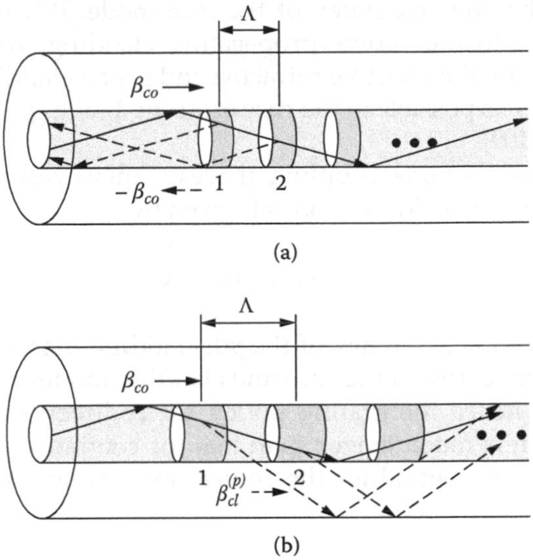

FIGURE 7.42
Illustration of mode couplings: (a) contradirectional coupling; (b) codirectional coupling.

diffracted radiation from the series of the perturbation should be in phase; that is, the following equation should be satisfied:

$$\beta_i - (\pm\beta_d) = \frac{2\pi}{\Lambda} m \tag{7.22}$$

where β_i and β_d, respectively, are the propagation constants for the incident and diffracted modes; Λ is the period of the grating; and m stands for an integer number. It is noteworthy that a minus sign before the propagation constant describes the case wherein the mode propagates in the $-z$ direction. The optical path difference between lights diffracted from the adjacent grating positions (see Figure 7.42) should be an integer multiple of the wavelength for a resonant coupling of modes, as in the case of a constructive interference in dielectric multilayers [108]. In most cases, first-order diffraction is dominant, and hence m is assumed to be unity [34]. Thus, the resonant wavelength can be obtained as follows:

$$\lambda = [n_{i,\text{eff}} - (\pm n_{\text{d,eff}})]\Lambda \tag{7.23}$$

In the case of a single-mode fiber, $n_{i,\text{eff}}$ and $n_{d,\text{eff}}$ can be the effective indices of both the core and cladding modes.

In the case of contradirectional coupling, the nominal Bragg wavelength of the core mode is given by

$$\lambda = 2n_{\text{co,eff}}\Lambda \tag{7.24}$$

where $n_{\mathrm{co,eff}}$ is the effective index of the core mode. When the core mode becomes coupled to the counterpropagating cladding mode, the $n_{d,\mathrm{eff}}$ in Eq. (7.23) is given by the effective refractive index of a cladding mode with a minus sign, which is possible in the case of strong Bragg gratings and blazed gratings [106,107,109].

In the case of codirectional coupling, the resonant wavelength for coupling between the core and cladding modes is given by

$$\lambda = (n_{\mathrm{co,eff}} - n_{\mathrm{cl,eff}}^{(p)})\Lambda \tag{7.25}$$

where $n_{\mathrm{cl,eff}}^{(p)}$ is the effective index of the pth cladding mode. The differences in effective indices between the core and cladding modes are much smaller than unity, and hence the grating period for codirectional coupling at a given wavelength is much longer than that for contradirectional coupling. Typically, the grating period for the codirectional coupling is hundreds of micrometers in length [110].

7.5.2 Coupled-Mode Theory

In coupled-mode theory, the perturbed permittivity as a function of space is written as

$$\varepsilon'(x,y,z) = \varepsilon_0(x,y) + \Delta\varepsilon(x,y,z) \tag{7.26}$$

where $\varepsilon_0(x, y)$ is the unperturbed part of the permittivity, and $\Delta\varepsilon(x, y, z)$ is the perturbed part of the permittivity and can vary periodically or aperiodically in the z direction—namely, along the fiber axis. The variation in permittivity is considered to be a perturbation that can couple the unperturbed eigenmodes of the given structure of media [34]. The unperturbed eigenmodes are determined by $\varepsilon_0(x, y)$, which are readily known. Based on the method of variation of constants, the perturbed mode can be expressed as an expansion in the eigenmodes of the unperturbed structure of media, where the expansion coefficients are assumed to be dependent on z. From the Lorentz reciprocity theorem [34], the general coupled-mode equation for a given wavelength λ is described by [97–99]

$$\nabla\cdot(E' \times H_p^* + E_p^* \times H') = -i\frac{2\pi c}{\lambda} E_p^* \cdot \Delta\varepsilon E' \tag{7.27}$$

where $p = 1, 2, 3, \ldots$, and c is the velocity of light in vacuum. The terms subscripted by p denote the pth eigenmode field, and the terms superscripted by a prime denote the resultant perturbed-mode field. $\Delta\varepsilon$ and * represent the permittivity perturbation and the complex conjugate, respectively. If it is assumed that there are N eigenmodes in the unperturbed structure of interest, the total number of the coupled-mode equations to be obtained becomes

N, which consist of independent equations based on each p. Details of the derivation of the coupled-mode equation can be found in references 97–99.

Let us suppose that the electric and magnetic fields of the pth eigenmode can be described by

$$\begin{pmatrix} E_p \\ H_p \end{pmatrix} = \begin{pmatrix} e_p(x,y)\exp(-i\beta_p z) \\ h_p(x,y)\exp(-i\beta_p z) \end{pmatrix} \tag{7.28}$$

where $e_p(x, y)$ and $h_p(x, y)$ are, respectively, the electric- and magnetic-field vectors normalized to a power flow of 1 W in the z direction, and β_p is the propagation constant in the z direction. The time-varying factor $\exp(i\omega t)$ is omitted for the sake of simplicity. In the case of guided modes in a single-mode fiber, the eigenmodes are described by the core and cladding modes. Assuming slowly varying amplitudes with the method of variation constants, as has been assumed in the coupled-mode approach, the perturbed fields E' and H' can then be described, without losing generality, as follows:

$$\begin{pmatrix} E' \\ H' \end{pmatrix} = \sum_q a_q(z) \begin{pmatrix} E_q \\ H_q \end{pmatrix} = \sum_q a_q(z) \begin{pmatrix} e_q(x,y)\exp(-i\beta_p z) \\ h_q(x,y)\exp(-i\beta_p z) \end{pmatrix} \tag{7.29}$$

That is, the perturbed electric or magnetic field is expressed as an expansion in the unperturbed eigenmodes with variation constants depending on z. By substituting Eqs. (7.28) and (7.29) into Eq. (7.27) and integrating both sides of Eq. (7.27) in the plane transverse to the z-direction, we can obtain the final result of the coupled-mode equation as

$$\frac{da_p(z)}{dz} = -i\frac{|\beta_p|}{\beta_p}\sum_q \kappa_{pq}(z)a_q(z)\exp[-i(\beta_q - \beta_p)z] \tag{7.30}$$

where $p = 1, 2, 3, \ldots$, and the coupling coefficient $\kappa_{pq}(z)$ is given by

$$\kappa_{pq}(z) = \frac{\pi c}{2\lambda}\iint e_p^*(x,y)\cdot\Delta\varepsilon(x,y,z)e_q(x,y)dx\,dy \tag{7.31}$$

Unless the grating structure is blazed (tilted), the nonvanishing $\kappa_{pq}(z)$ is obtained from the coupling between modes that have an identical order for azimuthal variation [106]. In most fiber gratings, the structure is composed of isotropic materials and hence the perturbed permittivity can be described by

$$\begin{aligned} \Delta\varepsilon(x,y,z) &= \varepsilon_0[n'^2(x,y,z) - n^2(x,y)] \\ &\cong 2\varepsilon_0 n(x,y)\Delta n(x,y,z) \qquad \text{if } \Delta n(x,y,z) \ll n(x,y) \end{aligned} \tag{7.32}$$

where $n(x, y)$ and $n'(x, y, z)$ are the refractive indices corresponding to the permittivity $\varepsilon_0(x, y)$ and $\varepsilon'(x, y, z)$, respectively. In the case where the materials exhibit anisotropy, the permittivity results in a tensor [34].

Consequently, it is possible to obtain the spectral response of the grating structure by solving Eq. (7.30). There are numerous modes even in a single-mode fiber, including a fundamental core mode and numerous cladding modes. However, in most cases of quasi-uniform fiber gratings, the mode coupling between the two modes is dominant in a specific region of interest. Thus, coupled-mode analysis is often described by a two-mode coupling [34]. In addition, the synchronous approximation is usually utilized in the two-mode coupling model [34,111]. That is, a rapidly oscillating term with z on the right-hand side of Eq. (7.30) can probably be neglected, since its contribution is insignificant over a distance that is much longer than the period of the index change. Note that the expression $\kappa_{pq}(z)$ is a generalized form without further assumption. When the synchronous approximation is applied, the permittivity perturbation is expanded in a Fourier series. Thus, a specific component of resonance is chosen for the two-mode coupling, but others are neglected because of their small contribution. However, when the effective index spacing of modes related to the coupling is sufficiently narrow or the grating structure is highly nonuniform, then multimode coupling should be taken into account [97–99].

In the next parts of this section, we present the coupled-mode approach on uniform sinusoidal gratings with the two-mode coupling for both contra-directional and codirectional couplings. We assume that the initially induced index change for the uniform sinusoidal gratings can be described by

$$\begin{aligned}\Delta n(x,y,z) &= \Delta n_{\mathrm{av}}(x,y) + \Delta n_{\mathrm{gr}}(x,y)\cos\left(\frac{2\pi}{\Lambda}z + \phi_{\mathrm{gr}}\right) \qquad (7.33)\\ &= \Delta n_{\mathrm{av}}(x,y) + \frac{\Delta n_{\mathrm{gr}}(x,y)}{2}\left\{\exp\left[-i\left(\frac{2\pi}{\Lambda}z + \phi_{\mathrm{gr}}\right)\right]\right.\\ &\quad \left. + \exp\left[i\left(\frac{2\pi}{\Lambda}z + \phi_{\mathrm{gr}}\right)\right]\right\}\end{aligned}$$

where Δn_{av} is the dc index change spatially averaged over a grating period, Δn_{gr} the amplitude of the sinusoidal index change, Λ the nominal period, and ϕ_{gr} the initial grating phase. If the index change is not a sinusoidal function, it is possible to expand the variation function as Fourier series and to choose the appropriate Fourier components [34].

In fact, the coupled-mode approach utilizes not the realistic index change but the effective index change in the z direction that is determined by calculating the overlap integral for the product of the realistic refractive index change and the corresponding electric fields [34]. Assuming two-mode coupling together with Eqs. (7.31)–(7.33), the coupling constants are calculated by

$$\kappa_{\sigma,i} = \frac{\pi\varepsilon_0 c}{\lambda}\iint e_i^*(x,y)\cdot n(x,y)\Delta n_{\text{av}}(x,y)e_i(x,y)dx\,dy \tag{7.34}$$

$$\kappa_{\sigma,d} = \frac{\pi\varepsilon_0 c}{\lambda}\iint e_d^*(x,y)\cdot n(x,y)\Delta n_{\text{av}}(x,y)e_d(x,y)dx\,dy \tag{7.35}$$

$$\kappa_{\xi} = \frac{\pi\varepsilon_0 c}{2\lambda}\iint e_i^*(x,y)\cdot n(x,y)\Delta n_{\text{gr}}(x,y)e_d(x,y)dx\,dy \tag{7.36}$$

where $\kappa_{\sigma,i}$, $\kappa_{\sigma,d}$, and κ_{ξ} respectively denote the self-coupling constants for the corresponding modes and the cross-coupling constant, and the subscripts i and d respectively signify the incident and diffracted modes. It is noteworthy that the z-directional varying factor is excluded differently from the generalized expression (7.31).

The self-coupling constant is equivalent to the change in propagation constant for the mode. Thus, the so-called effective index change of the mode due to the realistic index perturbation $\Delta n_{\text{av}}(x, y)$ and $\Delta n_{\text{gr}}(x, y)$ is determined by

$$\delta n_{\text{eff, av}} = \frac{\varepsilon_0 c}{2}\int\int e_i^*(x,y)\cdot n(x,y)\Delta n_{\text{av}}(x,y)e_i(x,y)dx\,dy \tag{7.37}$$

$$\delta n_{\text{eff, gr}} = \frac{\varepsilon_0 c}{2}\int\int e_i^*(x,y)\cdot n(x,y)\Delta n_{\text{gr}}(x,y)e_i(x,y)dx\,dy \tag{7.38}$$

Therefore, in the case of single-mode reflection gratings, the expressions for the coupling constants can be described by simpler expressions as follows:

$$\kappa_{\sigma} \equiv \kappa_{\sigma,i} = \kappa_{\sigma,d} = \frac{2\pi}{\lambda}\delta n_{\text{eff, av}} \tag{7.39}$$

$$\kappa_{\xi} = \frac{\pi}{\lambda}\delta n_{\text{eff, gr}} \tag{7.40}$$

Since the two modes of incidence and reflection are identical except for the propagating directions, $\kappa_{\sigma,i}$ equals $\kappa_{\sigma,d}$. As a result, this expression can be utilized as a convenient way to analyze FBGs, since the effective index and its change can be nominally estimated without calculating the precise values of the overlap integrals for modes. On the contrary, in the case of LPFGs, the two coupled modes that consist of core and cladding modes are not identical, and hence the corresponding coupling constants should be determined by proper calculation for modes and their overlap integrals [97–99,107]. Thus, the coupling constants for the case of cladding-mode couplings may be described as the following expressions:

$$\kappa_{\sigma,i} = \frac{2\pi}{\lambda}\delta n_{\text{eff, av}} \tag{7.41}$$

$$\kappa_{\sigma,d} = \frac{2\pi u_\sigma}{\lambda}\delta n_{\text{eff, av}} \tag{7.42}$$

$$\kappa\varepsilon = \frac{\pi u_\varepsilon}{\lambda}\delta n_{\text{eff, gr}} \tag{7.43}$$

where

$$u_\sigma = \frac{\int\int e_d^*(x,y)\cdot n(x,y)\Delta n_{\text{av}}(x,y)e_d(x,y)dx\,dy}{\int\int e_i^*(x,y)\cdot n(x,y)\Delta n_{\text{av}}(x,y)e_i(x,y)dx\,dy} \tag{7.44}$$

$$u_\xi = \frac{\int\int e_i^*(x,y)\cdot n(x,y)\Delta n_{\text{gr}}(x,y)e_d(x,y)dx\,dy}{\int\int e_i^*(x,y)\cdot n(x,y)\Delta n_{\text{gr}}(x,y)e_i(x,y)dx\,dy} \tag{7.45}$$

In general, the relation $u_\sigma < u_\xi < 1$ holds for most fiber gratings that have periodic index changes in the core region, since the field confinement of the cladding mode in the core region is much smaller than that of the core mode.

Common types of fiber gratings as classified by the variation of the induced index change along the fiber axis include uniform, apodized, chirped, phase-shift, and superstructure gratings [106]. The wide variety of optical properties based on the variation of the index change can be found in numerous articles for specific cases [112,113].

Contradirectional Coupling

When strongly coupled modes are propagating in opposite directions, the phenomenon is addressed as a *contradirectional coupling*. Assuming a sinusoidal index change, as defined in (7.33), and a two-mode coupling between modes that propagate in opposite directions, we can describe the resultant equations from the coupled-mode theory with the synchronous approximation by

$$\frac{da_i(z)}{dz} = -i\kappa_{\sigma,i}a_i(z) - i\kappa_\xi a_d(z)\exp(i\Delta\beta z - i\phi_{\text{gr}}) \tag{7.46}$$

$$\frac{da_d(z)}{dz} = i\kappa_{\sigma,d}a_d(z) + i\kappa_\xi^* a_i(z)\exp(-i\Delta\beta z + i\phi_{\text{gr}}) \tag{7.47}$$

where

$$\Delta\beta = \beta_i + \beta_d - \frac{2\pi}{\Lambda} \tag{7.48}$$

and $a_i(z)$ and $a_d(z)$ are the complex amplitude functions of the incident and reflected modes, respectively. The coupling coefficients $\kappa_{\sigma,i}$, $\kappa_{\sigma,d}$, and κ_ξ respectively denote the self-coupling constants for the corresponding modes and the cross-coupling constant, as defined in Eqs. (7.34)–(7.36).

The boundary conditions for contradirectional coupling are determined by the amplitudes of the incoming modes at both ends of the grating—that is, as $E_i(z)|_{z=0} = E_i(0)$ and $E_d(z)|_{z=L} = E_d(L)$, where L is the grating length. Together with these boundary conditions and the solution of the coupled Eqs. (7.46)–(7.48), the expression of the entire electric-field amplitudes can be described by

$$\begin{aligned} E_i(z) &= a_i(z)\exp(-i\beta_i z) \\ &= \left\{ \frac{\sigma_b \cosh\sigma_b(L-z) + i(\Delta\beta'/2)\sinh\sigma_b(L-z)}{\sigma_b\cosh\sigma_b L + i(\Delta\beta'/2)\sinh\sigma_b L} E_i(0) \right. \\ &\quad \left. + \frac{-ik_\xi \sinh\sigma_b z \exp[i(\gamma_b - \pi/\Lambda)L - i\phi_{gr}]}{\sigma_b\cosh\sigma_b L + i(\Delta\beta'/2)\sinh\sigma_b L} E_d(L) \right\} \\ &\quad \times \exp\left[-i(\gamma_b + \pi/\Lambda)z\right] \end{aligned} \tag{7.49}$$

$$\begin{aligned} E_d(z) &= a_d(z)\exp(i\beta_d z) \\ &= \left\{ \frac{-ik_\xi^* \sinh\sigma_b(L-z)\exp(i\phi_{gr})}{\sigma_b\cosh\sigma_b L + i(\Delta\beta'/2)\sinh\sigma_b L} E_i(0) \right. \\ &\quad \left. + \frac{\sigma_b\cosh\sigma_b z + i(\Delta\beta'/2)\sinh\sigma_b z}{\sigma_b\cosh\sigma_b L + i(\Delta\beta'/2)\sinh\sigma_b L} \exp[i(\gamma_b - \pi/\Lambda)L] E_d(L) \right\} \\ &\quad \times \exp[-i(\gamma_b - \pi/\Lambda)z] \end{aligned} \tag{7.50}$$

where

$$\Delta\beta' = \Delta\beta + \kappa_{\sigma,i} + \kappa_{\sigma,d} \tag{7.51}$$

$$\gamma_b = \frac{\beta_i - \beta_d + \kappa_{\sigma,i} - \kappa_{\sigma,d}}{2} \tag{7.52}$$

$$\sigma_b^2 = \kappa_\varepsilon^* \kappa_\xi - \left(\frac{\Delta\beta'}{2}\right)^2 \tag{7.53}$$

Therefore, the analytic expression of the contradirectional coupling in a uniform grating has been obtained. This formalism is applicable to reflection-type gratings that include, for example, FBGs and blazed gratings. For cases of strong Bragg gratings and blazed gratings, the mode coupling between the core and cladding modes can occur, and hence the diffracted mode is given by the cladding mode. If the grating structure is blazed, the cladding mode can be a mode with a different order for azimuthal variation from that of the core mode, due to the transversal distribution of the index change. The detailed approach for this can be found in Erdogan [106].

Codirectional Coupling

When strongly coupled modes are propagating in the same directions, the phenomenon is addressed as *codirectional coupling.* Assuming a sinusoidal index change as defined in Eq. (7.33) and a two-mode coupling between modes that propagate in the same directions, we can describe the resultant equations from the coupled-mode theory with the synchronous approximation by

$$\frac{da_i(z)}{dz} = -i\kappa_{\sigma,i}a_i(z) - i\kappa_\xi a_d(z)\exp(i\Delta\beta z - i\phi_{gr}) \tag{7.54}$$

$$\frac{da_d(z)}{dz} = -i\kappa_{\sigma,d}a_d(z) - i\kappa_\xi^* a_i(z)\exp(-i\Delta\beta z + i\phi_{gr}) \tag{7.55}$$

where

$$\Delta\beta = \beta_i - \beta_d - \frac{2\pi}{\Lambda} \quad (\beta_i > \beta_d) \tag{7.56}$$

and $a_i(z)$ and $a_d(z)$ are the complex amplitude functions of the incident and diffracted modes, respectively. The coupling coefficients $\kappa_{\sigma,i}$, $\kappa_{\sigma,d}$, and κ_ξ respectively denote the self-coupling constants for the corresponding modes and the cross-coupling constant, as defined in Eqs. (7.34)–(7.36).

The boundary conditions for codirectional coupling are determined by the amplitudes of the incoming modes at the entrance of the grating—that is, as $E_i(z)|_{z=0} = E_i(0)$ and $E_d(z)|_{z=0} = E_d(0)$. Along with these boundary conditions and the solution of the coupled Eqs. (7.54)–(7.56), the expressions of the entire electric-field amplitudes can be described by

$$\begin{aligned} E_i(z) &= a_i(z)\exp(-i\beta_i z) \\ &= \left\{\left(\cos\sigma_f z - i\frac{\Delta\beta'}{2\sigma_f}\sin\sigma_f z\right)E_i(0) - i\frac{\kappa_\xi}{\sigma_f}\sin\sigma_f z\exp(-i\phi_{gr})E_d(0)\right\} \\ &\quad \times \exp\left[-i(\gamma_f + \pi/\Lambda)z\right] \end{aligned} \tag{7.57}$$

$$E_d(z)=a_d(z)\exp(-i\beta_d z) \tag{7.58}$$

$$=\left\{-i\frac{\kappa_\xi^*}{\sigma_f}\sin\sigma_f z\exp(+i\phi_{gr})E_i(0)+\left(\cos\sigma_f z+i\frac{\Delta\beta'}{2\sigma_f}\sin\sigma_f z\right)E_d(0)\right\}$$

$$\times\exp\left[-i(\gamma_f-\pi/\Lambda)z\right]$$

where

$$\Delta\beta'=\Delta\beta+\kappa_{\sigma,i}-\kappa_{\sigma,d} \tag{7.59}$$

$$\gamma_f=\frac{\beta_i+\beta_d+\kappa_{\sigma,i}+\kappa_{\sigma,d}}{2} \tag{7.60}$$

$$\sigma_f^2=\kappa_\xi^*\kappa_\xi+\left(\frac{\Delta\beta'}{2}\right)^2 \tag{7.61}$$

Therefore, the analytic expression of the codirectional coupling in a uniform grating has been obtained. This formalism is applicable to transmission-type gratings that include, for example, LPFGs. In the case of LPFGs, the grating tilt has little effect different from the blazed grating in the reflection type, since the pitch of index change is sufficiently long so that the blazed edges are too short relative to the straight region.

Transfer Matrix Method for Nonuniform Gratings

The transfer matrix method has been widely used for almost-periodic grating analysis [96]. In this approach the waveguides are divided into short segments. In each segment the gratings are assumed to be periodic, and the transfer relation or matrix is composed of the results, based on the coupled-mode theory. The characteristics of almost-periodic gratings can then be obtained by multiplying each transfer matrix of a short segment. In many aperiodic but quasi-sinusoidal grating cases, the transfer matrix method is restricted to a two-mode coupling. However, this method provides sufficiently quantitative results in almost-periodic gratings, which include chirped, apodized, and superstructured gratings. The numerical approach will be very simple, provided the results from the coupled-mode theory are available and have been modified as a matrix form between the front and rear end planes of the individual segment.

Let us suppose that an almost-periodic grating is assumed to be divided into short segments, in which the grating structure is uniform and sinusoidal. The transfer matrix of the individual *k*th segment is assumed to be given by $\mathbf{T}_k$, which satisfies the following relation:

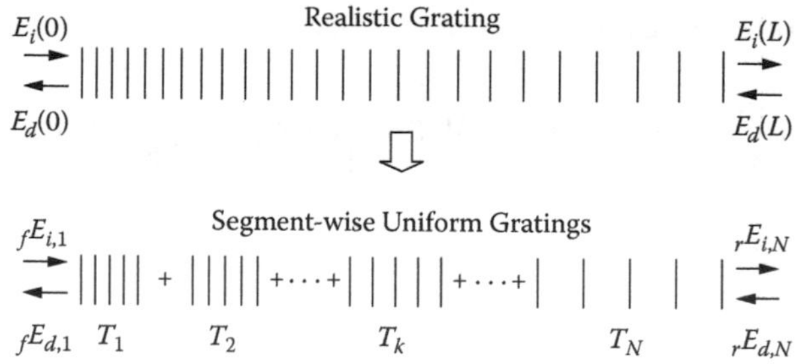

FIGURE 7.43
Illustration of transfer matrix method approach for an almost-periodic grating.

$$\begin{pmatrix} {}_rE_{i,k} \\ {}_rE_{d,k} \end{pmatrix} = \mathbf{T}_k \begin{pmatrix} {}_fE_{i,k} \\ {}_fE_{d,k} \end{pmatrix} \tag{7.62}$$

where the left-side subscripts f and r respectively indicate the front and rear end planes of the segment. The entire transfer relation through the segments can then be described by

$$\begin{pmatrix} {}_rE_{i,N} \\ {}_rE_{d,N} \end{pmatrix} = \mathbf{T}_N \ldots \mathbf{T}_{k+1}\, \mathbf{T}_k \ldots \mathbf{T}_1 \begin{pmatrix} {}_fE_{i,1} \\ {}_fE_{d,1} \end{pmatrix} \tag{7.63}$$

where N is the number of the total segments. As a result, together with the appropriate boundary conditions for the contradirectional coupling or the codirectional coupling as in the cases of Eqs. (7.49) and (7.50), or (7.57) and (7.58), the entire electric-field amplitudes can be obtained. The intuitive illustration of the transfer matrix method is shown in Figure 7.43.

After modifying the result from Eqs. (7.49) and (7.50), the components of the transfer matrix $\mathbf{T}_k$ for the contradirectional couplings can be described by

$$T_{k,(1,1)} = \left(\cosh \sigma_{b,k} L_k - i \frac{\Delta\beta'_k}{2\sigma_{b,k}} \sinh \sigma_{b,k} L_k \right) \exp[-i(\gamma_{b,k} + \pi/\Lambda_k) L_k] \tag{7.64}$$

$$T_{k,(1,2)} = -i \frac{\kappa_{\xi,k}}{\sigma_{b,k}} \sinh \sigma_{b,k} L_k \exp[-i(\gamma_{b,k} + \pi/\Lambda_k) L_k \phi_{gr,k}] \tag{7.65}$$

$$T_{k,(2,1)} = +i \frac{\kappa^*_{\xi,k}}{\sigma_{b,k}} \sinh \sigma_{b,k} L_k \exp[-i(\gamma_{b,k} - \pi/\Lambda_k) L_k + i\phi_{gr,k}] \tag{7.66}$$

$$T_{k,(2,2)} = \left(\cosh \sigma_{b,k} L_k + i \frac{\Delta\beta'_k}{2\sigma_{b,k}} \sinh \sigma_{b,k} L_k \right) \exp[-i(\gamma_{b,k} - \pi/\Lambda_k) L_k] \tag{7.67}$$

where the corresponding parameters are described in Eqs. (7.51)–(7.53), and the subscript k denotes the kth segment. In a similar manner, from Eqs. (7.57) and (7.58), the components of the transfer matrix $\mathbf{T}_k$ for the codirectional couplings are described by

$$T_{k,(1,1)} = \left(\cos\sigma_{f,k} L_k - i\frac{\Delta\beta'_k}{2\sigma_{f,k}} \sin\sigma_{f,k} L_k \right) \exp[-i(\gamma_{f,k} - \pi/\Lambda_k)L_k] \tag{7.68}$$

$$T_{k,(1,2)} = -i\frac{\kappa_{\xi,k}}{\sigma_{f,k}} \sin\sigma_{f,k} L_k \exp[-i(\gamma_{f,k} + \pi/\Lambda_k)L_k + i\phi_{gr,k}] \tag{7.69}$$

$$T_{k,(2,1)} = -i\frac{\kappa^*_{\xi,k}}{\sigma_{f,k}} \sin\sigma_{f,k} L_k \exp[-i(\gamma_{f,k} - \pi/\Lambda_k)L_k + i\phi_{gr,k}] \tag{7.70}$$

$$T_{k,(2,2)} = \left(\cos\sigma_{f,k} L_k + i\frac{\Delta\beta'_k}{2\sigma_{f,k}} \sin\sigma_{f,k} L_k \right) \exp[-i(\gamma_{f,k} - \pi/\Lambda_k)L_k] \tag{7.71}$$

where the parameters are described in Eqs. (7.59)–(7.61), except for the addition of the subscript k.

As a consequence, by repeatedly calculating the transfer matrices for the different segments and multiplying each of them, we can obtain the entire analysis. The coupled-mode theory approach with the synchronous approximation is valid only when the grating length is sufficiently longer than a one-period pitch. The aperiodic grating should be divided into segments with an appropriate segment number. The appropriate number of segment divisions is dependent on the degree of nonuniformity. For most slowly varying nonuniform gratings, a segment should be composed of hundreds of grating pitches [106]. This segment division is sufficient in the case of short-period gratings. However, in the cases of LPFGs, the number of the entire periodic sections results in a few tens, since, typically, the period is hundreds of micrometers, and the grating length is a few centimeters. Thus, it is probable that the coupled-mode theory approach with the synchronous approximation or the transfer matrix approach is capable of making numerical errors for a short-length grating or a highly nonuniform grating. In addition, it is possible that multimode couplings can occur in LPFGs, since the propagation constant difference between the cladding modes is very small, typically $\sim 10^{-4}$.

From this point of view, the coupled-mode theory approach that is not based on a synchronous approximation is required. The discretized coupled-mode theory approach has been developed as a generalized seminumerical approach for describing the multimode couplings in LPFGs or highly nonuniform gratings, which is not based on a synchronous approximation [97–99].

This method does not select a probable component from the Fourier series of the permittivity perturbation but fully estimates the realistic grating with sectionwise discretization. A numerical comparison between results by the conventional two-mode approach and the discretized approach shows that errors caused by the two-mode approach and the synchronous approximation are not negligible in the case of nonsinusoidal, long-period gratings. Detailed results can be found in references 97–99; however, this is outside the scope of this chapter.

7.5.3 Fiber Bragg Gratings

Fiber Bragg gratings (FBGs) are based on contradirectional couplings. In the case of a single-mode fiber, the propagating core mode is reflected into the identical core mode propagating in the opposite direction. In fact, it is possible for the core mode to be coupled to counterpropagating cladding modes in cases that include strong gratings and blazed gratings [106,107,109]. However, in most cases of moderate FBGs, the coupling of Bragg reflection is dominant compared to the cladding-mode couplings. Utilizing the expressions found in the first subsection of Section 7.5.2, useful properties of FBGs can be described as follows.

Assuming that the light is incident at $z = 0$ and that there is no incident light from the backside of the grating (i.e., that the boundary conditions are $E_i(z)|_{z=0} = E_i(0)$ and $E_d(z)|_{z=L} = 0$, the reflectivity of the FBG for the core-mode propagation is described by

$$R \equiv \left| \frac{E_d(0)}{E_i(0)} \right|^2 = \frac{\kappa_\xi^* \kappa_\xi \sinh^2 \sigma_b L}{\sigma_b^2 \cosh^2 \sigma_b L + (\Delta\beta'/2)^2 \sinh^2 \sigma_b L} \tag{7.72}$$

where the corresponding parameters are described in Eqs. (7.51)–(7.53).

The maximum reflectivity occurs at the wavelength that satisfies $\Delta\beta' = 0$, as follows:

$$R_{\max} = \tanh^2 |\kappa_\xi| L \tag{7.73}$$

$$\lambda_{\max} = 2(n_{\text{eff}} + \delta n_{\text{eff, av}})\Lambda = \left(1 + \frac{\delta n_{\text{eff,av}}}{n_{\text{eff}}}\right)\lambda_{B,0} \tag{7.74}$$

where $\lambda_{B,0} \equiv 2n_{\text{eff}}\Lambda$ is the nominal Bragg wavelength, n_{eff} is the unperturbed effective index of the core mode, and the self-coupling coefficient κ_σ is defined as in Eq. (7.39). When σ_b^2 is greater than zero, thereby producing a real value of σ_b, the power of the incident mode decays exponentially in the

z direction; that is, the power of the reflected mode grows exponentially in the opposite direction. Otherwise, no further decay or growth occurs, and, hence they evolve sinusoidally. Thus, the points at $\sigma_b^2 = 0$ can be defined as the band edges [106]. The band edges occur at the wavelengths

$$\lambda_{\text{edge}} = \lambda_{\max} \pm \frac{\delta n_{\text{eff,gr}}}{2n_{\text{eff}}} \lambda_{B,0} \tag{7.75}$$

and the reflectivity at the band edge is

$$R_{\text{edge}} = \frac{\kappa_\xi^* \kappa_\xi L^2}{1 + \kappa_\xi^* \kappa_\xi L^2} \tag{7.76}$$

Thus, the fractional bandwidth or the normalized bandwidth of a Bragg grating is described by [34,106]

$$\frac{\Delta\lambda_{\text{edge}}}{\lambda_{B,0}} = \frac{\delta n_{\text{eff,gr}}}{n_{\text{eff}}} \tag{7.77}$$

The reflection spectrum also consists of a series of sidelobes on both sides of the main band gap, as the phase mismatch increases. Mathematically, σ_b becomes purely imaginary when $(\Delta\beta'/2)^2$ is greater than $\kappa_\xi^* \kappa_\xi$ in Eq. (7.53), thereby producing an oscillatory decay outside the main band gap. Thus, the sideband reflection can be zero or a local maximum value based on $\sigma_b L$. In this case, the hyperbolic functions in Eq. (7.72) are to be modified into the corresponding triangular functions. The zero occurs at wavelengths

$$\lambda_{\text{sidelobe, zero}} = \frac{\lambda_{\max} \pm \dfrac{\delta n_{\text{eff, gr}} \lambda_{B,0}}{2n_{\text{eff}}} \sqrt{\left(\dfrac{p\lambda_{\max}}{\delta n_{\text{eff, gr}} L}\right)^2 + 1 - \left(\dfrac{p\Lambda}{L}\right)^2}}{1 - \left(\dfrac{p\Lambda}{L}\right)^2} \tag{7.78}$$

where $p = 1, 2, 3, \ldots$.

The peak reflectivities of the sidelobes are given by

$$R_{\text{sidelobe, peak}} = \frac{\kappa_\xi^* \kappa_\xi L^2}{(p + 1/2)^2 \pi^2 + \kappa_\xi^* \kappa_\xi L^2} \tag{7.79}$$

where $p = 1, 2, 3, \ldots$, which occurs at wavelengths

$$\lambda_{\text{sidelobe, peak}} = \frac{\lambda_{\max} \pm \frac{\delta n_{\text{eff},gr}\lambda_{B,0}}{2n_{\text{eff}}}\sqrt{\left(\frac{(p+1/2)\lambda_{\max}}{\delta n_{\text{eff, gr}}}\right)^2 + 1 - \left(\frac{(p+1/2)\Lambda}{L}\right)^2}}{1-\left(\frac{(p+1/2)\Lambda}{L}\right)^2} \tag{7.80}$$

where $p = 1, 2, 3, \ldots$.

As a result, another measurable bandwidth can be defined as the spectral width between first zeros [106]. Based on the strength of $|\kappa_\xi|L$, the bandwidth between first zeros from Eq. (7.78) can be found approximately as

$$\frac{\Delta\lambda_{\text{sidelobe, zero}}}{\lambda_{B,0}} \approx \frac{\delta n_{\text{eff, gr}}}{n_{\text{eff}}}\sqrt{\left(\frac{\lambda_{\max}}{\delta n_{\text{eff, gr}}L}\right)^2 + 1} \tag{7.81}$$

$$\approx \begin{cases} \frac{2}{L/\Lambda} & (|\kappa_\xi|L \ll \pi) \\ \frac{\delta n_{\text{eff, gr}}}{n_{\text{eff}}} & (|\kappa_\xi|L \gg \pi) \end{cases}$$

In the case of a weak grating so that $|\kappa_\xi|L \ll \pi$, the bandwidth is inversely proportional to L/Λ (i.e., to the number of grating periods). As a result, the reflection spectrum becomes a sinc-like function (see Figure 7.44). On the contrary, in the case of a strong grating so that $|\kappa_\xi|L \gg \pi$, the bandwidth is approximately identical to the fractional bandwidth in Eq. (7.77)—that is, the band gap edges are close to the first zeros. As a result, the main peak becomes a wide and flat squarelike function; however, the sidelobes also

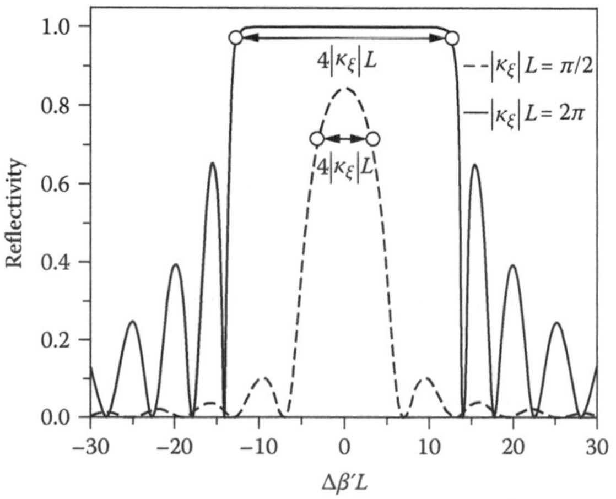

FIGURE 7.44
Reflection spectra of FBGs based on $\Delta\beta' L$ for $|\kappa_\xi| = \pi/2$ (dashed line) and $|\kappa_\xi|L = 2\pi$ (solid line).

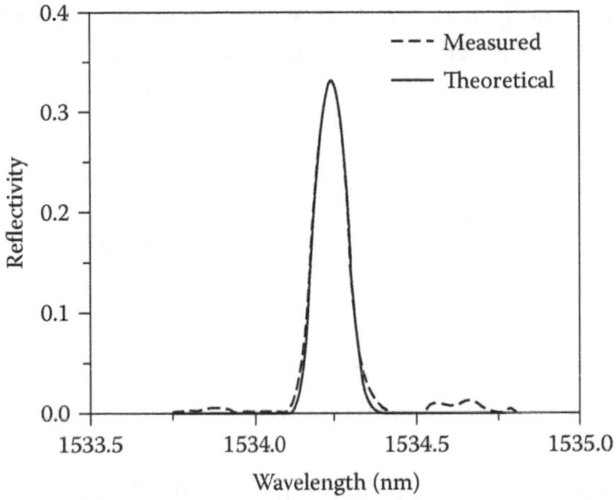

FIGURE 7.45
Measured (dashed line) and calculated (solid line) reflection spectra of a Gaussian-apodized FBG, wherein $\Lambda = 0.53$ μm, $L = 1$ cm, and $\delta n_{eff,av} = \delta n_{eff,gr} = 5.5 \times 10^{-4}$.

increase (see Eq. 7.80 and Figure 7.44). Examples of reflection spectra of FBGs are shown in Figure 7.44 and Figure 7.45.

Based on the coupled-mode theory, the basic properties of the uniform FBG have been derived. In the case of nonuniform fiber gratings, it is possible to analyze these by the transfer matrix method, examples of which are briefly discussed in the last part of this section.

7.5.4 Long-Period Fiber Gratings

Long-period fiber gratings (LPFGs) are based on codirectional couplings. In the case of a single-mode fiber, the propagating core mode is dominantly diffracted into cladding modes, which propagate in the same direction. In fact, the cladding modes are guided modes within the cladding boundary. However, they are leaky with respect to external perturbations (examples of which include jacket coatings or bending) and, as a result, are unable to propagate over a long distance. Thus, they give rise to band-rejection actions. Numerous cladding modes exist, even in a single-mode fiber, but one cladding mode that is most likely to be coupled with the core mode by the given grating period can be chosen in the regime of the two-mode coupling. Typically, the effective index difference between the core and cladding modes is ~10^{-3}, and, thus, the period of an LPFG is likely to be hundreds of micrometers, a so-called long period relative to that for an FBG. When the grating length is relatively short or the index change is nonuniform, a two-mode coupling approach may be insufficient, since the effective index difference between the adjacent cladding modes is, relatively, as small as ~10^{-4}. In that case, multimode coupling should be taken into account for more accurate evaluation [97–99]. Utilizing Eqs. (7.54)–(7.61), we can describe useful properties of LPFGs as follows.

Assuming that the light is incident at $z = 0$ and that there is no cladding-mode incidence at the entrance of the grating—that is, the boundary conditions are $E_i(z)|_{z=0} = E_i(0)$ and $E_d(z)|_{z=0} = 0$—the transmissivities of the core and cladding modes are respectively described by

$$T_{\infty} \equiv \left|\frac{E_i(L)}{E_i(0)}\right|^2 = \cos^2 \sigma_f L + \left(\frac{\Delta\beta'}{2\sigma_f}\right)^2 \sin^2 \sigma_f L \tag{7.82}$$

$$T_{\text{cl}} \equiv \left|\frac{E_d(L)}{E_i(0)}\right|^2 = \frac{\kappa_\xi^* \kappa_\xi}{\sigma_f^2} \sin^2 \sigma_f L \tag{7.83}$$

where the corresponding parameters are described in Eqs. (7.59)–(7.61). The diffracted light to the cladding mode is leaky with respect to external perturbations and, as a result, is evanescent after the grating. Thus, it is reasonable to define the transmissivity of an LPFG as the core-mode transmissivity. In the case that the external perturbations are excluded, it is possible for the cladding mode to propagate over a long distance and for it to be also recoupled into the core mode when it meets another LPFG [65,114].

The minimum transmissivity occurs at the wavelength that satisfies $\Delta\beta' = 0$ as follows:

$$T_{\min} = \cos^2 |\kappa_\xi| L \tag{7.84}$$

$$\lambda_{\min} = [\Delta n_{\text{eff}} + \delta n_{\text{eff,av}}(1-u_\sigma)]\Lambda = \left[1 + \frac{\delta n_{\text{eff,av}}(1-u_\sigma)}{\Delta n_{\text{eff}}}\right]\lambda_{F,0} \tag{7.85}$$

where $\lambda_{F,0} \equiv \Delta n_{\text{eff}}\Lambda$ is the nominal forward-diffraction wavelength, and Δn_{eff} is determined as the difference of the unperturbed effective indices between the core and cladding modes (i.e., $\Delta n_{\text{eff}} = n_{\text{eff},i} - n_{\text{eff},d}$). As the wavelength is detuned from $\lambda_{\min}$, the minimized transmission increases with side dips. The transmission can be unity when $\sigma_f L$ becomes an integer multiple of π as follows:

$$\lambda_{\text{unity}} = \frac{\lambda_{\min} \pm \dfrac{|u_\xi|\delta n_{\text{eff,gr}}\lambda_{F,0}}{\Delta n_{\text{eff}}}\sqrt{\left(\dfrac{p\lambda_{\min}}{|u_\xi|\delta n_{\text{eff,gr}} L}\right)^2 - 1 + \left(\dfrac{p\Lambda}{L}\right)^2}}{1 - \left(\dfrac{p\Lambda}{L}\right)^2} \tag{7.86}$$

where $p = 1, 2, 3, \ldots$ and $p > |\kappa_\xi| L/\pi$. Thus, the width between the first unities on either side of the main dip is usually defined as a measurable bandwidth

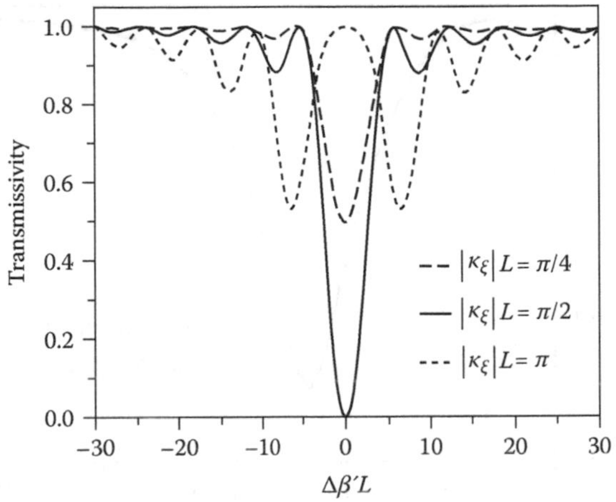

FIGURE 7.46
Transmission spectra of LPFGs based on $\Delta\beta'L$ for $|\kappa_\xi|L = \pi/4$ (dashed line), $|\kappa_\xi|L = \pi/2$ (solid line), and $|\kappa_\xi|L = \pi$ (dotted line).

[106]. In the case of $|\kappa_\xi|L < \pi$, the bandwidth between first unities from Eq. (7.86) can be approximated as

$$\frac{\Delta\lambda_{\text{unity}}}{\lambda_{F,0}} \approx \frac{2|u_\xi|\delta n_{\text{eff,gr}}}{\Delta n_{\text{eff}}}\sqrt{\left(\frac{\lambda_{\min}}{|u_\xi|\delta n_{\text{eff,gr}}L}\right)^2 - 1} \tag{7.87}$$

As opposed to contradirectional coupling, no band gap exists in the case of codirectional coupling. The power exchange between the two modes evolves based on the strength of $\sigma_f L$. The power exchange from the core mode to the cladding mode increases gradually with respect to $\sigma_f L$, and the maximum power exchange occurs at $\sigma_f L = \pi/2$. After that, until $\sigma_f L = \pi$, the direction of the power exchange becomes opposite (i.e., the core mode receives the power in return from the cladding mode). In this respect, the power exchange has an oscillatory evolution with the strength of $\sigma_f L$. For most cases of LPFGs, the strength of $\sigma_f L$ or $|\kappa_\xi|L$ in the range of interest is less than π. Examples of transmission spectra for LPFGs are shown in Figure 7.46 and Figure 7.47.

Based on the coupled-mode theory, the basic properties of the uniform LPFG have been derived. In the case of nonuniform LPFGs or pairs of LPFGs, it is possible to analyze these by the transfer matrix method, examples of which are discussed briefly in the last part of this section.

7.5.5 Examples of Nonuniform Fiber Gratings

In this subsection, we address several examples of nonuniform fiber gratings. For the case of contradirectional coupling, the properties of chirped fiber gratings are described. These are useful as dispersion compensators

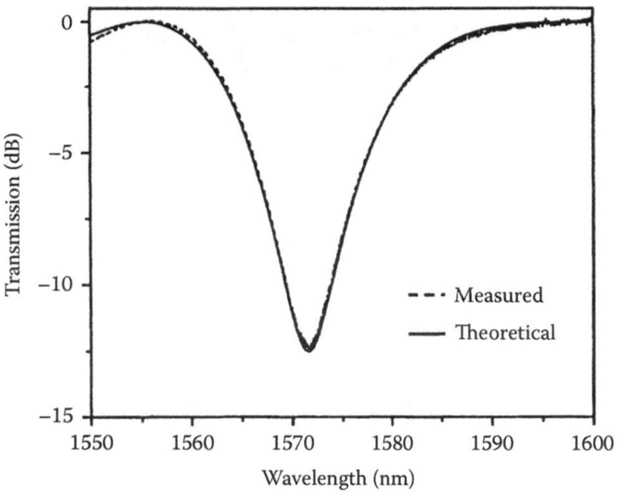

FIGURE 7.47
Measured (dashed line) and calculated (solid line) reflection spectra of a Gaussian-apodized LPFG, wherein $\Lambda = 500\ \mu m$, $L = 3$ cm, and $\Delta n_{av} = \Delta n_{gr} = 2.568 \times 10^{-4}$.

and integrating sensors [79,112], whose group-delay properties are discussed in the following with respect to different chirping methods. For the case of the codirectional coupling, the properties of phase-shifted gratings and cascaded gratings are described. It is possible to utilize their fringe patterns as bandpass filters and narrow-band, multichannel filters [114]. When their properties are combined with additional techniques, it becomes possible to use them as sensor demodulators, which is discussed in Section 7.3.7, and as all-optical switching devices [115].

Chirped Fiber Bragg Gratings

The chirped FBGs have a structure of monotonically increasing or decreasing local Bragg wavelengths through the gratings. The property gives rise to a broadband reflection relative to a uniform FBG and a dispersive group delay [78,79,112]. The chirped distribution of the local Bragg wavelengths can usually be obtained by one of two methods. One involves the use of sinusoidal index changes with a monotonically increasing or decreasing period plus an identical dc index change (i.e., by spatially varying Λ with a fixed $\delta n_{eff,av}$). The other is by monotonically increasing or decreasing the dc index change plus a sinusoidal index change with a uniform period (i.e., by spatially varying $\delta n_{eff,av}$) with a fixed Λ [112]. This can be easily seen from Eq. (7.74): It is possible for the local Bragg wavelength to be chirped by changing Λ or $\delta n_{eff,av}$ spatially through the grating. As a result, the local Bragg wavelength is given by

$$\lambda_B(z) = 2\left(n_{eff} + \delta n_{eff,av,0}\right)\Lambda(z) \tag{7.88}$$

or

$$\lambda_B(z)=2\left[n_{\text{eff}}+\delta n_{\text{eff,av}}(z)\right]\Lambda_0 \tag{7.89}$$

where $\delta n_{\text{eff,av,0}}$ or Λ_0 denotes a constant value for the corresponding parameter.

The first method requires UV-inscribing equipment to obtain chirped-period patterns that involve a chirped phase mask. This is simple but requires different phase masks in order to obtain differently distributed gratings. The second method requires UV laser scanning with intensity modulation to achieve a gradient of dc index change before UV-inscribing by a uniform phase mask [112]. The pre-illumination time is to be scheduled differently so as to obtain a spatially varying dc index change through the length of the fiber grating, since more UV illumination causes a higher dc index change in the core region. Here, the two cases can be described as follows.

The parameter $\Lambda(z)$ or $\delta n_{\text{eff,av}}(z)$ can vary spatially as a monotonically increasing or decreasing function. Thus, in the transfer matrix given by Eqs. (7.64)–(7.67), Λ_k or $\kappa_{\sigma,k}$ are determined in a sectionwise manner with appropriately discretized values based on the case of the method. It is noteworthy that the dc index change $\delta n_{\text{eff,av}}$ is related to κ_σ (see Eq. 7.39). Hereafter, we describe the dc index change in terms of κ_σ. By multiplying the entire transfer matrix, it is possible to obtain the spectral responses of the chirped gratings.

One feature of the chirped gratings is that they can give dispersive group delays for different wavelengths, which can be useful in dispersion compensation and intragrating sensing [79,112]. Dispersive properties of fiber gratings can be readily obtained from the derived results. The group delay of the reflected light from the grating can be determined by analyzing the phase factor of the amplitude for the reflected light. The group delay is defined as the constant phase time of a light propagation [78]. Thus, the group delay of the reflected light is given by

$$\tau_d \equiv \frac{d\phi_d}{d\omega}=-\frac{\lambda^2}{2\pi c}\frac{d\phi_d}{d\lambda} \tag{7.90}$$

where ϕ_d is the phase of the complex amplitude of the reflected light. Furthermore, the dispersion of the group delay can be determined by the rate of change of the delay with wavelength and is given by

$$\frac{d\tau_d}{d\lambda}=-\left(\frac{\lambda}{\pi c}\frac{d\phi_d}{d\lambda}+\frac{\lambda^2}{2\pi c}\frac{d^2\phi_d}{d\lambda^2}\right) \tag{7.91}$$

The parameter for $\Lambda(z)$ or $\kappa_\sigma(z)$ is to be defined as

$$\Lambda(z)=\Lambda_0+\delta\Lambda(z) \tag{7.92}$$

or

$$\kappa_\sigma(z)=\kappa_{\sigma,0}+\delta\kappa_\sigma(z) \tag{7.93}$$

where Λ_0 and $\kappa_{\sigma,0}$ are, respectively, the initial and mean values of the local period and the self-coupling constant. Assuming that the chirping variation is not so large relative to its initial value, the local Bragg wavelengths are respectively approximated by

$$\lambda_B(z)\approx\lambda_{B,0}+2n_{\text{eff}}\delta\Lambda(z) \tag{7.94}$$

or

$$\lambda_B(z)\approx\lambda_{B,0}+\frac{2n_{\text{eff}}\Lambda_0^2}{\pi}\delta k_\sigma(z) \tag{7.95}$$

where

$$\lambda_{B,0}=2(n_{\text{eff}}+\delta n_{\text{eff,av,0}})\Lambda_0 \tag{7.96}$$

It is of interest that the self-coupling constant variation is equivalent to the grating pitch change if multiplied by a factor Λ_0^2/π.

Assuming a linear chirp, the reflectivity and group-delay examples for both chirped gratings are shown in Figure 7.48. It can be seen that the two methods give nearly identical results.

Phase-Shifted and Cascaded Long-Period Fiber Gratings

Both the phase-shifted LPFG and cascaded LPFG have an anchoring section in the middle of the entire grating. The former consists of two sections of LPFGs where the second section starts with a phase shift with respect to the former grating sequence. On the contrary, the latter is a simple connection of two LPFGs with a bare fiber of an appropriate length. The bandpass filter has been demonstrated by Bakhti and Sansonetti with π-shift in the middle [116]. The pair of LPFGs has been presented by Dianov et al. [65], and the dependence of fringe spacing on the grating separation in an LPFG pair has been discussed by Lee and Nishii [114]. In general, both grating types can be composed of more than two different sections. Here, the two examples of the nonuniform LPFGs are described in the following.

The phase-shifted LPFG is readily analyzed by the transfer matrix given in Eqs. (7.68)–(7.71) with an appropriate determination of $\phi_{gr,k}$, which is the initial grating phase for section k. The entire grating can be divided into two uniform sections. Thus, the phase shift can be determined so that the second grating phase (i.e., $\phi_{gr,2}$) differs from the last phase of the first grating.

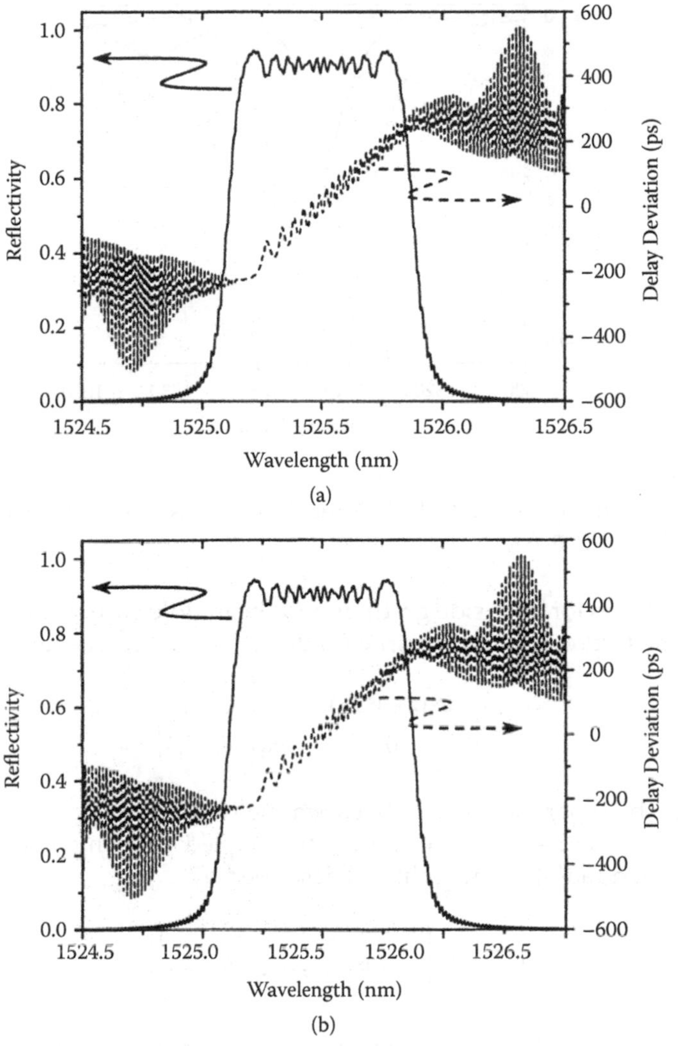

FIGURE 7.48
Reflection spectra and group delays of chirped FBGs by the transfer matrix method, wherein $\Lambda_0 = 0.527$ μm, $L \approx 5.27$ cm, $\delta n_{\mathrm{eff,av,0}} = \delta n_{\mathrm{eff,gr,0}} = 1 \times 10^{-4}$. The entire grating is divided into 200 segments in simulation: (a) linearly varying pitch (+0.05%); (b) linearly varying κ_σ (+723%).

The last grating phase is determined by the entire section length—that is, $2\pi \times$ (remainder of L_1 and Λ) + $\phi_{gr,1}$. The entire transfer relation is obtained by multiplication of the two transfer matrices. The numerical example for the phase-shifted LPFG is shown in Figure 7.49. At a phase shift of π, the rejection band splits with respect to the center of the nonphase-shifted one, which is similar to a bandpass filter [116].

The cascaded LPFG has a bare section in the middle that has no index change; hence, it can be divided into three sections: in other words, two

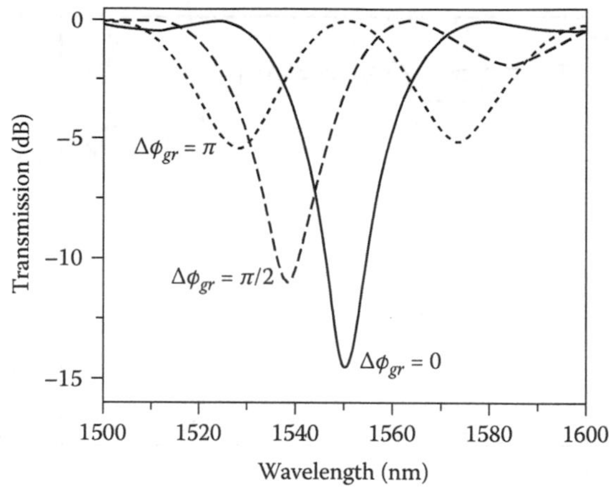

FIGURE 7.49
Transmission spectra of phase-shifted LPFGs based on the phase shift in the middle, wherein $\Lambda = 485\ \mu m$, $L = 2.425$ cm, $\Delta n_{eff} = 3 \times 10^{-3}$, and $n_{eff,gr} = 2 \times 10^{-4}$.

transfer matrices for both side gratings and the middle bare section. Since the middle bare section has no perturbed index change, it can be described by

$$T_S = \begin{pmatrix} \exp(-i\beta_i L_s) & 0 \\ 0 & \exp(-i\beta_d L_s) \end{pmatrix} \tag{7.97}$$

where L_s is the separation length between the two LPFGs.

The final transmission of the resonantly coupled cladding mode, which is to be a rejected radiation in reality, is described by

$$|t_{cl}|^2 \propto f_{cl}\left\{1+\cos\left(\frac{2}{\lambda}\Delta n_{eff} L_s + 2\varsigma\right)\right\} \tag{7.98}$$

where Δn_{eff} is the difference between the group indices of the core and the cladding modes. In addition, f_{cl} and ς are the inherent parameters of the individual LPFG, which are dependent on grating properties. Further derivation is omitted for the sake of simplicity. If the term for ς is neglected, in the case wherein the separation L_s is much longer than the LPFG length, the fringe spacing in a cascaded twin LPFG pair can be approximated by

$$\Delta\lambda_{fringe} \approx \frac{\lambda^2}{\Delta n_{eff} L_s} \tag{7.99}$$

Note that it is possible to control the fringe space by an appropriate separation L_s. Numerical examples for the cascaded LPFG are shown in Figure 7.50. The narrow-band fringe can be utilized in sensor demodulators, which are discussed in Section 7.3.7, and as all-optical switching devices wherein the

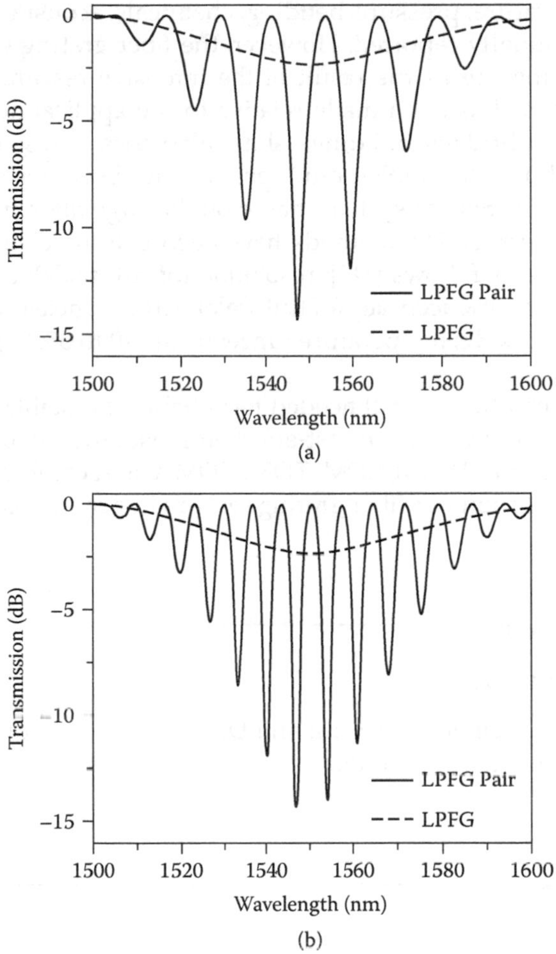

FIGURE 7.50
Transmission spectra of cascaded LPFGs based on the grating separation length, wherein Λ = 485 μm, $2L$ = 2.425 cm, $\Delta n_{eff} = 3 \times 10^{-3}$, $\delta n_{eff,av} = \delta n_{eff,gr} = 2 \times 10^{-4}$, and the two gratings are identical: (a) L_s = 5 cm; (b) L_s = 10 cm.

0.5-nm rejection bandwidth was obtained to enhance the nonlinear switching efficiency [115].

7.6 Conclusions

Fiber gratings are currently essential devices in optical communications and are becoming important devices for sensors as the result of extensive studies, which this chapter reviews. Many novel sensor systems for the measurement of strain, temperature, pressure, bending, chemicals, acoustic waves, etc. are still being continually reported. However, the fiber grating sensor technologies are now mature as the result of the intensive research to date, and considerable effort has been made relative to the application of sensors to real fields such as bridges, buildings, ships, airplanes, and so on. Many tests for quasi-distributed (multiple-point) sensing have been performed (see, for example, Rao [117]). Sensor systems based on the edge filter and Fabry–Perot tunable filter interrogation methods have been commercialized, and their performances are as follows [117]: resolution (of full scale): 0.01%; measurement range (strain): 1%; accuracy (of full scale): ±0.05%; measurable frequency (single channel): 1 kHz; temperature range: from –70 to 350°C; gauge length: >10 mm.

However, much effort is still needed for reliable and stable installment in real field situations and for the cost-effective sensor system development in multiplexings such as WDM, SDM, TDM, CDMA, and combinations thereof. These efforts will soon put fiber grating sensors in general use around us.

Acknowledgments

The authors thank Dr. Minho Song and Dr. Jaehoon Jung for valuable comments that improved this chapter.

References

1. K. O. Hill, Y. Fujii, D. C. Johnson, and B. S. Kawasaki, Photosensitivity in optical fiber waveguides: Application to reflection filter fabrication, *Appl. Phys. Lett.*, **32**, pp. 647–649, 1978.
2. G. Meltz, W. W. Morey, and W. H. Glenn, Formation of Bragg gratings in optical fibers by a transverse holographic method. *Optics Lett.*, **14**, pp. 823–825, 1989.
3. A. Othonos and K. Kalli, *Fiber Bragg Gratings—Fundamentals and Applications in Telecommunications and Sensing*, Artech House, Boston, 1999.
4. Y.-J. Rao, Fiber Bragg grating sensors: Principles and applications, in *Optical Fiber Sensor Technology*, Vol. 2, K. T. V. Grattan and B. T. Meggitt, eds., pp. 355–389, Chapman & Hall, London, 1998.
5. K. T. V. Grattan and B. T. Meggitt, eds., *Optical Fiber Sensor Technology*, Vol. 2, Chapman & Hall, London, 1998.
6. J. R. Dunphy, G. Meltz, and W. W. Morey, Optical fiber Bragg grating sensors: A candidate for smart structure applications, in *Fiber Optic Smart Structures*, E. Udd, ed., pp. 271–285, John Wiley & Sons, New York, 1995.

7. Y. N. Ning and B. T. Meggitt, Fiber Bragg grating sensors: Signal processing aspects, in *Optical Fiber Sensor Technology*, Vol. 2, K. T. V. Grattan and B. T. Meggitt, eds., pp. 381–417, Chapman & Hall, London, 1998.
8. S. M. Melle, K. Liu, and R. M. Measures, A passive wavelength demodulation system for guided-wave Bragg grating sensors, *IEEE Photonics Technol. Lett.*, **4**, 5, pp. 516–518, 1992.
9. A. B. Lobo Ribeiro, L. A. Ferreira, M. Tsvetkov, and J. L. Santos, All-fibre interrogation technique for fibre Bragg sensors using a biconical fibre filter, *Electronics Lett.*, **32**, 4, pp. 382–383, 1996.
10. M. Song, S. B. Lee, S. S. Choi, and B. Lee, Fiber laser strain sensor using an LPG (long period grating) Bragg wavelength demodulation filter, *The 2nd Optoelectronics and Communications Conf, (OECC '97) Technical Digest*, Seoul, Korea, pp. 676–677, July 1997.
11. M. A. Davis and A. D. Kersey, All-fibre Bragg grating strain-sensor demodulation technique using a wavelength division coupler, *Electronics Lett.*, **30**, 1, pp. 75–77, 1994.
12. Q. Zhang, D. A. Brown, H. Kung, J. E. Townsend, M. Chen, L. J. Reinhart, and T. F. Morse, Use of highly overcoupled couplers to detect shifts in Bragg wavelength, *Electronics Lett.*, **31**, 6, pp. 480–482, 1995.
13. J. Peupelmann and J. Meissner, Applications and field tests of a fibre Bragg grating sensor system, *Proc. 13th Int. Conf. Optical Fiber Sensors (OFS-13)*, Kyongju, Korea, *SPIE*, **3746**, pp. 470–473, 1999.
14. S. C. Kang, H. Yoon, S. B. Lee, S. S. Choi, and B. Lee, Real-time measurement for static and dynamic strain using a fiber Bragg grating and the ASE profile of EDFA, *Proc. 13th Int. Conf. Optical Fiber Sensors (OFS-13)*, Kyongju, Korea, *SPIE*, **3746**, pp. 530–533, 1999.
15. S. Kim, J. Kwon, S. Kim, and B. Lee, Temperature-independent strain sensor using a chirped grating partially embedded in a glass tube, *IEEE Photonics Technology Lett.*, **12**, 6, pp. 678–680, 2000.
16. A. D. Kersey, M. A. Davis, and T. Tsai, Fiber optic Bragg grating strain sensor with direct reflectometric interrogation, *Proc. Optical Fiber Sensors Conference (OFS-11)*, Sapporo, Japan, pp. 634–637, 1996.
17. V. Grubsky and J. Feinberg, Long-period fiber gratings with variable coupling for real-time sensing applications, *Optics Lett.*, **25**, 4, pp. 203–205, 2000.
18. R. W. Fallon, L. Zhang, A. Gloag, and I. Bennion, Identical broadband chirped grating interrogation technique for temperature and strain sensing, *Electronics Lett.*, **33**, pp. 705–706, 1997.
19. L. Zhang, R. W. Fallon, A. Gloag, I. Bennion, F. M. Haran, and P. Foote, Spatial and wavelength multiplexing architectures for extreme strain monitoring system using identical-chirped-grating-interrogation technique, *Proc. Optical Fiber Sensors Conference (OFS-12)*, Williamsburg, VA, pp. 452–455, 1997.
20. A. D. Kersey, M. A. Davis, H. J. Patrick, M. LeBlanc, K. P. Koo, C. G. Askins, M. A. Putnam, and E. J. Friebele, Fiber grating sensors, *J. Lightwave Technol.*, **15**, 8, pp. 1442–1463, 1997.
21. C. G. Askins, M. A. Putnam, G. M. Williams, and E. J. Friebele, Stepped-wavelength optical fiber Bragg grating arrays fabricated in line on a draw tower, *Optics Lett.*, **19**, pp. 147–149, 1994.

22. K. Usbeck, W. Ecke, V. Hagemann, R. Mueller, and R. Willsch, Temperature referenced fibre Bragg grating refractometer sensor for on-line quality control of petrol products, *Proc. 13th Int. Conf. Optical Fiber Sensors (OFS-13)*, Kyongju, Korea, *SPIE*, **3746**, pp. 163–166, 1999.
23. D. J. Webb, M. W. Hathaway, D. A. Jackson, S. Jones, L. Zhang, and I. Bennion, First in-vivo trials of a fiber Bragg grating based temperature profiling system, *J. Biomedical Optics*, **5**, 1, pp. 45–50, 2000.
24. S. Chen, Y. Hu, L. Zhang, and I. Bennion, Digital spatial and wavelength domain multiplexing of fiber Bragg grating based sensors, *Proc. Optical Fiber Sensors Conf. (OFS-12)*, Williamsburg, VA, pp. 448–451, 1997.
25. Y. Hu, S. Chen, L. Zhang, and I. Bennion, Multiplexing Bragg gratings combined wavelength and spatial division techniques with digital resolution enhancement, *Electronics Lett.*, **33**, 23, pp. 1973–1975, 1997.
26. M. A. Davis and A. D. Kersey, Fiber Fourier transform spectrometer for decoding Bragg grating sensors, *Proc. Optical Fiber Sensors Conf. (OFS-10)*, Glasgow, Scotland, pp. 167–170, 1994.
27. M. A. Davis and A. D. Kersey, Application of a fiber Fourier transform spectrometer to the detection of wavelength-encoded signals from Bragg grating sensors, *J. Lightwave Technol.*, **13**, 7, pp. 1289–1295, 1995.
28. D. A. Flavin, R. McBride, and J. D. C. Jones, Short optical path scan interferometric interrogation of a fibre Bragg grating embedded in a composite, *Electronics Lett.*, **33**, 4, pp. 319–321, 1997.
29. D. A. Flavin, R. McBride, and J. D. C. Jones, Short-scan interferometric interrogation and multiplexing of fibre Bragg grating sensors, *Optics Commun.*, **170**, pp. 347–353, 1999.
30. K. B. Rochford and S. D. Dyer, Demultiplexing of interferometrically interrogated fiber Bragg grating sensors using Hilbert transform processing, *J. Lightwave Technol.*, **17**, 5, pp. 831–836, 1999.
31. A. D. Kersey, T. A. Berkoff, and W. W. Morey, Multiplexed fiber Bragg grating strain-sensor system with a fiber Fabry–Perot wavelength filter, *Optics Lett.*, **18**, pp. 1370–1372, 1993.
32. S. T. Vohra, M. D. Todd, G. A. Johnson, C. C. Chang, and B. A. Danver, Fiber Bragg grating sensor system for civil structure monitoring: Applications and field tests, *Proc. 13th Int. Conf. Optical Fiber Sensors* (*OFS-13*), Kyongju, Korea, *SPIE*, **3746**, pp. 32–37, 1999.
33. P. J. Henderson, D. J. Webb, D. A. Jackson, L. Zhang, and I. Bennion, Highly multiplexed grating-sensors for temperature-referenced quasi-static measurements of strain in concrete bridges, *Proc. 13th Int. Conf. Optical Fiber Sensors* (*OFS-13*), Kyongju, Korea, *SPIE*, **3746**, pp. 320–323, 1999.
34. A. Yariv and P. Yeh, *Optical Waves in Crystals*, John Wiley & Sons, New York, 1984.
35. M. Volanthen, H. Geiger, M. G. Xu, and J. P. Dakin, Simultaneous monitoring fibre gratings with a single acousto-optic tunable filter, *Electronics Lett.*, **32**, pp. 1228–1229, 1996.
36. H. Geiger, M. G. Xu, N. C. Eaton, and J. P. Dakin, Electronic tracking system for multiplexed fibre grating sensors, *Electronics Lett.*, **32**, pp. 1006–1007, 1995.
37. M. G. Xu, H. Geiger, and J. P. Dakin, Modeling and performance analysis of a fiber Bragg grating interrogation system using an acousto-optic tunable filter, *J. Lightwave Technol.*, **14**, pp. 391–396, 1996.

38. M. G. Xu, H. Geiger, and J. L. Archambault, L. Reekie, and J. P. Dakin, Novel interrogating system for fibre Bragg grating sensors using an acousto-optic tunable filter, *Electronics Lett.*, **29**, pp. 1510–1511, 1993.
39. J. P. Dakin and M. Volanthen, Distributed and multiplexed fibre grating sensors, including discussion of problem areas, *IEICE Trans. Electronics*, **E83-C**, 3, pp. 391–399, 2000.
40. D. A. Jackson, A. B. Lobo Ribeiro, L. Reeckie, and J. L. Archambault, Simple multiplexing scheme for a fiber-optic grating sensor network, *Optics Lett.*, **18**, pp. 1192–1194, 1993.
41. G. P. Brady, S. Hope, A. B. Lobo Ribeiro, D. J. Webb, L. Reekie, J. L. Archambault, and D. A. Jackson, Demultiplexing of fibre Bragg grating temperature and strain sensors, *Optics Commun.*, **111**, pp. 51–54, 1994.
42. M. A. Davis and A. D. Kersey, Matched-filter interrogation technique for fibre Bragg grating arrays, *Electronics Lett.*, **31**, pp. 822–823, 1995.
43. S. C. Kang, S. Y. Kim, S. B. Lee, S. W. Kwon, S. S. Choi, and B. Lee, Temperature-independent strain sensor system using a tilted fiber Bragg grating demodulator, *IEEE Photonics Technol. Lett.*, **10**, 10, pp. 1461–1463, 1998.
44. A. D. Kersey, T. A. Berkoff, and W. W. Morey, High-resolution fibre-grating based strain sensor with interferometric wavelength-shift detection, *Electronics Lett.*, **28**, 3, pp. 236–238, 1992.
45. R. S. Weis, A. D. Kersey, and T. A. Berkoff, A four-element fiber grating sensor array with phase-sensitive detection, *IEEE Photonics Technol. Lett.*, **6**, 12, pp. 1469–1472, 1994.
46. K. P. Koo and A. D. Kersey, Bragg grating based laser sensors systems with interferometric interrogation and wavelength division multiplexing, *J. Lightwave Technol.*, **13**, pp. 1243–1249, 1995.
47. A. D. Kersey and T. A. Berkoff, Fiber-optic Bragg grating differential-temperature sensor, *IEEE Photonics Technol. Lett.*, **4**, 10, pp. 1183–1185, 1992.
48. A. D. Kersey, T. A. Berkoff, and W. W. Morey, Fiber-optic Bragg grating strain sensor with drift-compensated high-resolution interferometric wavelength-shift detection, *Optics Lett.*, **18**, 1, pp. 72–74, 1993.
49. A. Dandridge, A. B. Tveten, and T. G. Giallorenzi, Homodyne demodulation scheme for fiber optic sensors using phase generated carrier, *IEEE J. Quantum Electronics*, **18**, pp. 1647–1653, 1982.
50. Y. L. Lo, J. S. Sirkis, and C. C. Chang, Passive signal processing of in-line fiber etalon sensors for high strain-rate loading, *J. Lightwave Technol.*, **15**, pp. 1578–1585, 1997.
51. M. Song, S. Yin, and P. B. Ruffin, Fiber Bragg grating strain sensor demodulation with quadrature sampling of a Mach–Zehnder interferometer, *Appl. Opt.*, **39**, 7, pp. 1106–1111, 2000.
52. S. C. Kang, S. B. Lee, S. S. Choi, and B. Lee, A novel demodulation technique for the wavelength shift of fiber Bragg grating sensors using the I/Q signal processing scheme, *Proc. Conf. Lasers and Electro-Optics—Pacific Rim (CLEO/Pacific Rim '99)*, Seoul, Korea, pp. 135–136, Sept. 1999.
53. K. Kalli, G. P. Brady, D. J. Webb, D. A. Jackson, L. Zhang, and I. Bennion, Wavelength-division and spatial multiplexing using tandem interferometers for Bragg grating sensor networks, *Optics Lett.*, **20**, 24, pp. 2544–2546, 1995.

54. T. A. Berkoff and A. D. Kersey, Fiber Bragg grating array sensor system using a bandpass wavelength division multiplexer and interferometric detection, *IEEE Photonics Technol. Lett.*, **8**, 11, pp. 1522–1524, 1996.
55. G. A. Johnson, S. T. Vohra, B. A. Danver, K. Pran, G. B. Havsgard, and G. Wang, Vibration monitoring of a ship waterjet with fiber Bragg gratings, *Proc. 13th Int. Conf. Optical Fiber Sensors (OFS-13)*, Kyongju, Korea, *SPIE*, **3746**, pp. 616–619, 1999.
56. K. Pran. G. B. Havsgard, R. Palmstrom, G. Wang, G. A. Johnson, B. A. Danver, and S. T. Vohra, Sea-test of a 27 channel fibre Bragg grating strain sensor system on an air cushion catamaran, *Proc. 13th Int. Conf. Optical Fiber Sensors (OFS-13)*, Kyongju, Korea, *SPIE*, **3746**, pp. 145–148, 1999.
57. M. Song, B. Lee, S. B. Lee, and S. S. Choi, Interferometric temperature-insensitive strain measurement with different-diameter fiber Bragg gratings, *Optics Lett.*, **22**, 11, pp. 790–792, 1997.
58. S. B. Lee, Y. Liu, and S. S. Choi, Dual-wavelength fiber Bragg grating laser and its strained tuning characteristics, *Optoelectronics and Communication Conf.* (OECC '96), Chiba, Japan, pp. 352–353, July 1996.
59. M. Song, S. B. Lee, S. S. Choi, and B. Lee, Simultaneous measurement of temperature and strain using two fiber Bragg gratings embedded in a glass tube, *Optical Fiber Technol.*, **3**, pp. 194–196, 1997.
60. M. Song, S. B. Lee, S. S. Choi, and B. Lee, Dynamic-strain measurement with dual-grating fiber sensor, *Appl. Opt.*, **37**, 16, pp. 3484–3486, 1998.
61. P. M. Cavaleiro, F. M. Araujo, L. A. Ferreira, J. L. Santos, and F. Farahi, Simultaneous measurement of strain and temperature using Bragg gratings written in germanosilicate and boron-codoped germanosilicate fibers, *IEEE Photonics Technol. Lett.*, **11**, 12, pp. 1635–1637, 1999.
62. M. W. Hathaway, N. E. Fisher, D. J. Webb, C. N. Pannell, D. A. Jackson, L. R. Gavrilov, J. W. Hand, L. Zhang, and I. Bennoin, Combined ultrasonic and temperature sensor using a fibre Bragg grating, *Optics Commun.*, **171**, pp. 225–231, 1999.
63. R. Ashoori, Y. M. Gebremichael, S. Xiao, J. Kemp, K. T. V. Grattan, and A. W. Palmer, Time domain multiplexing for a Bragg grating strain measurement sensor network, *Proc, 13th Int. Conf. Optical Fiber Sensors (OFS-13)*, Kyongju, Korea, *SPIE*, **3746**, pp. 308–311, 1999.
64. Y. J. Rao, D. A. Jackson, L. Zhang, and I. Bennion, Dual-cavity interferometric wavelength-shift detection for in-fiber Bragg grating sensors, *Optics Lett.*, **21**, 19, pp. 1556–1558, 1996.
65. E. M. Dianov, S. A. Vasiliev, A. S. Kurkov, O. I. Medvedkov, and V. N. Protopopov, In-fiber Mach–Zehnder interferometer based on a pair of long-period gratings, *Proc. European Conf. Optical Commun.*, Gent, Belgium, pp. 65–68, 1997.
66. J. Jung, Y. W. Lee, and B. Lee, High-resolution interrogation technique for fiber Bragg grating strain sensor using long period grating pair and EDF, *The 14th Int. Conf. Optical Fiber Sensors (OFS-14)*, Venice, Italy, *SPIE*, **4185**, pp. 114–117, Oct. 2000.
67. G. A. Ball, W. W. Morey, and P. K. Cheo, Fiber laser source/analyzer for Bragg grating sensor array interrogation, *J. Lightwave Technol.*, **12**, pp. 700–703, 1994.
68. T. Coroy and R. M. Measures, Active wavelength demodulation of a Bragg grating fibre optic strain sensor using a quantum well electroabsorption filtering detector, *Electronics Lett.*, **32**, pp. 1811–1812, 1996.
69. S. H. Yun, D. J. Richardson, and B. Y. Kim, Interrogation of fiber grating sensor arrays with a wavelength-swept fiber laser, *Optics Lett.*, **23**, 11, pp. 843–845, 1998.

70. M. L. Dennis, M. A. Putnam, J. U. Kang, T.-E. Tsai, I. N. Duling, and E. J. Friebele, Grating sensor array demodulation by use of a passively mode-locked fiber laser, *Optics Lett.*, **22**, pp. 1362–1364, 1997.
71. M. A. Putnam, M. L. Dennis, J. U. Kang, T.-E. Tsai, I. N. Duling, and I. E. J. Friebele, Sensor grating array demodulation using a passively mode-locked fiber laser, *Technical Digest Optical Fiber Commun. Conf.*, Dallas, TX, Paper WJ4, pp. 156–157, Feb. 1997.
72. M. A. Putnam, M. L. Dennis, I. N. Duling III, C. G. Askins, and E. J. Friebele, Broadband square-pulse operation of a passively mode-locked fiber laser for fiber Bragg grating interrogation, *Optics Lett.*, **23**, pp. 138–140, 1998.
73. A. D. Kersey, A. Dandridge, and M. A. Davis, Low-crosstalk code-division multiplexed interferometric array, *Electronics Lett.*, **28**, pp. 351–352, 1992.
74. K. P. Koo, A. B. Tveten, and S. T. Vohra, Dense wavelength division multiplexing of fibre Bragg grating sensors using CDMA, *Electronics Lett.*, **35**, pp. 165–167, 1999.
75. P. K. C. Chan, W. Jin, and M. S. Demokan, Multiplexing of fiber Bragg grating sensors using subcarrier intensity modulation, *Optics Laser Technol.*, **31**, pp. 345–350, 1999.
76. P. K. C. Chan, W. Jin, J. M. Gong, and M. S. Demokan, Multiplexing of fiber Bragg grating sensors using an FMCW technique, *IEEE Photonics Technol. Lett.*, **11**, 11, pp. 1470–1472, 1999.
77. M. LeBlanc, S. Y. Huang, M. Ohn, R. M. Measures, A. Guemes, and A. Othonos, Distributed strain measurement based on a fiber Bragg grating and its reflection spectrum analysis, *Optics Lett.*, **21**, pp. 1405–1407, 1996.
78. A. M. Steinberg, P. G. Kwiat, and R. Y. Chiao, Measurement of single-photon tunneling time, *Physical Rev. Lett.*, **71**, pp. 708–711, 1993.
79. M. J. Marrone, A. D. Kersey, and M. A. Davis. Fiber sensors based on chirped Bragg gratings, *Optical Soc. America Annual Meeting*, Rochester, NY, Paper WGG5, Oct. 1996.
80. G. Duck and M. M. Ohn, Distributed Bragg grating sensing with a direct group-delay measurement technique, *Optics Lett.*, **25**, pp. 90–92, 2000.
81. S. Huang, M. M. Ohn, and R. M. Measures, Phase-based Bragg integrating distributed strain sensor, *Appl. Opt.*, **35**, pp. 1135–1142, 1996.
82. M. M. Ohn, S. Y. Huang, R. M. Measures, and J. Chwang, Arbitrary strain profile measurement within fibre gratings using interferometric Fourier transform technique, *Electronics Lett.*, **33**, pp. 1242–1243, 1997.
83 J. Azaña and M. A. Muriel, Reconstructing arbitrary strain distribution within fiber gratings by time-frequency signal analysis, *Optics Lett.*, **25**, pp. 698–700, 2000.
84. M. G. Xu, J. L. Archambault, L. Reekie, and J. P. Dakin, Discrimination between strain and temperature effects using dual-wavelength fiber grating sensors, *Electronics Lett.*, **30**, 1085–1087, 1994.
85. H. J. Patrick, G. M. Williams, A. D. Kersey, J. R. Pedrazzani, and A. M. Vengsarkar, Hybrid fiber Bragg grating/long period fiber grating sensor for strain/temperature discrimination, *IEEE Photonics Technol. Lett.*, **8**, 9, pp. 1223–1225, 1996.
86. W.-C. Du, X.-M. Tao, and H.-Y. Tam, Fiber Bragg grating cavity sensor for simultaneous measurement of strain and temperature, *IEEE Photonics Technol. Lett.*, **11**, 1, pp. 105–107, 1999.
87. W. Du, X. Tao, and H.-Y. Tam, Temperature independent strain measurement with a fiber grating tapered cavity sensor, *IEEE Photonics Technol. Lett.*, **11**, 5, pp. 596–598, 1999.

88. J. Jung, H. Nam, J. H. Lee, N. Park, and B. Lee, Simultaneous measurement of strain and temperature using a single fiber Bragg grating and an erbium-doped fiber amplifier, *Appl. Opt.*, **38**, 5, pp. 2749–2751, 1999.
89. J. Jung, N. Park, and B. Lee, Simultaneous measurement of strain and temperature by use of single fiber Bragg grating written in an erbium:ytter-bium-doped fiber, *Appl. Opt.*, **39**, 7, pp. 1118–1120, 2000.
90. R. Posey, Jr. and S. T. Vohra, An eight-channel fiber-optic Bragg grating and stimulated Brillouin sensor system for simultaneous temperature and strain measurement, *IEEE Photonics Technol. Lett.*, **11**, 12, pp. 1641–1643, 1999.
91. L. Zhang, Y. Liu, J. A. R. Williams, and I. Bennion, Enhanced FBG strain sensing multiplexing capacity using combination of intensity and wavelength dual-coding technique, *IEEE Photonics Technol. Lett.*, **11**, 12, pp. 1638–1640, 1999.
92. Y. Jeong, S. Back, and B. Lee, A self-referencing fiber-optic sensor for macro-bending detection immune to temperature and strain perturbations, *The 14th Intl. Conf. Optical Fiber Sensors (OFS-14)*, Venice, Italy, Oct. 2000.
93. K. P. Koo, M. LeBlanc, T. E. Tsai, and S. T. Vohra, Fiber-chirped grating Fabry–Perot sensor with multiple-wavelength-addressable free-spectral ranges, *IEEE Photonics Technol. Lett.*, **10**, pp. 1006–1008, 1998.
94. H. Kogelnik and C. W. Shank, Coupled wave theory of distributed feedback lasers, *J. Appl. Phys.*, **43**, pp. 2327–2335, 1972.
95. H. Kogelnik, Filter response of nonuniform almost-periodic structures, *Bell System Tech. J.*, **55**, pp. 109–126, 1976.
96. M. Yamada and K. Sakuda, Analysis of almost-periodic distributed feedback slab waveguides via a fundamental matrix approach, *Appl. Opt.*, **26**, pp. 3474–3478, 1987.
97. Y. Jeong and B. Lee, Long-period fiber grating analysis using generalized N × N coupled-mode theory by section-wise discretization, *J. Optical Soc. Korea*, **3**, pp. 55–63, 1999.
98. Y. Jeong and B. Lee, Nonlinear property analysis of long-period fiber gratings using discretized coupled-mode theory, *IEEE J. Quantum Electronics*, **35**, pp. 1284–1292, 1999.
99. Y. Jeong, *Linear and nonlinear characteristics of optical waves in waveguide gratings*, Ph.D. thesis, Seoul National University, Korea, Aug. 1999.
100. P. St. J. Russell, Bloch wave analysis of dispersion and pulse propagation in pure distributed feedback structures, *J. Modern Opt.*, **38**, pp. 1599–1619, 1991.
101. K. A. Winick, Effective-index method and coupled-mode theory for almost periodic waveguide gratings: A comparison, *Appl. Opt.*, **31**, pp. 757–764, 1992.
102. L. A. Weller-Brophy and D. G. Hall, Analysis of waveguide gratings: Application of Rouard's method, *J. Optical Soc. Amer. A*, **2**, pp. 863–871, 1985.
103. J. L. Frolik and A. E. Yagle, An asymmetric discrete-time approach for the design and analysis of periodic waveguide gratings, *J. Lightwave Technol.*, **13**, pp. 175–185, 1995.
104. C. Tsao, *Optical Fibre Waveguide Analysis*, Oxford University Press, New York, 1992.
105. D. Marcuse, *Theory of Dielectric Optical Waveguides*, 2nd ed., Academic Press, New York, 1991.
106. T. Erdogan, Fiber grating spectra, *J. Lightwave Technol.*, **15**, pp. 1277–1294, 1997.
107. T. Erdogan, Cladding-mode resonances in short- and long-period fiber grating filters, *J. Optical Soc. Amer. A*, **14**, pp. 1760–1773, 1997.
108. R. Guenther, *Modern Optics*, John Wiley & Sons, New York, 1990.

109. Y. Zhao and J. C. Palais, Fiber Bragg grating coherence spectrum modeling, simulation and characteristics, *J. Lightwave Technol.*, **15**, pp. 154–161, 1997.
110. A. M. Vengsarkar, P. J. Lemaire, J. B. Judkins, V. Bhatia, T. Erdogan, and J. E. Sipe, Long-period fiber gratings as band-rejection filters, *J. Lightwave Technol.*, **14**, pp. 58–65, 1996.
111. H. Kogelnik, Theory of optical waveguides, in *Guide-Wave Optoelectronics*, T. Tamir, ed., Springer–Verlag, New York, 1990.
112. K. O. Hill, F. Bilodeau, B. Malo, T. Kitagawa, S. Thériault, D. C. Johnson, and J. Albert, Chirped in-fiber Bragg gratings for compensation of optical-fiber dispersion, *Optics Lett.*, **19**, pp. 1314–1316, 1994.
113. R. Feced, M. N. Zervas, and M. A. Muriel, An efficient inverse scattering algorithm for the design of nonuniform fiber Bragg gratings, *IEEE J. Quantum Electronics*, **35**, pp. 1105–1115, 1999.
114. B. H. Lee and J. Nishii, Dependence of fringe spacing on the grating separation in a long-period fiber grating pair, *Appl. Opt.*, **38**, pp. 3450–3459, 1999.
115. Y. Jeong, S. Back, and B. Lee, All optical signal gating in cascaded long-period fiber gratings, *IEEE Photonics Technol. Lett.*, **12**, pp. 1216–1218, 2000.
116. F. Bakhti and P. Sansonetti, Realization of low back-reflection, wideband fiber bandpass filters using phase-shifted long-period gratings, *Optical Fiber Commun. Conf.*, Dallas, TX, Paper FB4, Feb. 1997.
117. Y. J. Rao, Recent progress in applications of in-fibre Bragg grating sensors, *Optics Laser Eng.*, **31**, pp. 297–324, 1999.

8

Fiber Optic Gyroscope Sensors

Paul B. Ruffin

CONTENTS

8.1 Introduction

The fiber optic gyroscope (FOG), which celebrated its 30th anniversary in 2006, represents the dominant solution in numerous applications of navigation, guidance, and stabilization, particularly in the 0.1 to 10°/hr range. The FOG offers unique advantages over the ring laser gyroscope (RLG), which include:

1. True solid-state device
2. No dithering required
3. Sensitivity can be increased by adding more fiber wraps
4. Tends itself to miniaturization
5. High reliability
6. Long lifetime
7. Quick to start

A brief summary of how FOG development has progressed over the past 30 years is provided, as is a simple description of the operation of the FOG. The three basic interferometric FOG (IFOG) configurations are described. Typical error sources, with emphasis on the largest error source (the "Shupe" effect) that tends to limit FOG performance when operating in thermally adverse environments, are examined. Parasitic phase shifts caused by time-varying environmental disturbances are discussed in detail. Compensation techniques (winding methods and coil designs), which significantly reduce bias uncertainty and noise in FOGs operating in thermally adverse environments, are presented. A description of a novel, symmetrical, crossover-free fiber-winding method, which greatly enhances gyroscope performance, is also provided. The effect of small radius bending on the polarization performance of single mode optical fibers is discussed. We also mention some of the current and future applications for FOGs followed by some concluding remarks.

8.2 Progression of Fiber Optic Gyroscope Development

The advent of the fiber optic gyroscope (FOG) dates back to the mid-1970s when Vali and Shorthill [1] demonstrated the first fiber optic rotation sensor. This breakthrough followed the pioneering efforts of R. B. Brown from the Navy Laboratory in 1968, who proposed a coil of optical fiber as a rotation sensor. Fringes were demonstrated in an optical fiber ring interferometer in 1975 using low-loss, single mode fiber. During the years to follow, a number of researchers and developers worldwide made the FOG concept become a reality [2–5]. A number of universities and industrial laboratories such as

McDonnell Douglas, Northrop-Grumman (Litton), Honeywell, Northrop, Singer, Lear Siegler, Martin Marietta, and others have investigated the FOG. Gyroscope bias errors of 0.01°/hr were being achieved in the laboratory by the early 1980s. Although the theoretical basis for FOG operation is published worldwide, the details of the design techniques and processes are not published in open literature due to proprietary restrictions.

The development of the FOG has flourished during the past 30 years. It has evolved from a laboratory experiment to the production floors, and thus into practical applications such as in navigation, guidance, and control of aircraft, missiles, automobiles, robots, and spacecraft. It has been in production by several companies for more than a decade. FOGs, which have replaced the RLG in a number of applications requiring 1.0°/hr performance accuracy, represent the prevalent solution in numerous applications of navigation, guidance, and stabilization in the better than 0.1°/hr regime (missiles, attitude heading and reference systems [AHRS], robotics, satellites, etc.). A great deal of effort has been made in the development of navigation-grade gyroscopes for aircraft and space applications with bias drift less than 0.01°/hr and scale factor of less than 10 parts per million (ppm). FOGs are currently used in the navigation system of aircraft such as the Boeing 777.

Tremendous progress has been made during the past decade in developing high-performance light source modules operating in the near-infrared region, integrated optics chips (IOC), environmentally robust packaging schemes for the sensor assembly, and automated precision techniques for the interconnection of miniature optical components and FOG subassemblies to meet the stringent performance requirements of space and submarine navigation systems applications [6–8]. The light source for IFOGs has progressed from the superluminescent diode (SLD) and the edge-emitting light-emitting diode (ELED) to the erbium-doped fiber superfluorescent fiber source (SFS) for increased power and wavelength stability over temperature [9,10].

Most recent efforts have been directed toward addressing issues associated with the development of low-cost, miniature, medium- to high-performance FOGs that can operate over adverse environments such as military environments. Attention is being given to performance and reliability aspects of ultraminiature single-mode fiber coil designs and the packaging of miniature FOG components. The packaging of miniature FOGs poses critical issues including where to place the fiber leads connecting each optical component in the circuit and how to package the depolarizers, in a depolarized IFOG, without degrading performance. A thermally symmetric, crossover-free winding technique has been developed to virtually eliminate polarization nonreciprocity error due to fiber crossovers and minimize time-varying thermal gradients.

Remaining issues include refinement of gyroscope designs (optical and electronics) to minimize environmental error sources for improved performance in adverse environments. Efforts are under way to reduce drift to 0.001°/hr for space applications and 0.1°/hr for ultraminiature FOGs.

8.3 Basic Operation of the Fiber Optic Gyroscope

8.3.1 Sagnac Effect

The physical phenomenon that explains the operation of an optical gyroscope is known as the Sagnac effect, named after French physicist Georges Sagnac [11–13]. A simple representation of the Sagnac effect is shown in Figure 8.1. The Sagnac interferometer acts as a nonreciprocal device where the light waves propagating in one direction of a loop under rotation are not equivalent to the light waves propagating in the opposite direction. Consider two light waves propagating in opposite directions around the ring interferometer shown in Figure 8.1, which is rotating at a rate Ω in the clockwise direction. The two light waves travel a different fiber length and take a different time to traverse the total length of fiber. The effective path lengths are $L_{cw} = 2\pi R + R\Omega t_{cw} = ct_{cw}$ and $L_{ccw} = 2\pi R - R\Omega t_{ccw} = ct_{ccw}$, where the transit times t_{cw} and t_{ccw} are

$$t_{cw} = 2\pi R/(c - R\Omega) \tag{8.1}$$

and

$$t_{ccw} = 2\pi R/(c + R\Omega) \tag{8.2}$$

for light waves traveling in the clockwise and counterclockwise directions, respectively. The free-space speed of light is denoted by c, and R is the radius of the ring. The transit time difference δt between the counterpropagation waves, in the case of N loops that enclose an area, $A = \pi R^2$ (or Sagnac area—SA), can be expressed as

$$\begin{aligned} \delta t &= t_{cw} - t_{ccw} \\ &= (4NA/c^2)^* \Omega \end{aligned} \tag{8.3}$$

The assumption is made that $c^2 >> R^2\Omega^2$. The resultant optical path length difference, δL, is $c^*\delta t$, or

$$\delta L = (4NA/c)^* \Omega \tag{8.4}$$

In the case of the analog or interferometric fiber optic gyroscope (IFOG), the Sagnac phase shift caused by a rotation can be expressed in terms of δL as

$$\begin{aligned} \delta\phi_s &= (2\pi/\lambda)^* \delta L \\ &= (8\pi NA/\lambda c)^* \Omega \end{aligned} \tag{8.5}$$

where λ is the wavelength of the free-space optical energy. Multiple wraps of fiber can be wound to significantly increase δL, thus improving the sensitivity. However, the optical attenuation tends to limit the length of fiber to several kilometers.

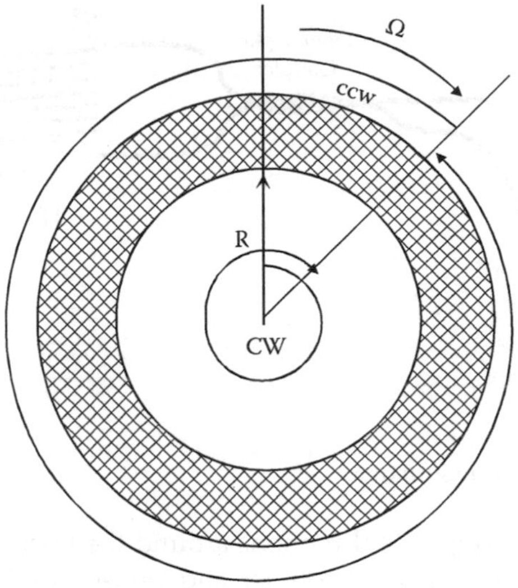

FIGURE 8.1
Illustration of Sagnac effect.

In the case of the digital or resonant fiber optic gyroscope (RFOG), the energy in the counterpropagating beams is coupled into the fiber loop at two different frequencies in the presence of a rotation. The relative frequency difference δf between the counterpropagating waves can be written in terms of δL as

$$\delta f / f = \delta L / L \tag{8.6}$$

where $f = c/\lambda$ and $L = 2\pi RN$ is the total distance traversed. The fundamental RFOG equation that relates δf to Ω is given by

$$\delta f = (2R/\lambda) * \Omega \tag{8.7}$$

Sagnac interferometers are highly sensitive measurement devices. The magnitude of the Sagnac effect can be realized from the following example. Consider a fiber optic ring interferometer of area $A = 100\ \text{cm}^2$, which experiences maximum Earth rotation rate $\Omega = 15°/\text{hr}$. Equation (8.4) suggests that $\delta L \approx 10^{-15}$ cm, which is very small. This requires phase detection with a resolution in the region of 10^{-8} radians for a 1-μm light source.

8.3.2 Basic Configuration

The IFOG is the object of discussion from this point. A basic IFOG configuration is shown in Figure 8.2. Light from a broadband source, such as a superluminescent diode (SLD), is projected into a 3-dB fiber optic coupler that splits the light into two waves. After traversing the coupler, the two light waves

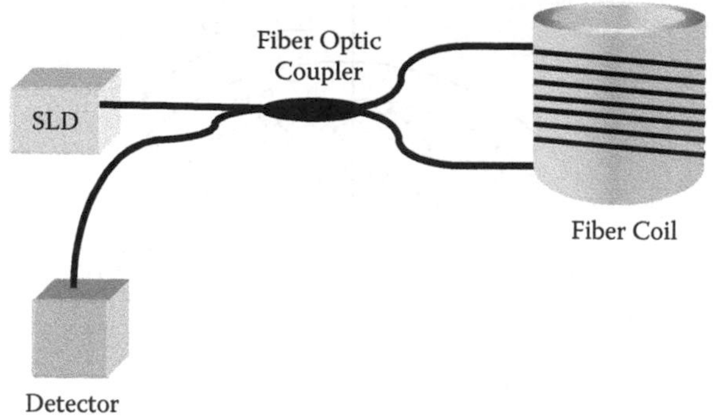

FIGURE 8.2
Schematic of basic IFOG configuration.

propagate equally in opposite directions around the fiber optic coil. The light waves interfere upon return to coupler and project a fringe pattern onto a photodetector.

The basic IFOG configuration is not reciprocal in the absence of rotation; that is, both light waves do not traverse identical paths upon recombination at the fiber optic coupler. The clockwise (CW) light wave experiences two reflections through fiber optic coupler, whereas the counterclockwise (CCW) light wave experiences two transmissions through coupler, which introduces a degree of nonreciprocity. Shaw and his research team at Stanford solved this problem of unintentional nonreciprocity in the basic IFOG configuration in 1981 [14,15].

8.3.3 Minimum Configuration

The Stanford group proposed a minimum configuration IFOG, as shown in Figure 8.3. A second coupler has been added after polarizer to ensure identical paths by equalizing intensity in CW and CCW waves. The polarizer, which also functions as a single-mode filter, ensures that the two light waves return to the first coupler in a single polarization, thus forming a fringe pat-

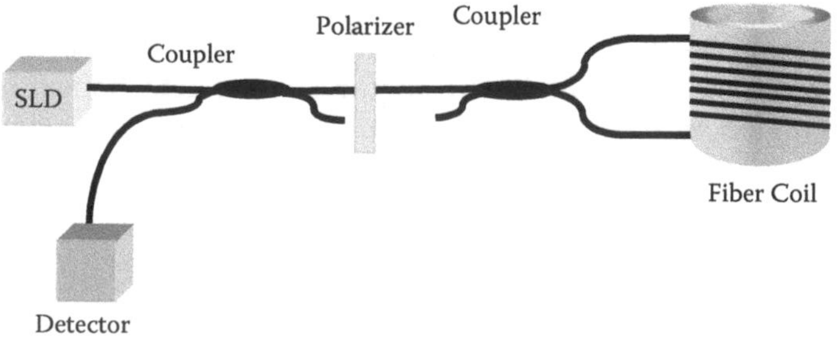

FIGURE 8.3
Schematic of minimum configuration IFOG.

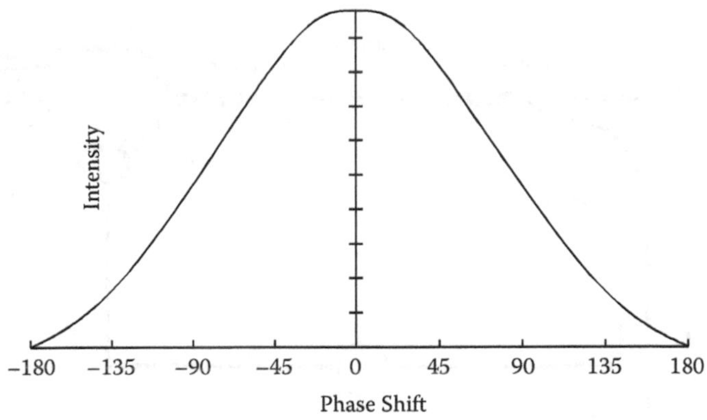

FIGURE 8.4
Optical intensity versus phase differences between interfering waves.

tern on the photodetector. The fringe shift ΔZ is written in terms of the Sagnac phase shift $\delta\phi_s$ expressed in Eq. (8.5) as $\Delta Z = \delta\phi_s/2\pi$ or

$$\Delta Z = (2RL/\lambda c) * \Omega = SF_o\Omega \tag{8.8}$$

where $SF_o = 2RL/\lambda c = \Delta S_o/\Delta S_i$ (ratio of change in output signal and change in input signal) is the open-loop optical scale factor. The characteristics of the scale factor depend on the stability of the light source.

In accordance with any two-wave interferometer, the intensity on the photodetector, which represents a mixture of the two light waves, varies as cosine of Sagnac phase $\delta\phi_s$ with its maximum value at zero as shown in Figure 8.4. This intensity is expressed as

$$I = I_o\left[1 + \cos(\delta\phi_s)\right] \tag{8.9}$$

where I_o is the mean value of the intensity. The detected intensity is used to calculate the rotation rate. In the case of no rotation, $\delta\phi_s = 0$, the light waves will combine in phase, which results in maximum intensity.

8.3.4 Open-Loop Biasing Scheme

In the presence of a rotation, the light waves travel different path lengths and mix slightly out of phase. The intensity is reduced due to the degree of destructive interference. The cosine function, which is symmetrical about zero, has its minimum slope there. For small rotation rates, it is impossible to determine the direction of rotation (CW or CCW) from the symmetrical aspect of Figure 8.4, where the slope is near zero. Furthermore, the gyroscope operating in this mode has minimum sensitivity near zero. Incorporating a dithering phase modulator with drive modulation capability asymmetrically in the loop (near one end of the coil) provides a means to introduce a

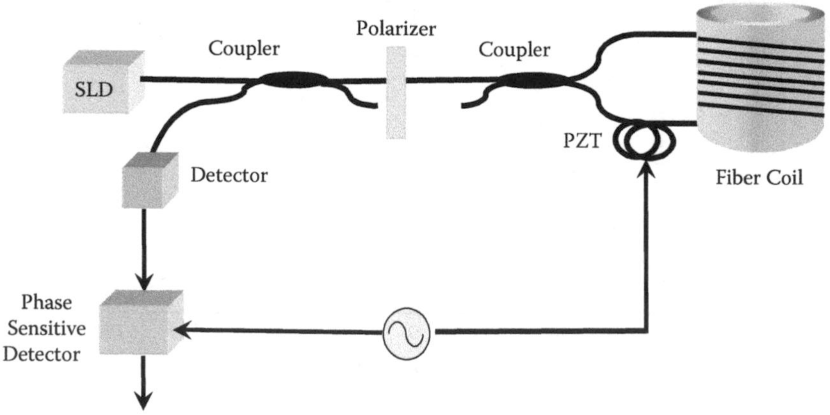

FIGURE 8.5
Schematic of IFOG with biasing phase modulator.

nonreciprocal phase shift to bias the gyroscope to its maximum sensitivity point [16,17]. This corrective measure solves both the low-sensitivity problem and the issue of ambiguous direction of rotation at low rotation rates.

The piezoelectric transducer (PZT) phase modulator shown in Figure 8.5 stretches the fiber at controlled rates via the application of a voltage. The phase modulator produces an optical shift as a result of the applied voltage. After exiting the coupler, the CCW wave or pulse encounters the phase modulator, which is fully stretched. After traversing the coil loop in a time ($\tau = n_f L/c(n_f)$ is the refractive index of the fiber), the wave returns to the coupler. The phase modulator is timed such that when the light wave propagating in the CW direction reaches the phase modulator the stretch has been relieved. Therefore, the light wave propagating in the CCW direction travels a longer distance. The two light waves experience a net nonreciprocal phase shift due to this path length difference. A schematic of the output of a gyroscope that is biased to operate at its maximum sensitivity point is shown in Figure 8.6 for zero input. When a phase modulator is used, the expression for the intensity on the photodetector is

$$I = I_o\left[1 + \cos(\delta\phi_s + \delta\phi_m)\right] \tag{8.10}$$

or with no rotation, $\delta\phi_s = 0$,

$$I = I_o\left[1 + \cos(\delta\phi_m)\right] \tag{8.11}$$

where the alternating bias phase shift is [17]

$$\delta\phi_m = 2\phi_m \sin(\omega_m \tau/2)\cos\left[\omega_m(t - \tau/2)\right] \tag{8.12}$$

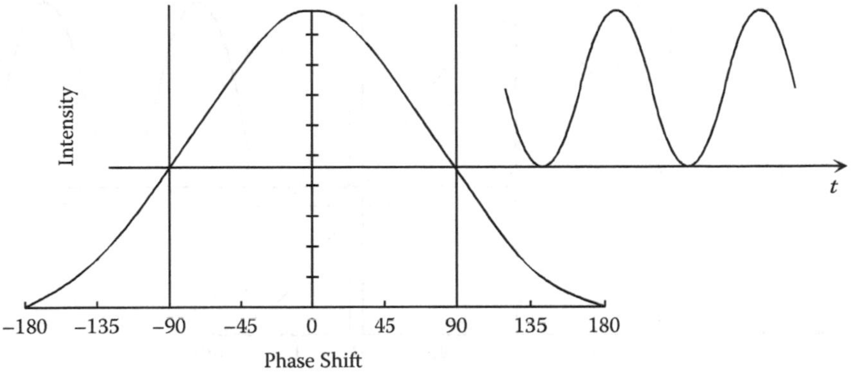

FIGURE 8.6
Optical intensity versus phase differences between interfering waves phase bias as a function of time with no rotation.

The modulation frequency is $\omega_m = 2\pi f_m$ and the amplitude of phase modulation is ϕ_m. The detector is synchronized to this alternating bias signal to permit detection of any variations in the output due to rotations. The phase shift is maximized at [18]

$$f_m = \frac{1}{2\tau} = \frac{c}{2nL} \tag{8.13}$$

An expression for the sensitivity is devised from Eq. (8.11) as

$$|dI/d\phi_m| = I_o \sin(\delta\phi_m) \tag{8.14}$$

It is obvious from Eq. (8.14) that a phase bias of 90° maximizes the sensitivity; that is, the gyroscope operates off its proper frequency when $\delta\phi_m = \pi/2$. For a 1-km length of fiber, the modulation frequency is 100 kHz. Shorter lengths of fiber require increased modulation frequencies for maximum sensitivity.

The intensity increases for a CCW rotation and decreases for a CW rotation as illustrated in Figure 8.7. An expression for the intensity at the photodetector of an IFOG biased to operate at maximum sensitivity, $\delta\phi_m = \pi/2$, in the presence of a rotation is derived from Eq. (8.10) as

$$I = I_o\left[1 + \sin(\delta\phi_s)\right] \approx I_o\left[1 - \delta\phi_s\right] \tag{8.15}$$

Open-loop IFOGs, which have moderate scale factor stability, have good bias stability and are basically immune to random noise; the open-loop IFOG, however, has limited dynamic range. The FOG becomes nonlinear for large rotation rates when operating in the open-loop configuration. It is obvious from Eq. (8.15) that the rotation rates are limited between $\delta\phi_s = \pm\pi/2$. The maximum rate in an open-loop IFOG operating at a wavelength of 1 μm with $A = 100\ \text{cm}^2$ and $N = 1000$ is approximately 200°/s.

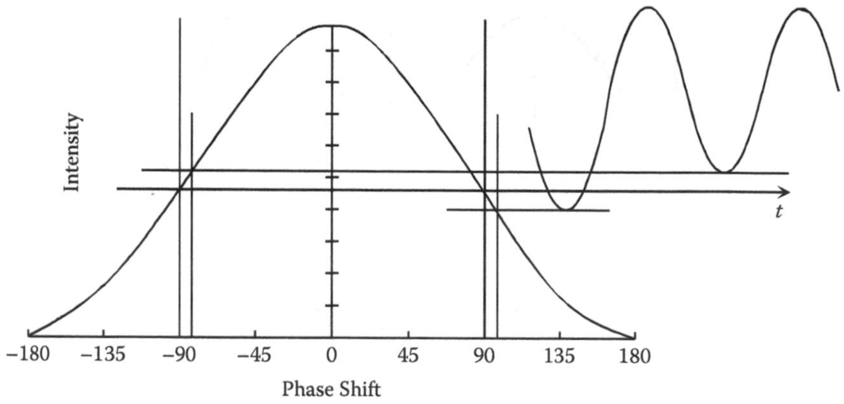

FIGURE 8.7
Optical intensity versus phase differences between interfering waves dynamic phase bias as a function of time with rotation.

8.3.5 Closed-Loop Signal Processing Schemes

Operating the gyroscope in a closed-loop configuration improves its performance. A number of IFOG researchers and developers have devised closed-loop signal processing schemes to null the output signal and continue to operate in the linear range [16,19–21]. These schemes include harmonic feedback, gated phase-modulation feedback, sinusoidal phase-modulation feedback, and incorporation of integrated-optic Bragg cell and Serredyne frequency-shifting elements. A loop-closure transducer, such as a phase-balancing element with square-wave modulation, as shown in Figure 8.8,

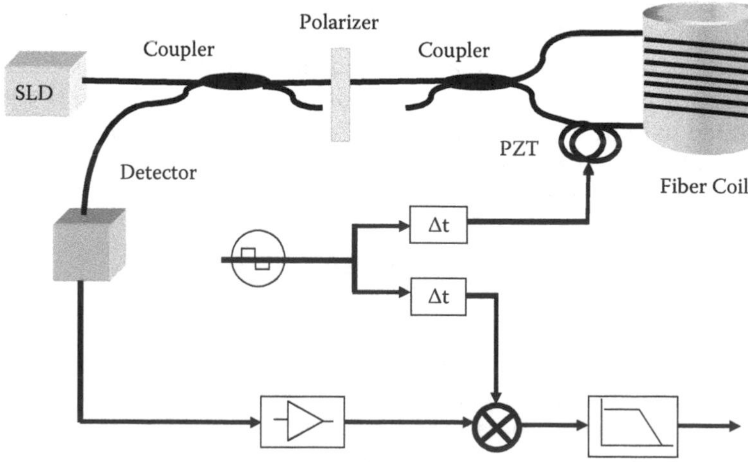

FIGURE 8.8
Closed-loop scheme with phase balancing element and square wave modulation.

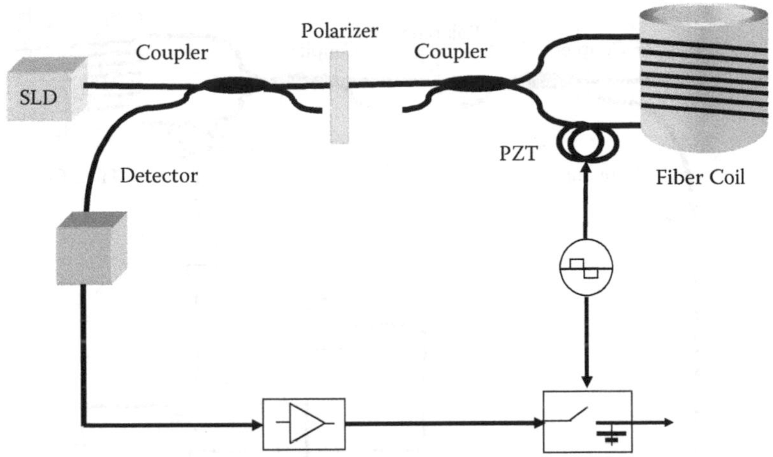

FIGURE 8.9
Closed-loop scheme with sample and hold demodulation and half wave detection.

can be used to keep the gyroscope operating at its null point (operational point at zero rotation) in the presence of a rotation. The open-loop signal is maintained at the null point by introducing an equal and opposite phase shift to compensate for a phase shift due to rotation. A measure of the feedback signal (to the phase modulator) required to null the output signal is proportional to the rotation. A digital output is possible for the case when a frequency shifter transducer is used.

The closed-loop configuration, which is immune to light source intensity fluctuations and gain instability in the detection electronics, increases the dynamic range of the gyroscope. The rotation rate is limited by the response of the loop-closure transducer in maintaining null operation. Figure 8.9 shows a configuration using a "sample and hold" demodulator with a square wave modulator and half-wave detection. Full-wave differential detection can be added as shown in Figure 8.10.

8.3.6 Fundamental Limit

The fundamental noise limit in the IFOG is set by photon shot noise—the random distribution of photons incident on the photodetector, which leads to random fluctuations in the detector output current, $i_s = \sqrt{(ei_o B_w)}$ [10,15]. The parameters e, i_o, and $B_w = 1/T$ are the electron charge, average current at detector, and the measurement bandwidth of the detection system, respectively. T is the sample or integration time. The resolution of the IFOG is the ratio of the minimum detectable rate of change in rotation angle caused by the uncertainty in detector output current and the angular rate.

An expression for the average number of photons arriving at the photodetector is written in terms of the average incident intensity, I_o, as the ratio of the incident energy and the energy of one photon as

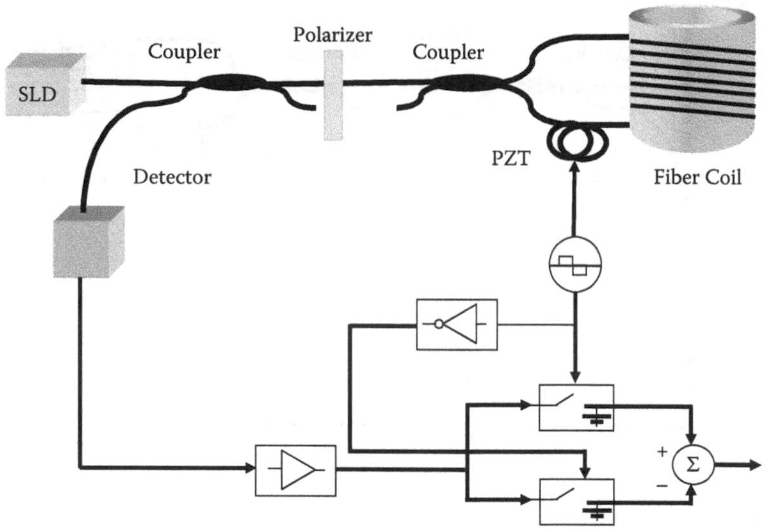

FIGURE 8.10
Closed-loop scheme with sample and hold demodulation and full wave differential detection.

$$N_p = I_o\lambda / B_w hc \tag{8.16}$$

An expression for phase noise, from Poisson statistics, is written as

$$\delta\phi_{noise} = 1/\sqrt{N_p} = \sqrt{(B_w hc / I_o\lambda} \tag{8.17}$$

An estimate of the theoretical limit for detection sensitivity of the IFOG is determined to be approximately 4×10^{-8} radians, for $\lambda = 1$ µm, $I_o = 100$ µW, and $B_w = 1$ Hz. In the case of a 10-cm-radius fiber optic ring interferometer containing 1 km of fiber, this corresponds to a shot noise equivalent rotation rate of $\Omega_{\min} \approx 0.01°/\text{hr}$ and angular random walk (ARW) of 10^{-4} deg/$\sqrt{\text{hr}}$.

8.3.7 Performance Accuracy and Parasitic Effects

Parasitic effects that cause drift—a time-varying zero offset in the output of IFOGs—limit the performance accuracy and must be minimized for high-performance operations. These noise sources include optical backscatter, light source instabilities, polarization noise, electro-optic effects, magneto-optic effects, thermal and stress gradients, and electronic noise. Noise sources directly related to gyroscope design and fiber parameters are discussed in this section. Phase-type bias error due to environmental perturbations will be discussed in detail later.

Rayleigh backscattering is caused by microscopic variations in the refractive index along the length of fiber due to inherent imperfections, splices, etc.

This can be in the form of coherent and incoherent backscatter and reflection noise. The incoherent component of the noise, which affects the light intensity at the photodetector, is a source of shot noise. This noise component, which does not cause a rotation error, contributes less than 1 dB to the shot noise. On the other hand, coherent noise adds coherently to the counterpropagating light waves, which is a coherent summation of light from individual scattering centers along the fiber. The combination with the counterpropagating light waves alters the phase between them. A low-coherence source such as an SLD is used to suppress this effect. Interference due to Rayleigh backscatter is averaged to zero when such a broadband source with low temporal coherence is used.

Light sources currently used in telecommunication operate at 1.55 μm. These sources, which typically produce greater than 10 mW of power with linewidths greater than 40 nm, have been adopted for FOGs. In the case of low coherence, broadband light source wavelengths beat against each other to cause intensity noise. Such light source instabilities lead to fluctuations (intensity and frequency) in the output signal. The intensity noise in semiconductor laser sources increases the shot noise by 1 to 2 dB.

8.4 IFOG Configurations

8.4.1 All-PM Fiber IFOG

A schematic of an all-fiber gyroscope configuration is shown in Figure 8.11. The design contains polarization-maintaining (PM) fiber, fiber optic components, light source, photodetector, and PM fiber-wound sensor coil. The PM fiber minimizes polarization noise and improves performance. A phase modulator is formed by wrapping a portion of one leg of the fiber around a piezoelectric cylinder and applying a voltage to stretch and relax the fiber at will. The phase of the light waves is modulated by the elasto-optic effect. The PZT is modulation-frequency limited. This configuration is used for low- to moderate-performance applications.

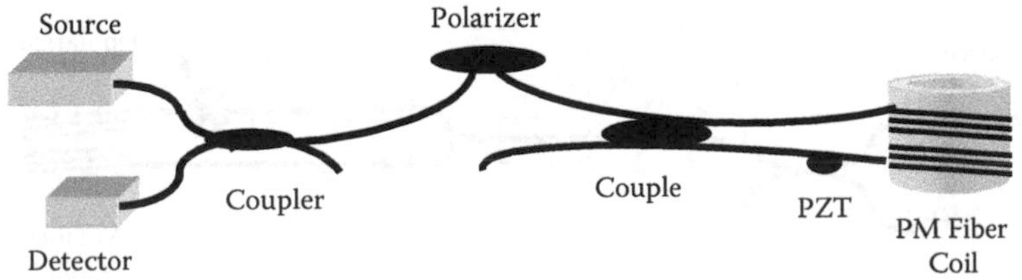

FIGURE 8.11
Schematic of all-fiber IFOG configuration.

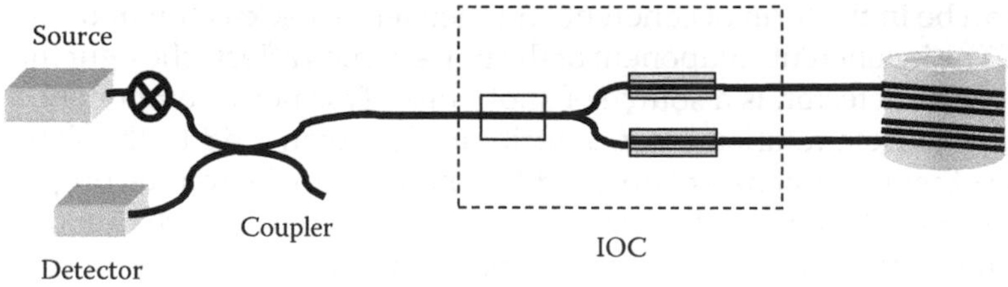

FIGURE 8.12
Schematic of integrated optics gyroscope configuration.

8.4.2 PM Fiber/Integrated Optics IFOG

A fiber/integrated optics gyroscope configuration is shown in Figure 8.12. An integrated optics chip (IOC) is incorporated into the FOG circuit to replace fiber optic components. Typically, the optical waveguide is fabricated using an annealed proton exchange in lithium niobate (LiNbO3). The fiber polarizer is eliminated due to the great polarization properties of the LiNbO3. The phase modulator in the integrated-optic component is constructed via positioning two electrodes adjacent to the waveguide. Inducing an electric field inside the waveguide modulates the phase of the light waves. The modulation frequency capability is substantially increased in the IOC. A y-branch beamsplitter is also integrated onto the chip. This design approach requires that fiber pigtails be attached to the chip. Attaching the fiber "pigtails" to the IOC could be time consuming. The IOC reduces the number of FOG components and provides superior modulation and polarization characteristics. The integrated optics gyroscope is necessary to achieve a mass producible FOG for small, low-cost, high-reliability devices. The cost of the PM fiber, typically 1000 m in a navigation grade gyroscope, is the dominant cost driver for the FOG.

8.4.3 Depolarized IFOG

A depolarized/single mode fiber optic gyroscope (D-FOG) configuration is shown in Figure 8.13. The light from the broadband source is linearly polarized via passing through the IOC before encountering the depolarizer. The depolarizer is required in the D-FOG to prevent signal fading and reduce

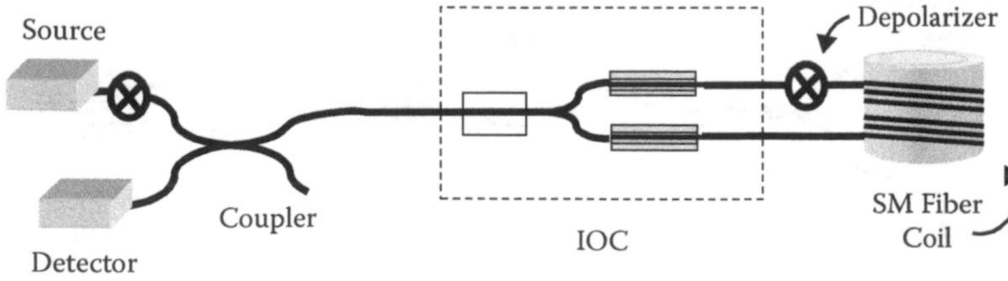

FIGURE 8.13
Schematic of depolarized/single mode fiber optic gyroscope configuration.

polarization error and magnetic field sensitivity via distributing the light evenly among all polarization states in the fiber coil loop [22]. The depolarized IFOG design, using low-cost single mode fiber, is typically more complex than the more expensive PM fiber approach. Significantly more fiber is required to achieve the same shot noise-limited performance as with the all-PM fiber configuration [22]. This is due to the loss resulting from adding depolarizers in which the fibers are spliced together at a 45° angle. The D-FOG approach is attractive for high-performance applications requiring moderate cost [23].

8.5 Phase-Type Bias Error

In practice, environmental effects can limit the rotation measurement accuracy of high-performance IFOGs. Environmental noise sources such as the Faraday effect and the "Shupe" effect introduce an optical intensity-induced nonreciprocal phenomenon. The magnitude of the environmental effects depends strongly on the way the FOG is packaged and employed during use. In the case of no rotation, in the presence of environmental perturbations, the intensity can be expressed as

$$I = I_o\left[1 + \cos(\delta\phi_m + \delta\phi_{env})\right] \tag{8.18}$$

where $\delta\phi_m$ is the phase modulation and $\delta\phi_{env} = \delta\phi_{pl} + \delta\phi_{fd} + \delta\phi_{kr} + \delta\phi_{sh}$ is the parasitic phase shift due to environmental perturbations: polarization nonreciprocity, the Faraday effect, the Kerr effect, and the Shupe effect. For best gyroscope performance, $\delta\phi_{env}$ must be minimized.

8.5.1 Polarization Nonreciprocity

Polarization mode coupling in the fiber coil can produce nonreciprocity effects. Coupling occurs between the two nearly degenerate modes in conventional single mode fiber, which typically supports two polarization modes with slightly different guided-wave propagation constants, in the presence of bends and lateral pressures on the fiber. The coupling points vary rapidly in a random manner in the presence of thermal and mechanical perturbations in the fiber coil. The use of PM fiber usually solves this problem.

8.5.2 Faraday Effect

The Faraday effect or magneto-optic effect in the glass fiber results when the plane of polarization of the light is rotated in the presence of a magnetic field. High-quality PM fiber and magnetic shielding have been used to minimize the Faraday effect [24–26].

8.5.3 Kerr Effect

The Kerr effect, which is a third-order, optical nonlinearity in glass fiber, results when the intensities of the counterpropagating light waves become unequal. Broadband light sources have been used to minimize the Kerr effect [27–29].

8.5.4 Shupe Effect

The Shupe effect, which is the largest error source in IFOGs, results when time-varying, spatial gradients arising from acoustic noise, vibration transients, and thermal and stress perturbations across the fiber optic sensing coil occur [30]. These external perturbations cause a time-varying refractive index change along the fiber length. Winding the sensing coil in such a manner that sections of the fiber coil equidistant from the two ends are placed as close together as possible minimizes the Shupe effect due to thermal transients. Cementing the fiber wraps in the coil pack typically solves the problem due to time-varying stress gradients. The remainder of this section is dedicated to addressing the issue of thermally induced nonreciprocity due to localized time-dependent thermal expansion in various portions of the fiber.

The nonreciprocity described by Shupe [30] causes a false rotation signal when the counterpropagating light waves encounter sections of fiber undergoing time-varying thermal gradients and no longer travel identical paths through the coil. The two counterpropagating waves in the IFOG sensing coil do not experience the same difference ΔT between the coil temperature and ambient temperature at a given instant in time, t. The IFOG research team at the Army Aviation and Missile Command conducted an in-depth investigation of the issue of thermal transients [31–34]. Nonreciprocity arises when the clockwise and counterclockwise light waves cross an infinitesimal region at a specified distance l from one end of the fiber (see diamond in Figure 8.14) at different times. Earlier, we discussed the phase shift $\delta\phi_s$ produced by a rotation rate Ω in an IFOG (see Eq. 8.5). An additive phase shift $\delta\phi_{sh}$ is produced in the gyroscope due to counterpropagating waves experiencing a different ΔT at a given instance in time. The phase shift due to time-dependent temperature fluctuations is given by [30–33]

$$\delta\phi_{sh} = B\int_0^L dl\left[\Delta T(l,t_1) - \Delta T(l,t_2)\right] \tag{8.19}$$

where $B = k(dn_f/dT + \alpha n_f)$, n_f is the refractive index, $k = 2\pi/\lambda$ is the wave propagation constant, and α is the thermal expansion coefficient of the fiber.

The time difference between the CW and CCW waves passing through the same point l in the fiber (see Figure 8.14) is

$$\delta t = t_1 - t_2 = \beta(2l - L)/\omega \tag{8.20}$$

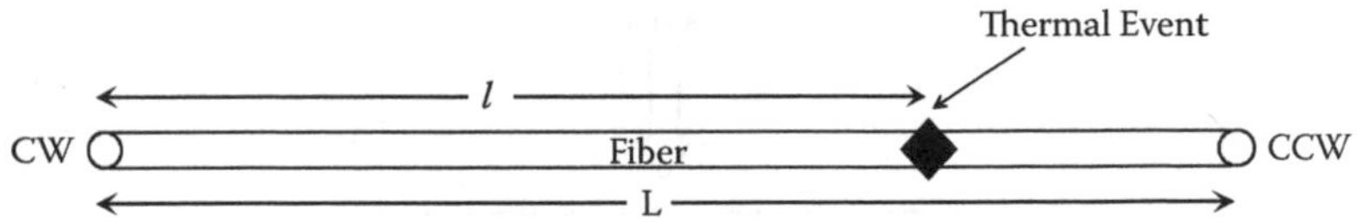

FIGURE 8.14
Illustration of time-dependent thermal gradient.

where $\beta = n_f k$ is the guided wave constant and $\omega = ck$ is the frequency of the light waves.

Equation (8.19) is simplified by a change of variable to yield an expression for the phase difference in the CW and CCW waves due to an elevated temperature difference $\Delta T(l)$ at the location l. Substituting Eq. 8.20 into Eq. 8.19 yields

$$\delta\phi_{sh} = B_T \int_0^{L/2} dl(2l - L)\left[\Delta\dot{T}(l) - \Delta\dot{T}(L-l)\right] \tag{8.21}$$

where $B_T = (c/n_f)B$. Shupe suggested that if $\Delta\dot{T}(l)$, the time derivation of ΔT, is made symmetric with respect to the point $L/2$, then $\delta\phi_{sh} = 0$.

Also, similar errors are produced by time-dependent mechanical stress [34]. Gyroscope error can be caused by time-dependent sinusoidal vibrations from a small segment of the fiber. This problem is practically solved via cementing the fiber pack and fiber leads.

Thermal perturbations can be caused by temperature events from both the axial and radial directions of the sensing coil as shown in Figure 8.15. The phase shift, $\delta\phi_{sh}$, can be separated into two parts to account for a thermal disturbance event in both the axial and radial directions. The temperature gradient difference, $[\Delta\dot{T}(l) - \Delta\dot{T}(L-l)]$, in Eq. 8.21 can be approximated as a sum of two components in axial (A) and radial (R) directions.

$$\Delta\dot{T}(l) - \Delta\dot{T}(L-l) = (\partial\dot{T}/\partial a)_A \Delta a + (\partial\dot{T}/\partial r)_R \Delta r \tag{8.22}$$

Equation 8.22 is used to estimate the temperature gradient difference between any two points of equal distance from the center point of a fiber wound in a coil pack. The resulting phase shift due to thermal events in the axial and radial directions of the fiber sensor coil is discussed in the next section.

8.6 Anti-Shupe Winding Methods

Typically, IFOG sensor coils contain 100 to 1000 m of fiber wound in multiple layers on 0.5- to 4-in.-diameter spools. The critical fiber-bending radius basically determines the minimum diameter of the sensor coil. The composition

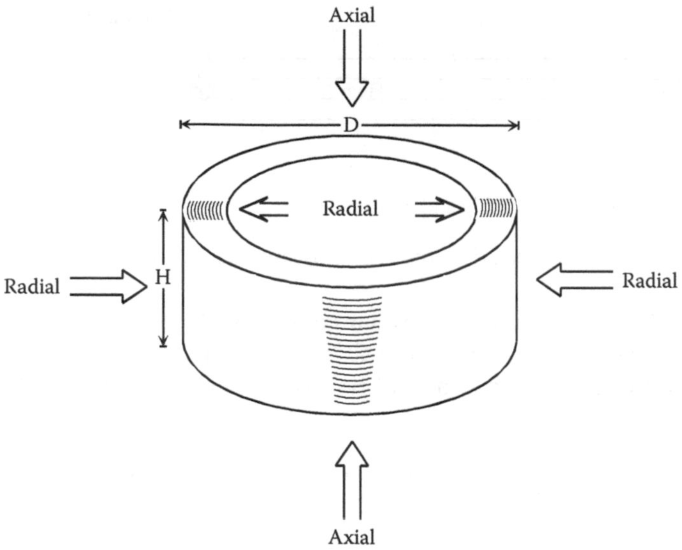

FIGURE 8.15
Schematic of fiber coil showing thermal flow from axial and radial directions.

and properties of the sensor coil constituent components have a significant impact on gyroscope performance. The conventional end-to-end or "simple" wind configuration is ideal for low-performance applications.

The conventional end-to-end ("simple"), precision-wind configuration is accomplished via winding the first layer of fiber, starting from one end of the coil next to a flange onto a cylindrical spool in the form of a gentle helix, which propagates with a pitch of one fiber diameter in the direction of the winding. The winding proceeds in a helix pattern as shown in Figure 8.16. Each fiber wrap is in contact with the preceding wrap. The direction of the pitch is changed during the transition from the first layer to the second layer when the direction of the winding is reversed. The fiber wraps of the second

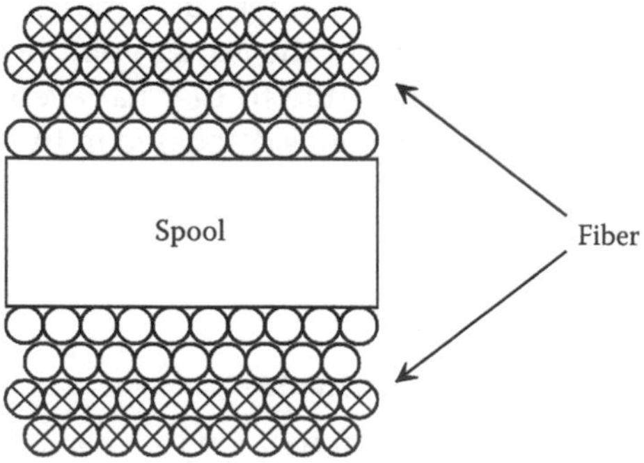

FIGURE 8.16
Schematic of simple end-to-end winding configuration.

layer are nested in the grooves formed by the underlying wraps, except in the crossover regions. In cross-section, the winding forms a hexagonal close pack pattern. During the transition the fiber is stepped back over a specified number of underlying fibers, then settles into a new groove. There is a crossover event for each fiber wrap. Since the pitch of the winding is reversed, the fiber in the second layer is forced from one groove and obliquely crosses over two wraps of fiber in the first layer before settling into the proper groove for precision winding. This process continues for all layers wound above. The open and the shaded circles in Figure 8.16 represent the inside half ($L/2$) and the outside half ($L/2$) of the fiber, respectively. The Shupe effect is most significant for the end-to-end winding configuration.

The linear transients effects discussed in the previous section are reduced by symmetrical winding and sufficient holding of the wound coil. Thermal gradients can arise from dynamic environments that gyroscopes are employed in as well as the way the sensing coils are packaged in the transport vehicle. Frigo [35] proposed several alternative winding techniques to reduce the time-varying thermal gradients present across the fiber optic sensing coil. The fiber must be wound such that points equidistance from the center point of the fiber loop lie in close mutual proximity in the pack. The research group at Litton Guidance and Control Systems devised an expression to estimate the Shupe effect caused by a given displacement of fiber sections from the center point [36,37]. The results suggested that the displacement must be less than 1 mm in order to obtain navigation grade performance, $\Omega = 0.01°/\text{hr}$.

A configuration of the dipole winding, proposed by Frigo [35], is shown in Figure 8.17. The fiber is wound according to the "simple" wind except the dipole winding is accomplished via winding in an alternating back-and-forth pattern from two different supply reels. The winding is initiated from the center of the coil by winding a continuous optical fiber that is supplied from two fiber-feed reels, each containing one-half the length, L, of fiber required for the sensor coil. The open and shaded circles in Figure 8.17 represent fiber from two different supply spools. The dipole winding technique reduces the

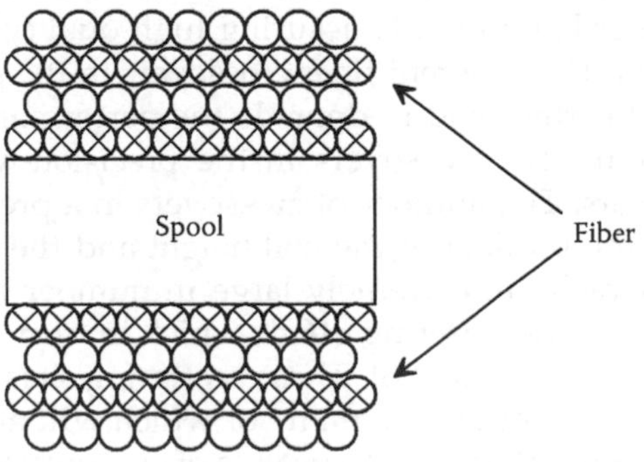

FIGURE 8.17
Schematic of dipole winding configuration.

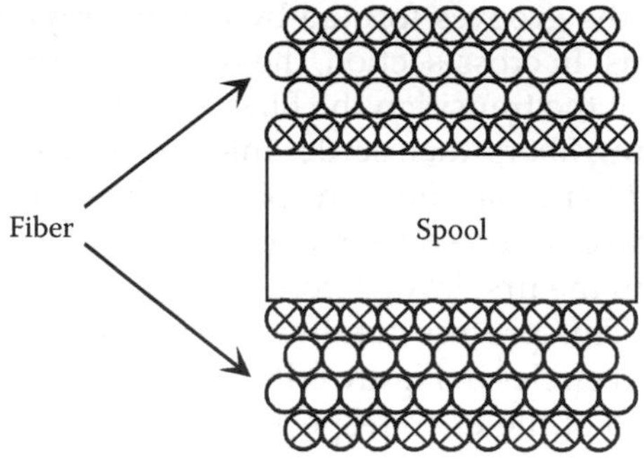

FIGURE 8.18
Schematic of quadrupolar winding configuration.

Shupe effect by a factor of approximately $1/2N_L$, where N_L is the number of fiber layers [31].

The center-to-end or quadrupolar winding, shown in Figure 8.18, has been adopted for moderate- to high-performance applications. This winding configuration features fiber layers wound in pairs, following the winding of the innermost layer. Each fiber wrap lies adjacent to the preceding fiber wrap. Each layer pair begins with a fiber segment on the opposite side of the innermost layer from the preceding layer pair to position fiber segments an equal distance from the center. After completing the winding of the first layer of any pair, the winding direction reverses at the spool flange. The fiber wraps of the second layer of any pair lie in the grooves formed by the fiber wraps beneath, except at the fiber crossovers due to the reverse winding fiber wraps moving one fiber diameter in the opposite direction. Alternating layer pairs are wound until winding is complete. This winding configuration virtually eliminates the radial component of a time-varying thermal gradient but only slightly reduces the effect of the axial component. The quadrupolar winding technique reduces the Shupe effect by a factor of approximately $1/N_L^2$ [31].

An alternate winding approach could be necessary for ultraminiature FOGs. The FOG optical components are inherently small, except perhaps the sensor coil. Reducing the size of the coil requires additional layers of fiber to maintain the fiber length while assuring high quality and performance stability of the FOG. The inherent fiber crossovers in the precision winding methods described earlier tend to degrade the performance of the D-FOG. Cross-coupling from fiber crossovers in the precision-wound coil causes "in-phase" bias errors. The number of crossovers in a precision-wound coil is proportional to the product of the coil height and the number of layers. Fiber crossovers, which are extremely large in number for ultraminiature coils, have a polarization-scattering effect, which in turn creates a polarization nonreciprocal (PNR) error [38]. These scattering sites provide for many (thousands) of birefringent delays, many of which will allow for recorrelation of PNR terms. Since there are thousands of crossovers in ultraminiature coils, these crossovers have a large effect on bias performance.

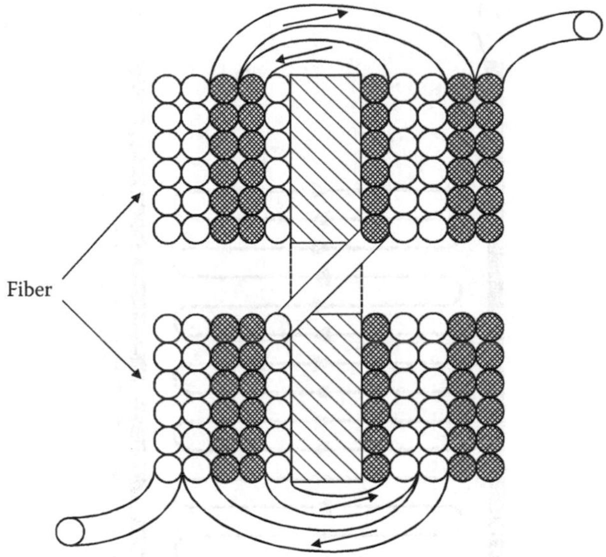

FIGURE 8.19
Schematic of thermally symmetric, crossover-free winding configuration.

A novel winding technique has been devised to eliminate fiber crossovers, as well as minimize the effect of time-varying thermal gradients. The significance of this development is twofold. First, the crossover-free (CF) method eliminates PNR bias errors for the depolarized FOG configuration, which will lessen the need for more complex, expensive electronics. Second, the winding technology virtually eliminates the radial component of thermal gradients in the coil pack. A schematic of the thermally symmetric, crossover-free winding technique is shown in Figure 8.19. The winding is accomplished according to the technique described in U.S. patent #5,781,301, "Thermally Symmetric, Crossover-Free Fiber Optic Sensor Coils and Method for Winding Them" [39]. The winding is initiated from the center of the coil as described in the dipole winding, except for the case of a flat-wind configuration. The first layer consists of compact spiral loops of fiber that is wound from the inside of the inner coil diameter to the outside of the outer coil diameter. The fiber is secured to a thin hollow disk via an adhesive that is applied to the disk prior to winding. A second spiral layer, which is a mirror image of the first spiral layer, is wound onto the opposite side of the thin disk. Subsequent spiral layers are wound such that the fiber loops positioned at equal distances from the center of the fiber optic coil are mirror images of the fiber loops on the opposite side of the disk. Ideally, this winding configuration completely eliminates the radial components of a thermal gradient because of the completely symmetric design configuration and virtually eliminates thermal gradients in the axial direction. This advanced winding technique improves volumetric efficiency and enhances the reliability of the

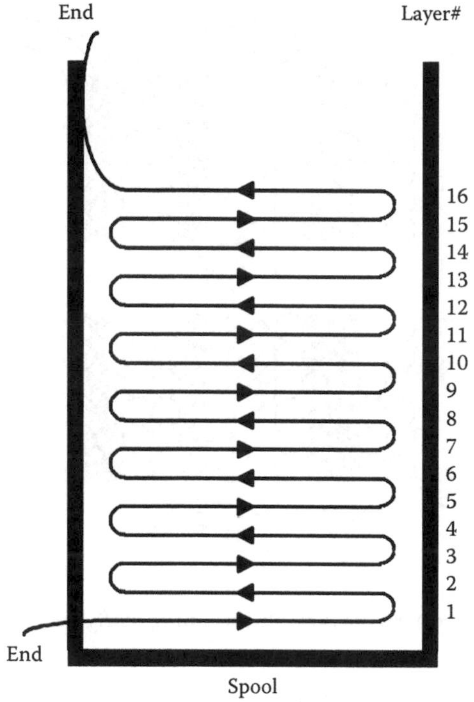

FIGURE 8.20
Simple end-to-end winding pattern for 16 fiber layers.

fiber sensor coil by eliminating the fiber crossovers, which are prime sites for scattering and fiber static fatigue failure.

Two-dimensional views, illustrating the difference between fiber layer positions in the "simple" or S-wind, the quadrupolar or Q-wind, and the thermally symmetric or TS-wind fiber coil packs, are provided in Figure 8.20, Figure 8.21, and Figure 8.22, respectively. Figure 8.20 shows an S-wind configuration for 16 layers of fiber. The first layer is wound directly onto a spool with subsequent layers numbered consecutively until the last layer, #16, is wound. Figure 8.21 shows a Q-wind configuration for 16 layers of fiber. The winding process is initiated with the fiber being wound from two supply spools beginning from the center of the fiber. The first layer is wound directly onto a spool using fiber from one of two fiber supply reels. The second and third layers, forming the first layer pair, are wound from the second supply reel. The winding proceeds via winding layer pairs in an alternating pattern from the two supply reels. The layer numbers in Figure 8.21 are a direct correspondence of the fiber layers in Figure 8.20. Figure 8.22 shows a TS-wind configuration for 10 spiral layers containing 16 spiral fiber loops. The solid and dashed lines represent fiber from two different supply reels, which is wound in alternating-layer pairs back and forth across a thin hollow disk, following the winding of the first layers on both sides of the disk.

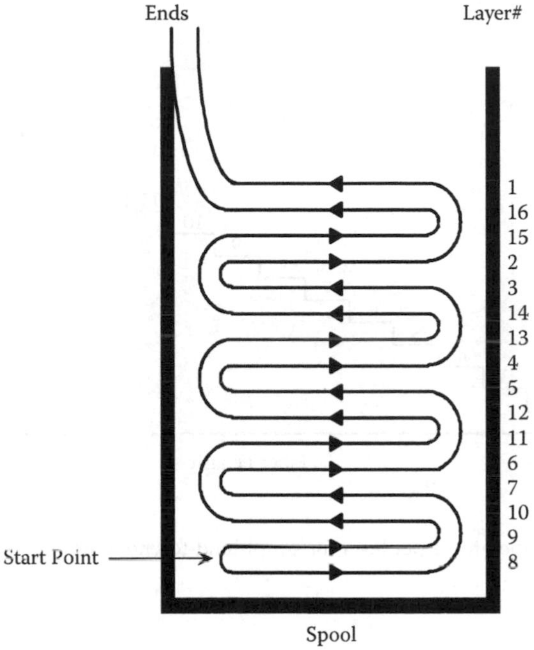

FIGURE 8.21
Quadrupolar-winding pattern for 16 fiber layers.

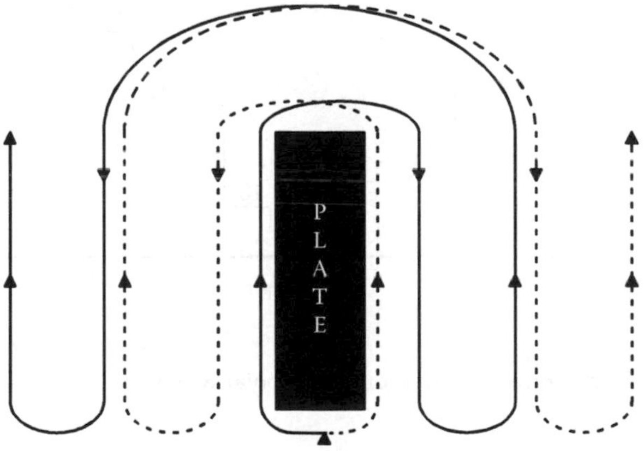

FIGURE 8.22
Thermally symmetric, crossover-free winding pattern for 10 spiral layers.

The corresponding graphs for temperature distribution throughout the fiber packs are illustrated in Figure 8.23, Figure 8.24, and Figure 8.25 for the S-wind, Q-wind, and TS-wind configurations, respectively. It is obvious from Figure 8.23 that the temperature difference between fiber layers equidistance

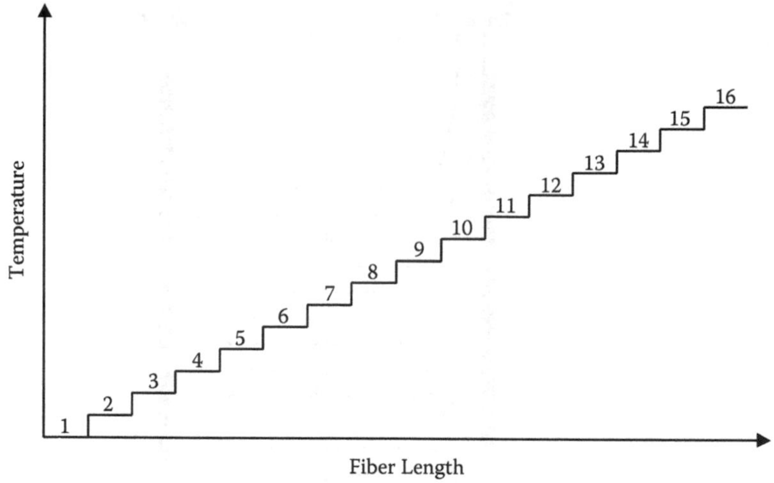

FIGURE 8.23
Plot of resulting temperatures per layer for simple end-to-end winding.

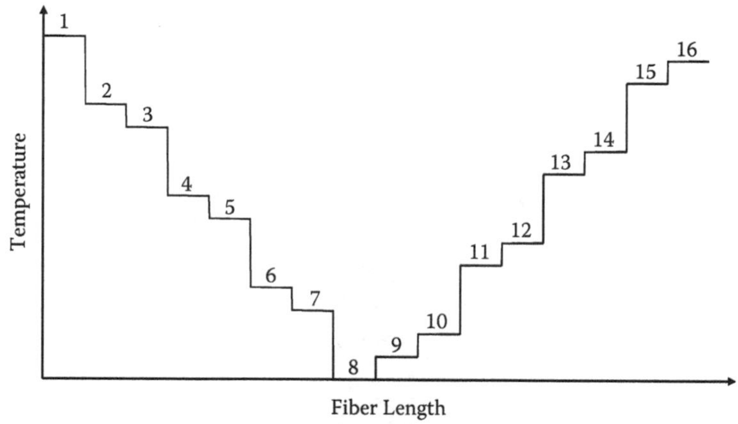

FIGURE 8.24
Plot of resulting temperatures per layer for quadrupolar winding.

from the center of the fiber is substantial for the S-wind configuration, with the minimum temperature difference being between layers #8 and #9 and increasing to a maximum between layers #1 and #16. The temperature difference, which is identical for all fiber layers equidistance from the center of the fiber, has been significantly reduced for the Q-wind as seen in Figure 8.24. The temperature difference between fiber layers equidistance from the center of the fiber is zero for the TS-wind as seen in Figure 8.25.

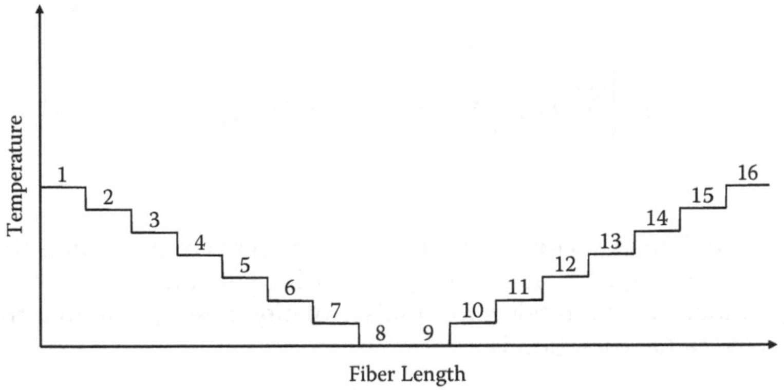

FIGURE 8.25
Plot of resulting temperatures per layer for thermally symmetric, crossover-free winding.

The phase shift $\delta\phi_A$ due to a time-varying temperature change in the axial direction is calculated in Eq. (8.21) by comparing $\Delta\dot{T}$ at two points of equal distance from the middle of the fiber. Incorporating a crude approximation,

$$(\partial\dot{T}/\partial a)_A \Delta a \approx (\Delta\dot{T})_A d/H, \tag{8.23}$$

allows us to write expressions for the phase change due to a thermal event in the axial direction as

$$\delta\phi_A = B_T(L^2/4)(d/H)*(\Delta\dot{T}) \tag{8.24}$$

where H is the height of the sensor coil, d is the diameter of the fiber, and $(\Delta\dot{T})_A$ is the temperature gradient between two specified points in the fiber sensor coil. Heat conduction models are used to determine $\Delta\dot{T}$ for a specified coil geometry [32]. Equation (8.5) is used to write an expression for a false rotation rate Ω_A due to a thermal event in the axial direction:

$$\Omega_A = \left[(\lambda c)/4\pi RL\right]*\left[\delta\phi_A\right] \tag{8.25}$$

Fiber diameters used in IFOGs range from 160 to 250 μm. Considering a fiber diameter of 200 μm (average), a temperature difference ΔT of 1°C/min in the axial direction of a 2.5-cm-diameter coil with a 1-in. height containing 200 m of fiber causes a rotation error of approximately 4°/hr for the S-wind configuration.

Expressions for the phase shift $\delta\phi_R$ due to a time-varying temperature change in the radial direction of the fiber coil is obtained from Eq. (8.21) as

$$\delta\phi_R(S-wind) = B_T\left\{\sum_{i=0}^{N_L-l}\left[\Delta\dot{T}(i+1)-\Delta\dot{T}(N_L-i)\right]\left[(2i+1)l_o^2 - N_L l_o^2\right]\right\} \tag{8.26}$$

and

$$\delta\phi_R(Q-wind)=B_T\left\{\left[\sum_{i=0}^{N_L-l}(-1)^i\left[\Delta\dot{T}(2i+1)-\Delta\dot{T}(2i+2)\right]\left[N_L-(2i+1)l_o^2-N_Ll_o^2\right]\right]\right\} \tag{8.27}$$

for the S-wind and Q-wind configurations, respectively. $N_L = mn$ is the total number of fiber layers, l_o is the length of each layer, and $m = 4$ for Q-wind configuration. The expression for a false rotation rate Ω_R due to a thermal event in the radial direction is

$$\Omega_R=\left[(\lambda c)/4\pi RL\right]*\left[\delta\phi_R\right] \tag{8.28}$$

In the case of the S-wind for $N_L = 16$, a false rotation rate caused by a temperature difference $\Delta\dot{T}$ of 1°C/min in the radial direction for a 200-m fiber coil is approximately 500°/hr. The error is reduced to 5°/hr for the Q-wind. The total rotation error due to Shupe effect is

$$\Omega_{sh}=\left[(\lambda c)/4\pi RL\right]*\left[\delta\phi_R+\delta\phi_A\right] \tag{8.29}$$

8.7 Geometrical and Polarization Effects in Crossover-Free IFOG Coils

8.7.1 Sagnac Area

The Sagnac area (SA; defined in Eq. 8.3), which is the total area circumscribed by the fiber loops, is substantially larger in the CF winding configuration as compared to the conventional winding configuration. In the conventional winding configuration, each fiber loop circumscribes an ideally equivalent area, whereas in the CF configuration subsequent loops of varying radii generate quadratically varying SA. A top view of a single spiral layer of fiber, which could represent the first layer in Figure 8.19, is shown in Figure 8.26 [39]. The area enclosed by the nth revolution of an Archimedean spiral is given by

$$A_n = (1/3)*(12n^2 - 12n + 4)\,\pi^3a^2 \tag{8.30}$$

and the area of a partial segment of the spiral is

$$A_{partial} = (\theta/2)*[r_1r_2 + 1/3(r_2 - r_1)^2], \text{ respectively [40,41].} \tag{8.31}$$

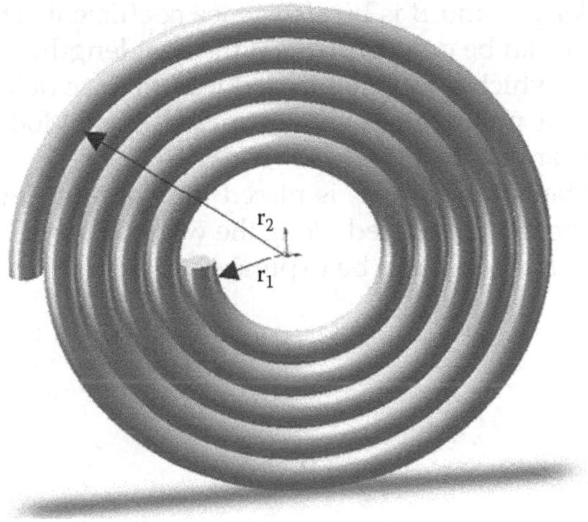

FIGURE 8.26
Top view of an Archimedean spiral layer of fiber in CF configuration.

n is the number of spiral rings, a is the intercycle spacing, r_1 and r_2 are the starting and ending radii for the partial cycle, and θ is the partial rotation angle. The total area encompassed by a single coil layer can be determined from Eqs. (8.30) and (8.31). A 150-m CF coil and 350-m CF coil encompass more than 5 and 10 times the area of the conventional coil, respectively.

8.7.2 Bending-Induced Birefringence

Single mode (SM) optical fibers are highly affected by external forces such as transverse pressure, temperature, and bending. Bending is the dominant factor in CF wind. Stress produced by bending an optical fiber induces birefringence. Birefringence increases as fiber radius is reduced. This bending stress modifies the refractive index of the fiber and in turn changes the propagation coefficient in any one of the two polarizations. To examine the effect of small radius bending on the birefringence of SM fiber, specific characteristics must be evaluated. Modal birefringence or beat length and mode-coupling parameter or extinction ratio or cross-talk (CT) are considered here for further discussion.

The inherent index of refraction difference, Δn, in SM fiber creates a difference in the propagation constants of two orthogonally polarized modes between the x and y components of the guided wave propagation constant, β_f. This difference is called the polarization birefringence or modal birefringence, $\Delta\beta$ [42]. $\Delta\beta$ is related to beat length, L_P, by [43]

$$L_p = \frac{\lambda}{B} = \frac{2\pi}{\Delta\beta} \tag{8.32}$$

where λ is wavelength and B is birefringence coefficient. The degree of the fiber birefringence can be determined via the beat length, which is the periodic distance over which birefringence induces a phase delay of 2π between the two orthogonal polarization modes in the fiber. Modal birefringence increases for shorter beat lengths.

When an SM fiber of radius, R_f, is placed in a circular loop of radius, R_s, modal birefringence is introduced along the wound fiber in the direction of winding. This birefringence can be expressed as [44]

$$\Delta\beta = \beta_x - \beta_y = -0.13\left(R_f/R_S\right)^2 \beta_f \tag{8.33}$$

where

$$\beta_f = \frac{\beta_x + \beta_y}{2}.$$

Using Eq. (8.32), B can be expressed as

$$B = \frac{\Delta\beta}{k} = \frac{\lambda}{L_P} \tag{8.34}$$

B varies between a maximum and a minimum value in the CF coil configuration, since the loop radius, R_s, continually varies in a multilayered Archimedean spiral. Using Eq. (8.34), predicted values for small curvature winding of a 125-μm diameter fiber are determined and illustrated in Table 8.1.

TABLE 8.1

Predicted B for Various SM Coil Radii

R	B for 125 μm Fiber
1 in.	1.14×10^{-6}
0.5 in.	1.82×10^{-5}
0.4 in.	2.84×10^{-5}
0.3 in.	5.06×10^{-5}
0.2 in.	1.14×10^{-4}
0.1 in.	4.55×10^{-4}

Typical values for the birefringence coefficient, B, of high-birefringent (PM) fibers are 10^{-4}. Therefore, based on predictions in Table 8.1, the SM fiber coil must have a radius less than 0.2 in. in conjunction with a fiber comprising 125 μm cladding before bending-induced birefringence becomes a design factor in D-IFOGs. Thus, the notion to replace the PM fiber in an all-PM fiber IFOG design with a low-cost SM fiber wound in a microcoil configuration becomes inconsequential.

8.7.3 Polarization Coupling

Ideally, if perfectly linearly polarized light is launched into a birefringent PM fiber with its plane exactly aligned to either of the fiber's principal axes, the light remains perfectly polarized as it propagates. However, in practice some of the light couples from the preferred or excited axis to the extinguished or coupled axis. The coupling between polarization states is due to the intrinsic properties of the optical fiber and to stresses outside the fiber such as mechanical and/or temperature stress.

When the axis of one polarizer is 90° to that of the second polarizer (analyzer), light transmittance through the pair is at a minimum. This is defined as extinction. The extinction is measured by launching linearly polarized light into one polarization mode and measuring the output power in that and the orthogonal mode. The extinction ratio or polarization cross-talk is calculated from Eq. (8.35) by [45]

$$\eta = 10\log\frac{P_y}{P_x} \tag{8.35}$$

where P_x and P_y are the powers of the excited and extinguished or coupled mode in the optical fiber, respectively. The extinction ratio or CT is usually expressed in decibels.

The polarization-maintaining ability or property of an SM fiber is characterized by the h-parameter, which is a measure of polarization cross-talk per unit length of fiber. An expression showing the relationship between the extinction ratio and the h-parameter is provided in Eq. (8.36) [43]:

$$\eta = 10\log\frac{P_y}{P_x} = 10\log\left(\tanh\left(hl\right)\right), \tag{8.36}$$

where l is the length of the fiber. The h-parameter has units of inverse meters and can also be derived from

$$h = 1/l\tanh^{-1}\left(P_y \middle/ P_x \right) \tag{8.37}$$

8.8 Applications of Fiber Optic Gyroscopes

The fiber optic gyroscope has reached a level of practical use in navigation, guidance, control, and stabilization of aircraft, missiles, automobiles, and spacecraft as shown in Figure 8.27. The FOG performance and design requirements (such as resolution, stable scale factor, maximum rate, frequency response, size, interface electronics, environment, etc.) have been scaled to fulfill a broad range of applications such as route surveying and mapping, well logging, self-guided service robots and factory floor robots, autonomous guided ground and air vehicles, tactical missiles, guided munitions, cannon-launched vehicles, smart bombs, and seeker, missile airframe and satellite antenna stabilization.

The open-loop FOG is best suited for low-cost applications such as gyrocompassing, attitude stabilization, and pitch and roll indicators, which require low- to moderate-performance accuracy. A number of corporations have developed low-performance FOGs for use in automotive applications.

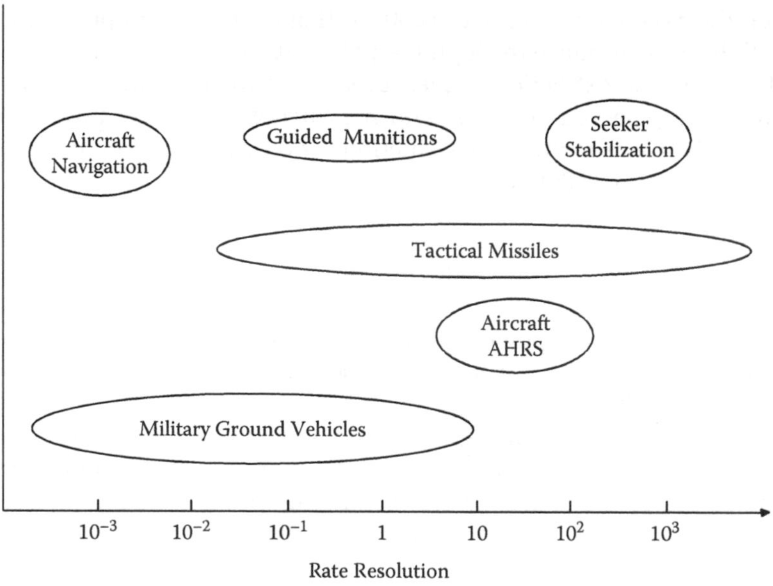

FIGURE 8.27
Application areas and drift rate requirements for fiber optic gyroscopes.

The FOG has been used for automobile navigation as a heading sensor in Nissan's luxury sedans since 1995 [46]. FOGs, which are ideal for GPS-aided systems, have been combined with GPS as part of vehicle location and navigation to provide accurate dead reckoning navigation during periods of GPS signal interruptions in parking garages, tunnels, tall buildings in urban areas, trees in rural areas, and physical obstructions.

The FOG has challenged the RLG in the medium-performance (0.1 to 10.0°/hr) regime. Commercial aircraft AHRS are one of the earliest medium performance applications of FOGs. An open-loop, PM fiber FOG, developed by Honeywell, was one of the first commercially available FOG units to go into production in the early 1990s [47]. This IFOG system is used as a backup unit in small commuter aircraft and the Boeing 777 aircraft to provide attitude and heading information [47,48]. The LN-200, comprising IFOGs, has been in production at Northrop-Grumman (Litton Guidance and Control Systems) since 1989 [49]. The LN-200 can be configured as an AHRS or an inertial measurement unit (IMU) for use in applications such as the Comanche helicopter, the AMRAAM missile guidance system, and the guided MLRS.

Closed-loop FOGs are required for high-performance applications. More than a decade ago, Northrop-Grumman (Litton Guidance and Control Systems) developed an integrated global positioning system/inertial navigation system (GPS/INS) using navigation-grade IFOGs as part of the GPS guided package (GGP) program under a DARPA contract [50]. Honeywell Technology Center has also developed navigation grade D-FOGs for use in tactical guidance and aircraft navigation [47].

The high reliability and environmental ruggedness features of the FOG make it attractive for use in space applications. Precision-grade (<0.001°/hr) FOGs have been developed by Honeywell Technology Center in support of deep-space missions and land and submarine navigation systems [47].

8.9 Conclusions

We have presented the basic operation of the FOG and described design criteria for best performance when operating under environmentally hostile conditions. Design considerations for optimal FOG performance include

1. An open-loop phase biasing scheme to maximize sensitivity that permits operation at low rotation rates and provides a direct indication of the direction of rotation
2. A low-coherence, broadband optical source to suppress bias uncertainty and noise due to optical backscatter
3. A closed-loop signal processing technique to improve performance and increase dynamic range
4. A high-quality PM fiber for the all-fiber and integrated optics design and depolarizers for a single mode fiber design to minimize polarization noise
5. An advanced Shupe winding technique to minimize any time-varying thermal gradients

The anti-Shupe compensation technique discussed in this chapter virtually eliminates radial thermal gradients. It also minimizes the axial thermal gradients, which are not specifically addressed in the typical quadrupolar-winding configuration. This winding technique offers a revolutionary advance in FOG technology.

The FOG has become a viable technology for navigation, guidance, and stabilization applications. The open-loop FOG, which covers medium performance applications, is used in automobile navigation, small commuter aircraft, the Boeing 777 aircraft, helicopters, and missile guidance systems. The closed-loop FOG, which covers high-performance applications, is slated for tactical guidance, aircraft navigation, land and submarine navigation, and deep-space missions.

References

1. V. Vali and R. W. Shorthill, Fiber ring interferometer, *Appl. Opt.*, **15**, p. 1099, 1976.
2. S. Ezekiel and H. J. Arditty, *Fiber-Optic Rotation Sensors*, Springer–Verlag, Berlin, 1982.
3. W. K. Burns, R. P. Moeller, C. A. Villaruel, and M. Abebe, Fiber-optic gyroscope with polarization-holding fiber, *Opt. Lett.*, **8**, pp. 540–542, 1983.
4. H. C. Lefevre, R. A. Bergh, and H. J. Shaw, All-fiber gyroscope with inertial-navigation short-term sensitivity, *Opt. Lett.*, **7**, pp. 454–456, 1982.
5. K. Bohm, P. Marten, K. Petermann, E. Weidel, and R. Ulrich, Low-drift using a superluminescent diode, *Elect. Lett.*, **17**, pp. 352–353, 1981.
6. G. A. Alphonse, J. C. Connolly, N. A. Dinkel, S. L. Palfrey, and D. B. Gilbert, Low spectral modulation high power output from a new AlGaAs superluminescent diode/optical amplifier structure, *Appl. Phys. Lett.*, **55**, 22, p. 2289, November 1989.
7. P. B. Ruffin and R. H. Smith, Fiber winding approaches for environmentally robust IFOG sensor coils, *SPIE Proc.*, **1792**, Components for Fiber Optic Applications, p. 179, September 1992.
8. Integrated Optical Circuit Consultant's SBIR Phase II Final Report, Rugged/low cost pigtailing approaches for LiNbO3 fiber optic gyroscope chips II, September 1995.
9. P. F. Wysocki, M. J. F. Digonnet, B. Y. Kim, and H. J. Shaw, Characterization of erbium-doped superfluorescent fiber sources for interferometer sensor applications, *IEEE J. Lightwave Technol.*, **12**, 3, pp. 550–567, March 1994.
10. A. Lawrence, *Modern Inertial Technology: Navigation, Guidance, and Control*, 2nd ed., Springer–Verlag, New York, Inc., 1992.
11. G. Sagnac, L'ether luminex demonte par l'effect du vent relatif d'ether dans un interferometre en rotation uniforme, *C. R. Acad. Sci.*, 95, pp. 708–710, 1913.
12. G. Sagnac, Sur la preuve de la realite de l'ether lumineux par l'experience de l'interferographe tournant, *C. R. Acad. Sci.*, 95, pp. 1410–1413, 1913.
13. P. Hariharan, Sagnac or Michaelson–Sagnac interferometer? *Appl. Opt.*, **14**, 10, pp. 2319–2321, 1975.
14. R. A. Bergh, H. C. Lefevre, and H. J. Shaw, All-single-mode fiber-optic gyroscope, *Opt. Lett.*, **6**, 4, pp. 198–200, April 1981.
15. R. A. Bergh, H. C. Lefevre, and H. J. Shaw, Overview of fiber-optic gyroscopes, *IEEE J. Lightwave Technol.*, **LT-2**, pp. 91–107, 1984.
16. K. Bohm and K. Petermann, Signal processing schemes for the fiber-optic gyroscope, *SPIE Proc.*, **719**, Fiber Optic Gyros: 10th Anniversary Conference, p. 36, 1986.
17. R. P. Moeller, W. K. Burns, and N. J. Frigo, Open-loop output and scale factor stability in a fiber-optic gyroscope, *J. Lightwave Technol.*, **7**, 2, pp. 262–269, Feb. 1989.
18. H. C. Lefevre, J. P. Bettini, S. Vatoux, and M. Papuchon, Progress in optical fiber gyroscopes using integrated optics, *NATO/AGARD Conf. Proc.*, **383**, 9A-1, 9A-13, 1985.
19. B. Y. Kim, H. C. Lefevre, R. A. Bergh, and H. J. Shaw, Harmonic feed-back approach to fiber gyroscope scale factor stabilization, *Proc. 1st Conf. Optical Fiber Sensors*, pp. 136–137, London, 1983.
20. B. Y. Kim and H. J. Shaw, Gated phase-modulation feedback approach to fiber-optic gyroscopes, *Opt. Lett.*, **9**, p. 263–265, 1984.

21. J. L. Davis and S. Ezekiel, Closed-loop, low-noise fiber-optic rotation sensor, *Opt. Lett.*, **6**, pp. 505–507, 1981.
22. J. Blake, B. Szafraniec, J. Feth, and K. Dimond, Progress in low cost interferometric fiber optic gyroscopes, *SPIE Proc.*, **1694**, pp. 188–192, 1992.
23. B. Szafraniec, J. Feth, R. Bergh, and J. Blake, Performance improvements in depolarized fiber gyros, *SPIE Proc.*, **2510**, Fiber Optic Laser Sensors XIII, pp. 37–48, 1995.
24. K. Hotate and K. Tabe, Drift of an optical fiber gyroscope caused by the Faraday effect: influence of the Earth's magnetic field, *App. Opt.*, **25**, pp. 1086–1092, 1986.
25. K. Bohm, K. Petermann, and E. Weidel, Sensitivity of a fiber optic gyroscope to environmental magnetic fields, *Opt. Lett.*, **6**, pp. 180–182, 1982.
26. J. N. Blake, Magnetic field sensitivity of depolarized fiber optic gyros, *SPIE Proc.*, **1367**, Fiber Optic and Laser Sensors VIII, pp. 81–86, 1990.
27. R. A. Bergh, H. C. Lefevre, and H. J. Shaw, Compensation of the optical Kerr effect in fiber-optic gyroscopes, *Opt. Lett.*, **7**, 6, pp. 282–284, June 1982.
28. K. Takiguchi and K. Hotate, Method to reduce the optical Kerr-effect-induced bias in an optical passive ring-resonator gyro, *IEEE Photonics Tech. Lett.*, **4**, 2, Feb. 1992.
29. R. A. Bergh, B. Culshaw, C. C. Cutler, H. C. Lefevre, and H. J. Shaw, Source statistics and the Kerr effect in fiber-optic gyroscopes, *Opt. Lett.*, **7**, pp. 563–565, 1982.
30. D. M. Shupe, Thermally induced nonreciprocity in the fiber-optic interferometer, *App. Opt.*, **19**, 5, pp. 654–655, 1980.
31. P. B. Ruffin, C. M. Lofts, C. C. Sung, and J. L. Page, Reduction of nonreciprocity in wound fiber optic interferometers, *Opt. Eng.*, **33**, 8, pp. 2675–2679, Aug. 1994.
32. C. M. Lofts, P. B. Ruffin, M. Parker, and C. C. Sung, Investigation of effects of temporal thermal gradients in fiber optic gyroscope sensing coils, *Opt. Eng.*, **34**, 10, pp. 2856–2863, Oct. 1995.
33. J. Sawyer, P. B. Ruffin, and C. C. Sung, Investigation of effects of temporal thermal gradients in fiber optic gyroscope sensing coils, Part 2, *Opt. Eng.*, **36**, 1, pp. 29–34, Jan. 1997.
34. P. B. Ruffin, C. C. Sung, and R. Morgan, Analysis of temperature and stress effects in fiber optic gyroscopes, Fiber Optic Gyros: 15th Anniversary Conference, *SPIE Proc.*, **1585**, pp. 293–299, Sep. 1991.
35. N. J. Frigo, Compensation of linear sources of non-reciprocity in Sagnac interferometers, Fiber Optic Laser Sensors I, *SPIE Proc.*, **412**, pp. 268–271, 1983.
36. A. Cordova, R. Patterson, J. Rahu, L. Lam, and D. Rozelle, Progress in navigation grade FOG performance, *SPIE Proc.*, **2837**, pp. 207–217, 1996.
37. G. Pavlath, The LN200 fiber gyro based tactical grade IMU, *Proc. Guidance, Navigation Control, AIAA*, pp. 898–904, 1993.
38. P. B. Ruffin, J. S. Baeder, and C. C. Sung, Study of ultraminiature sensing coils and the performance of a depolarized interferometric fiber optic gyroscope, *Opt. Eng.*, **40**, 4, pp. 605–611, Apr. 2001.
39. U.S. Patent Number: 5,781,301, 1998.
40. A. Lompado, M. S. Ktanz, J. S. Baeder, L. C. Heaton, and P. B. Ruffin, Geometrical and polarization analyses of crossover-free fiber optic gyroscope sensor coils, *SPIE Proc.*, **63140E**, 2006.
41. B. Hart, Math 406 homework assignment, University of Illinois at Urbana Champaign.

42. T. Okoshi, Single polarization single mode optical fibers, *IEEE J. Quantum Electron.*, **QE-17**, 6, pp. 879–884, June 1981.
43. J. Noda, K. Okamoto, and Y. Sasaki, Polarization-maintaining fibers and their applications, *J. Lightwave Technol.*, **LT 4**, 8, pp. 1071–1089, Aug. 1986.
44. J. S. Baeder and P. B. Ruffin, Microsensor coils for miniature fiber optic gyroscopes, *SPIE Proc.*, **5560**, pp. 219–227, Oct. 2004.
45. M. Tsubokawa, N. Shibata, and S. Seikai, Evaluation of polarization mode coupling coefficient from measurement of polarization mode dispersion, *J. Lightwave Technol.*, **LT-3**, 4, pp. 850–853, Aug. 1985.
46. T. Kamagai et. al., Fiber optic gyroscopes for vehicle navigation systems, Fiber Optic Laser Sensors XI, *SPIE Proc.*, **2070**, Sep. 1993.
47. G. A. Sanders, B. Szafraniec, R.-Y. Liu, C. Laskoskie, and L. Strandjord, Fiber optic gyros for space, marine and aviation applications, *SPIE Proc.*, **2837**, pp. 61–71, 1996.
48. G. A. Sanders and B. Szafraniec, Progress in fiber-optic gyroscope applications II with emphasis on the theory of depolarized gyros, AGARD/NATO Conference Report on Optical Gyros and Their Applications, *AGARDougraph*, **339**, pp. 11/1–42, 1999.
49. G. A. Pavlath, Productionization of fiber gyros at Litton Guidance and Control Systems, *SPIE Proc.*, 1585, pp. 2–5, 1991.
50. R. A. Patterson, E. L. Goldner, D. M. Rozelle, N. J. Dahlen, and T. L. Caylor, IFOG technology for embedded GPS/INS applications, *SPIE Proc.*, **2837**, pp. 113–123, 1996.

9

Optical Fiber Hydrophone Systems

G. D. Peng and P. L. Chu

CONTENTS

Optical fiber hydrophone systems have been under research and development for many years. Various sensor head designs and system architectures have been proposed and intensively studied. Some have been tested in the field and have demonstrated excellent performance. However, a number of challenges remain for this particular technology to move into commercial applications such as geophysical and seismic hydrophone arrays. This chapter presents a review of the basic principles and techniques of fiber hydrophone systems and their potential applications. We focus on key issues such as the interferometer configuration interrogation/demodulation scheme, multiplexing architecture, polarization-fading mitigation, and system integration, including component selection and sensor head design. We also address recent advances in sources and detectors, fiber optical amplifiers, wavelength division multiplexing (WDM) components, optical isolators, and circulators. These significant developments in optical fiber communications will greatly impact the cost effectiveness and system performance of optical fiber hydrophones.

9.1 Introduction

Optical fiber sensing has great application potential in most fields of modern science and technology, such as industrial manufacturing, civil engineering, military technology, environmental protection, geophysical survey, oil exploration, and medical and biological technologies. Many types of optical fiber sensors—rotation, temperature, strain, stress, vibration, acoustic, and pressure—have been under intensive research and development for more than 20 years.

A *fiber hydrophone* is an acoustic sensor using optical fiber as the sensing element. Many of its features make it a very promising alternative to the conventional piezoelectric ceramic sensor. These features include high sensitivity, a large dynamic range, and freedom from electromagnetic interference. Development of the optical fiber hydrophone began in the late 1970s (see, for example, Shajenko et al. [1] and Price [2]). A wide range of sensing schemes based on measuring the optical fiber's (1) intensity (amplitude), (2) frequency, (3) polarization, or (4) phase change of light transmitted has been developed. The most promising of these schemes are those based on *optical interferometry*—measuring the phase change of light induced by a particular measurand of interest [3]. Interferometric optical fiber hydrophones made with very high sensitivity and quite a large dynamic range compared to conventional piezoelectric ceramic devices, in addition to their immunity to electromagnetic interference and downlead insensitivity to disturbances, have been built [4]. A number of educational institutions and government research laboratories have worked on optical fiber hydrophones (e.g., references 5–9). The U.S. Naval Research Laboratories (NRL) and Plessey (now Thomson Marconi Sonar) of Great Britain are two leading groups in the research and application of optical fiber hydrophone systems and arrays [10–12].

Most fiber hydrophone systems have been developed with only one sensing head or one single channel for research purposes. For practical applications, it is often necessary to develop fiber hydrophone systems or arrays with a large number of sensing/signal channels. Only highly multiplexed fiber hydrophone systems can possibly satisfy this requirement. For this reason channel multiplexing was the main subject of fiber hydrophone research in the 1990s. Various fiber hydrophone multiplexing schemes were intensively developed and tested by the NRL [13] and by the U.K. Defense Evaluation and Research Agency [14]. These schemes are driven by the prospects of myriad underwater warfare applications. A multiplexed fiber hydrophone array with 64 channels was demonstrated by Kersey et al. of the NRL [15]. For geophysical applications such as seismic exploration for oil reserves, for example, a surface vessel towed streamer or ocean-bottom cable array can have several hundreds (sometimes thousands) of signal channels [16,17]. In these applications, sensor arrays with a large number of channels are required for the acquisition of detailed two-, three-, and even four-dimensional information on targeted areas [18]. The number of signal channels achievable in a fiber hydrophone array system is highly dependent on system sensitivity requirements, the cross-talk between sensing channels, multiplexing/demultiplexing techniques, and interrogation schemes.

In Section 9.2 we describe the basic interferometric configurations for fiber hydrophones. We present interrogation, or demodulation, schemes in Section 9.3 and a variety of multiplexing techniques for fiber hydrophones in Section 9.4. Sensor head designs and polarization-fading mitigation techniques are discussed in Sections 9.5 and 9.6, respectively. Finally, in Section 9.7, we examine the component and system aspects and application potentials of fiber hydrophone technology.

9.2 Basic Fiber Optic Hydrophone Configurations

It is now widely accepted that in order to achieve sufficiently high sensitivity, the fiber optic hydrophone must be an interferometric type, which measures the phase change introduced in the sensing arm by the measurand. Over the years, many different interferometer architectures have been proposed and investigated for hydrophone systems. The most important ones can be categorized into four fundamental configurations: the Mach–Zehnder interferometer, Michelson interferometer, Sagnac interferometer, and Fabry–Perot interferometer. These architectures can also be classified into two types: transmission type (represented by the Mach–Zehnder interferometer) and reflection type (represented by the Michelson interferometer).

9.2.1 Mach–Zehnder Interferometer

This is one of the main interferometric schemes widely used in early fiber hydrophone systems. This configuration represents a typical transmissive

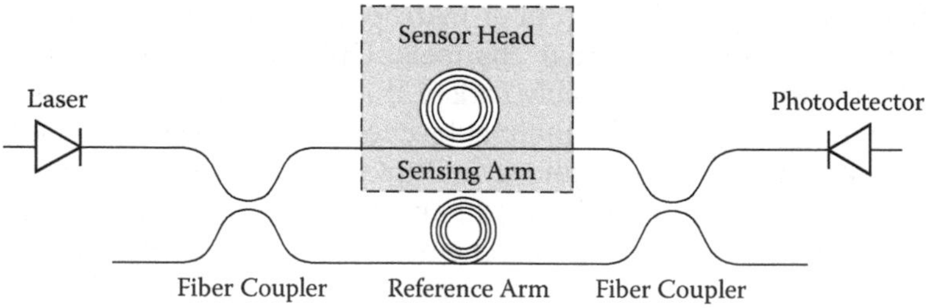

FIGURE 9.1
Schematic diagram of a fiber hydrophone based on a Mach–Zehnder interferometer.

fiber hydrophone. It can be used to build a transmission-type sensor array. In the configuration shown in Figure 9.1, light from a laser is split into two beams by the first fiber coupler, one entering the sensing arm and the other the reference arm. In a single sensor case, the sensor head in the sensing arm is immersed in the sensing environment (the "wet" end, the shaded area in Figure 9.1). The reference arm (and the delay or balancing coil, if any) provides the phase reference, and it can stay with all other parts of the system onboard (the "dry" end). At the output end of the second fiber coupler, these two beams are combined and sent to a photodetector, which produces a current resulting from the interference of the signal and reference beams at the receiver circuit. The phase change the sensing beam experiences is recovered. The extraction of the phase change information can be achieved by a variety of methods, which we describe later.

This configuration suffers from the following disadvantages: (1) signal fading due to an unstable bias point; (2) source phase noise translated into intensity noise; and (3) polarization fading. Techniques to overcome these problems are described in later sections.

Balanced and Unbalanced Path Schemes

Looking into various interferometric configurations of the fiber hydrophone, two possible path schemes are evident: balanced and unbalanced. In a balanced hydrophone scheme, the signal path and the reference path are made nearly the same. For the MZI scheme shown in Figure 9.1, the path balance can be achieved by choosing the length of the reference head. This scheme is very popular because it is usually not sensitive to the phase noise of light sources and does not require high-coherence (narrow linewidth) sources. An unbalanced path scheme is simpler. Some interrogation schemes such as the phase-generated carrier (PGC, to be discussed later) require substantial path imbalance in the MZI. The path imbalance causes the phase noise of a laser output to convert into the intensity noise and hence deteriorate the performance of the sensor system. Another disadvantage of the unbalanced scheme is that it would have to use laser sources with a very long coherence

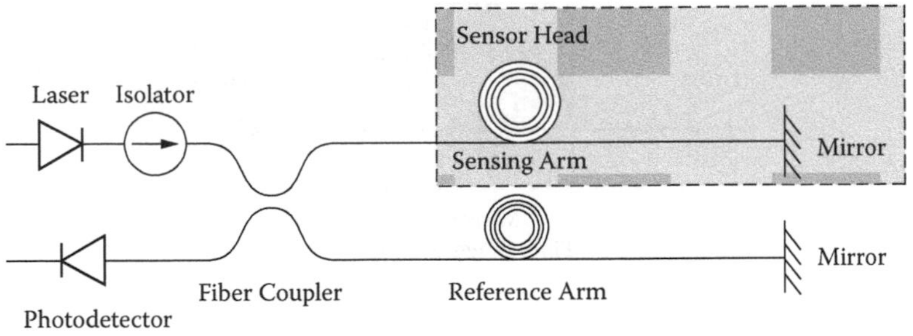

FIGURE 9.2
Schematic diagram of a fiber hydrophone based on a Michelson interferometer.

length (which is essentially longer than the path difference, typically 50 ~ 150 m). This could be a stringent requirement for a light source.

9.2.2 Michelson Interferometer

As shown in Figure 9.2, a Michelson interferometer is essentially a folded Mach–Zehnder interferometer with a pair of mirrors reflecting the signal and reference at the center of two arms, respectively. This configuration is a typical reflection type of fiber hydrophone, appropriate for building a reflective sensor array. To avoid the reflected signals being fed back into the light source, which usually affects the source output, an optical isolator should be used.

The Michelson interferometer has two advantages over the Mach–Zehnder interferometer:

1. Both the transmitter and receiver electronics are located at the same end. In this case, a link to the sensor head, which could be remotely located from the system "dry end," is achieved by a single fiber. With a Mach–Zehnder interferometer, two fibers are needed: one for the light signal from the source to the sensor head and another for the signal from the head to the detector.
2. The sensing beam goes through the sensor head twice, thus making it theoretically twice as sensitive as the Mach–Zehnder interferometer.

9.2.3 Fabry–Perot Interferometer

In a fiber sensor configuration based on the Fabry–Perot cavity shown in Figure 9.3, a light signal is reflected backward and forward within the cavity many times. The output will be a maximum when all the exit beams are in phase and will drop quickly when they are slightly out of phase. Thus, a small phase change introduced in the cavity can result in a large change in the output intensity.

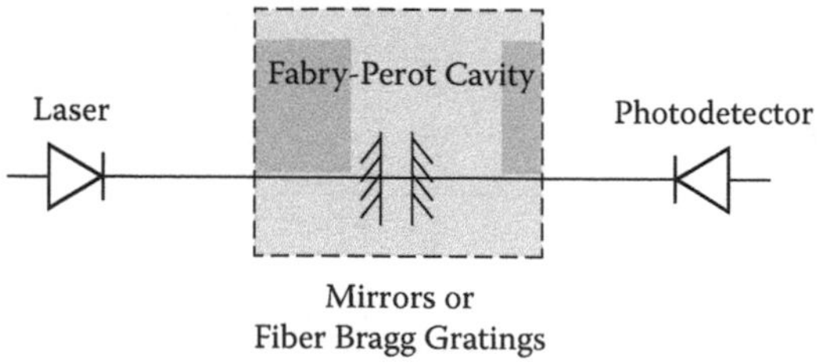

FIGURE 9.3
Schematic diagram of a fiber hydrophone based on a Fabry–Perot interferometer.

This interferometer has two advantages: (1) high sensitivity—a hydrophone constructed in this manner has been shown to have a sensitivity of –122 dB re 1 V/μPa; (2) compactness—the Fabry–Perot cavity can be constructed out of the two end surfaces of a broken fiber or a continuous fiber with a pair of in-fiber Bragg gratings. However, its main drawback is its small dynamic range.

9.2.4 Sagnac Interferometer

In this interferometer configuration (Figure 9.4), light from the source enters the middle port of a 3 × 3 fiber coupler and is split into a clockwise beam (CW beam) and a counterclockwise beam (CCW beam). The CW beam travels through the delay loop first and then through the sensor head, while the CCW beam travels through the sensor head and then through the delay loop. Because of the nonsymmetrical positioning of the sensor head, the phase difference the measurand induces in the sensor head is now translated into the phase difference between the two beams as they meet again at the 3 × 3 coupler. This phase difference in turn is converted into intensity modulation when the beams are combined by the coupler.

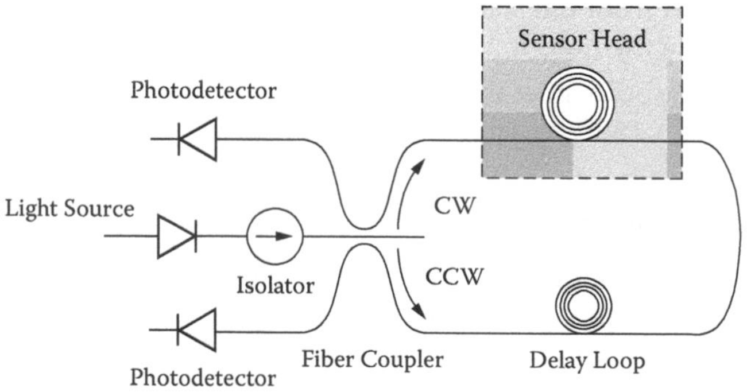

FIGURE 9.4
Schematic diagram of a fiber hydrophone based on a Sagnac interferometer.

The advantages of the Sagnac interferometer are as follows:

1. The phase noise of the optical source does not translate into intensity noise as in a Mach–Zehnder interferometer.
2. It can use a broadband superfluorescent optical source instead of a narrow linewidth laser source.
3. Polarization fading can be minimized.

One main disadvantage of this scheme is its intrinsic insensitivity to low frequencies. Another is its complication in multiplexing. Researchers at Stanford University worked on fiber acoustic sensor arrays based on Sagnac interferometers [19,20].

9.3 Interrogation (Demodulation) Techniques

Here we briefly introduce some of the conventional interrogation (demodulation) schemes used in fiber hydrophones.

9.3.1 Phase-Generated Carrier (PGC)

This scheme was first proposed by Dandridge et al. [21]. This scheme allows time division multiplexing (TDM) or frequency division multiplexing (FDM) to be easily implemented [22]. The output from an interferometer is typically represented by a sinusoidal function of the phase difference between the signal and reference, $\theta(t)$. This phase difference contains the relevant information of the measurand of interest. At the optical detector, the interferometer output produces an optical current given by

$$i(t) = A\cos\theta(t) + B \tag{9.1}$$

where A and B are parameters related to the signal intensity and detector sensitivity. The phase-generated carrier technique introduces a carrier term of frequency ω_0 into the phase factor $\theta(t)$:

$$i(t) = A\cos[\phi_m \cos\omega_0 t + \phi(t)] + B \tag{9.2}$$

where $\phi(t)$ contains the phase information that we want to extract. $\phi_m \cos\omega_0 t$ is the phase-generated carrier term with ϕ_m the modulation index. Now i has many harmonic components from the phase modulation. The amplitude of its nth harmonic is associated with a Bessel function of the nth order, with the argument ϕ_m. For example, the fundamental component is

$$2AJ_1(\phi_m)\sin\phi(t)\cos\omega_0 t \tag{9.3}$$

and the second harmonic is

$$-2AJ_2(\phi_m)\cos\phi(t)\cos 2\omega_0 t \tag{9.4}$$

To extract $\phi(t)$, we first subtract the amplitudes of the fundamental and second harmonics from $i(t)$ by frequency mixing, $i(t)$ cos ω0t and $i(t)$ cos 2ω0t, respectively. The frequency-mixed signals are passed through a lowpass filter, respectively, and this process produces the amplitudes of the fundamental and second harmonics: $2AJ1(\phi_m)$ sin $\phi(t)$ and $-2AJ2(\phi_m)$ cos $\phi(t)$. Now we differentiate both the fundamental and second harmonic components with respect to time, and thus we have the signal corresponding to the fundamental harmonic

$$2AJ_1(\phi_m)\cos\phi(t)\dot{\phi}(t) \tag{9.5}$$

and that corresponding to the second harmonic

$$2AJ_2(\phi_m)\sin\phi(t)\dot{\phi}(t) \tag{9.6}$$

They are cross-multiplied by the original harmonics respectively and the products are subtracted from each other. Mathematically, the process can be represented by

$$\begin{aligned} &2AJ_1(\phi_m)\sin\phi(t)2AJ_2(\phi_m)\sin\phi(t)\dot{\phi}(t) \\ &\quad-(-2A)J_2(\phi_m)\cos\phi(t)2AJ_1(\phi_m)\cos\phi(t)\dot{\phi}(t) \\ &= 2A^2J_1(\phi_m)J_2(\phi_m)\dot{\phi}(t) \end{aligned} \tag{9.7}$$

Thus, the output is directly proportional to the derivative of the phase $\phi(t)$. A simple integration step would finally produce the phase $\phi(t)$ and thus provide the measurand's relevant information.

Various methods can be used to realize the phase-generated carrier interrogation. A simple technique is directly modulating the laser, as shown in Figure 9.5, where the PGC interrogation scheme is applied to retrieve the phase induced in the sensor head in a Mach–Zehnder interferometric type of fiber hydrophone.

Here we drive the laser source with a sinusoidal signal at frequency ω0. In this method, the laser wavelength has not been changed. Only its light intensity has been modulated by ω0. The optical length of the sensing path should be significantly different from that of the reference path (i.e., $L1 \neq L2$) so that $\phi_m \neq 0$. This scheme is usually called *phase-generated carrier (PGC) homodyne.*

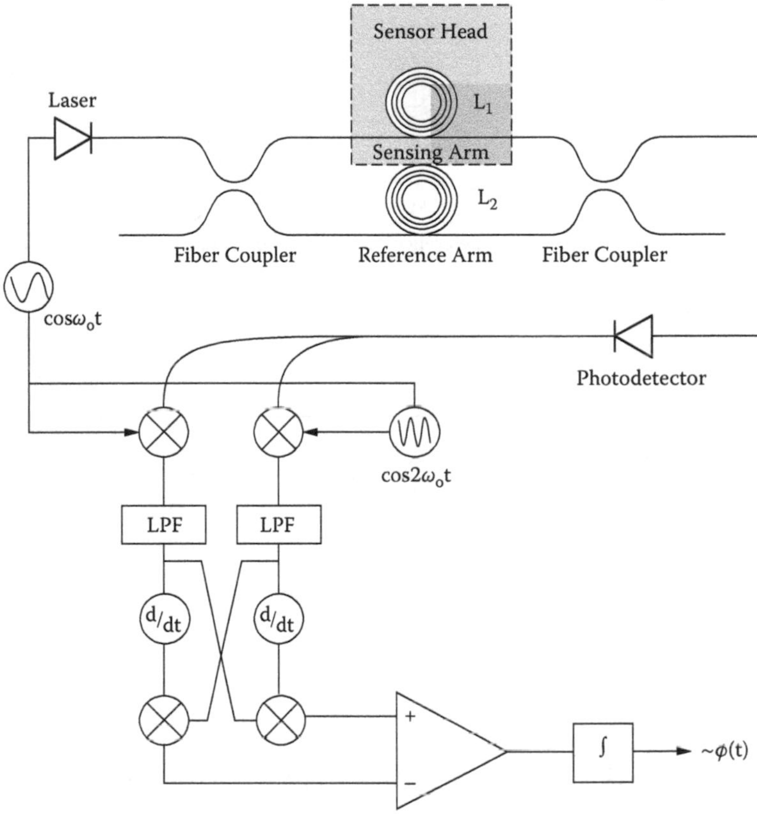

FIGURE 9.5
A fiber hydrophone using the phase-generated carrier homodyne technique.

A slightly different scheme is the *PGC heterodyne*, in which the wavelength of the laser source is changed by driving it with a ramp voltage at repetition rate ω_0. It can be shown that the electrical signal output from the photodetector is given by

$$i(t) = A\cos[m\omega_0 t + \phi(t)] \tag{9.8}$$

where m is an integer dependent on the laser modulation depth. To recover $\phi(t)$, we pass the signal through a bandpass filter followed by a phase-locked loop and an integrator.

9.3.2 Phase-Modulated Compensator (PMC)

This scheme was first proposed by Al-Chalabi et al. [23,24]. In the scheme shown in Figure 9.6, the laser source has a short coherence length, shorter than the path length difference between L_1 and L_2 in the sensor head. In this way, there is no interference between the sensing signal and the reference

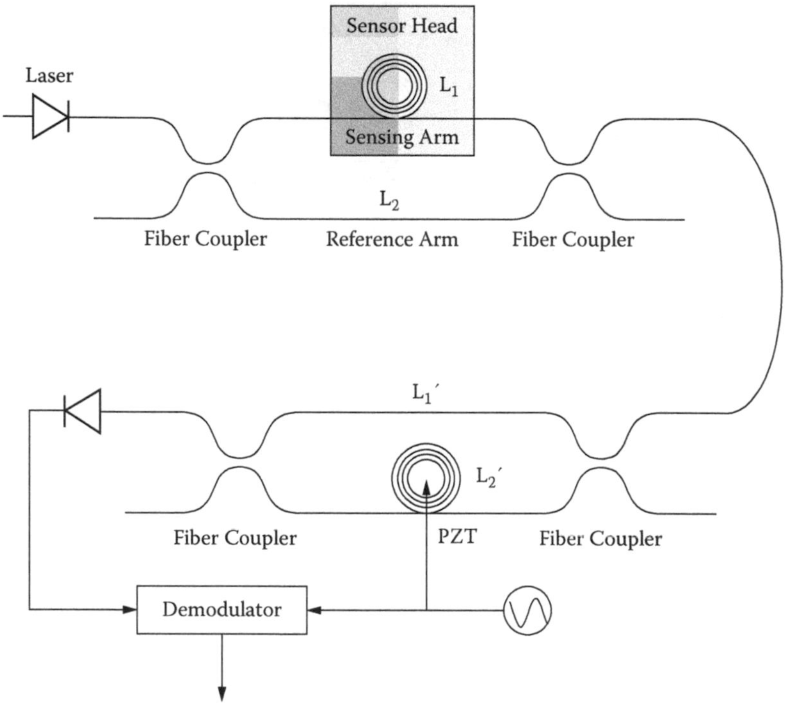

FIGURE 9.6
A fiber hydrophone using the phase-modulated compensator (PMC) technique.

signal in the sensing interferometer. Both signals are then transmitted to the receive interferometer whose two arms have path lengths L'_1 and L'_2. One of these arms can be phase modulated by an external source so that a portion of the sensing signal can interfere with the reference signal, thus giving rise to an interference effect at the photodetector. It can be shown that the detector output is

$$i(t)=\frac{i_0}{8}\left\{2+\cos\left[\frac{2\pi n v}{c}(\Delta L(t)-\Delta L'(t))\right]\right\} \tag{9.9}$$

where v is the emission frequency of the laser source and $\Delta L = |L_1 - L_2|$, $\Delta L' = |L'_1 - L'_2|$.

Similar to PGC, carrier components are generated (by the phase modulator) in this scheme and the phase induced by acoustic signal can be retrieved by appropriate demodulation electronics. This method is suitable for the reduction of phase noise of the laser source. This is particularly attractive because it allows us to use a short coherence-length light source for a sensor head made of long sensing fiber. This means a partially coherent source could be used for interferometric sensors with long sensing arm lengths. For this scheme to work properly, two conditions must be satisfied simultaneously:

$$L_c \gg |\Delta L - \Delta L'| \tag{9.10}$$

and

$$L_c \ll \Delta L \approx \Delta L' \tag{9.11}$$

The first condition ensures that effective interference between the signal and reference could be achieved with high contrast and that significant source-related phase noise could be avoided. The second condition guarantees that the residual contrast, resulting from the interference of components not carrying useful sensing information, is kept sufficiently low. Hence, the signal-to-noise characteristics of a PMC interferometric sensor system will be critically determined by how well both conditions are satisfied.

9.3.3 Frequency-Modulated Continuous Wave (FMCW)

This method derives from the radar ranging principle by ramping the frequency of the radar source by Giles et al. [25]. There are a number of variations of this technique. Usually, sweeping the optical carrier over a certain frequency range or simply ramping the driving current of a diode light source, with a triangular or sawtooth waveform, can produce appropriate FMCW signals.

As shown in Figure 9.7, the laser diode source is driven by a sawtooth current i. This generates a change $d\nu$ of the laser frequency. When this light reaches the interferometer with unequal optical path length $\Delta L(= L1 - L2)$

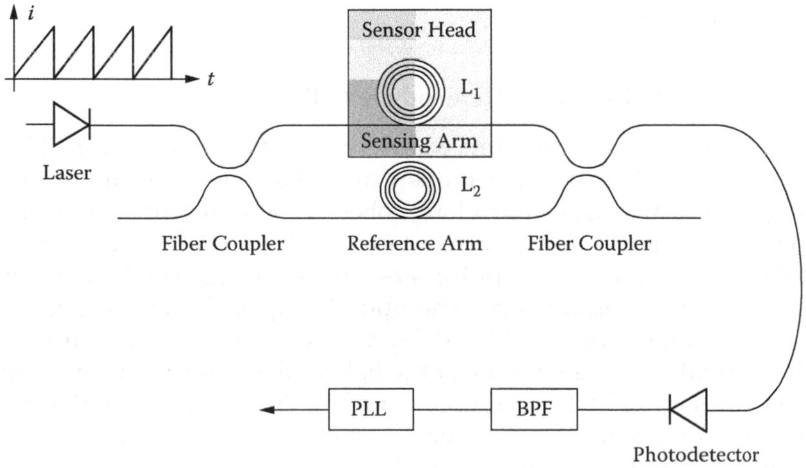

FIGURE 9.7
An MZI-based fiber hydrophone using the frequency-modulated continuous-wave (FMCW) technique.

between the two arms, a change $d\phi$ of the phase at the interferometric output is obtained in the following manner:

$$d\phi = 2\pi n\,\Delta L\frac{dv}{c} \tag{9.12}$$

The rate of this phase shift is related to the fringe shift of the output and is generated by the current ramp rate di/dt:

$$\frac{d\phi}{dt} = \frac{dv}{dt}\frac{2\pi n\Delta L}{c} = \frac{di}{dt}\left(\frac{dv}{di}\right)\frac{2\pi n\,\Delta L}{c} \tag{9.13}$$

The resulting photodetector signals derived from the optical outputs have complex spectra, composed of components at the fundamental and harmonics of the ramp frequency. The distribution of energy in each of these frequency components depends on the total phase excursion of the fringe pattern during the ramp period T. Experimentally, it was found to be possible to concentrate most of the power in the fundamental of the ramp frequency by adjusting di/dt such that the Mach–Zehnder fringe pattern was driven over one full fringe during each ramp period—that is, $(d\phi/dt)\,T = 2\pi$. Bandpass filtering around the fundamental of the ramp frequency then produces a strong carrier, which is free from the distortion caused by the higher harmonics associated with the ramp flyback. The instantaneous phase of this recovered carrier is equivalent to that between the arms of the interferometer. Random and signal-induced optical phase shifts in the interferometer thus directly phase modulate the carrier. Final recovery of the signal may then be achieved using conventional FM demodulation techniques through the phase-locked loop.

9.3.4 Differential Delayed Heterodyne (DDH)

This pulsed interrogation scheme was developed by Dakin et al. [26,27]. This scheme allows TDM to be easily implemented. As shown in Figure 9.8, in this technique, optical pulses from a long coherence-length source are launched into a remote, unbalanced two-arm interferometer. The input pulse sequence is arranged such that pairs of pulses are repetitively launched into the input fiber, as shown in Figure 9.8(a). The optical frequencies of the two pulses are f_1 and f_2, respectively, and the delay τ between the pulse pair is equal to the differential propagation time of the light in the two arms. Consequently, as shown in Figure 9.8(b), a component of the first input pulse that passes through the long arm of the interferometer overlaps a component of the second pulse, which passes through the short arm at the interferometer output. These optical components mix and produce a burst of heterodyne carrier signal of frequency $f_1 - f_2$ at the detector. Due to the repetitive input pulsing, a continuous series of pulsed heterodyne signals is generated at the detector,

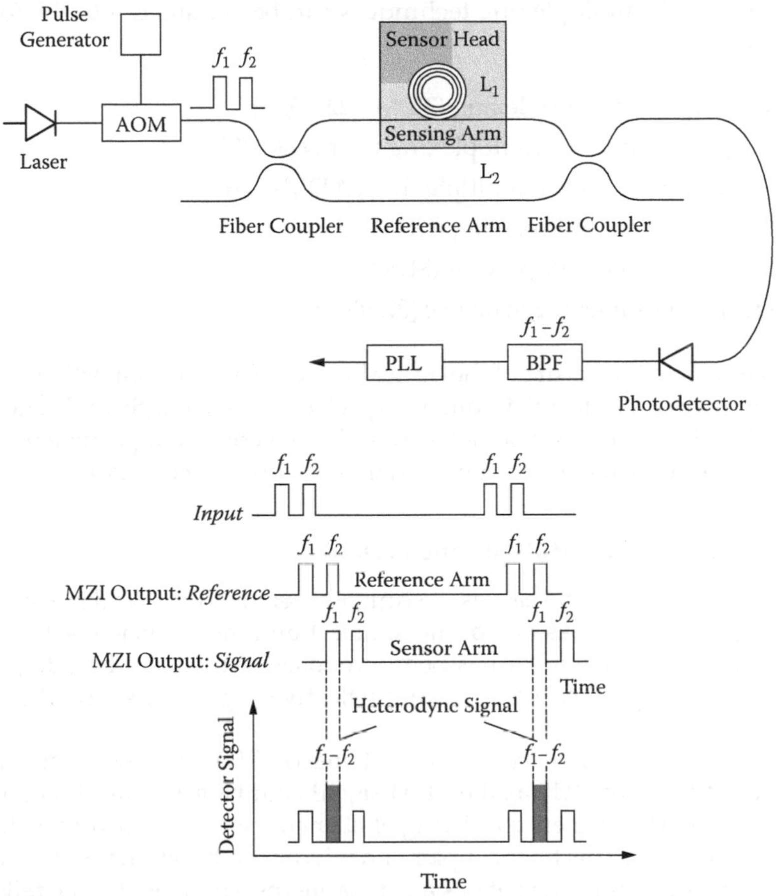

FIGURE 9.8
An MZI-based fiber hydrophone using the differential delayed heterodyne technique.

which can be filtered to produce a steady carrier and is then processed in a conventional manner using standard FM/PM demodulators.

In the Photonics and Optical Communications Group at UNSW, we also have been developing all-optical fiber hydrophones based on DDH [28,29].

9.4 Multiplexing Techniques

Multiplexing is to incorporate a number of sensors in one or a pair of transmission optical fibers. There are many different multiplexing schemes. Each scheme is not only crucial to the number of sensors that can be incorporated, but is also important in determining the sensor's frequency response, dynamic range, noise floor, cross-talk, and cost effectiveness.

In general, the multiplexing techniques can be classified into the following types:

1. Time division multiplexing (TDM) [22,30,31]
2. Frequency division multiplexing (FDM) [32,33]
3. Wavelength division multiplexing (WDM) [34]
4. Code division multiplexing (CDM) [35,36]
5. Space division multiplexing (SDM)
6. Hybrid multiplexing schemes [37,38]

All trace back to the nature of the optical source. They are respectively based on the time (in pulse form), frequency spectrum, wavelength, and coherence of the light beam. It must be noted that the selection of a particular multiplexing scheme also depends on the demodulation scheme used.

9.4.1 Time Division Multiplexing (TDM)

This scheme is one of the earliest techniques demonstrated and still receives considerable attention as one of the simplest and most efficient schemes. A wide range of architectures has been proposed, which can be categorized into two main types: transmissive and reflective. Figure 9.9 shows the transmissive arrangement.

Figure 9.10 shows two types of reflectometric TDM arrangements. In the first, shown in Figure 9.10(a), the TDM signal channels are based on units of Michelson interferometers consisting of fiber couplers and mirrors coated on the fiber end faces. This fiber coupler and mirror combination is not very efficient in terms of light usage. It has a very large insertion loss for the reflected signal. An improved arrangement is to use a fiber Bragg grating instead, as shown in Figure 9.10(b). These fiber Bragg gratings reflect a small fraction of the optical power back, without the insertion loss problem associated with

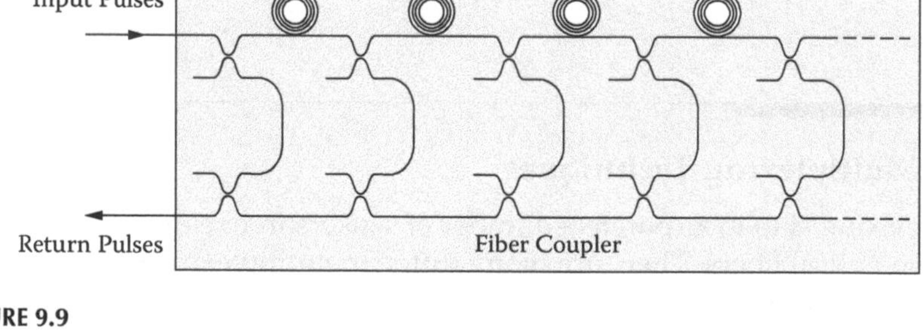

FIGURE 9.9
A schematic diagram of a transmissive TDM hydrophone array.

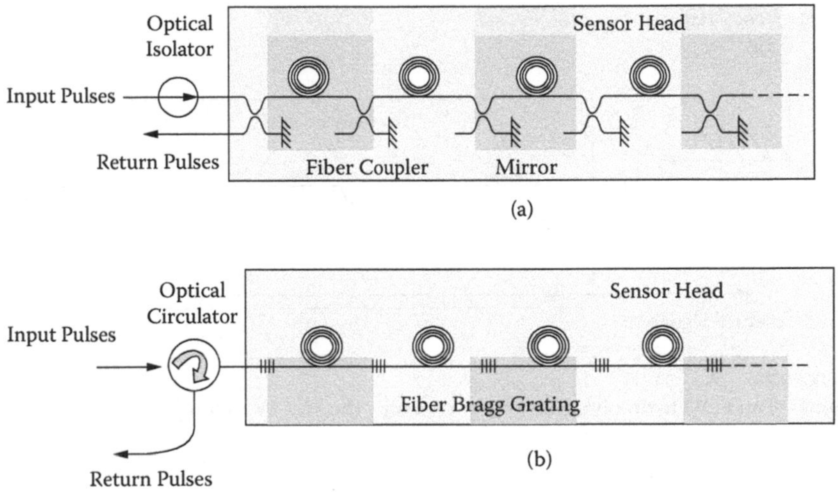

FIGURE 9.10
Schemes of reflective TDM hydrophone arrays.

a fiber coupler. The low reflection of grating also ensures that the multiple reflection effects are minimized.

In all these arrangements, interrogation is usually by means of a heterodyne technique. In the TDM scheme, a switched CW laser or pulsed laser is used. The pulse width is typically 1 ms or less, depending on the amount of multiplexing. The system output will be a train of pulses, each corresponding to a single hydrophone channel. Each pulse is then electronically gated and the acoustic signal demodulated.

9.4.2 Frequency Division Multiplexing (FDM)

An early demonstration of an FDM hydrophone array was a 16-channel vertical line array deployed by Yurek et al. at NRL [32]. The FDM uses the FMCW scheme for demodulation. Each interferometer in the array has a different optical path length difference ΔL so that a different phase change $d\phi$ is induced through the following relation, which we explained before:

$$d\phi = 2\pi \Delta L \frac{dv}{c} \tag{9.14}$$

As shown in the section on FMCW, as the drive current of the optical source ramps up, a fringe shift is produced. A different ΔL in a different interferometer therefore induces a different fringe shift. Hence, tracking these fringes by means of separate phase-locked loops allows the signal from different interferometers to be extracted. Figure 9.11 shows the realization of such a scheme in which the interferometers are connected in series.

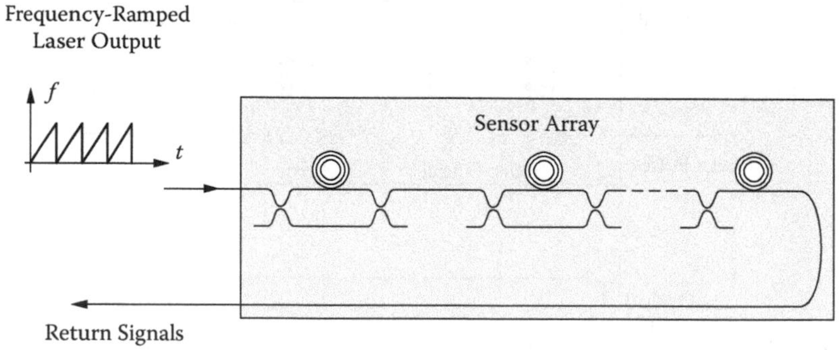

FIGURE 9.11
A layout of an FDM hydrophone array incorporating the FMCW technique.

9.4.3 Wavelength Division Multiplexing (WDM)

In this scheme many beams of light of different wavelengths are launched into the sensor system. Each beam may set up one sensor array based on one of the multiplexing and demodulation schemes discussed previously. WDM is particularly useful in addition to other multiplexing techniques such as TDM and FDM. Thus, N number of wavelengths sets up N arrays sharing most of the optical components. If there are n sensors in a single array, then the total number of sensors in the N-wavelength WDM system will be $n \times N$. This is a very powerful technique in increasing the number of sensors. Another advantage is that the WDM system in telecommunications is fast becoming a mature technology. Components required for 30 or more WDM channels are readily available in the market.

The basic optical components required for WDM are

1. Wavelength filters, usually in the form of fiber Bragg gratings
2. Optical add–drop multiplexers (OADMS) to direct the desired wavelength to a particular channel of sensor arrays
3. Lasers with different wavelengths
4. Optical fiber amplifiers and pump lasers. Since efficient amplifiers are erbium-doped fiber amplifiers (EDFAs), the wavelengths to be used will reside in the 1550-nm window.
5. Optical multiplexers to combine all the beams of different wavelengths from the laser sources and launch them into a single transmission fiber
6. Optical demultiplexers to split the received beams into different receivers

Figure 9.12 shows a scheme of multiplexing N wavelengths λ_1, λ_2, and λ_n into an N-channel fiber hydrophone array. This WDM hydrophone array is

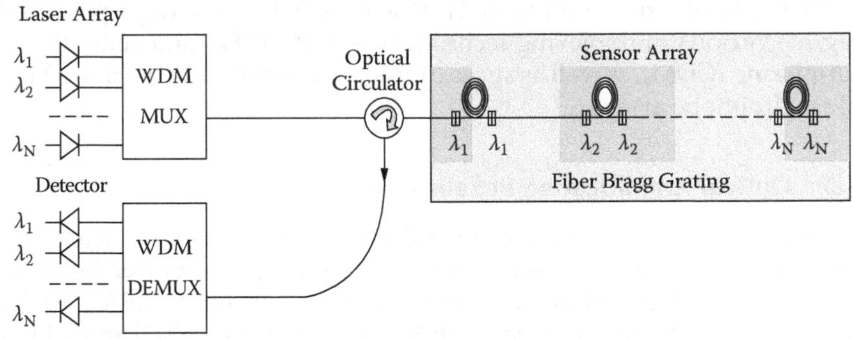

FIGURE 9.12
A schematic diagram of an *N*-channel WDM hydrophone array.

based on a Michelson interferometer configuration using fiber Bragg grating pairs as reflection elements. The state-of-the-art WDM technology can provide the wavelength spacing between channels as small as 0.4 nm (frequency spacing 50 GHz) or less at the 1550-nm region. It is possible to have the crosstalk between the two wavelength channels less than –40 dB.

9.4.4 Hybrid Multiplexing

Either TDM or FDM alone cannot provide sufficient channel count for many applications [34]. The number of sensors that can be incorporated into one array depends on the multiplexing and demodulation techniques used. For example, the DDH/TDM scheme allows about 60 sensors in one array. If 30 WDM wavelengths are used, the total number of sensor channels in one complete system is 30 × 60 = 1800. This number would be enough for towed array applications. In Figure 9.13 we show a schematic arrangement

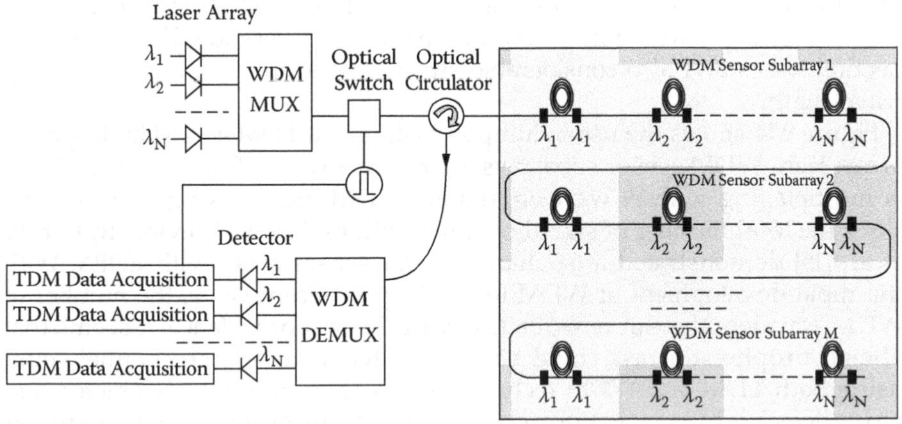

FIGURE 9.13
A schematic diagram of an *N* × *M* channel TDM–WDM hydrophone array.

of a hydrophone array with both TDM and WDM. It is always possible to integrate various multiplexing techniques—FDM, TDM, and code division multiplexing (CDM), as well as space division multiplex (SDM)—into a large-scale hydrophone array.

9.4.5 Optical Amplification in Sensor Arrays

For fiber hydrophone arrays with a significant number of sensor heads, optical amplification will be required for compensating light power lost in the array by splicing, branching, and bending in sensor heads [39,40]. Mainly two techniques can currently be used: Raman amplification [41] and erbium-doped fiber amplification [15]. Both are state-of-the-art optical telecommunication techniques that could produce sufficiently high optical gain. The main difference between them is that Raman optical amplification has distributed gain while erbium-doped fiber amplification usually has a lumped optical gain. An advantage of Raman amplification is having a gain distribution of up to tens of kilometers, providing the possibility to online compensate the loss along the whole hydrophone array.

Erbium-Doped Fiber Amplification (EDFA)

The use of EDFAs in a hydrophone array has three advantages. A single EDFA can (1) amplify all the wavelength channels within its gain bandwidth, (2) reduce noise, and (3) reduce the effect of the optical source's noise phase. EDFAs may be arranged at the beginning of the array (*preamplification*) or at the end of the array (*postamplification*), or both. It is also possible to have "distributed" EDFA—inserting short lengths of erbium-doped fiber between two successive rungs of the array. In this case, only one pump source is needed at the sending end. However, this scheme does not seem particularly attractive because the splicing loss between an EDFA fiber and a normal fiber is usually high. Too many short sections of small-gain EDFAs stretched over a long sensor array would make the pump efficiency very low, due to the connection losses and also considerable attenuation at the usual 980-nm pump wavelength.

Figure 9.14 shows the use of lumped EDFAs in a TDM–WDM hydrophone array. Here WDM sensor subarrays are multiplexed in TDM series. If WDM is multiplexing with N wavelengths and TDM multiplexing with M time slots, the total number of signal channels will be $N \times M$. Previously, Kersey et al. [15] demonstrated a 64-channel TDM sensor array with EDFA. With the rapid development of WDM technology in recent years, the number of WDM wavelengths can now readily reach 32 or more. Hence, it seems that fiber hydrophone arrays could be constructed with a large channel count using both TDM and WDM techniques. A variation of interferometer configurations and interrogation schemes may be incorporated into a similar type of system arrangement. As an example, the figure shows a TDM–WDM based on the DDH scheme. The lights from different wavelength sources are

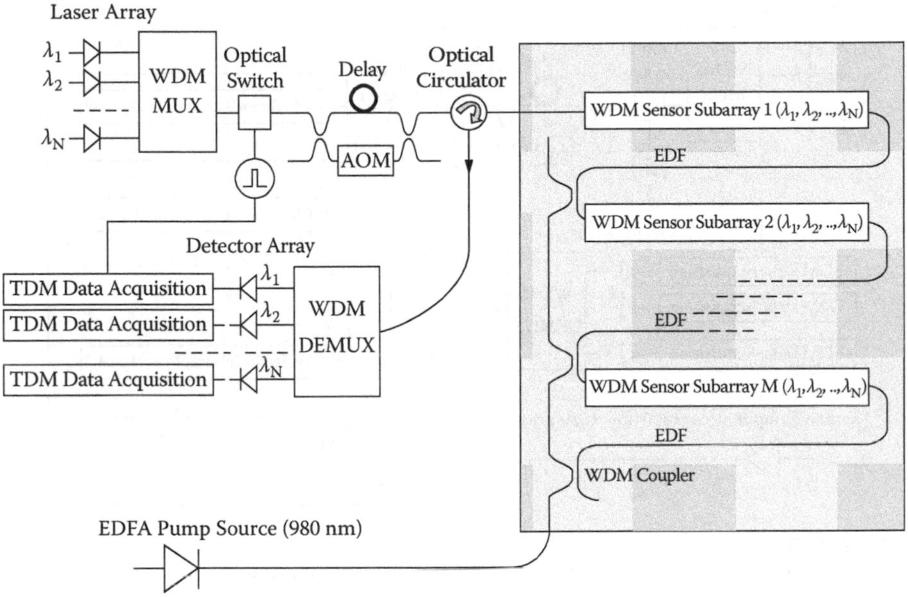

FIGURE 9.14
Schematic arrangement using EDFA in a sensor array.

converted into pulse trains, and then they enter a structure that produces frequency-shifted and time-delayed (twin) pulse trains.

Raman Amplification

The advantage of Raman optical amplification is that its pump wavelengths are readily within the low-loss window of optical fibers [42]. So it is more suitable for amplification at the remote end of fibers where the optical signals are the weakest. Moreover, the very much uniformly distributed gain is more desirable to avoid other nonlinear optical effects and it can improve the optical signal-to-noise ratio. Figure 9.15 shows a scheme using distributed optical Raman amplification in a sensor array. Here the whole fiber sensor array is Raman amplified both forward (preamplification) and backward (postamplification) using a fiber coupler and a single pump source. The main disadvantage of the Raman amplifier is that the gain bandwidth is not very broad. Therefore, if many wavelengths are used in a WDM system, a few pumps of different wavelengths or one pump with multiple wavelengths will be needed.

9.5 Sensor Head Design

Which hydrophone head design is appropriate for a given application depends largely on the required sensitivity, the dynamic range, and the frequency

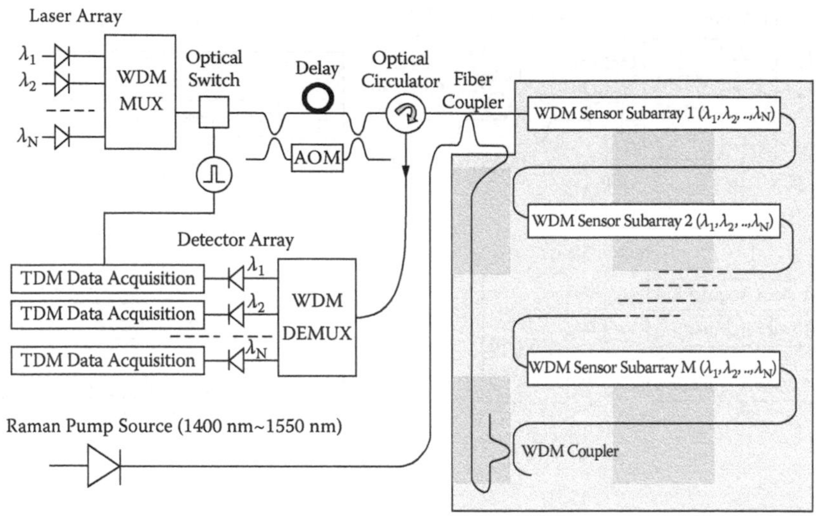

FIGURE 9.15
Schematic arrangement using Raman amplification in a sensor array.

response range, as well as the operational environment. Other parameters are also relevant to head design, such as sensing directivity and acceleration sensitivity. The directivity of the sensor is determined by the configuration of the winding of the fiber on the mandrill. If the fiber is wound into a thin coil, its directivity is omnidirectional. If it is wound into a long (compared to the wavelength of acoustic wave), thin cylinder, it is most sensitive in the direction transverse to the cylinder axis.

In interferometric fiber hydrophone, the change in phase of the light is proportional to the strain produced from stretching or compressing the optical fiber by acoustic wave (pressure). Phase sensitivity is the most important parameter of a sensor head. The sensitivity of a fiber hydrophone head depends on several factors: (1) the material constants such as Young's modulus, Poisson ratio, and the stress-optic coefficient of fiber; (2) the sensor head material and structure; and (3) the length of fiber. Formulas exist to calculate this sensitivity when these parameters are known. Silica glass optical fiber itself is typically not sensitive to acoustic wave, due to its very large Young's modulus, and has very limited phase sensitivity. So various head designs have to be developed to increase the sensitivity.

The popular design of a fiber hydrophone head has been the mandrill, which is a length of fiber wound on a hollow air-backed or foam-filled cylinder. The reference head is usually wound on a solid mandrill to reduce its sensitivity to sound or pressure waves. The difference in sensitivity produces the signal corresponding to the sound waves. A mandrill design is advantageous in keeping the head size small and streamlined, while simultaneously achieving high sensitivity by increasing the length of optical fiber wrapped on the mandrill. By selecting proper mandrill materials and structure and fiber winding scheme, one can substantially suppress the undesirable longitudinal acceleration sensitivity.

Additional sensitivity may be achieved by using a Michelson interferometer configuration rather than a Mach–Zehnder interferometer (MZI) or by using a push–pull design in an MZI where both arms of the interferometer are used for sensing. Both techniques will add 6 dB of sensitivity.

Usually, the mandrills are made of solid plastic or air-backed plastic or metal [4]. In a solid mandrill head, fiber is wound under tension on a polymer solid mandrill such as nylon or Teflon. Its sensitivity is proportional to the reciprocal of the bulk modulus of the mandrill material. In an air-backed mandrill head, fiber is wrapped on a highly compliant, thin metal-walled air-backed mandrill. An air-backed plastic mandrill could provide substantially higher sensitivity—however, its sensitivity could be affected by hydrostatic pressure and thus be depth dependent. Hence, its operation depth could be limited. Highly sensitive air-backed hydrophones may reach their plastic deformation limit before they reach their operating depth.

Nonmandrill hydrophones such as flexural disks, prolate spheroids, and coated fiber designs also consist of optical fiber closely contacted to a compliant medium; thus, the sensitivities of these devices depend on the same factors as the mandrill [43].

Many applications require that the hydrophones be operational in a changing temperature and hydrostatic pressure environment. Hydrophone deformation, due to either temperature or hydrostatic pressure, can be a problem. One way to deal with large temperature gradients is to build the hydrophones in the push–pull configuration. Because both arms of the interferometer in this type of hydrophone respond in the same way to a temperature change, the thermal effect would cancel.

It is worth noting that polymer optical fiber technology has made significant progress in recent years [44,45]. New polymer optical fibers based on perfluorinated polymers have been made with very low loss [46]. Fiber Bragg gratings have also been made on polymer optical fibers [47,48]. The Young's modulus of polymer fiber materials is typically 30 times less than that of glass. Hence, for strain-related optical sensors such as fiber hydrophone, polymer optical fibers are intrinsically 30 times more sensitive to acoustic wave than silica glass fibers [49]. The use of polymer optical fiber in a hydrophone could revolutionize the sensor head design because of the much better acoustic compatibility between water and fiber.

9.6 Polarization-Fading Mitigation Techniques

Polarization fading refers to the polarization-related signal fading that has been one of the major obstacles in the development of interferometric sensors [50].

This fading comes from the effects of fluctuation in fiber birefringence under the influence of environmental changes. For the sensing beam and the reference beam in the interferometer to produce the phase-related interference, they should have the same polarization. Because of the birefringence fluctuation, the fibers in the two arms cannot maintain the same polarization all the time. This is especially serious for fiber hydrophones that usually have long lengths of fibers in each sensor head. The fluctuation in turn causes a random fading of the detected signal. In the worst case, when the two interfering lights are nearly orthogonal in polarization, the signal could be totally lost. As a result, the system signal-to-noise ratio could seriously deteriorate.

Several methods have been proposed to deal with the polarization-fading problem (e.g., references 51–54). They include

1. Polarization-maintaining fibers and devices
2. Polarization scrambling
3. Polarization switching
4. Faraday rotation mirror
5. Tri-state polarization diversity detection
6. Active polarization tracking at receiver

The most direct—and also most expensive—solution to this problem is to construct fiber hydrophones using polarization-maintaining optic components and fibers. However, this approach has the following disadvantages: (1) its performance can be degraded by the finite extinction ratios of the fiber couplers and the splices, and (2) the fiber is usually much more expensive than standard nonpolarization-maintaining fibers.

Hence, alternative techniques have been sought and developed. All these techniques try to mitigate the problem by introducing a certain polarization diversity in the source (polarization scrambling), modulator (polarization switching or scanning), sensor head (Faraday rotation mirror), or receiver (tri-state polarization mask, polarization tracking, etc.).

Figure 9.16 shows one of our schemes for mitigating the fading problem. Here our scheme is based on the input polarization-switching technique proposed by Cranch and Nash [51]. This system is a TDM–MZI hydrophone array based on DDH with path balancing. A train of rectangular pulses of one polarization (for example, x-polarization) is launched by an optical source into a fiber coupler, which splits the pulse train into two. One of the pulse trains (the top branch of the coupler output in Figure 9.16) is switched by the polarization optical switch. The switch rotates the polarization of every second pulse in this train by 90° (e.g., conversion into y-polarization from x-polarization), while leaving the first pulse unchanged. Hence, this train is essentially interleaved by optical pulses with the original polarization (e.g., x-pol) and its orthogonal polarization (e.g., y-pol).

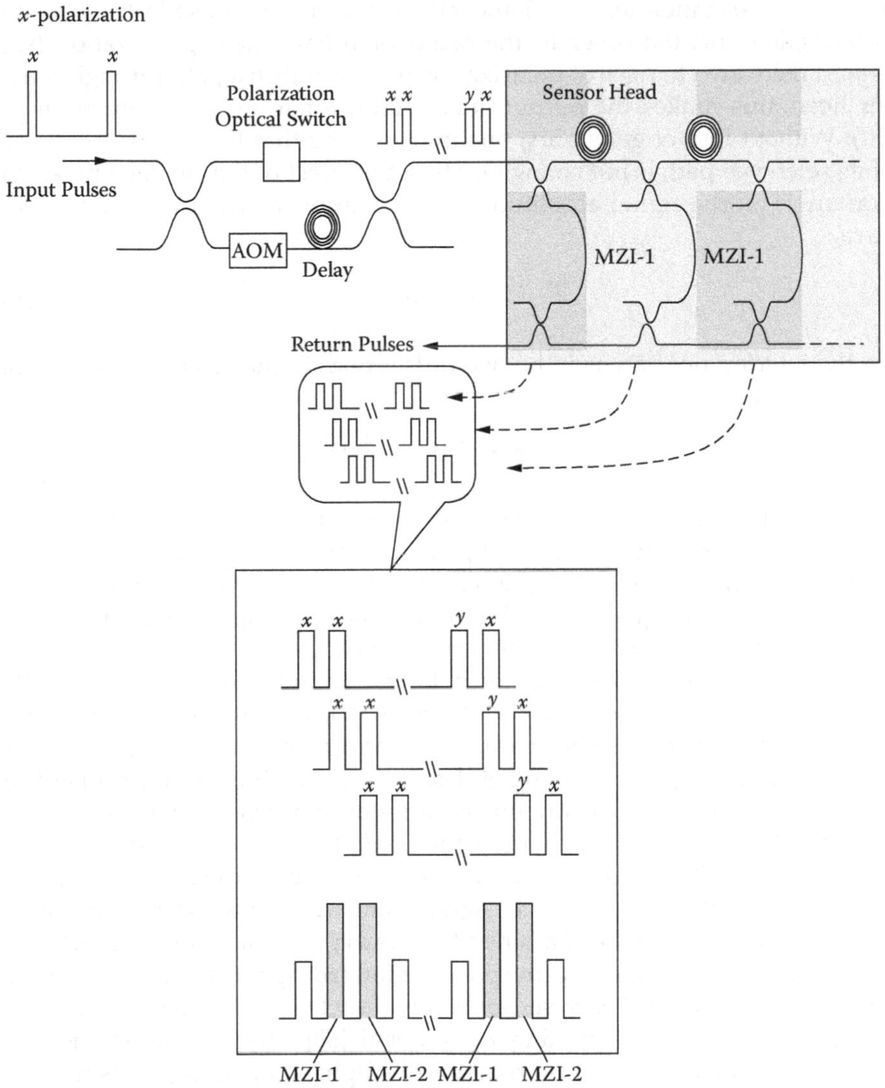

FIGURE 9.16
Mitigation of polarization-induced signal fading in a hydrophone array by polarization-switching approach.

The other pulse train is frequency shifted by an acousto-optic modulator and also delayed by the path-balancing fiber coil in the bottom branch of the fiber coupler. The two trains of optical pulses are merged into one and then launched into the sensor array via a second fiber coupler. In the signal detection, each channel is collecting signals from the interference of succession pulses, as shown in Figure 9.16. This means that the signal will be retrieved based on the interference between the two pulses both in the original polarization (e.g., x–x) and also between one pulse in the original polarization and the other in the orthogonal (switched) polarization (e.g., x–y). If there are changes of polarization effects within a path through a sensor head, this makes the output polarization rotate randomly by an angle $\Theta(t)$. Without loss of generality, one may assume that the polarization state of its reference path is not changed. Thus, the effect of the changed polarization angle on the signal amplitude of this channel can be expressed in two parts:

$$\vec{E}_s \cdot \vec{E}_r \sim E_x E_x \cos\left[\Theta(t)\right] \tag{9.15}$$

for the interference between the two pulses both in the original polarization (e.g., x–x) and

$$\vec{E}_s \cdot \vec{E}_r \sim E_y E_x \sin\left[\Theta(t)\right] \tag{9.16}$$

for the interference between one pulse in the original polarization and the other in the orthogonal (switched) polarization (e.g., x–y). For a simple case with equal amplitudes, $E_x = E_y$, we could easily remove the polarization fading by adding the square of each part to produce a signal amplitude independent of the random polarization rotation $\Theta(t)$.

One simple scheme to counter the polarization-induced signal fading is the *tri-state mask technique* proposed by Frigo et al. [52]. In this scheme, the combined beam from the reference beam and the sensing beam is expanded and launched into a mask consisting of three polarizer films arranged in appropriate angles. In this way, any polarization of light from the incident beam can be split into three polarization components. A quadrant photodetector situated behind the tri-state polarization mask can convert the three polarization lights into three electrical signals that are then separately amplified. These signals can then be combined to always give a maximum output.

Another approach to overcome polarization fading is the use of the *Faraday rotation mirror* (FRM) at each sensor head in Michelson interferometer-based sensor systems, demonstrated by Kersey et al. [53]. The *Faraday effect* denotes the nonreciprocal polarization rotation of a light signal as it passes through an optical medium within a magnetic field. The FRM incorporates a Faraday rotator in front of the mirror. It is this element that provides the nonreciprocal 45° rotation of the state of polarization each time the light passes through it. Situated at the end of an optical fiber, the Faraday rotation mirror is designed to rotate a signal's state of polarization (SOP) by 45° twice: once when the light enters, and again when the light is reflected back into the fiber. Since the Faraday effect is nonreciprocal, the SOP is rotated by 90° with respect to the original signal. These rotations, applied in combination with a reversal of the polarization state's handedness upon reflection at the mirror interface, yield a state perpendicular to the original signal. In this way, any SOP fluctuations that occur along the fiber anywhere between the source and the FRM are exactly compensated and their unwanted effects neutralized.

9.7 Component and System Issues

The two key issues of practical applications are invariably system performance and cost. They in turn are largely determined by the availability and quality of components and system integration techniques. The rapid advance of optical communications technology has brought a range of cost-effective, high-performance, reliable photonic components and system techniques to the market. A large proportion of the technology developed is equally applicable to optical fiber sensing. This will definitely have a significant impact on the field application of fiber hydrophone systems. In the past few years, a number of enabling techniques essential to fiber hydrophone arrays have been developed, including

Optical isolator, circulator, and switch
Fiber amplifier (EDFA)
DWM or DWDM mux and demux
Fiber Bragg grating
Fiber coupler and connector
Optical source and detector arrays

All these component and system techniques can be used in fiber hydrophone system designs. Fiber Bragg gratings as in-line devices, for example, are very compact, low insertion loss, and low cost. The use of erbium-doped amplification or Raman optical amplification will compensate the light attenuation in a sensor array, which in turn allows a large number of sensor heads to be incorporated. Also, the WDM or DWDM technology, which allows more than 100 wavelengths transmitted over one fiber, can be used to boost the number of signal channels to more than 1000, incorporating with the TDM or FDM technique. However, the advantages are not only the new technologies but also the huge economies of scale that result from products made for the telecommunications industry.

Since the early years of optical fiber sensor research and development, the hydrophone system has been one of the most challenging topics. A great deal of effort has put into fiber hydrophone system development, mainly focusing on a variety of naval and military applications. These applications, which include seabed arrays (e.g., ocean-bottom cables [OBCs]), fixed and fast deployable systems, towed arrays behind surface ships [55] or submarines (kinds of streamers, but smaller in scale), and sculled mounted arrays, are mainly concerned with submarine warfare for area surveillance and warship tracking. Excellent demonstrations of fiber hydrophone systems have been carried out in ocean bottom, undersea, towed array, submarine, and terrestrial environments [56]. However, these techniques have yet to be tested to see whether or not they are viable for industrial and civil applications that are usually much more sensitive to system cost rather than superior performance.

In fact, fiber hydrophone systems could find many more important, non-military applications. For example, fiber hydrophone systems could be deployed in the ocean, on the seabed, and down the boreholes or oil wells for industrial and scientific purposes. In the ocean, hydrophone arrays can be used as seismic streamers for oil exploration. In this application, long line arrays of hydrophones are towed behind a surface vessel for seismic survey. An airgun sends acoustic waves to the seabed and underground and the hydrophone array receives the acoustic waves reflected from the underground. Thus, the underground structure and geophysical properties could be revealed by analyzing the reflected signals. This process is often useful for finding or monitoring oil reserves under the seabed. On the seabed, fiber hydrophone arrays could be deployed to monitor seismic activities underground or marine activities aboveground.

Industrial applications of hydrophone systems are emerging. For example, Litton Guidance and Control Systems develops a borehole system with 96 channels based on a 96-fiber hydrophone array [57]. These hydrophones are frequency division multiplexed using six tunable lasers that each provide optical power for a set of 16 sensors. The fiber hydrophones in boreholes listen to underground sound waves to locate and monitor oil and gas reserves. The advantages of optical fiber hydrophones in this application include avoiding electronics that hardly survive the corrosive fluids and high temperature (150°C or higher) of the deep underground. The optical fiber can withstand temperatures of 500°C or higher.

9.8 Final Remarks

Experts in the field had long believed that sooner or later the era of practical applications of fiber hydrophone systems would come. However, the progress has been a bit slow. The related industries and market economics have yet to act in favor of the technology. The availability of low-cost, high-performance photonic devices and components has been a key issue. Also, system packaging and integration techniques may be further developed. Nevertheless, by combining the advanced architecture designs with the latest development of fiber component and system techniques, optical fiber hydrophone system technology is now fast approaching a stage where it is ready to enter the market currently dominated by the well-established piezoelectric ceramics.

References

1. P. Shajenko, J. P. Flatley, and M. B. Moffett, On fiber-optic hydrophone sensitivity, *J. Acoust. Soc. Amer.*, **64**, 5, pp. 1286–1288, 1978.
2. H. L. Price, On the mechanism of transduction in optical fiber hydrophones, *J. Acoust. Soc. Amer.*, **66**, 4, pp. 976–979, 1979; M. B. Moffett, P. Shajenko, J. W. Frye, R. N. Thurston, H. L. Price. Comments on "On the mechanism of transduction in optical fiber hydrophones," *J. Acoust. Soc. Amer.*, **67**, 3, pp. 1071–1075, 1980.
3. J. P. Dakin and B. Culshaw, eds., *Optical Fiber Sensors; Principles and Components*, Artech House, Boston, 1988.
4. T. G. Giallorenzi, Optical technology in naval applications, *Optics and Photonics News*, pp. 23–36, April 1999.
5. W. B. Spillman, Jr. and D. H. McMahon, Frustrated-total-internal-reflection multimode fiber-optic hydrophone, *Appl. Opt.*, **19**, 1, pp. 113–116, 1980.
6. J. Jarzynski, R. Hughes, T. R. Hickman, and J. A. Bucaro, Frequency response of interferometric fiber-optic coil hydrophones, *J. Acoust. Soc. Amer.*, **69**, 6, pp. 799–808, 1981.
7. T. K. Stanton, Noise-equivalent pressure of a single-fiber interferometric acoustic sensor, *J. Acoust. Soc. Amer.*, **69**, 1, pp. 311–312, 1981.
8. D. A. Jackson, A. D. Kersey, D. Corke, and J. D. C. Jones, Psuedo-heterodyne detection scheme for optical interferometer, *Electron. Lett.*, **18**, pp. 1081–1083, 1982.
9. M. L. Henning, S. W. Thornton, R. Carpenter, W. J. Stewart, J. P. Dakin, and C. A. Wade, Optical fiber hydrophones with down lead insensitivity, *First Int. Conf. Fibre Sensors*, London, pp. 23–27, 1983.
10. J. P. Dakin and C. A. Wade, Optical fiber hydrophone array—recent progress, *Proc. OFS'84*, Stuttgart, pp. 375–379, 1984.
11. A. D. Kersey, A. Dandridge, and A. B. Tveten, Time-division multiplexing of interferometric fiber sensors using passive phase-generated carrier interrogation, *Opt. Lett.*, **12**, pp. 775–777, 1987.
12. A. D. Kersey and J. P. Dakin, eds., *Distributed and Multiplexed Fiber Optic Sensor III*, Proc. SPIE 2071, 1993.
13. A. D. Kersey, Multiplexed interferometric fiber sensors, presented at the *7th Int. Optical Fiber Sensors Conf.*, Sydney, Jan. 1990.
14. P. J. Nash, G. A. Cranch, L. K. Cheng, D. de Bruijn, and I. Crowe, 32-element TDM optical hydrophone array, *Proc. SPIE*, **3483**, pp. 238–242, 1998.
15. A. D. Kersey, A. Dandridge, A. R. Davis, C. K. Kirdendall, M. J. Marrone, and D. G. Gross, 64-element time-division multiplexed interoferometric sensor array with EDFA telemetry, *Proc. Optical Fiber Comm., OFC'96*, pp. 270–271, 1996.
16. M. H. Houston, B. N. P. Paulsson, and L. C. Knauer, Fiber optic sensor systems for reservoir fluids management, *Proc. Annual Offshore Technology Conf.*, Houston, TX, **3**, pp. 185–193, 2000.
17. E. B. Wooding, K. R. Peal, and J. A. Collins, The ORB ocean-bottom seismicdata logger. *Sea Technology*, **39**, 8, pp. 85–89, Aug. 1998.
18. P. Kristiansen and P. Christie, Monitoring Foinaven reservoir: Advances in 4-D seismic, *World Oil*, **220**, 11, pp. 71–74, 1999.
19. B. I. Vakoc, M. J. F. Digonnet, and G. S. Kino, A folded configuration of a fiber Sagnac-based sensor array, *Optical Fiber Technology*, **6**, 4, pp. 388–399, 2000; B. J. Vakoc, Folded sagnac sensor array, U.S. Patent, no. 6,034,924, March 7, 2000.

20. B. I. Vakoc, M. J. F. Digonnet, and G. S. Kino, A novel fiber-optic sensor array based on the Sagnac interferometer, *J. Lightwave Tech.*, **17**, 11, pp. 2316–2326, 1999.
21. A. Dandridge, A. B. Tveten, and T. G. Giallorenzi, Homodyne demodulation scheme for fiber optic sensors using phase generated carrier, *IEEE J. Quantum Electron.*, **QE-18**, p. 1647, 1982.
22. A. D. Kersey, A. Dandridge, and A. B. Tveten, Time-division multiplexing of interferometric fiber sensors using passive phase-generated carrier interrogation, *Opt. Lett.*, **12**, pp. 775–777, 1987.
23. S. A. Al-Chalabi, B. Culshaw, and D. E. N. Davies, Partially coherent sources in interferometric fiber sensors, *Proc. 1st Intl. Conf. Optical Fiber Sensors*, London, **221**, pp. 132–135, 1983.
24. S. A. Al-Chalabi, B. Culshaw, D. E. N. Davies, I. P. Giles, and D. Uttam, Multiplexed optical fiber interferometers: An analysis based on radar systems, *IEE Proc., Part J: Optoelectronics*, **132**, 2, pp. 150–156, 1985.
25. I. P. Giles, D. Uttam, B. Culshaw, and D. E. N. Davies, Coherent optical fiber sensors with modulated laser sources, *Electronics Lett.*, **19**, 1, pp. 14–15, 1983.
26. J. P. Dakin, Multiplexed and distributed optical fiber sensor systems, *J. Phys. E: Sci. Instrum.*, **20**, pp. 954–967, 1987.
27. J. L. Brooks, B. Moslehi, B. Y. Kim, and H. J. Shaw, Time-domain addressing of remote fiber-optic interferometric sensor arrays, *J. Lightwave Tech.*, **15**, pp. 1014–1023, 1987.
28. P. L. Chu, T. W. Whitbread, and P. M. Allen, An all fiber hydrophone, *Proc. 13th ACOFT*, pp. 241–244, 1988.
29. P. L. Chu, T. W. Whitbread, and P. M. Allen, Trade-off between sensitivity and dynamic range of reflectometric pulsed interferometric fiber sensors, *Proc. 7th Optical Sensor Conf.*, pp. 229–232, 1990.
30. S. C. Huang, W. W. Lin, and M. H. Chen, Time-division multiplexing of polarization-insensitive fiber-optic Michelson interferometric sensors, *Opt. Lett.*, **20**, pp. 1244–1246, 1995.
31. A. D. Kersey, A. Dandridge, and K. L. Dorsey, Transmissive serial interferometric fiber sensor array, *J. Lightwave Tech.*, **7**, 5, pp. 846–854, 1989.
32. A. M. Yurek, *Proc. OFC'93*, Florence, Italy, May 4–6, 1993.
33. A. Dandridge, and A. D. Kersey, Multiplexed interferometric fiber sensor arrays, *Proc. SPIE*, 1586, pp. 176–183, 1991.
34. A. D. Kersey, Array topologies for implementing serial fiber Bragg grating interferometer arrays, U.S. Patent No. 5, 987,197, Nov. 1999.
35. A. D. Kersey, A. Dandridge, and M. A. Davis, Low-cross-talk code-division multiplexed interferometric array, *Electronics Lett.*, **28**, 4, pp. 351–352, 1992.
36. C. P. Jacobson and H. K. Whitesel, Code division multiplexing of optical fiber sensors for shipboard applications, *Proc. Ship Control Systems Symp.*, **1**, pp. 381–390, 1997.
37. P. J. Nash and G. A. Cranch, Multichannel optical hydrophone array with time and wavelength division multiplexing, *Proc. SPIE*, **3746**, pp. 304–307, 1999.
38. Y. J. Rao, A. B. Lobo Ribeiro, D. A. Jackson, L. Zhang, and I. Bennion, In-fiber grating sensing network with a combined SDM, TDM, and WDM topology, *Proc. Lasers and Electro-Optics Soc. Annual Meeting*, p. 244, 1996.

39. C. W. Hodgson, J. L. Wagener, M. J. F. Digonnet, and H. J. Shaw, Optimization of large-scale fiber sensor arrays incorporating multiple optical amplifiers. Part I. Signal-to-noise ratio, Optimization of large-scale fiber sensor arrays incorporating multiple optical amplifiers. Part II: pump power, *J. Lightwave Tech.*, **16**, 2, pp. 218–231, 1998.
40. W.-W. Lin, S.-C. Huang, J.-S. Tsay, and S.-C. Hung, System design and optimization of optically amplified WDM–TDM hybrid polarization-insensitive fiber-optic Michelson interferometric sensor, *J. Lightwave Tech.*, **18**, 3, pp. 348–359, 2000.
41. D. M. Spirit, L. C. Blank, S. T. Davey, and D. L. Williams, Systems aspects of Raman fibre amplifiers, *IEE Proc., Part J: Optoelectronics*, **137**, 4, pp. 221–224, 1990.
42. E. M. Dianov, A. A. Abramov, M. M. Bubnov, A. M. Prokhorov, A. V. Shipulin, G. G. Devjatykh, A. N. Guryanov, and V. F. Khopin, 30dB gain Raman amplifier at 1.3[μ]m in low-loss high GeO2-doped silica fibres, *Electronics Lett.*, **31**, 13, pp. 1057–1058, 1995.
43. B. J. Flaskerud and J. B. Kreijger, Fiber optic flexural disk hydrophone design and system evaluation, Thesis, Naval Postgraduate School, 1990.
44. Y. Koike and T. Ishigure, Bandwidth and transmission distance achieved by POF, *IEICE Trans. Elect.*, **E82-B**, pp. 1287–1295, 1999.
45. G. Giaretta, W. White, M. Wegmueller, and R. V. Yelamaty, 11 Gb/sec data transmission through 100m of perfluorinated graded-index polymer optical fiber. *Tech. Digest, OFC'99*, p. 3, 1999.
46. N. Yoshihara, Performance of perfluorinated POF, *Polymer Optical Fiber Conf. '97*, paper Tub-2, Sept. 1997; N. Tanio and Y. Koike, What is the most transparent polymer?, *J. Polymer* **32**, pp. 43–50, 2000.
47. G. D. Peng, Z. Xiong, and P. L. Chu, Photosensitivity and gratings in dye-doped polymer optical fibres, *Optical Fiber Tech.*, **5**, 242–251, April 1999.
48. Z. Xiong, G. D. Peng, B. Wu, and P. L. Chu, Highly tunable single-mode polymer optical fiber grating, *IEEE Photonic Tech. Lett.*, **11**, 3, pp. 352–354, March 1999.
49. G. D. Peng and P. L. Chu, Polymer optical fiber photosensitivities and highly tunable fiber gratings, *Fiber and Integrated Optics*, special issue on *Fiber and Integrated Optic Gratings: Fundamentals, Devices and Applications*, **19**, 4, pp. 277–293, 2000.
50. A. D. Kersey, M. J. Marrone, and A. Dandridge, Experimental investigation of polarisation-induced fading in interferometric fiber sensor arrays, *Electronics Lett.*, **27**, 7, pp. 562–563, 1991.
51. J. T. Ahn and B. Y. Kim, Polarisation switching approach to the suppression of polarisation induced signal fading in fiber-optic sensor array, *Proc. 10th Intl. Optical Fiber Sensor Conf.*, Glasgow, Scotland, p. 502, 1994.
52. N. Frigo, A, Dandridge, and A. B. Tveten, Technique for eliminating polarisation fading in fibre interferometers, *Electronics Lett.*, **20**, 8, pp. 319–320, 1984.
53. A. D. Kersey, M. J. Marrone, and M. A. Davis, Polarization-insensitive fiber optic Michelson interferometer, *Electron. Lett.*, **27**, pp. 518–520, 1991.
54. B. Y. Kim, Fiber optical time-division-multiplexed unbalanced pulsed interferometer with polarization fading compensation, U.S. Patent No. 5,173,743, Dec. 22, 1992.
55. A. Caiti, S. M. Jesus, and A. E. Kristensen, Geoacoustic seafloor exploration with a towed array in a shallow water area of the Strait of Sicily, *IEEE J. Oceanic Engineering*, **21**, pp. 355–366, 1996.

56. G. A. Cranch and P. J. Nash, High multiplexing gain using TDM and WDM in interferometric sensor arrays, *Proc. SPIE*, **3860**, pp. 531–537, 1999.
57. F. Su, Developing large-scale multiplexed fiber optic arrays for geophysical applications, an interview with Mark Houston (Litton) and Philip Nash (DERA), *OE Report*, p. 5, Sept. 2000.

10

Applications of Fiber Optic Sensors

Y. J. Rao and Shanglian Huang

CONTENTS

10.1 Introduction

Fiber optic sensors (FOSs) have been subject to considerable research for the past 25 years or so since they were first demonstrated about 30 years ago [1,2]. Early applications were focused on military and aerospace uses during the late 1970s and early 1980s. Fiber optic gyroscopes and acoustic sensors are examples, and they are widely used today. With the increase in the popularity of FOSs in the 1980s, a great deal of effort was made toward the commercialization of FOS, in particular intensity-based sensors. In the 1990s, new fiber optic sensor technologies emerged, such as in-fiber Bragg grating (FBG) sensors [3,4], low-coherence interferometric sensors [5,6], and Brillouin scattering distributed sensors [7]. Dramatic advances in the field of FOSs have been made due to the appearance of these new technologies, which have enhanced the practical applicability of FOSs significantly.

Apart from the well-known advantages of FOSs—such as electromagnetic interference (EMI) immunity, small size, good corrosion resistance, and ultimate long-term reliability—more advantages of FOSs based on new sensing mechanisms have been identified. For example, FBG sensors offer a number of distinguishing advantages over other implementations of FOSs, including direct absolute measurement, low cost, and unique wavelength-multiplexing capability. These new sensing technologies have formed an entirely new generation of sensors offering many important measurement opportunities and great potential for diverse applications. In fact, we have seen that rapid progress in FOS applications has been made in recent years as a result of the feeding of advanced fiber optic sensing technologies innovated. Hence, it is considered useful to update the reader with the state-of-the-art progress in FOS applications. This chapter provides an overview of FOS applications in terms of application areas with emphases on newly emerged FBG sensors. Following the introduction, other sections deal with selected applications:

1. Large composite and concrete structures
2. The electrical power industry
3. Medicine
4. Chemical sensing
5. The gas and oil industry

10.2 Applications to Large Composite and Concrete Structures

When compared with traditional electrical strain gauges used for strain monitoring of large composite or concrete structures, FOSs have several distinguishing advantages, including

1. A much better invulnerability to electromagnetic interference, including storms, and the potential capability of surviving in harsh environments, such as in nuclear power plants [8]
2. A much less intrusive size (typically 125 μm in diameter—the ideal size for embedding into composites without introducing any significant perturbation to the characteristics of the structure)

3. Greater resistance to corrosion when used in open structures, such as bridges and dams
4. A greater capacity of multiplexing a large number of sensors for strain mapping along a single fiber link, unlike strain gauges, which need a huge amount of wiring
5. A higher temperature capacity with a widely selectable range
6. A longer lifetime, which could probably be used throughout the working lifetime of the structure (e.g., >25 years) as optical fibers are reliable for long-term operation over periods greater than 25 years without degradation in performance

These features have made FOSs very attractive for quality control during construction, health monitoring after building, and impact monitoring of large composite or concrete structures [9]. Since the uses of FOSs in concrete was first suggested in 1989 [10] and the demonstration of embedding a fiber optic strain sensor in an epoxy–fiber composite material was reported in 1989 [11], a number of applications of FOSs in bridges, dams, mines, marine vehicles, and aircraft have been demonstrated.

10.2.1 Bridges

One of the first monitoring demonstrations for large structures using FOSs was a highway bridge using carbon fiber-based composite prestressing tendons for replacement of steel-based tendons to solve the serious corrosion problem [12]. Because composite materials are not well proven in their substitution for steel in concrete structures, there is considerable interest in monitoring the strain and deformation or deflection, temperature, or environmental degradation within such types of composite structures using an integrated fiber optic sensing system. FBG sensors could be suitable for achieving such a goal. An array of FBGs has been adhered to the surface of a composite tendon, and the specially protected lead-in/out optical fibers egress through recessed ports in the side of the concrete girders, as shown in Figure 10.1.

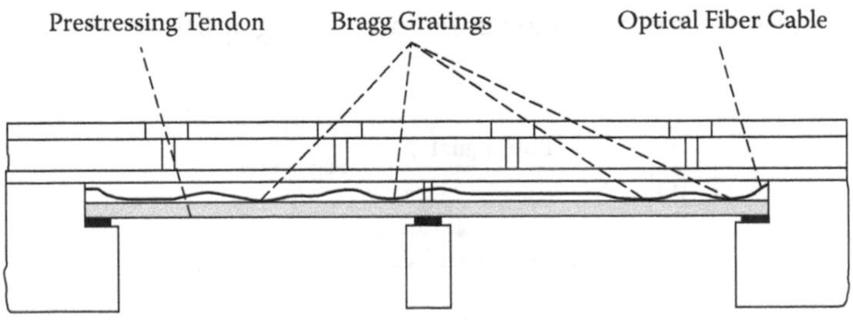

FIGURE 10.1
Schematic diagram of FBG sensor locations for strain monitoring a bridge.

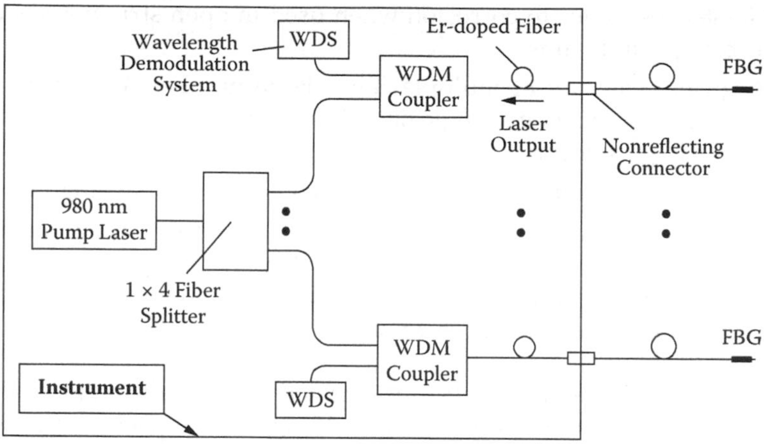

FIGURE 10.2
Schematic diagram of Bragg grating fiber laser sensor demodulation system.

However, if the FBG sensors could be embedded into the composite tendons during their manufacture, excellent protection for the sensors and their leads would be provided. This has been done recently [13]. A strain-decoupled FBG temperature sensor was installed within each girder to allow for correction of thermally induced strain. A four-channel demodulation system, as shown in Figure 10.2, was developed based on the combination of the linear filter method and an Er-doped fiber laser used for enhancement of the small reflective signal levels from the FBG sensors. In this arrangement a length of Er-doped optical fiber pumped by a semiconductor laser operating at 980 nm serves as the fiber laser whose wavelength is tuned by the sensing FBG, and the wavelength shift of the sensing FBG induced by strain change is detected via a bulk linear filter that converts the wavelength shift into intensity change, as shown in Figure 10.3 [14]. The measurement range and resolution of this interrogation system are 5 mε and 1 με, respectively. An accuracy of ~ ±20 με was demonstrated, which is mainly limited by the Er-doped fiber laser frequency jitter. The maximum measurement bandwidth is about 700 Hz. The transient strain change and static loading associated with passing and parking a 21-ton truck on the bridge were demonstrated, indicating the potential for possible traffic monitoring applications.

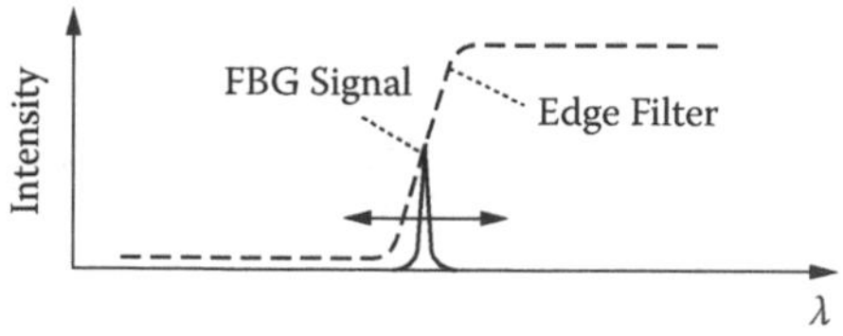

FIGURE 10.3
Principle of the edge filter method.

Also, the same research group embedded such a system into a concrete bridge. Two similar FBG sensor systems using a long-period fiber grating [15] and a chirped fiber grating [16] as wavelength discriminating elements for demodulating the sensor output were used for replacement of the bulk linear filter and were field-tested for strain monitoring of concrete bridges in the late 1990s. These two approaches provide an all-fiber, robust design. In order to obtain more detailed information about the strain distribution in a bridge structure due to damage, an FBG sensor system with up to 60 FBGs has been embedded into a quarter-scale bridge model by the U.S. Naval Research Laboratory [17]. This system, with a typical response time of 0.1 s, is well suited for static strain mapping but not for dynamic strain measurement, due to the scanning-speed constraint of the Fabry–Perot tunable filter used for the wavelength-shift measurement.

The preliminary results obtained from these demonstrations are quite encouraging. However, the typical resolution of 1 $\mu\varepsilon/\sqrt{\text{Hz}}$ is not adequate for traffic usage; for example, the 21-ton truck generated only a strain level of ~20 με [12]. The resolution would need to be improved by a factor of at least 10. Recently, a new approach using FBG to form the reflectors in a Fabry–Perot interferometer, interrogated by low-coherence interferometry to minimize the interferometric phase noise, was demonstrated to achieve high sensitivity for dynamic strain measurement [18]. The configuration of such a system is shown in Figure 10.4, which combines FBGs with an all-fiber Fabry–Perot interferometric sensor (FFPI) formed by writing two FBGs with the same central wavelength on a length of fiber. The reflectivities of the two FBGs are selected as ~30 and ~100%, respectively, in order to obtain the maximum fringe visibility. For a single-sensor design, the phase change of the FFPI is used for high-sensitivity dynamic strain measurement while the wavelength

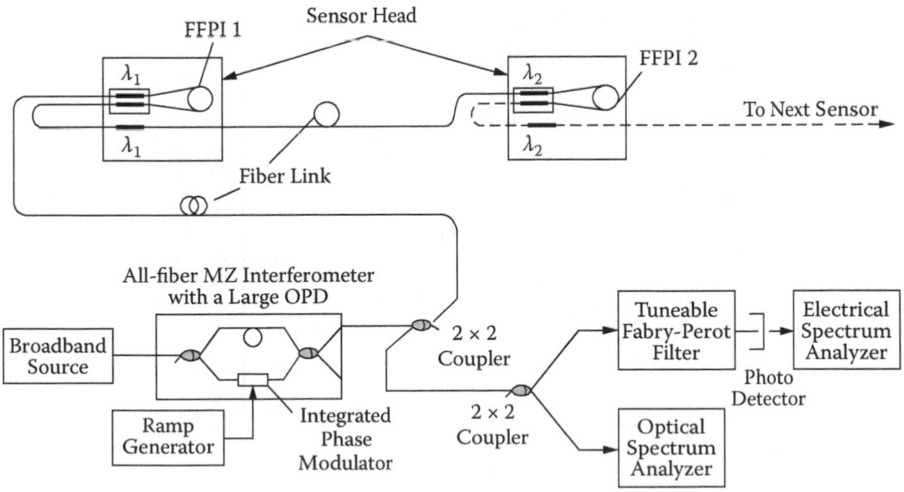

FIGURE 10.4
Schematic diagram of the wavelength division multiplexed FBG–FFPI sensor system.

shift of one of the two FBGs protected against strain is used for correction of thermal apparent strain. The wavelength difference between the two FBGs caused by a temperature gradient would be a problem for practical applications as it would degrade the visibility of the interferometric signal. This undesired effect could be reduced by simply selecting the two FBGs with a larger linewidth. Also, as the gauge length of the sensor is normally much less than the size of the structure to be monitored, the wavelength difference caused could be negligible due to small environmental temperature gradients between the two FBGs.

Static strain monitoring is simply obtained by directly measuring the wavelength shift of another FBG sensor that is arranged in tandem near the FFPI and has a different central wavelength to the FBGs in the FFPI, although it can also be achieved by the identification of the central fringe position of the interferometric signal when the FFPI is interrogated with a scanned local receiving interferometer [19,20]. This FFPI–FBG combination allows simultaneous measurement of three different parameters—static strain, temperature, and transient strain. Multiple FFPI–FBG sensor pairs are wavelength multiplexed for facilitating quasi-distributed measurement.

An experimental system has been demonstrated that included two 1-m-long FFPIs with central wavelengths of 1531 and 1534 nm and an FBG with a central wavelength of 1555 nm. The results for static strain and temperature measurement are shown in Figure 10.5 and Figure 10.6, respectively. A static strain resolution of better than 1 με over a range of 5 mε, a temperature sen-

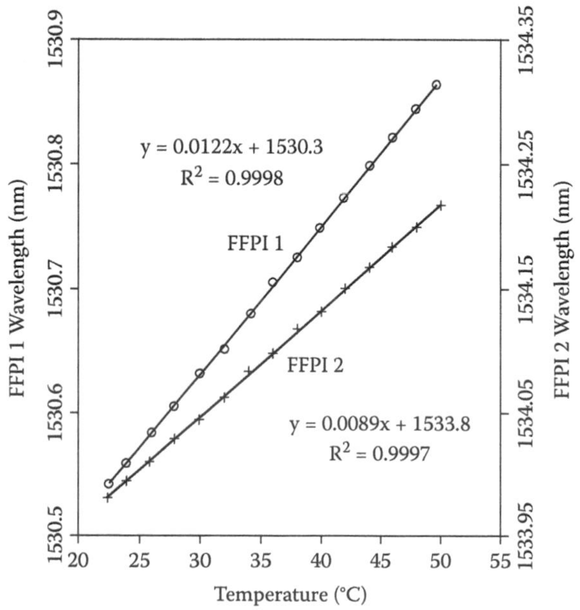

FIGURE 10.5
Variation of FFPI wavelength with temperature.

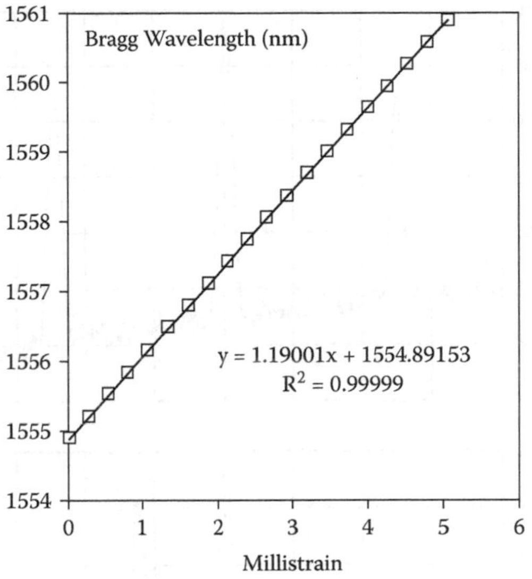

FIGURE 10.6
Strain measurement results with an FBG centered at 1555 nm.

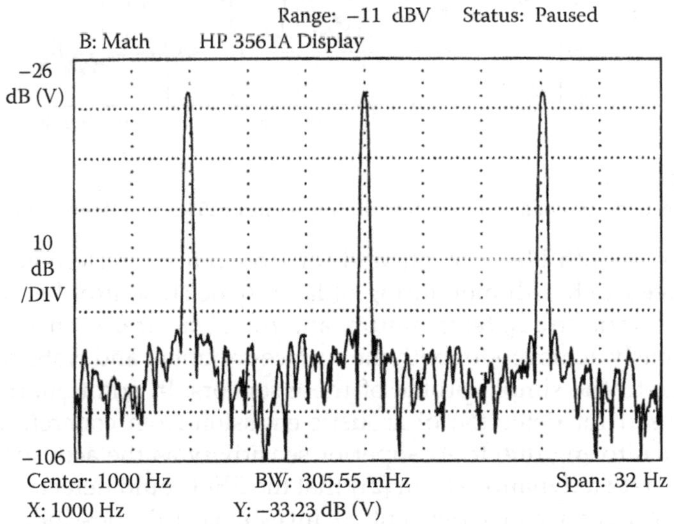

FIGURE 10.7
Experimental results of low-frequency, dynamic strain measurement with FFPI.

sitivity of 0.1°C, and a dynamic strain sensitivity of better than $1\text{n}\varepsilon/\sqrt{\text{Hz}}$ have been obtained. Figure 10.7 shows the result for a 10-Hz low-frequency dynamic strain with a heterodyne carrier frequency at 1 kHz. The measured system cross-talk is less than 50 dB. This sensor system, combining the

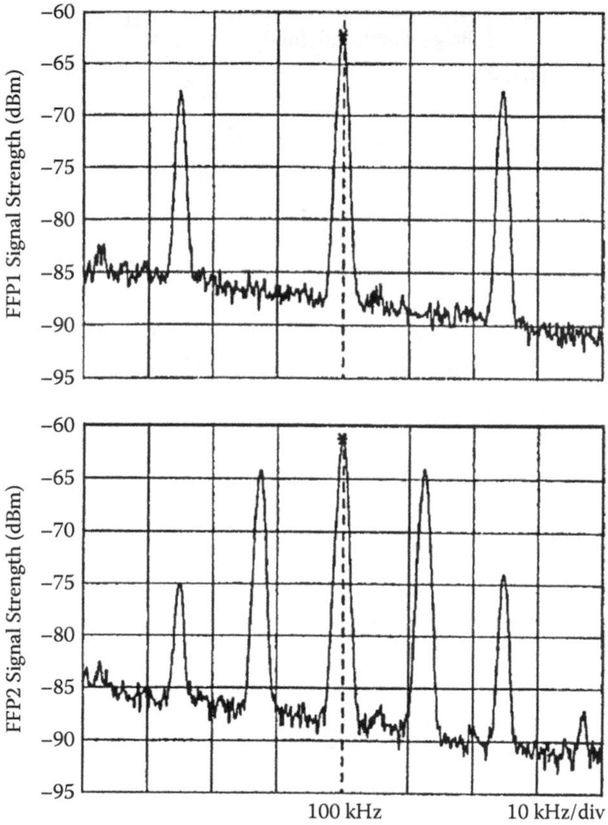

FIGURE 10.8
FFPI high-frequency, dynamic strain spectra. Top: 0.4 με at 25 kHz; bottom: 6 με at 12 kHz.

advantages of both FBG sensors and low-coherence interferometry, would be well suited to health monitoring of large-scale structures because quasi-distributed static strain, temperature, and transient strain sensing could be simultaneously achieved for both the surface-mounted and embedded applications due to the simple profile of these sensors. In addition, this system could be used for detection of acoustic emission from concrete cracks for damage monitoring due to its superior sensitivity as the acoustic emission would generate a dynamic strain [21] that the FFPI could detect.

An experiment was carried out to investigate this possibility, and the preliminary results show that sub-nε sensitivity can be achieved for high-frequency dynamic strain signal at frequencies of tens of kilohertz (see Figure 10.8, where the heterodyne carrier frequency is 100 kHz). For the dynamic strain application, parallel wavelength division multiplexing (WDM) filters could be used for replacement of the tunable Fabry–Perot filter in order to achieve real-time interrogation of each sensor. Furthermore, in order to enhance the sensitivity for dynamic strain measurement, the reflectivity of the first FBG in the Fabry–Perot cavity can be selected as ~100% so that a fiber Bragg grating laser is formed [22]. Such a fiber laser sensor can achieve a dynamic strain sensitivity of up to $10^{-5} n\varepsilon/\sqrt{Hz}$ [23].

Another example of the application of the FBG sensor to bridges is for distributed load monitoring of carbon fiber reinforced polymer cables used in a cable-stayed suspension road bridge during construction and its behavior during traffic [24]. The interrogation system based on a charge coupled device (CCD) spectrometer with a calibrated lamp as the wavelength reference has a resolution of 1 με, but response time is slow due to the speed limitation of signal processing with software. An FBG sensor system with the combination of a two-mode fiber and an FBG has also been proposed for such an application, where simultaneous strain and temperature measurement could be achieved [25].

Fiber optic low-coherence interferometry [5] is also applied for strain mapping of concrete bridges. An example of such a system is that developed by Inaudi and coworkers, as shown in Figure 10.9 [26]. An all-fiber Michelson interferometer with an optical path difference (OPD) between the two arms is used as a sensing interferometer (SI). Strain leads to a change in OPD of

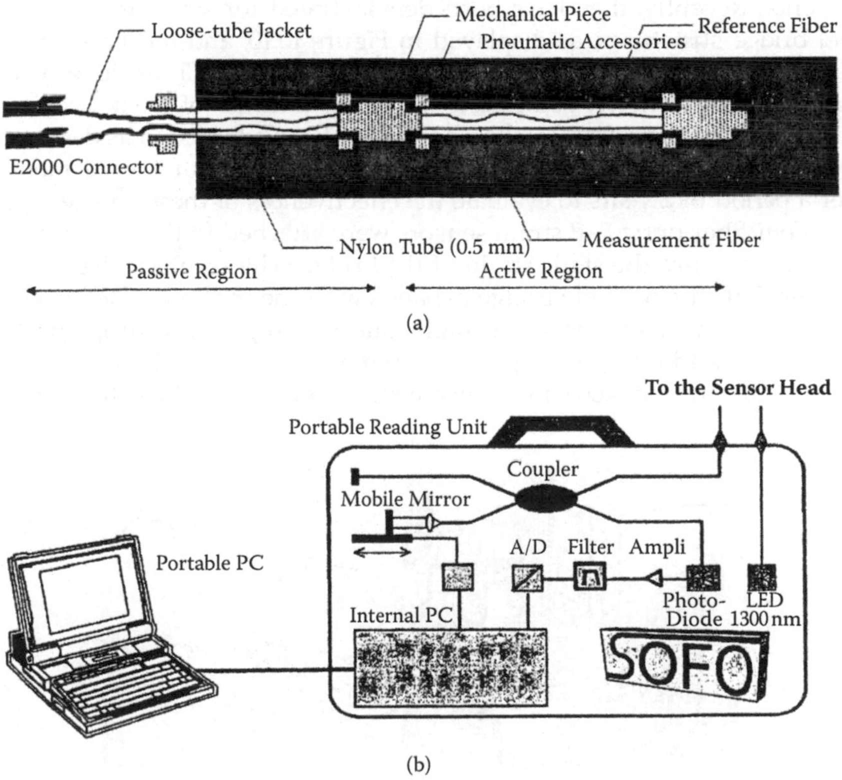

FIGURE 10.9
Fiber optic low-coherence interferometric sensing system for bridges. (a) Sensor head; (b) interrogation instrument.

the interferometer as one of the arms is surface attached to or embedded into the host concrete structure while the other arm, used as a temperature reference, is placed loose in a tube nearby. This OPD change is detected by a local receiving interferometer (LRI), which is also a Michelson interferometer with a scanning mirror. Due to the short coherence length of the light source, interferometric fringes are observed only when the OPD of the SI matches that of the LRI. So the OPD of the SI can be determined accurately by means of the displacement of the motorized mirror in the LRI. As each measurement takes about 10 s, this system is ideal for long-term monitoring of bridge deformations but is unsuitable for dynamic strain measurement, such as vibration. To date, more than 2000 sensors have been installed on tens of concrete bridges in Switzerland. This work demonstrates a very good example of using FOSs for practical applications.

In addition, short fiber optic strain gauges with a typical length of a few centimeters have been used for bridge monitoring, such as extrinsic fiber Fabry–Perot (F–P) strain sensors [27]. This type of sensor is not very sensitive to temperature change; hence, temperature compensation for thermally induced strain error could be achieved easily using a moderate temperature reference. Recently, they have been demonstrated for strain monitoring of steel bridge structures, as displayed in Figure 10.10, and a concrete bridge called Hongcaofang Crossroads Bridge in Chongqing, China, as shown in Figure 10.11. This newly completed bridge has a length of 210 m over seven spans. As a number of new technologies have been adopted in its design and construction, the performance of the bridge has to be measured monthly over a period of 2 years to evaluate the effectiveness of these new technologies. Four fiber optic F–P strain sensors were attached to the centers of two spans to measure the static strain of the bridge. The results in Figure 10.12 indicate that this concrete bridge expands with the increase of temperature in daytime and concentrates at night due to temperature dropping. The strain peaks and troughs are just in accordance with the values of temperature at 3:00 p.m. and 4:00 a.m., respectively. In order to evaluate the accuracy

FIGURE 10.10
Photograph of the steel bridge structures with the fiber Fabry–Perot strain sensors.

FIGURE 10.11
Photograph of the Hongcaofang crossroads concrete bridge in Chongqing, China.

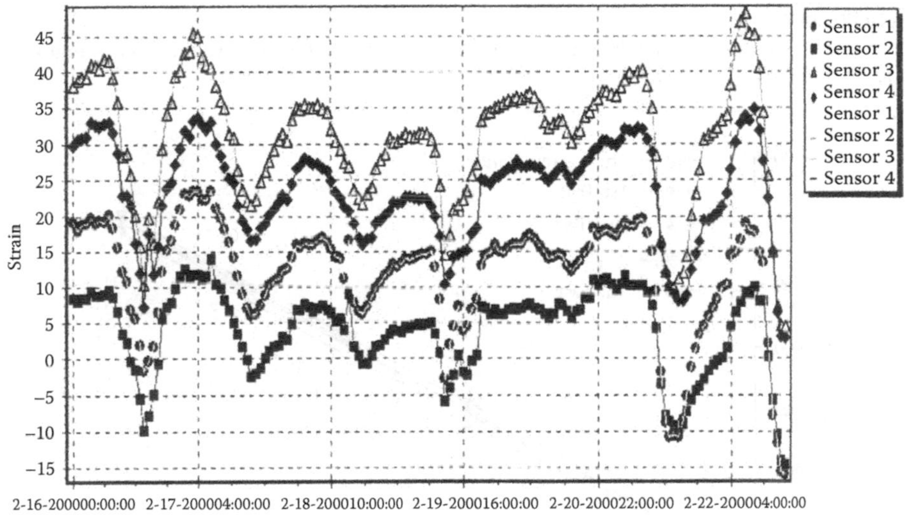

FIGURE 10.12
Experimental results of static strain measurement for the Hongcaofang Bridge, February 16–22, 2000.

and repeatability of the fiber F–P strain sensor, two experiments have been carried out based on a standard cantilever calibration setup with the strain gauge as a reference. The results are shown in Figure 10.13 and Figure 10.14, respectively. It can be seen that the fiber F–P strain sensor works very well.

10.2.2 Dams

Dams are probably the biggest structures in civil engineering; hence it is vital to monitor their mechanical properties during and after construction in order to ensure the construction quality, longevity, and safety of the dam. FOSs are ideal for health monitoring applications of dams due to their

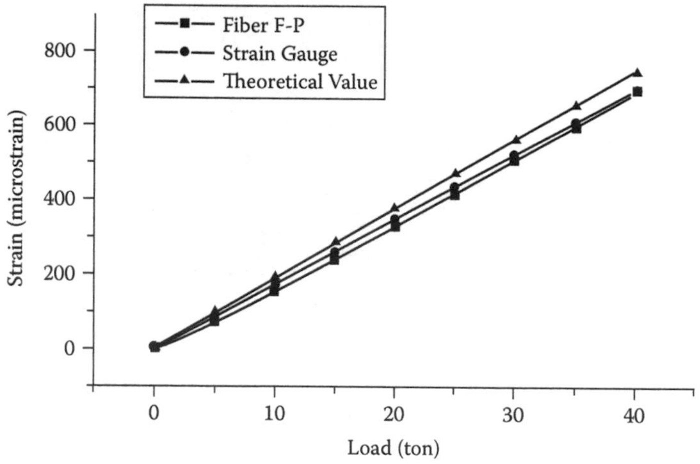

FIGURE 10.13
Comparison of theoretical values and experimental results obtained with the strain gauge and the fiber F–P strain sensor.

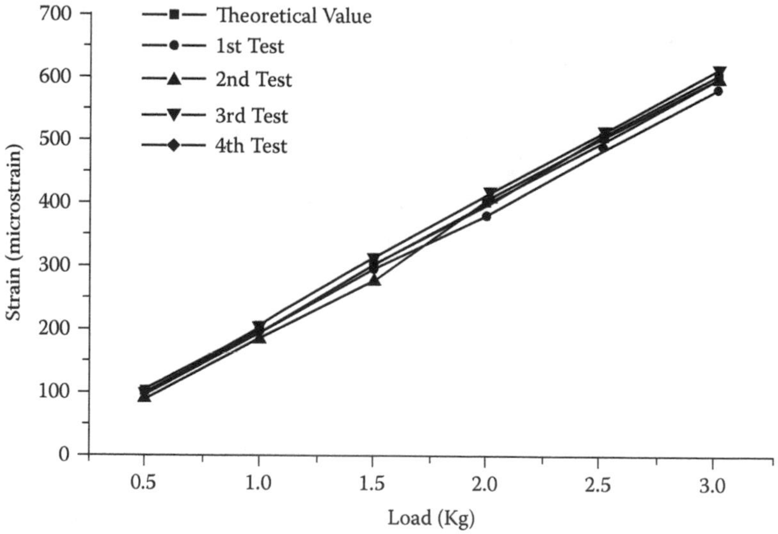

FIGURE 10.14
Repeatability test results for the fiber F–P strain sensor based on a standard cantilever setup.

excellent ability to realize long-range measurement. Truly distributed FOSs are particularly attractive as they normally have tens of kilometers measurement range with meter spatial resolution. A distributed temperature sensor has been demonstrated for monitoring concrete setting temperatures of a large dam in Switzerland [28]. This monitoring is of prime importance as the

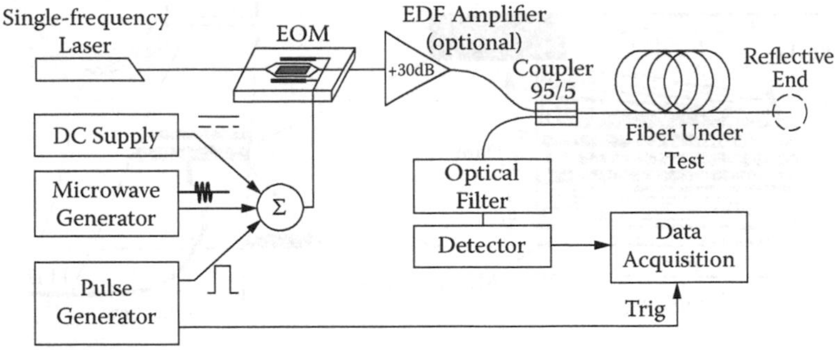

FIGURE 10.15
Schematic diagram of the distributed temperature sensing system based on stimulated Brillouin scattering.

density and microcracks are directly related to the maximum temperature the concrete experiences during the setting chemical process.

The sensor system is shown in Figure 10.15. The sensing mechanism used here is called *stimulated Brillouin scattering*, which is a unique parametric interaction offering a simultaneous sensitivity to temperature and strain. The Brillouin process in an optical fiber couples, through an acoustic wave, two counterpropagating light beams, which are frequency shifted by an amount dependent on the optical and elastic properties of silica. A Brillouin-based distributed sensor makes use of the temperature or strain dependence of the Brillouin shift. In practice, two beams—the pump and the Stokes waves—are launched into both ends of the fiber. The measurement is derived from the acquisition of the transmitted pump or Stokes signal (referred to as the *Brillouin loss* or *gain method*, respectively) as a function of the pump/Stokes frequency shift. This reveals a Lorentzian profile, and its central frequency is a linear function of the temperature and strain. Positional information, which requires that at least one of the beams be pulsed, is obtained through a standard time-delay analysis. The spatial resolution of the sensor is fixed by the duration of the pulse but is normally limited to about 1 m by the finite response time of the Brillouin interaction.

Normally, two highly stable, single-frequency lasers are required to achieve accurate measurement. The system shown in Figure 10.15 overcomes this drawback, and a single laser source is used for both pumping and probing. The key device is a Mach–Zehnder electro-optic modulator, which plays two roles at the same time—pulsing of the clockwise (CW) light from the laser to generate the pump signal and frequency-tuning and measurement of the probe signal via a microwave signal applied to the modulator. A spatial resolution of 1 m and a temperature accuracy of 1°C have been obtained with this system, although a temperature accuracy of 0.25°C can be obtained at the expense of longer measurement time. This system has been used for concrete setting temperature distribution in a concrete slab with dimensions of 15 m (*L*) × 10 m (*W*) × 3 m (*H*). These concrete slabs are used for raising the height

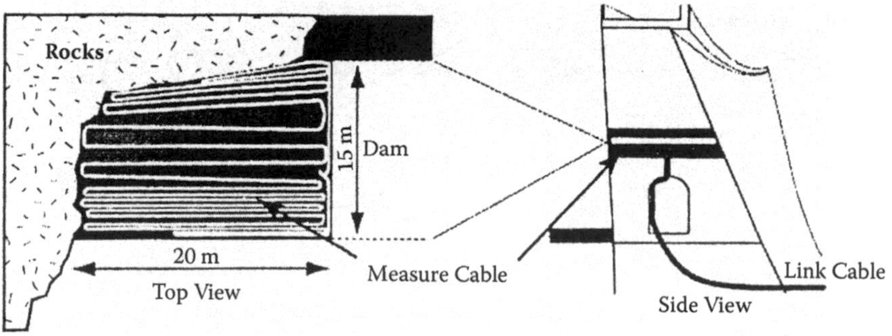

FIGURE 10.16
Layout of an optical communication cable inside the concrete slab.

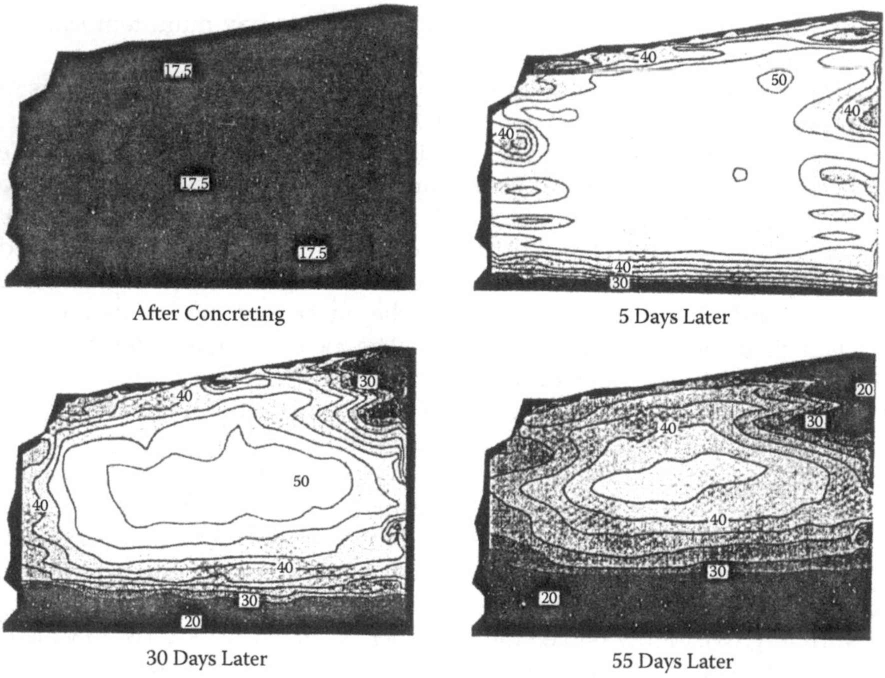

FIGURE 10.17
Temperature distribution during the setting of the concrete slab.

of the dam in order to increase the power capability of the associated hydroelectric plant. The layout of an optical communication cable inside the slab is shown in Figure 10.16, which gives a two-dimensional temperature distribution of the whole slab area. The fiber cable is installed during the concrete pouring. Figure 10.17 shows the temperature distribution over the slab at different times after concreting. It reveals that the temperature at the central area of the slab can be as high as 50°C, and it takes many weeks for this region to cool down.

10.2.3 Mines

Measurement of load and displacement changes in underground excavations of mines and tunnels is vital for safety monitoring. Multiplexed FBG sensor systems could replace the traditional electrical sensors, such as strain gauges and load cells, which cannot be operated in a simple multiplexed fashion and in a very hazardous environment with strong electromagnetic interference generated by excavating machinery. An FBG sensor system based on a broadband Er-doped fiber source and a tunable Fabry–Perot filter has been designed for long-term static displacement measurement in the ultimate roof of the mining excavations and in the hanging wall of the ore body's mineshaft [29]. A specially designed extensometer with a mechanical-level mechanism can cope with the large displacements of up to a few centimeters applied to the extensometer by controlling the overall strain change of the FBG to be less than 1%. This system is currently undergoing its field test.

10.2.4 Marine Vehicles

Advanced composite materials are currently finding an increased interest in marine vehicle design and construction as the introduction of new composite materials can reduce hull weight considerably and is especially attractive for fast vehicles. It is necessary to obtain a complete characterization of the behavior of such structures in order to achieve an optimum use of material for reinforcement and cost-effective construction. Approximately 100 sensors are required for monitoring bending moments, shear force, and slamming force at various positions of a vehicle model, and the test results are transferred to a full-scale vessel by appropriate scaling. The FBG sensor may be an ideal candidate for such a specified application. An FBG system based on the use of a dynamic locking distributed feedback (DFB) laser for wavelength-shift detection induced by strain has been demonstrated for measurement of the bending moment at the middle of a catamaran model, as shown in Figure 10.18 [30].

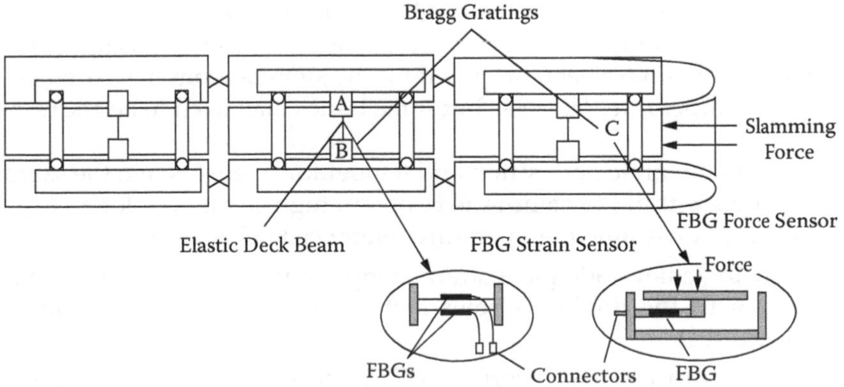

FIGURE 10.18
Schematic of a marine vehicle model with FBG strain and force sensors.

Two FBGs are mounted at the top and bottom of a stainless-steel beam that is a part of the model. An FBG positioned in the wet deck between the two hulls of the model is used for measurement of the slamming force generated by sea waves. The major advantage of using a DFB laser is its high S/N and the major disadvantage is limited wavelength tuning range (<1 nm). A limited range to resolution of ~500:1 has been demonstrated, which is less than that of most other interrogation schemes [4].

Advanced FBG systems with improved performance and capacity of multiplexing a very large number of FBGs are needed for this application. Recently, a 16-channel WDM FBG dynamic strain sensing system with interferometric detection was demonstrated for approaching such a goal [31]. A strain resolution of $< 10\,\text{n}\varepsilon/\sqrt{\text{Hz}}$ with a bandwidth of up to 5 kHz was achieved. This system has been successfully used for the slamming-load tests of a composite panel used in the design and fabrication of an air-cushion catamaran ship.

10.2.5 Aircraft

Advanced composite materials are now routinely used for manufacturing engineering structures such as aerospace structures (e.g., parts of airplane wings). Compared with metallic materials, advanced composite materials can have higher fatigue resistance, lighter weight, higher strength-to-weight ratio, the capability of obtaining complex shapes, and no corrosion. Hence, the use of composite materials with embedded FBG systems can lead to a reduction in weight, inspection intervals, and maintenance cost of aircraft—and, consequently, to an improvement in performance [32]. However, there is a major challenge in realizing *real-time* health and usage monitoring in service with an onboard sensor system. A distributed FBG sensor system could be ideally suitable for such an application. Because FBG sensors are sensitive to both strain and temperature, it is essential to measure strain and temperature simultaneously in order to correct the thermally induced strain for static strain measurement. A simple and effective method often used is to employ an unstrained temperature reference FBG, but this approach is not suitable in all cases; for example, for FBG sensors embedded in composites. A number of approaches have been proposed for the simultaneous measurement of strain and temperature [4,33,34], but some issues may need to be addressed:

1. How to integrate both strain and temperature sensors into the same location within the composite to obtain high spatial resolution (e.g., when an unstrained temperature reference FBG is used)
2. How to achieve adequate strain–temperature discrimination accuracy with the dual-parameter method—in particular, in a multiplexed system
3. How to process the distorted FBG signal if the FBG experiences a nonuniform strain along its length, which is likely to occur when the applied strain is large

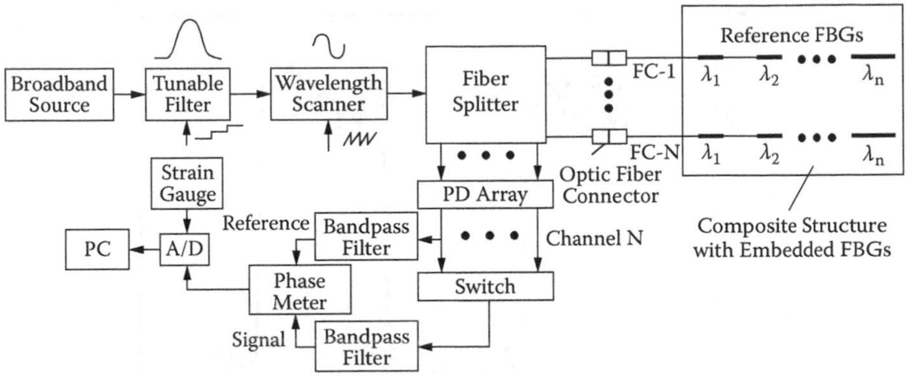

FIGURE 10.19
Principal diagram of SDM–WDM topology for dynamic strain monitoring of a composite specimen.

Strictly speaking, local and accurate temperature compensation for static strain measurement with FBG sensors embedded in composites is still a key problem to be solved. At this stage it may be easier to measure dynamic strain rather than static strain as temperature variations are normally much slower than dynamic strain changes and hence would not affect the measurement accuracy. Dynamic strain measurement can be used for vibration and impact mode analysis of the structure, which is highly desirable in practice.

Figure 10.19 shows a schematic diagram of a multiplexing system capable of measuring the two-dimensional dynamic strain distribution of a carbon fiber composite specimen used for aircraft [35]. This system combines spatial division multiplexing (SDM) with WDM. This SDM–WDM topology with a tunable wavelength filter (TWF) and an interferometric wavelength scanner (IWS) combines the advantages of the series and parallel multiplexing topologies and hence can create an efficient two-dimensional distributed FBG sensor network. Both the TWF and the IWS are located immediately after the source rather than in the front of the detector to allow FBG elements with similar wavelengths on all-fiber lines to be interrogated simultaneously. In operation, the TWF is used for selection of specified FBG channels with different center wavelengths along a single fiber line, whereas the IWS, arranged after the TWF, is used to achieve both high-resolution and high-speed wavelength-shift measurement.

A step strain change was applied to the composite to simulate a periodic transient strain. The results obtained from the FBGs embedded in the composite were in good agreement with those from conventional strain gauges used as a strain reference (see Figure 10.20). This system has a potential multiplexing capacity of up to a few hundred of FBGs with a strain resolution of sub-με and a bandwidth of up to megahertz. Hence, it could also be used for possible detection of acoustic emission either from material degradation (e.g., matrix cracking, fiber break) or from impact of foreign bodies with sufficient impact energy to cause damage and thus to achieve both strain and

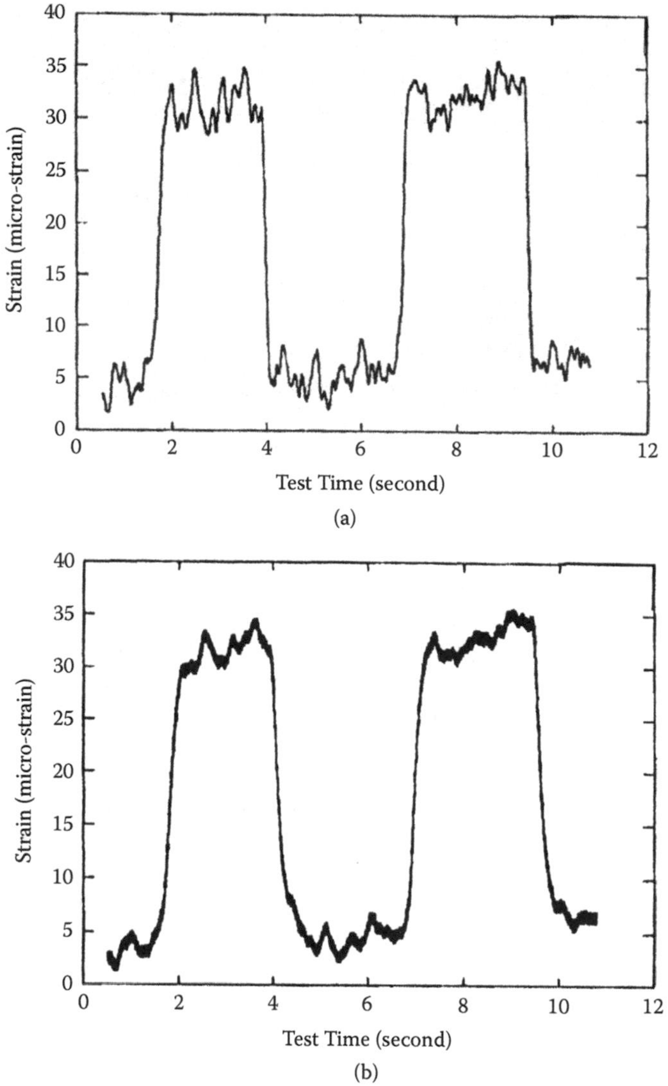

FIGURE 10.20
Results of dynamic strain measurement of composites with embedded FBGs. (a) FBG output; (b) strain gauge output.

damage detection with the same system. For such an application, FBGs have been combined with conventional piezoelectric sensors to provide a simpler solution, but piezoelectric sensors suffer from electromagnetic interference and shorter lifetimes [36].

In general, for FBGs embedded in composites it is ideal to measure all the multiple axes of strain and also temperature at multiple locations, as transverse strain loading may be on the order of the longitudinal strain for many applications. Recently, a technique based on the combination of an

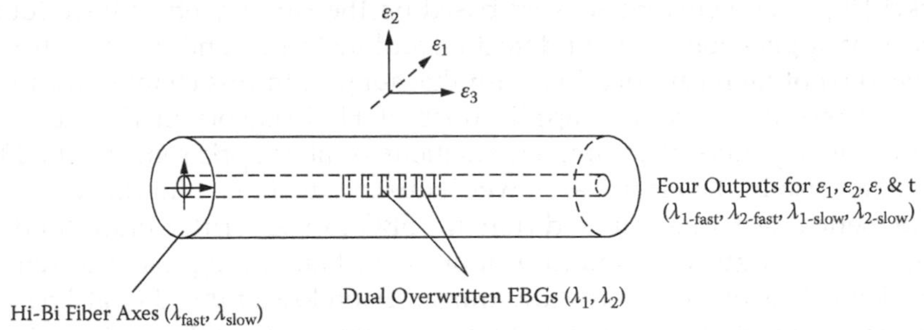

FIGURE 10.21
Schematic diagram of a multi-axis strain and temperature sensor based on dual FBGs overwritten onto a Hi-Bi fiber.

FBG and an in-line fiber Fabry–Perot sensor was demonstrated for simultaneous measurement of axial and transverse strain [37]. As the in-line FP sensor is insensitive to transverse strain, it can be used for axial strain measurement directly and the transverse strain is then obtained via subtracting the axial strain from the output of the FBG, which is intrinsically sensitive to both axial and transverse strain. Furthermore, a sophisticated technique of using dual overlaid FBGs written onto polarization-preserving fiber has been proposed to address the subject of simultaneous three-axis strain and temperature measurement, as shown in Figure 10.21 [38]. Four effective FBGs are generated with this procedure, which includes two FBGs at widely separated central wavelengths and two FBGs corresponding to each polarization axis of the fiber. The result is that there are four equations in four unknowns that relate the three axes of strain and temperature to the four grating wavelengths. This method could be very effective if adequate strain–temperature discrimination accuracy could be achieved.

Many research groups have demonstrated the "simple" operation of FBGs embedded in large composite or concrete structures for strain measurement; however, further work may be needed in order to realize a cost-effective, multifunctional FBG sensor multiplexing system that is able to measure static strain, temperature, and dynamic strain simultaneously with adequate resolution and accuracy. With such a type of system, the FBG sensor would be able to realize its full potential and perhaps dominate the market for health monitoring of large composite and concrete structures in the future.

10.3 Applications to the Electric Power Industry

Electric current measurement using FOSs in 1977 is probably the earliest application of FOSs in the electric power industry [39]. After more than two decades of development, fiber-optic current sensors recently entered the market [40]. These current sensors based on the Faraday effect have found important applications in fault detection and metering, and there is already a great deal of literature dealing with this subject. In this chapter, we focus on some recent progress in applications of FBG sensors in the electrical power industry. Like other implementations of fiber optic sensors, the FBG is ideal for use in the electrical power industry due to its immunity to electromagnetic interference. In addition, the FBG can be written onto standard 1.55-μm-wavelength telecommunication fiber; hence, long-distance remote operation is feasible due to the low transmission loss of the fiber at 1.55 μm. Loading of power transmission lines, winding temperature of electrical power transformers, and large electrical currents have been measured with the FBG sensor.

10.3.1 Load Monitoring of Power Transmission Lines

An excessive mechanical load on electrical power transmission lines, which may be caused by heavy snow, for example, may lead to a serious accident. In particular, for those lines located in, for example, mountainous areas, there is no easy access for inspection. Therefore, an online measurement system is needed to monitor the changing load on the power line. A multiplexed FBG system with more than 10 sensors distributed over a distance of 3 km has been demonstrated, as shown in Figure 10.22 [41]. The load change is simply converted into strain via a metal plate attached to the line and onto which the FBG is bonded. Obviously, many more sensors are required for such an application. WDM may no longer be able to cope with the significant increase in sensor number due to the limited bandwidth of the light source;

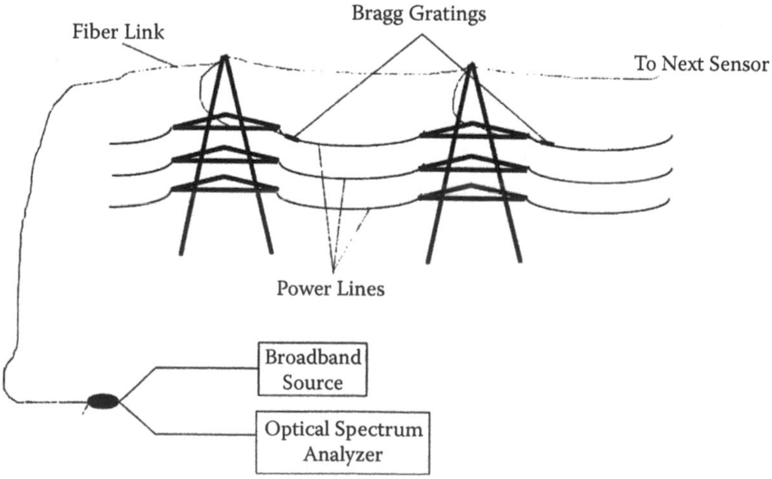

FIGURE 10.22
Schematic diagram of FBG load monitoring system for power transmission lines.

however, TDM could be used to improve the multiplexing capacity considerably. As the distance between adjacent FBGs is large, high-speed modulation and demodulation would not be required. Overall, this is an excellent example of applying FBG sensors for long-distance remote monitoring in harsh environments.

10.3.2 Winding Temperature Measurement

Knowledge of the local temperature distribution present in high-voltage, high-power equipment, such as generators and transformers, is essential in understanding their operation and in verifying new or modified products. Defective or degraded equipment can be detected by continuously monitoring the variations in the winding (the "hot spot") temperature, which reflects the performance of the cooling system. The FBG sensor has been demonstrated for such an application where the winding temperature of a high-voltage transformer is measured with two 1550-nm FBGs as sensing and reference elements interrogated via a standard optical spectrum analyzer [42]. A measurement accuracy of ±3°C has been achieved for long-term monitoring. If WDM is added into this system, *real-time* distributed measurement can be achieved.

10.3.3 Electric Current Measurement

Optical fiber sensors exploiting the Faraday effect have been intensively researched and developed for measurement of large currents at high voltages in the power distribution industry for more than a decade [43]. However, problems associated with induced linear birefringence, temperature, and vibration have limited the application of this technique. An alternative method is to measure the large current indirectly by using a hybrid system consisting of a conventional current transformer (CT) and a piezoelectric element (PZ). The CT converts the current change into a voltage variation, and then this voltage change is detected by measuring the deformation of the PZ using an FBG sensor [44].

An interferometric wavelength-shift detection method has been exploited for the detection of wavelength shift induced by the current, and a current resolution of $0.7\,\mathrm{A}/\sqrt{\mathrm{Hz}}$ over a range of up to 700 A has been obtained with a good linearity. More recently, the resolution has been further improved by replacement of the FBG with an FBG-based Fabry–Perot interferometer formed by arranging two FBGs with the same central wavelengths along a length of fiber, as shown in Figure 10.23 [45]. Figure 10.24(a) and Figure 10.24(b) show the experimental results for measurement of a 50-Hz, 12 A current and the performance of the sensor, respectively; from Figure 10.24(b), one can see that good linearity has been obtained. Compared with expensive CTs used in industry, these approaches offer a much lower cost alternative as the sophisticated electrical insulation is no longer required.

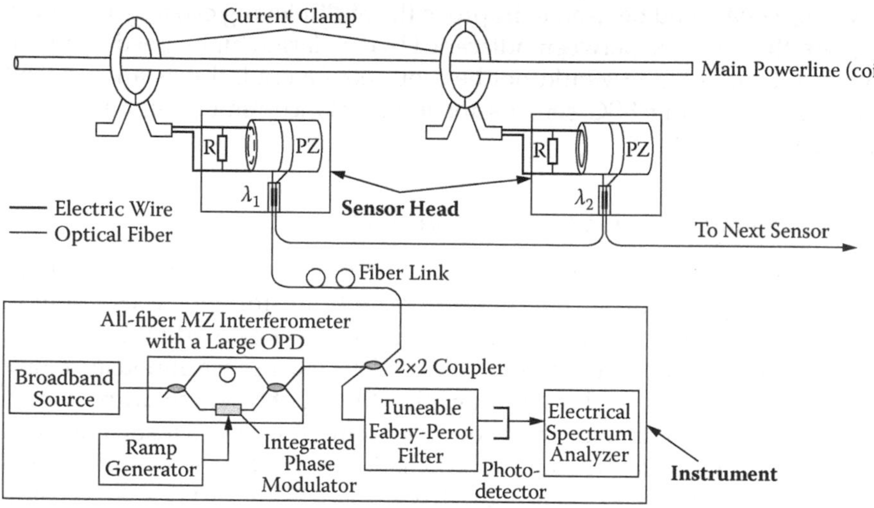

FIGURE 10.23
Schematic diagram of the multiplexed FBG-based Fabry–Perot current sensor system.

10.4 Applications to Medicine

The majority of commercial sensors widely used in medicine are electrically active and hence not appropriate for use in a number of medical applications—in particular, in high microwave/radiofrequency fields or ultrasound fields or laser radiation associated with hyperthermia treatment—due to local heating of the sensor head and the surrounding tumor due to the presence of metallic conductors and electromagnetic interference of currents and voltages in the metallic conductors, resulting in erroneous readings. Fiber optic sensors can overcome these problems as they are virtually dielectric. A range of miniature fiber optic sensors based on intensity modulation has been successfully commercialized in recent years [46]. Generally speaking, these sensors are all point sensors that can only provide readings over a small volume in the human body. Although passive multiplexing of these point sensors is possible, it is difficult to achieve in practice due to limitations on the probe size. By using the unique multiplexing property of the FBG sensor it is possible to realize quasi-distributed sensor systems with a single fiber link. A number of temperature and ultrasound sensing systems have been demonstrated to date.

10.4.1 Temperature

A novel FBG temperature sensor system has been demonstrated, as shown in Figure 10.25, in which high-resolution detection of the wavelength shifts induced by temperature changes is achieved using drift-compensated

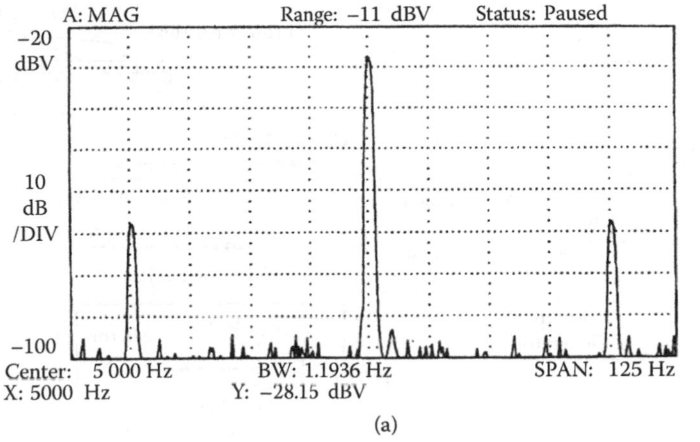

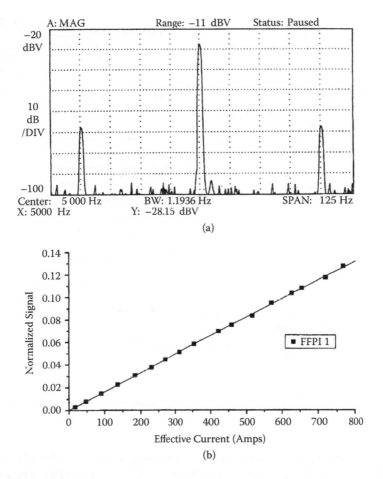

FIGURE 10.24
Results of current measurement using FBG-based Fabry–Perot interferometry. (a) Spectrum of carrier at 5 kHz and current signal for an effective current of 12 A; (b) normalized sideband signal amplitude as a function of effective current at 50 Hz.

interferometric detection while the return signals from the FBG sensor array are demultiplexed with a simple monochromator, which offers cross-talk free WDM [47]. A "strain-free" probe shown in the inset of Figure 10.25 is designed by enclosing the FBG sensor array in a protection sleeve. The inner diameter of the sleeve is very small (typically 0.5 mm); hence, it can support large transverse stress without transferring it to the fiber. The inner diameter is selected to be a few times larger than the diameter of the fiber, so that the inner surface of the sleeve does not contact the fiber under the maximum transverse stress condition. The fiber link is connected to the processing unit via a fiber connector, making the probe disposable and interchangeable. A resolution of 0.1°C and an accuracy of ±0.2°C over a temperature range of 30–60°C have been achieved, as shown in Figure 10.26, which meet or exceed the requirements for many medical applications.

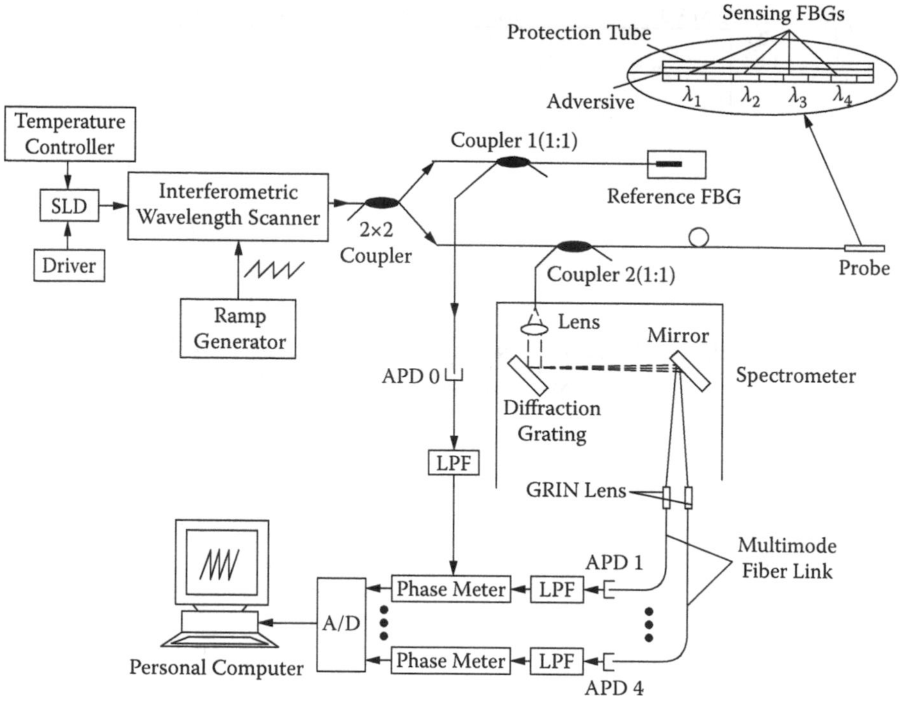

FIGURE 10.25
Schematic diagram of the multichannel FBG temperature sensor system. SLD: superluminescent diode; APD0–APD4: avalanche photodiodes: LPF: low-pass filter, A/D: analog-to-digital converter.

Another FBG temperature sensor system using a tunable Fabry–Perot filter (FPF) has been developed for achieving both wavelength-shift detection and WDM simultaneously, making the system simple and compact, as shown in Figure 10.27 [48]. The performance of the probe with four sensing FBGs is tested by placing it in a container inside an NMR machine with a high magnetic field of ~4.7 T. In-situ measurement is achieved by using an ~25 m long fiber line to link the instrument and the NMR machine. A resolution of 0.1°C and an accuracy of ±0.5°C over a temperature range of 25–60°C have been achieved. Figure 10.28 shows the experimental results obtained by cycling the FBGs from water at room temperature to ice. All the FBG sensors work well, and the transitions caused by suddenly changing the temperature of the FBGs are readily seen.

The spatial resolution is mainly determined by the thermal cross-talk between adjacent FBGs due to heat transfer along the connecting fiber (here the length of each FBG is shorter than the interval between adjacent FBGs). It may not be easy to determine the exact values of the spatial resolution by theory due to the relative complexity of the specific heat transfer conditions involved, such as the heat convection between the fiber and the surrounding

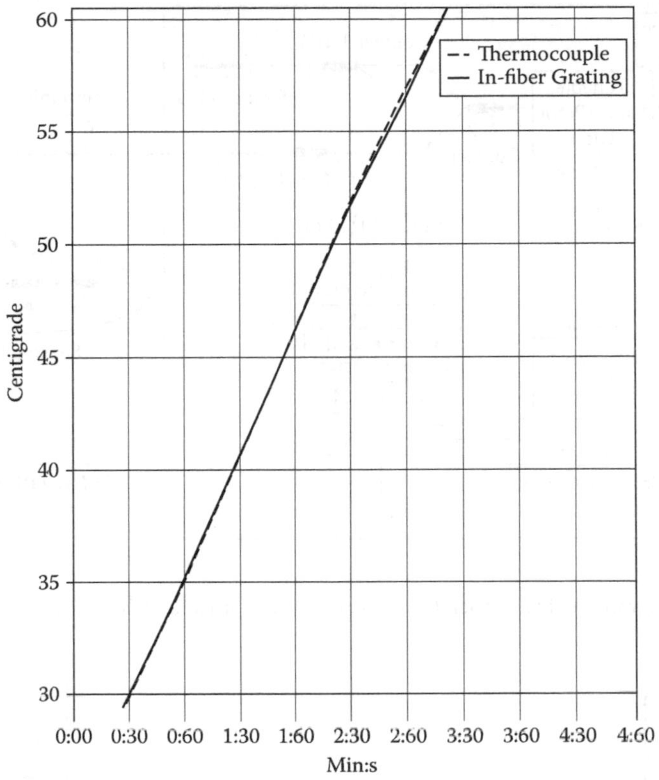

FIGURE 10.26
Experimental comparison between the FBG sensor and the thermocouple sensor.

medium. However, experiments could be used to measure the spatial resolution for specified experimental conditions. For example, an experiment has been carried out to determine the spatial resolution for FBGs exposed in air, in which three 1-mm-long FBGs with an interval of 5 mm were scanned through by a heated metal wire with a diameter of 0.2 mm at a constant temperature of ~42°C, as shown in Figure 10.29. It can be seen from the results in Figure 10.30 that the minimum spatial resolution without thermal cross-talk is approximately between 7 and 8 mm.

A more portable FBG system using a simple CCD spectrometer has also been employed. Compared with the two systems mentioned before, this approach is more simple and inexpensive [49,50]. The arrangement is similar to that shown in Figure 10.25, but the graded-index (GRIN) array has been replaced by a linear CCD array, so the Michelson interferometer is no longer required. The spectrum of the FBGs is recorded by a fast A/D converter, and the centroid of each FBG can be determined by fitting the FBG profile using a standard Gaussian function. A resolution of 0.1°C and an accuracy of ±0.2°C over a temperature range of 20–50° have been achieved, as shown

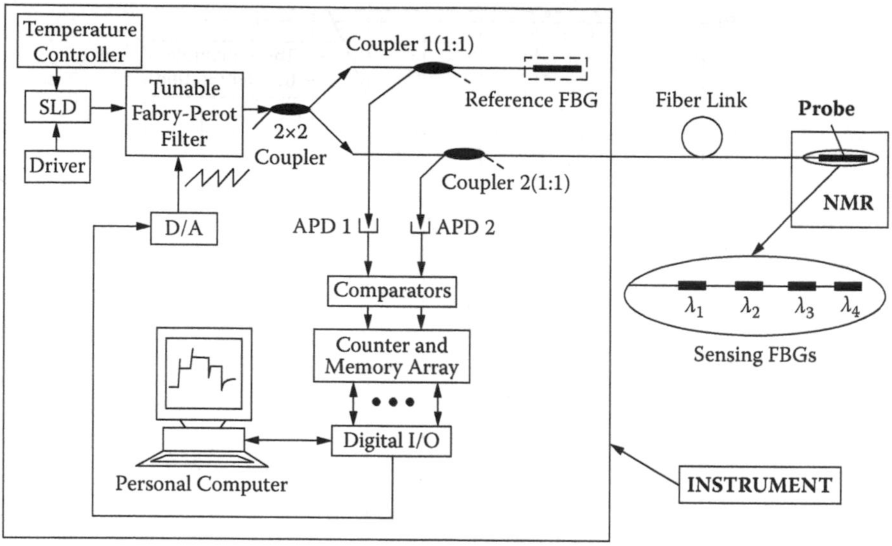

FIGURE 10.27
Schematic diagram of FBG temperature sensor system with a tunable FP filter.

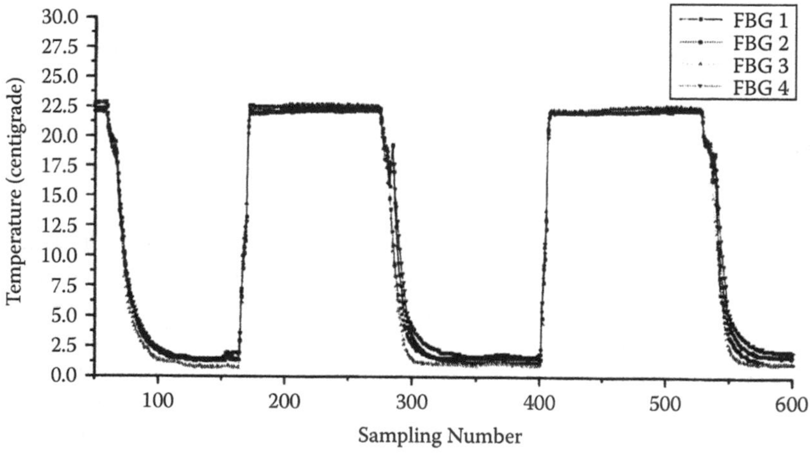

FIGURE 10.28
Results of temperature measurement with FBGs operated in a nuclear magnetic resonance (NMR) machine.

in Figure 10.31(a) and Figure 10.31(b), respectively. The measurement speed is relatively slow (a few seconds for each sensor) due to the limitation of the current signal processing speed; it can be improved in the future.

The FBG sensor can also be used for the measurement of the heart's efficiency based on the flow-directed thermodilution catheter method in which doctors inject patients with a cold solution to measure their heart's blood

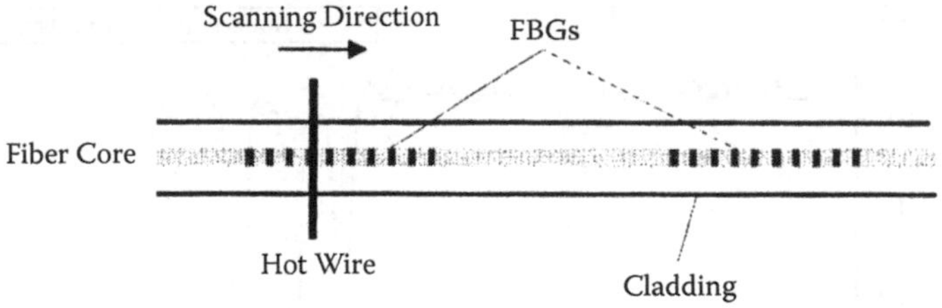

FIGURE 10.29
Schematic of experimental setup for thermal cross-talk test for FBG sensors scanned by a heated metal wire.

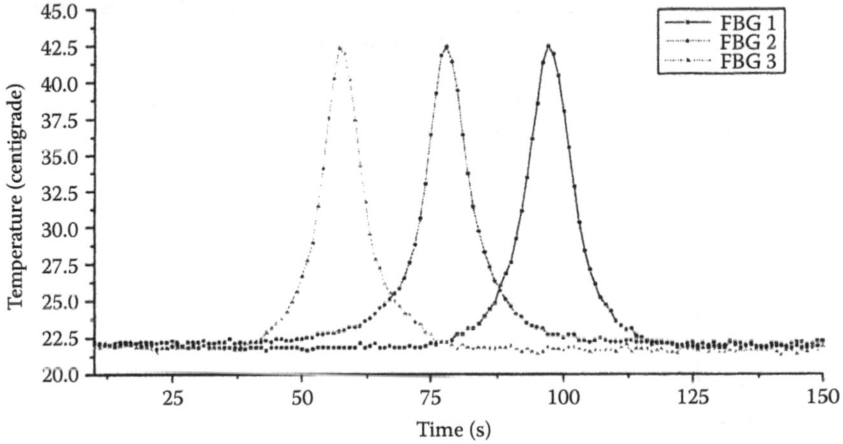

FIGURE 10.30
Experimental results of thermal cross-talk test for FBG sensors.

output. A flow-directed thermodilution catheter is inserted into the right atrium of the heart, allowing the solution to be injected directly into the heart for measurement of the temperature of the blood in the pulmonary artery. By combining temperature readings with pulse rate, doctors can determine how much blood the heart pumps. Such a type of catheter with an FBG sensor has been demonstrated for replacement of a conventional catheter with a thermistor or a thermocouple [51]. To simulate the change of blood flow due to the size change of the blood vessel when the pump rate is kept constant, a clamp is used to squeeze the tubing, as shown in Figure 10.32. The results for both clamped and unclamped cases with a pump rate of 108 rotations/min are shown in Figure 10.33(a) and Figure 10.33(b). It can be seen that for both the FBG sensor and the thermocouple the time delays have changed from ~0.9 to ~1.1 s due to the partial clamping, corresponding to a relative change of ~10%. Preliminary results show that the optical outputs are in good agreement with the electrical thermocouple. It is also found that as the FBG is

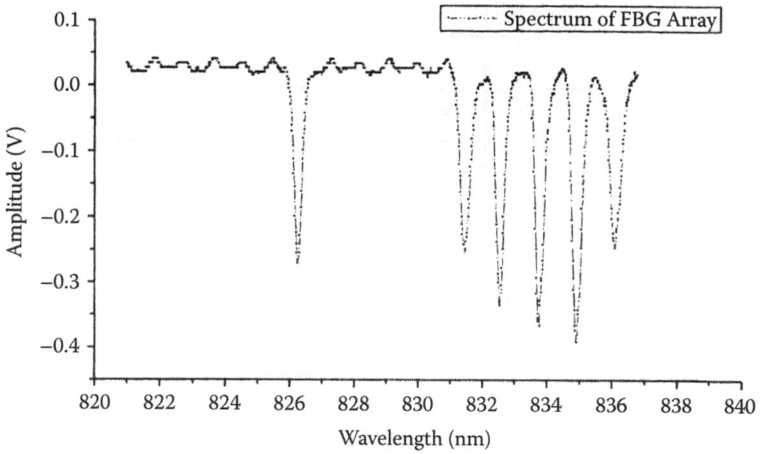

(a)

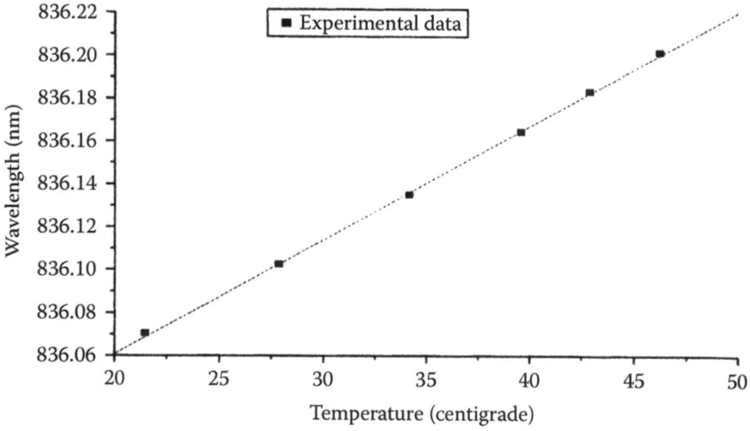

(b)

FIGURE 10.31
Results of the FBG temperature sensor system based on a CCD spectrometer. (a) Spectrum of sensing and reference FBGs; (b) experimental results.

longer than electrical sensors, more accurate measurement is achieved due to smoothing of temperature profiles.

10.4.2 Ultrasound

Similar to temperature monitoring for the assessment of the operating safety of high-RF or microwave fields mentioned previously, an ultrasound sensor is required to monitor the output power from diagnostic ultrasound equipment used for a range of medical applications (including ultrasound surgery, hyperthermia, and lithotripsy). Piezoelectric devices are the most common

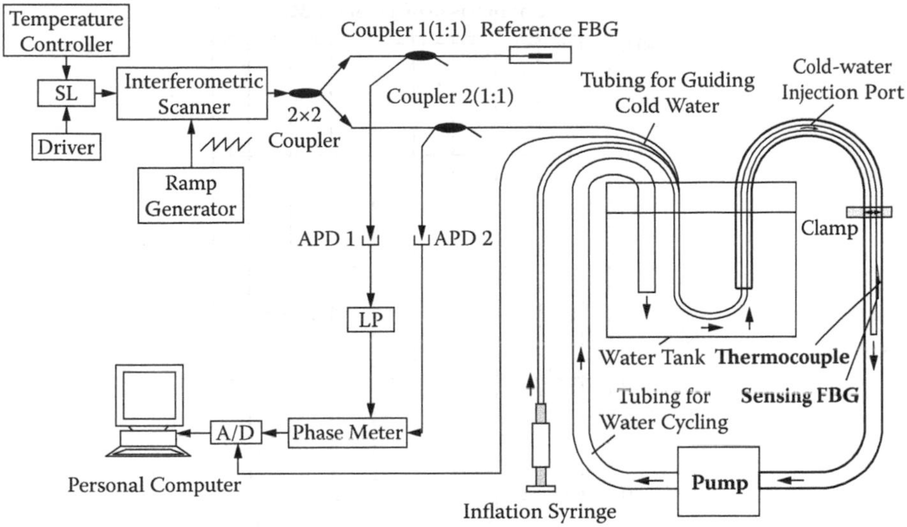

FIGURE 10.32
Schematic diagram of the FBG-based flow-directed thermodilution catheter system. SLD: superluminescent diode; APD1 and APD2: avalanche photodiodes; LPF: low-pass filter; A/D: analog-to-digital converter.

sensors but suffer from a susceptibility to electrocomagnetic interference and signal distortion and from the difficulty of direct determination of ultrasound fields in vivo due to the limitation of the probe size. The FBG sensor can overcome these problems and is able to measure the ultrasound field at several points simultaneously due to its unique multiplexing capability.

An FBG sensing system based on interferometric detection with heterodyne signal demodulation has been demonstrated for detection of a focused ultrasound field at a frequency of ~2 MHz [52]. The maximum frequency range measurable is mainly limited by the length of the FBG used (typically 4–10 mm) as the wavelength of the high-frequency ultrasound waves needs to be smaller than the length of the FBG. Apart from the wanted wavelength modulation induced by the ultrasound, an undesired amplitude modulation of the output signal has been observed. This is because compressional standing waves set up by the ultrasound in the fiber only partially modulate the grating (as their wavelength is less than the length of the grating); the grating is subject to a nonuniform strain and hence leads to regions of the grating acting as spectral filters for the back-reflected light from other regions of the grating. As a consequence, an FBG with a length of 1 mm has been used to reduce this amplitude modulation effect [53]. Except for the region where the 1-mm FBG is located, all other regions of the fiber near the ultrasound field need to be desensitized to eliminate the standing waves in the fiber; this is achieved simply by jacketing the fiber. A pressure resolution of $\sim 10^{-3}\ \mathrm{Atm}/\sqrt{\mathrm{Hz}}$ has been demonstrated.

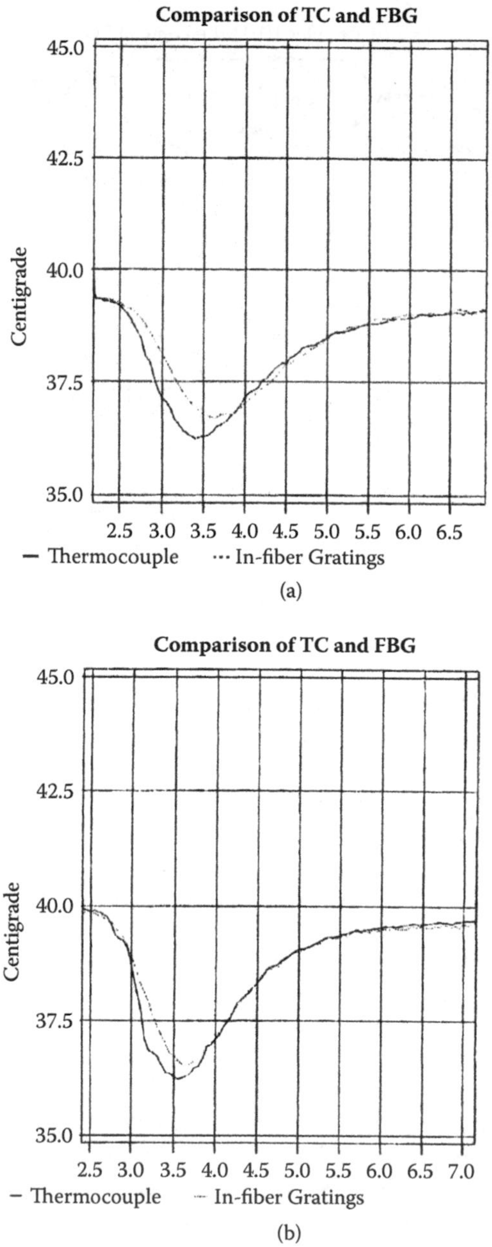

FIGURE 10.33
Experimental results of simulation for blood-vessel blocking test. (a) With clamping of the tubing; (b) without clamping of the tubing.

10.5 Applications to Chemical Sensing

Similar to applications of FOSs in the electric power industry, FOSs have found widespread applications for chemical sensing, in particular those based on multimode fibers used for light transmission between the sensor head and the data processing unit [54]. In environmental and pollution control, FOSs are being developed for sensing gases such as methane, ammonia, sulfur dioxide, nitrogen oxides, and hydrocarbon and for monitoring seawater and drinking water. Biomedical applications include in-vivo sensors for pH, O_2, and CO_2 levels in blood, glucose, and cholesterol sensors. In the chemical industry, the nonconductive and passive nature of FOSs makes them safe in hazardous environments, while their flexibility allows access to remote or dangerous areas. In recent years, chemical FOSs based on the evanescent wave coupling, where interaction with a chemical or intermediate dye occurs within the evanescent field region of the fiber, have emerged and attracted considerable research interest [55]. Side-polished fiber and D-fiber chemical sensors have been demonstrated for the detection of pH, O_2, methane, hydrocarbon, and others. The FBG sensor is also used for chemical sensing based on the fact that the central wavelength of an FBG varies with refractive index change—that is, chemical concentration change—via the evanescent field interaction between the FBG and the surrounding chemical.

An approach based on an FBG written onto an etched D-fiber has been demonstrated [56], and recently a modified version based on a side-polished fiber configuration was reported as a refractive index sensor, allowing fast online measurements of chemicals, such as carbon hydrides in the petrol industry [57]. Because of the constraint of current interrogation techniques with a typical resolution of 1 pm for static wavelength-shift measurement, the sensitivity obtained is much lower (up to 10^{-5}) compared with other fiber-optical techniques, such as the interferometric method with a sensitivity of up to 10^{-8}. A new type of fiber grating called *long-period grating* (LPG) has been discovered to be more sensitive to the refractive index change of the material around the grating cladding when compared with FBGs [58]. As LPGs couple light from the forward-propagating mode into several forward-propagating cladding modes, as shown in Figure 10.34, any variation on the refractive index of the

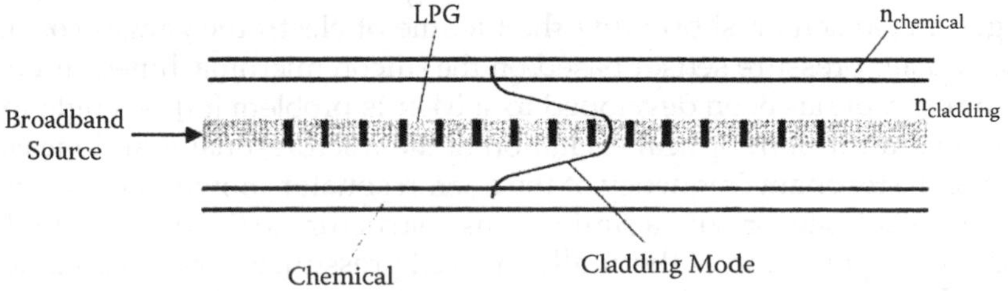

FIGURE 10.34
Schematic diagram of the long-period grating chemical sensor.

material around the cladding modifies the transmission spectrum properties to generate loss peaks. A sensitivity of up to 10^{-7} is possible with further improvement of the interrogation resolution. As LPGs are also sensitive to temperature change, temperature compensation is required. The concentration of a number of chemicals, including ethanol, hexanol, methylcyclohexane, hexadecane [59], $CaCl_2$, and $NaCl_2$ [60], has been tested with LPGs. In principle, any chemical with its loss peak lying in the refractive index range from 1.3 to 1.45 can be detected using such an LPG technique.

10.6 Applications to the Oil and Gas Industry

FOSs could be ideal for applications in the oil and gas industry due to their inherent advantages, such as being intrinsically safe, immune to EMI, workable at high temperature, capable of multiplexing, and minimally invasive. Of these advantages, the multiplexed or distributed sensing feature is particularly attractive for downhole applications, where monitoring of a parameter or parameters at many spatial locations through the wellbore is essentially necessary. The major applications of FOSs in the oil and gas industry concentrate on sensing a wide range of parameters, such as pressure, temperature, vibration, flow, and acoustic fields for downhole oil and gas reservoir monitoring, in both retrievable and permanently installed systems. Significant progress has been made in recent years, especially in pressure and temperature measurement.

10.6.1 Pressure Sensing

It has been found that hydrocarbon accumulations in relatively small reservoirs are at an increasing depth. In order to improve recovery efficiency of these reservoirs, it is of importance to obtain primary reservoir data on a continuous basis—that is, online information for downhole pressure, downhole temperature, and downhole flow velocity. The typical temperature range for these measurements is between 150 and 250°C. Such a high environmental temperature has caused problems in applying traditional electronic downhole gauges, in particular, shortening the lifetime of electronic gauges considerably. A point pressure sensor based on the silicon micromachined resonator sensor concept has been developed to solve this problem [61]. A single-mode fiber is used for both optical excitation of the microresonator and measurement of the resonant frequency of the microresonator shown in Figure 10.35. A temperature sensor with a similar sensor structure to the pressure probe is used for compensation of thermally induced pressure errors. An accuracy of 0.2 bar has been achieved. This system has demonstrated its excellent long-term stability via a 5-year test without any maintenance and recalibration. Such a sensor was successfully commercialized recently.

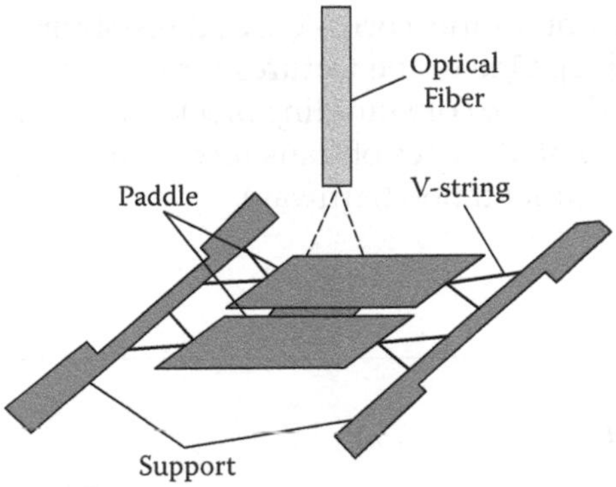

FIGURE 10.35
Schematic diagram of the fiber optic pressure sensor based on a silicon micromachined resonator.

The remaining problem for this type of sensor system is the multiplexing of a number of sensors. Time division multiplexing (TDM) and code division multiplexing (CDM) have been demonstrated for multiplexing silicon microresonator sensors [62,63], but it is very difficult to multiplex a large number of sensors due to limitations in S/N or cross-talk. Recently, a novel approach based on FBGs written onto side hole single-mode fibers was reported to increase the multiplexing capability significantly [64]. Under high-pressure conditions, dual effective Bragg reflective wavelengths are generated from a single FBG wavelength due to stress-induced bi-refringence in the core as there are two air holes in the cladding. By detection of the peak-to-peak spectral separation of the two wavelengths, it is found that the side hole fiber FBG pressure sensor has a much higher sensitivity over the normal single-mode fiber FBG pressure sensor, and, more interestingly, its temperature sensitivity is much lower instead. This may allow straightforward temperature compensation in practical applications.

10.6.2 Temperature Sensing

An effective technique to improve production efficiency in oil recovery is to use steam flood wells, where steam is injected into a well to heat the trapped oil. The temperature profile of the heat flow from the borehole to reservoir is of importance for optimization of the oil recovery. Distributed temperature sensors based on the Raman backscattering have been widely deployed for this application. The spatial resolution is typically 1–3 m, limited by the "width" of optical pulses from the laser source. The measurement range can be up to 10 km. The main drawbacks for the distributed temperature sensor are (1) slow data processing (typically a few minutes) due to the inherent weak nature of the Raman signal, and (2) it is only suitable for temperature monitoring. New FBG sensors are under development to solve these problems, and it is anticipated that distributed FBG sensor systems for measurement of multiparameters, such as pressure, temperature, vibration, acoustic fields, flow, oil and water ratio, corrosion, and resistivity, could become possible in the future [65]. One of the features for such types of systems is that the FBG can be used as the core building block for a number of wavelength-encoded sensors so that all types of transducers are compatible with a single form of surface instrumentation hardware.

10.7 Discussion

In this chapter, recent progress in applications of FOSs to large composite and concrete structures, the electrical power industry, medicine, chemical sensing, and the oil and gas industry has been reviewed with emphases on FBG sensors. Recent applications have concentrated on the strain mapping of large composite and concrete structures in a surface-attached or embedded fashion, and this may lead to the development of a major market for FOSs if cost-effective FOS systems could become available. FOSs based on FBGs and low-coherence interferometry have provided a good example for bridge monitoring. Distributed FOSs will play a more important role in health monitoring of large structures, such as dams, due to their unique feature of a very long working range. For the embedding mode of operation in composites, the correction of thermally induced strain is still a difficult problem for high-accuracy static strain measurement in practice due to limitation of either the spatial resolution or strain/temperature discrimination accuracy, or possible nonuniform strain; this needs to be solved in the future. Also, simultaneous measurement of multiaxis strain and temperature in composites remains as a major challenge for the optical fiber sensors community, due to its complexity.

Fiber optic electric current sensors based on the Faraday effect have successfully entered the market eventually after more than two decades of development. A number of electric field or voltage sensors are under the stage of field tests. Preliminary studies have shown that the FBG sensor is a promising technique for use in the electrical power industry, and there is no doubt that more application examples can be foreseen, due to the capability of FBGs to be used in high-voltage environments. For medical applications, extrinsic FOSs based on multimode fiber transmission have matured. A number of FBG temperature and ultrasound sensor systems have been developed. With further engineering, it is anticipated that these FBG systems could be used for in vivo measurement of temperature and/or ultrasound. In chemical sensing, FOSs based on evanescent wave coupling are still under development. FBG and LPG sensors represent another active application area where quasi-distributed measurement is required.

References

1. B. Culshaw and J. P. Dakin, eds., *Optical Fiber Sensors*, Vols. 3 and 4, Artech House, Boston, 1996.
2. K. T. V. Grattan and B. T. Meggitt, eds., *Optical Fibre Sensor Technology II: Devices and Technology*, Chapman and Hall, London, 1998.
3. W. W. Morey, G. Meltz, and W. H. Glenn, Fibre optic Bragg grating sensors, *Proc. SPIE*, **1169**, pp. 98–107, 1989.
4. Y. J. Rao, Review article: In-fibre Bragg grating sensors, *Meas. Sci. and Technol.*, **8**, pp. 355–375, 1997.
5. K. T. Grattan and B. T. Meggitt, eds., *Optical Fibre Sensor Technology*, Chapman and Hall, London, pp. 269–310, 1995.
6. Y. J. Rao and D. A. Jackson, Review article: Recent progress in fibre-optic low-coherence interferametry, *Meas. Sci and Technol.*, 7, pp. 981–999, 1996.
7. X. Bao, J. Dhliwayo, N. Heron, D. J. Webb, and D. A. Jackson, Experimental and theoretical studies on a distributed temperature sensor based on Brillouin scattering, *J. Lightwave Technol.*, **13**, pp. 1340–1346, 1995.
8. R. D. Townsend and N. H. Taylor, Fibre optic monitoring of temperature and strain along insulated pipework at high temperature, *Proc. Euromaintenance '96*, 1996.
9. U. Udd, ed., *Fibre-Optic Smart Structures*, Wiley, New York, 1995.
10. A. Mendez, T. E. Morse, and E. Mendez, Applications of embedded optical fibre sensors in reinforced concrete buildings and structures, *Proc. SPIE*, **1170**, pp. 60–69, 1989.
11. R. M. Measures, Smart structures in nerves of glass, *Prog. Aerosp. Sci.*, **26**, pp. 289, 1989.
12. R. M. Measures, A. T. Alavie, R. Maaskant, M. Ohn, S. Karr, and S. Huang, Bragg grating structural sensing system for bridge monitoring, *Proc. SPIE*, **2294**, pp. 53–60, 1994.
13. R. M. Measures, R. Maaskant, T. Alaview, R. M. Measures, G. Tadros, S. H. Rizkalla, and A. G. Thakurta, Fibre-optic Bragg gratings for bridge monitoring, *Cement Concrete Composites*, **9**, pp. 21–23, 1997.
14. S. M. Melle, K. Liu, and R. M. Measures, A passive wavelength demodulation system for guided-wave Bragg grating sensors, *IEEE Photon. Technol. Lett.*, **4**, pp. 516–518, 1992.
15. J. Meissner, W. Nowak, V. Slowik, and T. Klink, Strain monitoring at a prestressed concrete bridge, *Proc. of 12th Int. Conf. Optical Fibre Sensors*, Williamsburg, VA, pp. 408–411, 1997.
16. W. L. Schulz, E. Udd, J. M. Seim, and G. E. McGill, Advanced fibre grating strain sensor systems for bridges, structures, and highways, *Proc. SPIE*, **3325**, pp. 212–221, 1998.
17. A. D. Kersey, M. A. Davis, T. A. Berkoff, D. G. Bellemore, K. P. Koo, and R. T. Jones, Progress towards the development of practical fibre Bragg grating instrumentation systems, *Proc. SPIE*, **2839**, pp. 40–63, 1996.
18. Y. J. Rao, P. J. Henderson, D. A. Jackson, L. Zhang, and I. Bennion, Simulations strain, temperature and vibration measurement using a multiplexed in-fibre Bragg grating/fibre-Fabry–Perot sensor system, *Electron. Lett.*, **33**, 2063–2064, 1997.

19. J. P. Dakin, W. Ecke, M. Rothhardt, J. Schauer, K. Usbeck, and R. Willsch, New multiplexing scheme for monitoring fibre optic Bragg grating sensors in the coherence domain, *Proc. OFS'12*, pp. 31–34, 1997.
20. J. P. Dakin, V. Foufelle, S. J. Russell, O. Hadeler, W. Ecke, E. Geinitz, J. Schauer, and R. Willsch, Sensor network for structural strain and high hydraulic pressure, using optical fibre grating pairs interrogated in the coherence domain, *Proc. 13th Int. Conf. Optical Fibre Sensors*, Kyongju, Korea, pp. 157–160, 1999.
21. D. A. Jackson and J. D. C. Jones, Optical fibre interferometry, in B. Culshaw and J. P. Dakin, eds., *Optical Fibre Sensors II*, Chap. 10, pp. 359–361, Academic Press, San Diego, 1989.
22. G. A. Ball, W. W. Morey, and W. H. Glenn, Standing-wave monomode erbium fibre laser, *IEEE Photon. Tech. Lett*, **3**, pp. 613–615, 1991.
23. K. P. Koo and A. D. Kersey, Bragg grating-based laser sensor systems with interferometric interrogation and wavelength division multiplexing, *J. Lightwave Technol.*, **13**, pp. 1243–1249, 1995.
24. U. Sennhauser, R. Bronnimann, and P. M. Nellen, Reliability modelling and testing of optical fibre Bragg sensors for strain measurements, *Proc. SPIE*, **2839**, pp. 64–75, 1996.
25. P. A. Robertson and B. P. Ludden, A fibre optic distributed sensor system for condition monitoring of synthetic ropes, *Proc. IEEE Colloquium on Optical Techniques for Smart Structures* and *Structural Monitoring*, London, Feb. 1997.
26. D. Inaudi and S. Vurpillot, Structure monitoring by curvature analysis using interferometric fibre optic sensors, *Smart Mater and Struct.*, **7**, pp. 199–208, 1997.
27. V. Bhatia, M. E. Jones, J. L. Grace, K. A. Murphy, R. O. Claus, J. A. Greene, and T. A. Tran, Applications of "absolute" fibre optic sensors to smart materials and structures, *Proc. OFS'10*, pp. 171–174. 1994.
28. L. Thevenaz, M. Facchini, A. Fellay, P. Robert, D. Inaudi, and B. Dardel, Monitoring of large structures using distributed Brillouin fiber sensing, *Proc. OFS'13*, pp. 345–348, 1999.
29. P. Ferdinand, O. Ferragu, J. L. Lechien, B. Lescop, V. Marty. V. S. Rougeault, G. Pierre, C. Renouf. B. Jarret, G. Kotrolsios, V. Neuman, Y. Depeursings, J. B. Michel, M. V. Uffelen, Y. Verbandt, M. R. H. Voet, and D. Toscano, Mine operating accurate stability control with optical fibre sensing and Bragg grating technology: The BRITE-EURAM *STABILOS* project, *J. Lightwave Technol.*, **13**, pp. 1303–1313, 1995.
30. D. R. Hjelme, L. Bjerkan, S. Neegard, J. S. Rambech, and J. V. Aarsnes, Application of Bragg grating sensors in the characterization of sealed marine vehicle modes. *Appl. Opt. 1997*, **36**, pp. 328–326, 1997.
31. S. T. Vohra, M. A. Davis, A. Dandridge, C. C. Chang, B. Althouse, H. Patrick, M. Putnam, T. Tsai, G. Wang, P. O. Baalerud, G. B. Haavsgard, and K. Pran, Sixteen channel WDM fibre Bragg grating dynamic strain sensing system for composite panel slamming tests, *Proc. 12th Int. Conf. Optical Fibre Sensors*, Williamsburg, VA, paper #PDP3-1, 1997.
32. P. D. Foote. Fibre Bragg grating strain sensors for aerospace smart structures, *Proc. SPIE*, **2361**, pp. 162–166, 1994.
33. J. D. C. Jones, Review of fibre sensor technique for temperature-strain discrimination, *Proc. 12th Int. Conf. Optical Fibre Sensors*, Williamsburg, VA, pp. 35–39, 1997.

34. T. Liu, G. F. Fernando, L. Zhang, I. Bennion, Y. J. Rao, and D. A. Jackson, Simulations strain and temperature measurement using a combined fibre Bragg grating/extrinsic Fabry–Perot sensor, *Proc. 12th Int. Conf. Optical Fibre Sensors*, Williamsburg, VA, pp. 40–43, 1997.
35. Y. J. Rao, D. A. Jackson, L. Zhang, and I. Bennion, Strain sensing of modern composite materials with a spatial/wavelength multiplexed fibre grating network, *Opt. Lett.*, **21**, pp. 683–685, 1996.
36. J. M. Menendez, S. Diaz-Carrillo, C. Pardo de Vera, and J. A. Guemes, Embedded piezoelectrics and optical fibres as strain and damage sensors of large composite structures, *Proc. 17th Conf. Aerospace Materials*, Paris, 1997.
37. X. D. Jin, J. S. Sirkis, and V. S. Venkateswaran, Simultaneous measurement of two strain components in composite structures using embedded fibre sensors, *Proc. 12th Int. Conf. Optical Fibre Sensors*, Williamsburg, VA, pp. 44–47, 1997.
38. E. Udd, D. Nelson, and C. Lawrence, Multiple axis strain sensing using fibre gratings written onto birefringent single mode optical fibre, *Proc. 12th Int. Conf. Optical Fibre Sensors*, Williamsburg, VA, pp. 354–357, 1997.
39. A. J, Rogers, Optical-fibre current measurement, *Int. J. Optoelectronics*, **3**, pp. 391–407, 1988.
40. T. Bosselmann, Magneto- and electro-optic transformers meet expectations of power industry, *Proc. OFS'12*, pp. 111–114, 1997.
41. Y. Ogawa, J. I. Iwasaki, and K. Nakamura, A multiplexing load monitoring system of power transmission lines using fibre Bragg grating, *Proc. 12th Int. Conf. Optical Fibre Sensors*, Williamsburg, VA, pp. 468–471, 1997.
42. T. E. Hammon and A. D. Stokes, Optical fibre bragg grating temperature sensor measurements in an electrical power transformer using a temperature compensated fibre Bragg grating as a reference, *Proc. 11th Int. Conf. Optical Fibre Sensors*, Sapporo, Japan, pp. 566–569, 1996.
43. G. W. Day, M. N. Deeter, and A. H. Rose, Faraday effect sensors: A review of recent progress, *Proc. SPIE*, **PM07**, pp. 11–26, 1992.
44. P. J. Henderson, N. E. Fisher, and D. A. Jackson, Current metering using fibre-grating based interrogation of a conventional current transformer, *Proc. 12th Int. Conf. Optical Fibre Sensors*, Williamsburg, VA, pp. 186–189, 1997.
45. Y. J. Rao, P. J. Henderson, N. E. Fischer, D. A. Jackson, L. Zhang, and I. Bennion, Wavelength-division-multiplexed in-fibre Bragg grating Fabry–Perot sensor system for quasi-distributed current measurement, *Proc. Annual Conf. Applied Optics and Optoelectronics (IOP)*, Brighton, pp. 99–104, 1998.
46. B. Trimble, Fifty thousand pressure sensors per year: A successful fibre sensor for medical applications, *Proc. OFS'9*, pp. 457–462, 1993.
47. Y. J. Rao, D. J. Webb, D. A. Jackson, L. Zhang, and I. Bennion, In-fibre grating temperature sensor system for medical applications, *J. Lightwave Technol.*, **15**, pp. 779–785, 1997.
48. Y. J. Rao, D. J. Webb, D. A. Jackson, L. Zhang, and I. Bennion, In-situ temperature monitoring in NMR machines with a prototype in-fibre Bragg grating sensor system, *Proc. 12th Int. Conf. Optical Fibre Sensors*, Williamsburg, VA, pp. 646–649, 1997.
49. C. G. Atkins, M. A. Putman, and E. J. Friebele, Instrumentation for interrogating many-element fibre Bragg grating sensors, *Proc. SPIE*, **2444**, pp. 257–264, 1995.

50. A. Ezbiri, S. E. Kanellopoulos, and V. A. Handerek, High resolution instrumentation system for demodulation of Bragg grating aerospace sensors, *Proc. 12th Int. Conf. Optical Fibre Sensors*, Williamsburg, VA, pp. 456–459, 1997.
51. Y. J. Rao, D. J. Webb, D. A. Jackson, L. Zhang, and I. Bennion, In-fibre Bragg grating flow-directed thermodilution catheter for cardiac monitoring, *Proc. 12th Int. Conf. Optical Fibre Sensors*, Williamsburg, VA, pp. 354–357, 1997.
52. N. E. Fisher, S. F. O'Meill, D. J. Webb, C. N. Pannell, D. A. Jackson, L. R. Gavrilov, J. W. Hand, L. Zhang, and I. Bennion, Response of in-fibre Bragg gratings to focused ultrasound fields, *Proc. 12th Int. Conf. Optical Fibre Sensors*, Williamsburg, VA, pp. 190–193, 1997.
53. N. E. Fisher, D. J. Webb, C. N. Pannell, D. A. Jackson, L. R. Gavrilov, J. W. Hand, L. Zhang, and I. Bennion, Probe for measuring ultrasound fields using short in-fibre Bragg gratings, *Proc. SPIE*, **3555**, pp. 451–456, 1998.
54. O. S. Wolfoeis, ed., *Fibre Optic Chemical Sensors*. Vols. I & II, CRC Press, Boca Raton, FL, 1991.
55. G. Stewart and W. Johnstone, Evanescently coupled components, in B. Culshaw and J. P. Dakin, eds., *Optical Fiber Sensors*, Vol. 3, ch. 3, Artech House, Boston, 1996.
56. G. Meltz, S. J. Hewlett, and J. D. Love, Fibre grating evanscent-wave sensors, *Proc. SPIE*, **2836**, pp. 342–350, 1996.
57. W. Ecke, K. Usbeck, V. Hagemann, R. Mueller, and R. Willsch, Chemical Bragg grating sensor network basing on side-parallel optical fibre, *Proc. SPIE*, **3555**, pp. 457–466, 1998.
58. V. Bhatia and A. M. Vengsarkar, Optical fibre long-period grating sensors, *Opt. Lett.*, **21**, pp. 692–694, 1996.
59. Z. Zhang, J. S. Sirkis, Temperature-compensated long period grating chemical sensor, *Proc. 12th Int. Conf. Optical Fibre Sensors*, Williamsburg, VA, pp. 294–297, 1997.
60. R. Falciai, A. G. Mignani, and A. Vannini, Optical fibre long-period gratings for the refractometry of aqueous solutions, *Proc. SPIE*, **3555**, pp. 447–450, 1998.
61. P. Eigenraam, B. S. Douma, and A. P. Koopman, Applications of fibre optic sensors and instrumentation in the oil and gas industry, *Proc. OFS'13*, pp. 602–607, 1999.
62. Y. J. Rao, D. Uttamchandani, and B. Culshaw, Passive multiplexing of silicon microresonator sensors using a high speed fibre optic TDM system, *Int. J. Optoelectronics*, **8**, pp. 77–85, 1993.
63. Y. J. Rao, D. Uttamchandani, and B. Culshaw, Spread spectrum technique for passive multiplexing of reflective frequency-out fibre optic sensors, *Opt. Comm.*, **96**, pp. 214–217, 1993.
64. R. J. Schroeder, T. Yamate, and E. Udd, High pressure and temperature sensing for the oil industry using fibre Bragg gratings written onto side hole single mode fibre, *Proc. OFS'13*, pp. 42–45, 1999.
65. A. D. Kersey, Optical fibre sensors for downwell monitoring applications in the oil and gas industry, *Proc. OFS'13*, pp. 326–331, 1999.

11

Fiber Optic Bio and Chemical Sensors

Shizhuo Yin, Chun Zhan, and Paul B. Ruffin

CONTENTS

11.1 Introduction

Fiber optic bio and chemical sensors have been widely studied for the past 10 years. The availability of new types of optical fibers (such as IR fibers, photonic crystal fibers), new types of sources (such as solid-state light-emitting diodes [LEDs]) and detectors (such as highly sensitive point and array detectors), new sensing materials, and new technologies (such as nanotechnology) has rapidly advanced the performance and enabled new applications of bio and chemical sensors.

Fiber optic sensors have intrinsic advantages over the electronic counterpart due to their small footprint, low cost, rapid sensing speed, real-time monitoring capability, high sensitivity and selectivity, the possibility of distributed measurement, and immunity to electromagnetic interferences. Furthermore, many optical sensor materials (such as certain optical polymers) are biocompatible, which makes them excellent choices for in situ biosensing.

Fiber-optic bio and chemical sensors are based on the transmission of light along the fiber to the site of the analyte. The key components of fiber sensors include light sources (such as LEDs and lasers), detectors (photodiodes), fibers, and sensing materials. Light-intensity measurements can be read directly or can be interpreted by electrical circuitry. There are two groups of fiber-optic bio and chemical sensors according to constructions: extrinsic and intrinsic sensors. For extrinsic sensors, fibers are simply used to guide light to and from the region where the light beam is influenced by the measurand. In this case, the fiber is not used for the sensing function. On the other hand, for intrinsic sensors, the interaction with the analyte occurs within the optical fiber element; that is, the optical fiber structure is modified and the fiber itself plays an active role in the sensing function. Many parameters, like phase, polarization and intensity of the output light, may be modulated.

A chemical sensor is defined as a device responding to a particular analyte in a selective way through a chemical reaction, and it can be used for the qualitative or quantitative determination of the analyte.[1] Biosensors are a subset of chemical sensors, which can be broadly defined as any measuring device that can sense a biological element. In general, there are many types of sensing mechanisms that can be applied to bio sensors, which include absorbance, reflectance, fluorescence, fluorescence quenching, evanescent wave, surface plasmon resonance, and luminescence. In real-world application, different types of sensing mechanisms and technologies have been developed for serving different sensing purposes and sensing different elements, such as toxins, pesticides, explosives, nucleic acids, antibodies, and other small molecules.

Furthermore, sensing mechanisms based on amplitude change or phase change are called intensity-modulated sensors and phase-modulated sensors, respectively. Intensity modulation is relatively simple and even an incoherent light source can be employed, whereas phase-modulated sensors require coherent lasers as light source, which offers higher sensitivity but also higher cost and more sensitive to the ambient factors such as temperature fluctuation and vibrations.

The practical challenges[2,3] of real-world applications may include (1) multi-analyte sensing ability; (2) interference from ambient light, background absorbance, or Raman scattering of the fiber; and (3) limited dynamic range in comparison with electronic counterpart. Furthermore, irreversibility may be a problem if the reagent is expensive.

This chapter presents the basics of fiber optic bio and chemical sensors, which include sensing mechanism, operation principles, classifications, transduction methods, applications, and recent developments and current trends.

11.2 Principle of Operation

Glass fibers are the most commonly used fibers in the field of bio and chemical sensing because they can transmit light in the visible and near infrared regions. However, for applications in the UV region, quartz (i.e., pure silica) fiber is needed. Furthermore, remarkable development of various infrared optic fibers has further extended the wavelength of measurement to 2- to 15-μm regions,[4] where many molecules have strong absorption bands. Therefore, the sensitivity and the accuracy of the sensors are dramatically improved.

11.2.1 Optical Fundamentals

Optical Fibers

Optical fibers consist of a core with a refractive index n_1 surrounded by a cladding with a lower refractive index n_2. For the conventional fiber (except the photonic crystal fiber), the light transmission in optical fiber is based on

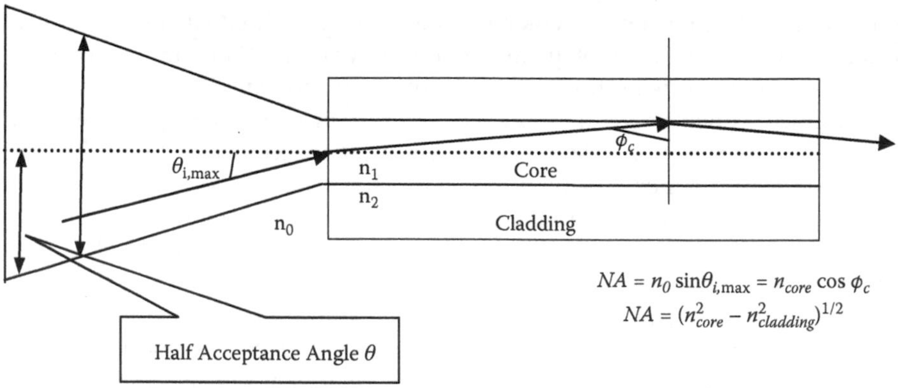

FIGURE 11.1
An illustration of light guiding in optical fiber.

the principle of total internal reflection (TIR), which requires the refractive index of the core (n_1) to be higher than that of the cladding (n_2)—that is, $n_1 > n_2$. As illustrated in Figure 11.1, important fiber parameters include (a) critical angle ϕ_c, which is defined by the ratio between the cladding and the core refractive indices, as given by

$$\sin\phi_c = n_2/n_1\,; \tag{11.1}$$

(b) the acceptance cone angle, $\theta_{i,max}$, which depends on the refractive indices of the core, the clad, and the ambient refractive index, n_0,

$$\sin\theta_{i,\max} = \frac{\sqrt{(n_1^2 - n_2^2)}}{n_0}\,; \tag{11.2}$$

and (c) the numerical aperture (NA), which defines the fiber's light collection efficiency and is related to the acceptance cone's angle as:

$$NA = n_0 \sin\theta_{i,\max} \tag{11.3}$$

All these parameters are critically important when designing the fiber optic bio and chemical sensors.

Evanescent Wave

When the incident light is reflected from an interface at an angle greater than the critical angle, the total internal reflection occurs. However, its intensity does not abruptly decay to zero at the interface and a small portion of light penetrates into the reflecting medium. This penetrated electromagnetic field is called the evanescent wave, as illustrated in Figure 11.2. Since the amplitude of evanescent wave decays exponentially with the distance, the penetration depth

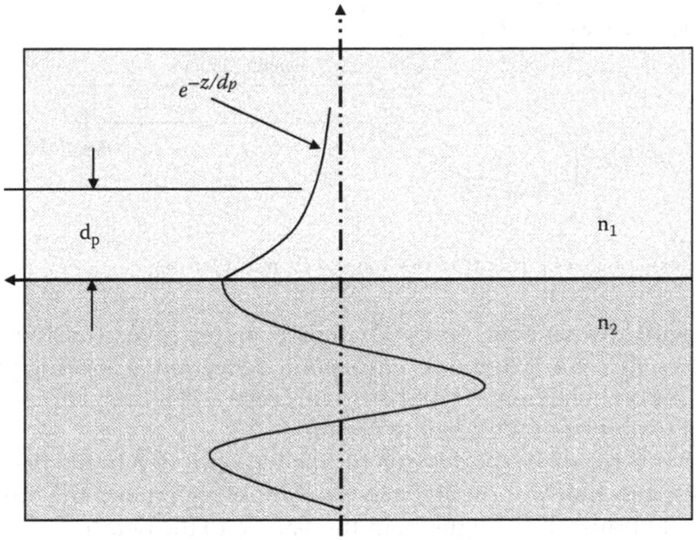

FIGURE 11.2
An illustration of exponential decay of evanescent field.

(d_p) is defined as the distance required for the electric field amplitude to fall to 1/e (0.37) of its value at the interface, which is a function of both the wavelength of the light and the angle of incidence, as mathematically given by[5]

$$d_p = \frac{\lambda}{4\pi[n_1^2 \sin^2\theta - n_2^2]^{1/2}} \tag{11.4}$$

where λ is the wavelength of the transmitted light, θ is the incident angle at the core/cladding interface, and n_1, n_2 are the refractive indices (RI) of the core and cladding, respectively.

Penetration depth is a very important parameter that needs to be considered when designing evanescent wave-based bio and chemical sensors, to be discussed in detail in later sections.

11.2.2 Optrode-Based Fiber Optic Biosensors (Bio-Optrode) and Evanescent Wave Fiber Optic Biosensors

Bio-Optrode

The word "optrode" is a combination of the words "optical" and "electrode" and refers to fiber optic devices that can measure the concentration of a specific chemical or a group of chemicals.[6] The basic structure of an optrode is composed of a source fiber and a receiver fiber that is connected to a sensing fiber by a special connector as illustrated in Figure 11.3.[6] To achieve sensing capability, the tip of the sensing fiber is usually coated with a sensing material, such as by the dip coating procedure. The chemicals to be sensed may

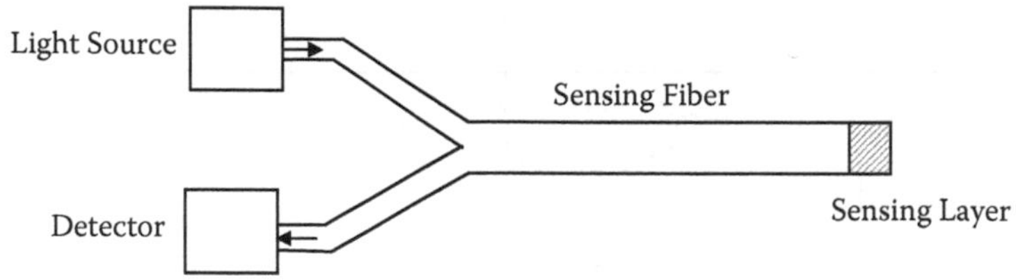

FIGURE 11.3
A configuration of fiber optic optrode for bio and chemical sensing.

interact with the sensing tip by changing one or several of the following parameters: its refractive index, absorption, reflection, scattering properties; or polarization behaviors. The fiber in this case acts as a light pipe transmitting light to and from the sensing region.

There are three different designs for the bio-optrode.[6] In the first case, one fiber is used to transmit the light to the sample region and the other fiber is employed to transmit the light from the sample to the detector. In the second design, bifurcated fibers are harnessed. The third design utilizes a different configuration, in which the sample is put in the central region of the fiber and the signals are collected by the multiple fibers surrounding the central region.

Evanescent Wave Fiber Optic Biosensors

Optrodes use the light transported to the end of the fiber to generate a signal at or near the fiber end, whereas evanescent wave sensors utilize the electromagnetic component of the reflected light at the side surface between the fiber core and the fiber cladding.

The evanescent wave can interact with analytes within the penetration depth; thus, by immobilizing biological material within this region, the absorption of propagating light or generation of fluorescence during the binding of analytes can be detected. The major advantage of using evanescent wave is the ability to couple light out of the fiber into the surrounding medium, which offers a large interaction surface, as depicted in Figure 11.4. Therefore, higher sensitivity can be achieved.

Examples of evanescent field based sensors include attenuated total reflection (ATR) type sensors and total internal reflection fluorescence (TIRF) sensors. In ATR, the evanescent wave produces a net flow of energy across the reflecting surface in the surrounding medium. The energy transfer can lead to attenuation in reflectance that depends on the absorption of the evanescent waves. In TIRF, when the evanescent light selectively excites a fluorophore, the fluorescence emitted by the fluorophore can be coupled back into the fiber and guided all the way to the detector.

To further enhance sensor sensitivity and reduce noise, D-shaped optical fibers[7] or holey fibers[8] have also been employed recently for evanescent wave based biosensors.

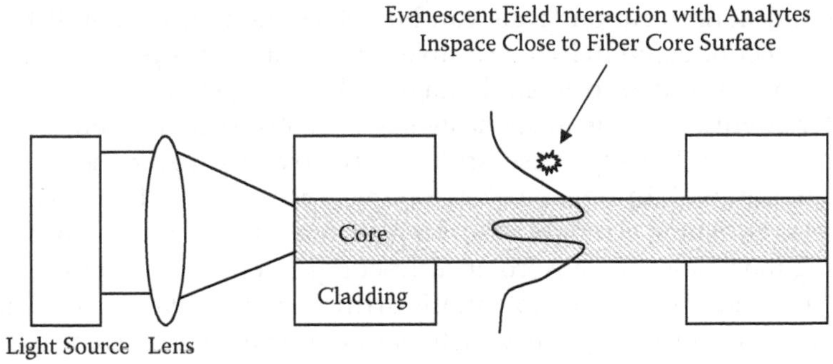

FIGURE 11.4
A schematic of biosensor based on fiber coupled evanescent wave.

11.2.3 Optical Transducers

The optical transducer of a biosensor is referred to convert the observed biochemical change into a measurable optical signal. According to the types of transduction methods, fiber optical sensor can be classified into the following groups: (1) direct absorption, (2) fluorescence, and (3) chemiluminescence and bioluminescence.

Absorption

The simplest optical bio and chemical sensors use absorptions to determine changes in the concentration of analytes.[2] The sensor works by sending light through an optical fiber to the bio sample; the amount of light absorbed by the analyte is determined by measuring the light coupled out via the same fiber or a second optical fiber. To ensure the stability of the sensing region, the biological material is immobilized at the distal end of the optical fibers, which enable one either to produce or extract the analyte that absorbs the light.[9]

From a physics point of view, absorption is a process in which light energies are absorbed by an atom or a molecule. Based on the Lambert–Beer law (usually referred to as Beer's law),[10] the intensity of transmitted light (I) through a uniform absorption medium can be mathematically described by the following formula[10]:

$$I = I_0 \exp^{-\varepsilon C \Delta x}, \tag{11.5}$$

where I_0 denotes the incident light intensity, ε is the extinction coefficient, C represents the concentration of the absorption analyte, and Δx is the thickness (or length) of the absorption medium. Since the absorption is usually wavelength dependent and different species may have different absorption spectra, by measuring the absorption spectra via fiber optic sensor, different species and concentration levels can be determined. Note that, if the measurement is conducted in living samples, the light source power will need to be carefully controlled to avoid damaging the living samples.

As aforementioned, the major advantages of absorption-based sensors are that they are simple, easy to use, and cost effective. However, they also suffer some fundamental limitations. For example, the relatively few bio processes produce or consume strongly absorbing chromophores and many naturally occurring materials may absorb or diffract light in the visible range, causing increased backgrounds and potentially artificial results. Moreover, water has a high background absorption, which may be problematic for some molecules and in some light wavelength ranges—for example, in 975-nm IR regime.[11]

Fluorescence

Fluorescence is commonly used in bio-optrodes. Fluorescence occurs when molecules absorb light at one wavelength and then emit light at a longer wavelength, as illustrated in Figure 11.5.[10] Since the excitation and emission occur only at distinct energy levels, each fluorescent molecule has a unique fluorescence spectral fingerprint, which is very important for the sensor application.

Generally speaking, since there are few intrinsically fluorescent biological molecules, fluorescent tags are commonly used for generating fluorescent light, increasing the signal-to-noise ratio, and allowing specific signals to be clearly distinguished from background. Fluorescent intensity and fluorescent lifetime are two important measurants for the sensor applications.

Intensity. The analyte concentration level can be determined by measuring the increase or decrease in fluorescence intensity. The fluorescence intensity-based sensor is most commonly used in immunosensors and/or affinity-binding type sensors. Furthermore, quenching of fluorescence intensity is commonly used in both affinity-based and biocatalytic fiber optic sensors.

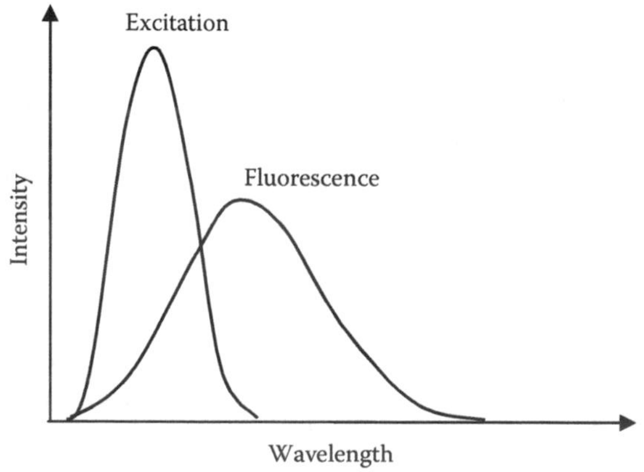

FIGURE 11.5
An illustration of fluorescent light emission.

In the affinity-based case, the analyte itself quenches fluorescence emission. On the other hand, in the biocatalytic sensor case, enzymatic production or consumption of a quenching species is employed for the signal transduction.

Again, to improve the sensitivity and the selectivity of the sensors, some biosensors have used not only a change in fluorescence intensity for detection, but also a shift in fluorescence emission wavelengths. It is more accurate to differentiate signal from emission wavelength change than the intensity change because the accuracy of the intensity change measurement can be compromised by optical variations, fiber movement, or light scattering from those caused by analyte recognition.

Lifetime. The fluorescence lifetime is another commonly used sensing measurand, which is defined as the average amount of time that a fluorophore stays in the excited state between the photon absorption and the fluorescence emission. Mathematically, the fluorescent intensity as a function of decay time is described by

$$I(t) = I_0 e^{-t/\tau} \tag{11.6}$$

where I_0 is the intensity at initial time, t is time, and τ is the lifetime, which is also defined as the fluorescent intensity to decay to 1/e of its initial value. Due to the existence of multiple fluorescent decaying processes, Eq. (11.6) can be rewritten in a more general form, as given by[6]

$$I(t) = I_0 \Sigma \alpha_i e^{-t/\tau_i}, \tag{11.7}$$

where α_i denotes the coefficient for the ith fluorescent decay process. Since the fluorescent lifetime can be very short (e.g., in the nanosecond regime), sophisticated high-speed electronics are usually required. The major advantages of using fluorescent lifetime as the sensing measurand include (1) the measurement is independent of analyte concentration, and (2) measurement is not affected by leaching or bleaching of the fluorophore. One way to avoid using the high-speed electronics is to measure fluorescence lifetime in frequency domain. In the frequency-domain measurement approach, the sinusoidally modulated light is used to excite the fluorescent molecule; the resulting emission light also oscillates at the same exciting frequency. However, because of the finite lifetime of fluorescence, the emission light is phase shifted with respect to the excitation light. Since the amount of phase shift is directly related to the lifetime, the lifetime can be determined by measuring this phase shift.

Chemiluminescence and Bioluminescence

Chemiluminescence is similar to fluorescence. The difference is that chemiluminescence occurs by exciting molecules with a chemical reaction (usually occurring by the oxidation of certain substances such as oxygen or hydrogen peroxide), whereas fluorescence occurs by exciting molecules via light. Thus, in the case of chemiluminescence, no external source of light is required to initiate the reaction that eliminates the need of light source for the sensor application. Chemiluminescence is usually involved with two steps, as illustrated by[6]

$$A \xrightarrow{k_1} A', \qquad \textit{exciting step} \tag{11.8a}$$

$$A' \xrightarrow{k_2} h\nu, \qquad \textit{emission step} \tag{11.8b}$$

where k_1 is the excitation rate, and k_2 is the decay rate.

A chemiluminescence-based sensor is a commonly used chemical for generating light signal in many bio-optrodes. The reaction between luminol and HO produces a luminescence signal and this reaction is also catalyzed by certain ions or molecules. To enhance the sensing capability, chemiluminescence and fluorescence techniques are usually combined. For example, an antigen is labeled with luminol and an antibody is labeled with a fluorescent compound. When the antigen is in contact with the antibody, the chemiluminescence emission from the antigen can excite the fluorescent light from the antibody. Thus, sensing the interaction between the antigen and the antibody can be realized.

Bioluminescence is simply chemiluminescence occurring in living organisms, which represents a biological chemiluminescent reaction process.[10] Many organisms produce bioluminescence for signaling, mating, prey attracting, food hunting, and self-protection. For example, a very familiar example of high-efficiency bioluminescence is the firefly. The ratio of the number of photons produced for a given number of molecules is as high as 0.9. Since the bioluminescence is generated via biological reaction processes, the certain biological process can be sensed by detecting the bioluminescence.

Surface Plasmon Resonance (SPR)

Surface plasmon resonance is a unique optical transduction method, which has been commercially employed for optical biosensors. Basically, SPR is an evanescent electromagnetic field generated at the metal surface (e.g., silver or gold) by coupling the exciting light into the metal surface.[12] To effectively generate SPR, proper exciting wavelength is needed.

The basic SPR apparatus is referred to as the Kretschman prism arrangement,[13] as illustrated in Figure 11.6. In the device, first, a thin film of metal (usually a 400–500 Å thick gold or silver film) is coated on the prism, and a biosensing layer containing an immobilized biorecognition element is also coated on the metal surface. When the light of an appropriate wavelength interacts with the dielectric–metal interface at the proper angle, it can induce the electron plasmatic resonance at the metal surface. Under this situation, the photon energy is largely transferred to the resonant energy of

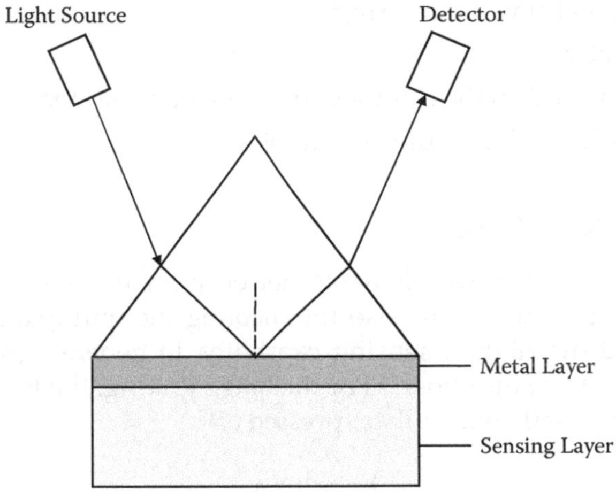

FIGURE 11.6
A schematic diagram of the Kretschmann prism-based surface plasmon resonance sensor.[13]

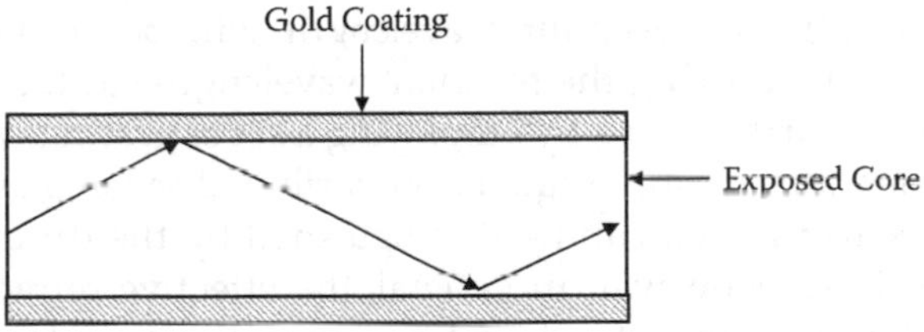

FIGURE 11.7
Surface plasmon resonance sensors based on optical fiber configuration.

SPR so that the reflection light from the metal film will greatly be attenuated. Thus, one can observe a sharp minimum of light reflectance when the angle of incidence is at this proper resonant angle. Since the resonance angle depends on several factors—the wavelength of the incident light, the metal, and the nature of the media in contact with the surface—the nature of the media can be sensed by measuring this resonance angle.[13]

Besides Kretschman prism based optical sensors, surface plasmons have also been used to enhance the sensing signal and the sensitivity for other types of spectroscopic measurements, including fluorescence, Raman scattering, and second harmonic generation. Furthermore, in addition to the prism coupling, SPR sensors can also be based on optical fibers or integrated optical waveguides,[14,15] as illustrated in Figure 11.7.

In biosensor application, SPR has been successfully used to detect DNA or proteins by measuring the changes in the local index of refraction upon adsorption of the target molecule to the metal surface. The major advantages of SPR sensors include

high-speed, real-time monitoring
high sensitivity
label-free, enabled analysis for a wide range of bio systems
requiring only small amounts of samples

Fiber Grating Based Sensors

Fiber gratings are effective elements not only for enhancing the sensing sensitivity and selectivity but also for enabling the multiparameter, multifunctional, and distributed sensing capability. In general, gratings can be photoinduced into a silica fiber.[16] For the Bragg grating, the Bragg resonance condition can be mathematically expressed as[17]

$$\lambda_B = 2n_{eff}\Lambda \tag{11.9}$$

where n_{eff} is the effective refractive index of the fiber, Λ is the grating pitch, and λ is the resonant reflected Bragg wavelength.

When there is a change in refractive index or grating period (e.g., induced by the measurand), the resonant wavelength will be shifted (based on Eq. 11.9). Thus, by measuring the resonant wavelength shift, the measurand can be measured. Furthermore, by employing a set of gratings with different resonant wavelengths, multiple agents or distributed sensing can be realized because these sensing data can be distinguished by the different resonant wavelength regimes. However, in general, the effective refractive index of the fundamental mode of a standard fiber is practically independent of the refractive index of the surrounding medium. In order to enable the ambient sensing capability, it is important to make optical modes penetrate evanescently into the surrounding media so that there is an interaction between the guided light field and the bio and chemical agents to be sensed. Many methods have been proposed to optimize this interaction,[18–22] such as using blazed gratings[18] or long period gratings,[19] etching the fiber close to the core diameter to increase the sensitivity, or side polishing the fiber. A recent work by Chryssis et al. demonstrated that the minimum detectable refractive index resolution could be as small as 7.2×10^{-6} when the fiber diameter was etched to ultrathin (i.e., 3.4 μm).[22] This exciting result shows the outstanding sensing capability of fiber grating based sensors.

Besides the chemical sensing, the high sensitivity offered by grating based sensors can also be used for biological sensing. An important example is DNA sensing. In the operation, the probe DNA is attached on the sensing area, and the detection is based on the immobilization of target DNA. The combination between the target DNA and the complementary probe DNA induces a refractive index change, which can be sensed by the shift of resonant wavelength of the Bragg grating. Besides sensing the refractive index change, the sensing mechanism can also be based on grating period change such as the grating expansion induced by the specific polymer coating.[23]

11.2.4 Performance Factors

The performance of fiber sensors can be quantitatively evaluated by the following important performance factors.

Selectivity

The selectivity is defined as the ability to discriminate between the target analyte and the background analytes, which are the most important characteristics of bio and chemical sensors. The ideal bio and chemical sensor will only respond to changes in concentration of the target analyte, and will not be influenced by the presence of other chemical species.

Calibration

To ensure sensor accuracy, it is usually necessary to make periodic calibrations at regular intervals during the sensing process.

Background Signal

Usually, a sensor signal will have some background level. To optimize the sensor performance, these background signals need to be subtracted. In addition to the previously discussed performance factors, many other factors such as the dynamic response, the temperature dependence, and the stability and biocompatibility are important factors that also need to be considered for evaluating bio/chemical sensors.

11.2.5 Key Optical Components

A fiber optic sensor is usually composed of the following key optical components.

Light Source

The commonly used light sources include LEDs, lasers, tungsten lamps, xenon lamps, broadband supercontinuum generation, etc. The key requirements for bio and chemical sensors include the power, the wavelength, the stability, the size, and the cost.

Light Detectors

A light detector is an indispensable component for bio/chemical sensors. In general, the signals for bio/chemical samples are weak. Thus, high-sensitivity detectors such as photomultiplier tubes (PMTs) and avalanched photodiodes are usually used as light detectors. Besides point detectors, charge coupled devices (CCDs) are also often used in bio and chemical sensing fields, especially for imaging usage. Furthermore, optical spectrum analyzers (OSAs) are used for spectral measurement and analysis (such as measuring the wavelength shift of Bragg grating based fiber optic sensors).

11.2.6 Types of Biosensors

Based on the sensing mechanism, biosensors may be classified into the following categories: (1) catalytic-based biosensors, and (2) bioaffinity-based optrodes.

Catalytic Biosensors

The function of catalytic biosensors is realized by recognizing and binding of an analyte followed by a catalyzed chemical conversion of the analyte from a nondetectible form to a detectible form. The reaction progress of the biocatalysis can be monitored by detecting the rate of formation of a product, the disappearance of a reactant, or the inhibition of the reaction. Many types of biomaterials can be biocatalysts, including an isolated enzyme, a microorganism, a subcellular organelle, or a tissue slice.

Enzymes belong to natural proteins, which have the capability to transform a specific substrate molecule into a product without being consumed in the reaction. Since most enzymes do not have intrinsic optical property, changes that can be used to indicate interaction with the analyte and chemical transducers (such as O_2, pH, CO_2, NH_3, etc.) are frequently employed in fiber-optic enzymatic biosensors for indicating the change of analyte concentration in the sample. The commonly used enzymes in optical biosensors include (1) oxidases and oxidoreductases that catalyze the oxidation of compounds using oxygen or NAD, (2) esterases that produce acids, (3) decarboxylases that produce CO_2, and (4) deaminases that produce NH_3.

A successful example of catalytic biosensor is measuring the glucose level by using glucose oxidase. The enzyme oxidase catalyzes the conversion of glucose to gluconic acid and H_2O_2 via the reaction with O_2, as given by

$$\text{Glucose} + O_2 + (\text{Glucose oxidase}) \rightarrow \text{Gluconic acid} + H_2O_2 \qquad (11.10)$$

Thus, by monitoring the generated H_2O_2, or gluconic acid, or the amount of consumed oxygen, the concentration of glucose can be measured. In the operation, the monitoring of the amount of consumed oxygen can be realized by employing O_2-sensitive ruthenium-based dyes.[24,25] With increasing concentrations of glucose (increased consumption of O_2), O_2-mediated quenching of fluorescence is ameliorated, resulting in an increase in fluorescence. On the other hand, the amount of generated H_2O_2 can be detected by using chemiluminescence indicators.

Since in vivo monitoring of glucose is needed for millions of persons suffering from diabetes, this type of biosensor has a huge commercial potential.

Bioaffintity-Based Optrodes

Bioaffinity biosensors are based on affinity interactions by separating an individual or selected range of components from complex mixtures of biomolecules. Two typical examples of bioaffinity biosensors are immunoassay optical fiber-based biosensors and nucleic acid optical fiber-based biosensors.

Immunoassay optical fiber-based biosensors. Immunoassay optical biosensors are based on the optical signals generated by antibody–antigen binding.[6] The binding can generate different types of optical signals such as a fluorescent optical label, a refractive index change directly, etc. The major advantage of this type of sensor is high selectivity because the nature of antibodies makes them very powerful sensing elements for recognizing their binding partners.

Immunoassays can be implemented in one of three modes: direct, competitive, and sandwich. In the direct mode, the antigen (for instance, a naturally fluorescent compound) is incubated with excess amounts of an immobilized antibody. The measured signal is directly proportional to the amount of antigen present. In the competitive mode, the detection is based on competition for the antibody binding site between the antigen sample (unlabeled) and an externally added fluorescent-labeled antigen. For example, in one configuration of this scheme, the unlabelled and labeled analytes compete for binding to recognition molecules immobilized on the surface of the waveguide; the decrease in fluorescent signal will be proportional to the amount of unlabeled species in the mix. In the sandwich mode, a second recognition species is required. To simplify the operation, antibodies are often used as both capture and tracer elements. From the application point of view, sandwich mode has proven to be effective for the detection of bacteria, viruses, proteins, and protozoa.

Nucleic acid optical fiber-based biosensors. Nucleic acid optical fiber biosensors are implemented by using the affinity of single-stranded DNA to form double-stranded DNA with complementary sequences. The applications of this type of sensor include the detection of chemically induced DNA damage and the detection of microorganisms through the hybridization of species-specific sequences of DNA.[26]

11.2.7 Immobilization Issue

In biosensors, to ensure the consistency of the measurement and to avoid the movement of biosamples, the immobilization is a critical issue that needs to be considered. There are three common methods of immobilization: (1) adsorption, (2) entrapment, and (3) covalent bonding, as illustrated in Figure 11.8.[27]

The absorption approach itself includes using physical adsorption and chemical adsorption. Although both methods involve adsorbing the sensing material onto a solid surface or polymer matrix, physical adsorption is usually weak and the chemical adsorption is much stronger due to the formation of covalent bonds. The major advantages of absorption method are

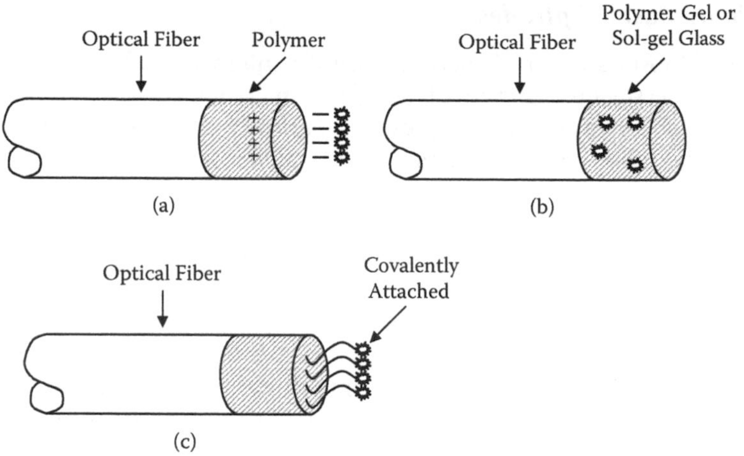

FIGURE 11.8
A schematic illustration of immobilization methods.[27] (a) Adsorption immobilization, (b) entrapment immobilization, (c) covalent binding.

simplicity, absence of a clean-up step, and highly reproducible. However, accuracy of this technique may be affected by changes in the medium pH or by changes in other ion concentrations.

Entrapment immobilization involves the physical entrapment of sensing biomolecules. In the operation, biomolecules are mixed with a monomer solution. When the monomer solution is polymerized, the polymer gel can trap the biomolecules. This immobilization approach can be strong; however, it may have a slow diffusion rate of analytes.

The third approach is covalent binding, in which a functional group is bound in the biomaterial to the support matrix. In general, immobilization methods that employ covalent bonding agents are applicable only to proteins (enzymes, antibodies, etc). The covalent binding is a stable immobilization approach that can avoid problems of protein leaching from the support. The drawback is that covalent linkage could lead to the loss of one or more functional binding sites.

Finally, we would like to point out that no matter which immobilization method is used, to achieve best performance, the immobilization procedure should be optimized in terms of signal intensity, selectivity, and sensitivity by choosing proper materials (e.g., fluorophores) and methods depending on the application.

11.3 Selected Applications of Bio and Chemical Sensing

Bio and chemical sensing have tremendous applications. Due to the page limitation, in this section, we will only introduce several typical applications done by different groups around the world, including the measurements of pH, gas, ions, etc.

11.3.1 Measurement of pH

The measurement of pH is fundamental to many chemical applications, which can be realized by using certain dyes as pH indicators. For example, methyl red is a suitable pH dye that has a distinctive visible spectrum with well separated maxima for its acidic and basic forms.

11.3.2 Measurement of Gases

Many types of gases, such as CO_2, O_2, NH_3, and SO_2, can be detected by fiber optic chemical sensors. For example, CO_2 concentration can be determined by using a pH probe that includes a gas permeable membrane containing a hydrogen carbonate buffer. Oxygen can be detected by using its fluorescence quenching feature. Furthermore, NH_3 can be sensed by using oxazine perchlorate dye, and SO_2 (down to 84-ppm level) can be detected by using benzofluoranthene.[28] These chemical gas sensors can be very useful for industrial pollutant monitoring and control.

11.3.3 Ion Measurement

Fiber optic chemical sensors can also be used to measure ions such as Cl^-, Br^-, I^-, and Na^+ via fluorescence quenching. Acridinium or quinidinium is a frequently used fluorescent reagent for this purpose.[29]

11.3.4 Explosives Detection

Fiber optic chemical sensors can also be used for explosive detection by using the mid-IR spectrum. For example, thin polymer films can be coated onto a mid-IR transmissive chalcogenide fiber. The explosive chemicals can be detected when they partition into the polymer film and absorb IR radiation at characteristic wavelengths.

11.3.5 Other Chemical Compound and Environmental Sensors

Besides the chemicals mentioned in previous sections, many other chemical compounds can also be detected by using fiber optic chemical sensors. For example, nitroaromatics in water can also be sensed by using a mid-IR spectrum based fiber-optic sensor and ozone can be detected by using an ultraviolet (UV) evanescent wave sensor.[30] In an ozone sensor, the sensing probe is made of gas-permeable silicone cladding; the output signal shows a linear response to ozone over the range of 0.02–0.35 vol% with a response time of about 1 min. Furthermore, pesticide analysis can be realized by using surface plasmon-resonance (SPR) biosensors. For example, a gold-coated waveguide

SPR sensor can be used to monitor the attachment of biotin–avidin layers to the surface of the sensor in an aqueous environment, which enables the detection of the pesticide simazine.[31] Finally, there are numerous other types of fiber optic chemical sensors for monitoring acid, wastes, ground water, toxicity, and heavy metals.[32–34]

11.3.6 Clinic Sensors

The rapid development of bio-optrodes has also enabled the potential clinical applications, which can have a great impact from both the healthcare and economic points of view. For example, an optical fluorescence biosensor has been investigated for blood gas analysis, which can be used to determine blood gases and pH via fluorescence measurements.[35] Fiber optic biosensors have also been employed for free cholesterol determination. The sensor is based on the use of a cholesterol oxidizer and oxygen transduction. The signal is obtained by the oxygen sensitive complex immobilized in the bioactive layer. Hence, the signal can be correlated to the cholesterol concentration.[36] Furthermore, fiber optic biosensors have also been used for monitoring drug delivery. In this application, the drug dissolution level is monitored by using UV/visible spectral analysis.[37] Finally, in situ glucose monitoring is conducted by using the quenching effect of the luminescence of ruthenium diimine complex.[38]

11.4 Recent Developments and Future Trends

Fiber optic bio and chemical sensors are fast growing technologies. In particular, with the recent advent in nanotechnology, advanced signal detection (such as single molecule fluorescent emission) and processing, higher sensitivity, selectivity, and multi-agent detection capability are enabled. In this section, we will briefly review several recent developments in this field.

11.4.1 Nano Bio-Optrodes

One of the most exciting advances in sensing development is to realize sensing in the nanoscale, which enables one to monitor biomolecule concentrations inside a single living cell and thereby leads to a better understanding of many cellular processes. One type of nano bio-optrode is fabricated by pulling optical fibers to tapered shape with typical distal ends around 20–80 nm, and then immobilizing the bio indicator (e.g., pH-sensitive dye or fluorescent labels) is added on the fiber tip.

11.4.2 Multi-Analyte Sensing

Another recent development is multi-analyte sensing, which is very important for clinical, environmental, and industrial analysis. One approach is to use discrete sensing regions, each region containing different biosensing elements. The sensing regions can be formed on the end of fiber imaging fiber by using photopolymerization techniques.[39] For example, a multi-analyte imaging fiber sensor can detect pH, CO_2, and O_2 simultaneously. The sensing element is based on covalently immobilizing fluorescent indicators within polymer matrices via photopolymerization, resulting in a region of analyte sensing at the fiber's distal end. The sensing ranges for O_2, CO_2, and pH are 0–100%, 0–10%, and 5.5–7.5, respectively.[40] To achieve a large number of fiber channels, instead of bundling multiple individual fibers together, an imaging fiber array that consists of thousands of optical fibers is used, in which each individual fiber channel maintains its ability to carry its own light signal from one end of the fiber to the other. By attaching a sensing material to the individual fiber's distal end, an array that contains thousands of sensing elements can be constructed. To ensure good contact, microwells are fabricated on the end of each individual fiber channel by selectively etching the fiber cores so that high aspect ratio microwell array can be built on the imaging fiber tip, and the sensing elements are prepared by immobilizing fluorescent indicators to the microsphere surface. This type of device has been successfully used to detect multiple drugs, digoxin and theophylline, simultaneously.[41]

11.4.3 Other Advanced Developments

New Material Development

Conventional optical fibers used in sensing applications are silica fibers, which have good transmission in visible and near IR spectral regions but are opaque in the longer wavelength infrared region (i.e., >2-μm wavelength). Since molecules have strong absorptions in the mid-IR (2–15 μm) region, the sensitivity and accuracy can be notably improved by extending the wavelength of measurement into the mid-infrared region. Thus, one active topic in this field is to develop infrared fibers that have good transmission in this mid-IR transmission window. A variety of IR fibers are being developed, including fluoride glass fibers, chalcogenide glass fibers, halide crystalline fibers, and sapphire crystalline fibers.[4] Another approach is to use hollow waveguides or light pipes, which not only can transmit the light but also can function as a sensing chip for advanced chemical sensing. Another trend in the new material development is to find new materials that exhibit selective sensitivity for a specific material to be measured.

New Sensing Schemes

A new sensing scheme is to use microstructured optical fiber (MOF). In sensing, the characteristic micron-sized holes that run along the length of MOF can be filled with various fluids containing the species to be sensed. Since the species can be in contact with the mode field propagating through the fiber over long interaction lengths, it provides large overlap of mode field of reaction materials and greatly enhanced sensitivity.[42] Furthermore, multiple holes also enable the highly sensitive multi-agent sensing capability.

Methods for Enhancing Sensing Sensitivity

The sensitivity of the sensor can be greatly enhanced by integrating the surface plasmon resonance with the long interaction fiber structure. In the operation, silver–gold alloy nanoparticles are coated on the fiber surface. Ag–Au alloy offers a better performance than the metal-host based fiber optic SPR sensor.[43] A D-type fiber biosensor based on SPR technology and heterodyne interferometry is also presented.[44] The device is made of a single-mode optical fiber in which half the core is polished away and a thin-film layer of gold is deposited on the polished surface. Instead of measuring the light intensity as used in traditional SPR techniques, the phase-difference variations are measured. Thus, a very high sensitivity (2×10^{-6} for refractive index measurement) is achieved.[44]

Another method to increase the sensitivity is to use tapered optical fiber. Since the evanescent field interaction can be significantly increased by using the tapered structure, a substantial increase in sensor sensitivity can be realized by employing the tapered fiber. For example, Maraldo et al.[45] recently demonstrated that a tapered fiber sensor could be effectively used for rapid assessment to determine the presence of bacteria by growth. As cells grew on the tapered surface, the evanescent scatter and absorption increased, which caused transmission to decrease. Thus, by monitoring the transmission changes, the bacteria growth rate can be detected.

Besides using the surface plasmon resonance, localized surface plasmon resonance (LSPR) was also recently investigated by Chau et al.[46] The sensitivity of this reflection-type sensor is comparable with transmission-based analogues but offers a faster response time, smaller pixel size, and capability of simultaneous LSPR sensing and surface-enhanced Raman scattering.

In addition to using Bragg grating (as aforementioned), long period gratings (LPGs) are also used to enhance the sensitivity and selectivity. For example, a reflection mode LPG was employed as a transducer for immunosensing.[47] In the operation, LPG partially couples incoming light from the core to a cladding mode, which is reflected by the mirror at the end of the fiber and recombined into the core mode by the same LPG. Thus, there is interference between the forward and the backward propagated modes. Since this interference pattern can be very sensitive to the external perturbation, a very sensitive sensor can be built.

11.5 Conclusion

In conclusion, this chapter provides a brief introduction on fiber optic bio and chemical sensors. First, we explained the fundamentals of optical fiber sensors, including (1) principles of operation, (2) classifications, and (3) transduction methods. Second, we presented typical applications of chemical and bio sensors to pH measurement, chemical gas sensing, and biological agent detection. Third, we reported the recent developments and future trends such as nano bio-optrode and multi-agent sensing. Throughout the chapter, the advantages and limitations of different types of sensors were also discussed, which might be helpful for readers to determine proper sensors for their specific needs. Bio and chemical sensors are growing areas. Every day, there are new sensors or new applications emerging. We believe that with the rapid advent of bio and nano technologies, fiber optic bio and chemical sensors will have a bright future.

References

1. R. W. Catterall, *Chemical Sensors,* Oxford University Press, Oxford, UK, 1997.
2. M. Mehrvar, C. Bis, J. M. Scharer, M. Moo-Young, and J. H. Luong, Fiber-optic biosensors—Trends and advances, *Anal. Sci.,* **16**, pp. 677–692, 2000.
3. M. D. Marazuel and M. C. Moreno-Bondi, Fiber-optic biosensors—An overview, *Anal. Bioanal. Chem.,* **372**, pp. 664–682, 2002.
4. M. Saito and K. Kikuchi, Infrared optical fiber sensors, *Opt. Rev.,* 4, 5, pp. 527–538, 1997.
5. N. J. Harrick, *Internal Reflection Spectroscopy,* Wiley, New York, p. 327, 1967.
6. F. S. Ligler and C. A. Rowe Taitt, *Optical Biosensors: Present and Future,* Elsevier Science, B. V., 2002.
7. W. Jin, G. Stewart, M. Wilkinson, B. Culshaw, F. Muhammad, S. Murray, and J. O. W. Norris, Compensation for surface contamination in a D-fiber evanescent wave methane sensor, *J. Lightwave Technol.,* **13**, 6, 1995.
8. Y. L. Hoo, W. Jin, H. L. Ho, D. N. Wang and R. S. Windeler, Evanescent-wave gas sensing using microstructure fiber, *Opt. Eng.,* **41**, 1, pp. 8–9, January 2002.
9. D. L. Wise and L. B. Wingard, *Biosensors with Fiberoptics,* Humana Press, Clifton, NJ, pp. 1–300, 1991.
10. D. G. Buerk, *Biosensors—Theory and Applications,* Technomic Publishing Company, Inc., Lancaster PA, 1995.
11. http://www.dartmouth.edu/~etrnsfer/water.htm.
12. L. M. Lechuga, A. Calle, and F. Prieto, Optical sensors based on evanescent field sensing. Part I. Surface plasmon resonance sensors, *Quím Anal.,* **19**, 54–60, 2000.
13. D. R. Purvis, D. Pollard-Knight, and P. A. Lowe, Biosensors based on evanescent waves. In G. Ramsay (ed.), *Commercial Biosensors,* Wiley, New York, 1998.
14. R. L. Earp and E. D. Raymond, Surface plasmon resonance. In G. Ramsay (ed.), *Commercial Biosensors,* Wiley, New York, 1998.
15. J. Homola, S. S. Yee, and G. Gauglitz, Surface plasmon resonance sensors: Review, *Sensor Actuator B-Chem,* **54**, pp. 3–15, 1999.
16. K. O. Hill, Y. Fujii, D. C. Johnson, and B. S. Kawasaki, Photosensitivity in optical fiber waveguides: Application to reflection filter fabrication, *Appl. Phys. Lett.,* **32**, p. 647, 1978.

17. A. D. Kersey, M. A. Davis, H. J. Patrick, M. LeBlac, K. P. Koo, C. G. Askins, M. A. Putnam, and E. J. Friebele, Fiber grating sensors, *J. Lightwave Technol.*, **15**, pp. 1442–1463, Aug. 1997.
18. G. Meltz, W. W. Morey, and J. R. Dunphy, Fiber Bragg grating chemical sensor, *Chemical, Biochemical, and Environmental Fiber Sensors III, Proc. SPIE,* **1587**, pp. 350–361, 1991.
19. V. Bhatia, and A. M. Vengsarkar, Optical fiber long-period grating sensors, *Opt. Lett.*, **9**, 21, pp. 692–694, 1996.
20. A. Asseh, S. Sandgren, H. Ahlfeldt, B. Sahlgren, R. Stubbe, and G. Edwall, Fiber optical Bragg grating refractometer, *Fiber Integr. Opt.*, **17**, pp. 51–62, 1998.
21. K. Schroeder, W. Ecke, R. Mueller, R. Willsch, and A. Andreev, A fiber Bragg grating refractometer, *Meas. Sci. Technol.*, **12**, pp. 757–764, 2001.
22. A. N. Chryssis, S. M. Lee, S. B. Lee, S. S. Saini, and M. Dagenais, High sensitivity evanescent field fiber Bragg grating sensor, *IEEE Photonics Technol. Lett.*, **17**, 6, 2005.
23. G. B. Tait, G. C. Tepper, D. Pestov, and P. M. Boland, Fiber Bragg grating multifunctional chemical sensor, *Proceedings of SPIE—The International Society for Optical Engineering,* **5994**, Chemical and Biological Sensors for Industrial and Environmental Security, p. 599407, 2005.
24. W. Zhong, P. Urayama, and M-A. Mycek, Imaging fluorescence lifetime modulation of a ruthenium-based dye in living cells: The potential for oxygen sensing *J. Phys. D: Appl. Phys.* **36**, pp. 1689–1695, 2003.
25. P. S. Grant and M. J. McShane, Development of multilayer fluorescent thin film chemical sensors using electrostatic self-assembly, *IEEE Sensors Journal,* **3**, 2, April 2003.
26. U. J. Piunno, R. H. E. Krull, M. J. D. Hudson, and H. Cohen, Fiber optic biosensor for fluorimetric detection of DNA hybridization, *Anal. Chim, Acta,* **288**, p. 205, 1994.
27. B. Mattiasson, Immobilization methods. In B. Mattiasson (ed.), *Immobilized Cells and Organelles,* vol. I., CRC Press, Inc., Boca Raton, FL, pp. 3–26, 1983.
28. B. Eggins, *Biosensors: An Introduction,* Wiley, New York, 1996.
29. B. R. Eggins, *Chemical Sensors and Biosensors,* John Wiley & Sons, Inc., New York, 2002.
30. R. A. Potyrailo, S. E. Hobbs, and G. M. Hieftje, Optical waveguide sensors in analytical chemistry: Today's instrumentation, applications and trends for future development, *Anal. Chem.*, **48**, p. 456, 1998.
31. R. D. Harris, B. J. Luff, J. S. Wilkinson, R. Wilson, D. J. Schiffrin, J. Piehler, A. Brecht, R. A. Abuknesha, and C. Mouvet, Integrated optical surface plasmon resonance biosensor for pesticide analysis, *IEEE Colloquium on Optical Techniques for Environmental Monitoring* (1995/182), p. 6, London, 30 Oct. 1995.
32. K. J. Kuhn and J. T. Dyke, A renewable-reagent fiber-optic sensor for measurement of high acidities, *Anal. Chem.*, **68**, 2890, 1996.
33. R. B. Thompson, Z. Ge, M. Patchan, C. Chin-Huang, and C. A. Fierke, Fiber optic biosensor for Co(II) and Cu(II) based on fluorescence energy transfer with an enzyme transducer, *Biosens. Bioelectron.*, **11**, p. 557, 1996.
34. CMSTCP (Characterization, Monitoring & Sensor Echnology Crosscutting Program), Technology Catalog-HaloSnif- Optic Spectrochemical Sensor, http://www.cmst.org/cmst/Tech_Cat.text/6.1.html (Oct. 1998).
35. V. M. Owen, Optical fluorescence biosensoring application and diversification—A case history, *Biosens. Bioelectron.*, **11**, 1, pp. v–vii, 1996.

36. M. D. Marazuela, B. Cuesta, M. C. Moreno-Bondi, and A. Quejido, Free cholesterol fiber-optic biosensor for serum samples with simplex optimization, *Biosens. Bioelectron.*, **12**, p. 233, 1997.
37. P. J. Gemperline, J. Cho, B. Baker, B. Batchelor, and D. S. Walker, Determination of multicomponent dissolution profiles of pharmaceutical products by in situ fiber-optic UV measurements *Anal. Chim. Acta*, **345**, p. 155, 1997.
38. A. D. Neubauer, D. Pum, U. B. Sleyler, I. Klmant, and O. S. Wolbeis, Fibre-optic glucose biosensor using enzyme membranes with 2-D crystalline structure, *Biosens. Bioelectron.*, **11**, 3, p. 317, 1996.
39. G. L. Bowlin and G. Wnek, *Encyclopedia of Biomaterials and Biomedical Engineering*, Marcel Dekker, New York, 2004.
40. J. A. Ferguson, B. G. Healey, K. S. Bronk, S. M. Barnard, and D. R. Walt, Simultaneous monitoring of pH, CO_2 and O_2 using an optical imaging fiber, *Anal. Chim. Acta*, **340**, p. 123, 1997.
41. F. Szurdoki, K. L. Michael, and D. R. Walt, A duplexed microsphere-based fluorescent immunoassay, *Anal. Biochem.*, **291**, pp. 219–228, 2001.
42. F. M. Cox, A. Argyros, and M. C. J. Large, Liquid-filled hollow core microstructured polymer optical fiber, *Optics Express*, **14**, 9, pp. 4135–4140, 2006.
43. A. K. Sharma and B. D. Gupta, Fibre-optic sensor based on surface plasmon resonance with Ag–Au alloy nanoparticle films, *Nanotechnology*, **17**, pp. 124–131, 2006.
44. M-H. Chiu, S-F. Wang, and R-S. Chang, D-type fiber biosensor based on surface-plasmon resonance technology and heterodyne interferometry, *Optics Letters*, **30**, 3, p. 233, 2005.
45. D. S. Maraldo, P. Shankar, M. Mohana, and R. Mutharasan, Measuring bacterial growth by tapered fiber and changes in evanescent field, *Biosensors and Bioelectronics*, **21**, 7, pp. 1339–134, 2006.
46. L-K. Chau, Y-F. Lin, S-F. Cheng, and T-J. Lin, Fiber-optic chemical and biochemical probes based on localized surface plasmon resonance, *Sensors and Actuators, B: Chemical*, **113**, 1, pp. 100–105, 2006.
47. D. W. Kim, Y. Zhang, K. L. Cooper, and A. Wang, Fiber-optic interferometric immunosensor using long period grating, *Electronics Letters*, **42**, pp. 324–325, 2006.

Index

A

B

C

D

G

J

K

L

M

N

O

P

Q

R

S

T

U

V

W

Y

Milton Keynes UK
Ingram Content Group UK Ltd.
UKHW021907071024
449327UK00022B/1636

9 780367 387563